The DOG ^{and} Its GENOME

The DOG and Its GENOME

COLD SPRING HARBOR MONOGRAPH SERIES

The Lactose Operon
The Bacteriophage Lambda
The Molecular Biology of Tumour Viruses
Ribosomes
RNA Phages
RNA Polymerase
The Operon
The Single-Stranded DNA Phages
Transfer RNA:
    Structure, Properties, and Recognition
    Biological Aspects
Molecular Biology of Tumor Viruses, Second Edition:
    DNA Tumor Viruses
    RNA Tumor Viruses
The Molecular Biology of the Yeast *Saccharomyces:*
    Life Cycle and Inheritance
    Metabolism and Gene Expression
Mitochondrial Genes
Lambda II
Nucleases
Gene Function in Prokaryotes
Microbial Development
The Nematode *Caenorhabditis elegans*
Oncogenes and the Molecular Origins of Cancer
Stress Proteins in Biology and Medicine
DNA Topology and Its Biological Effects
The Molecular and Cellular Biology of the Yeast *Saccharomyces:*
    Genome Dynamics, Protein Synthesis, and Energetics
    Gene Expression
    Cell Cycle and Cell Biology
Transcriptional Regulation
Reverse Transcriptase
The RNA World
Nucleases, Second Edition
The Biology of Heat Shock Proteins and Molecular Chaperones
Arabidopsis
Cellular Receptors for Animal Viruses
Telomeres
Translational Control
DNA Replication in Eukaryotic Cells
Epigenetic Mechanisms of Gene Regulation
C. elegans II
Oxidative Stress and the Molecular Biology of Antioxidant Defenses
RNA Structure and Function
The Development of Human Gene Therapy
The RNA World, Second Edition
Prion Biology and Diseases
Translational Control of Gene Expression
Stem Cell Biology
Prion Biology and Diseases, Second Edition
Cell Growth: Control of Cell Size
The RNA World, Third Edition
The Dog and Its Genome

The DOG and Its GENOME

EDITED BY

Elaine A. Ostrander
National Human Genome Research Institute
National Institutes of Health

Urs Giger
Section of Medical Genetics
School of Veterinary Medicine
University of Pennsylvania

Kerstin Lindblad-Toh
Broad Institute of Massachusetts Institute of Technology
and Harvard University

COLD SPRING HARBOR LABORATORY PRESS
Cold Spring Harbor, New York

The Dog and Its Genome

Monograph 44
©2006 by Cold Spring Harbor Laboratory Press
All rights reserved
Printed in the United States of America

Publisher	John Inglis
Acquisition Editor	Alex Gann
Production Manager	Denise Weiss
Project Coordinator	Inez Sialiano
Production Editor	Patricia Barker
Desktop Editor	Danny deBruin
Interior Book Designer	Denise Weiss and Emily Harste
Cover Designer	Mike Albano

Front Cover Artwork: Composite by Heidi Parker. Original photographs courtesy of Purina.

Library of Congress Cataloging-in-Publication Data

The dog and its genome / edited by Elaine A. Ostrander, Urs Giger, Kerstin Lindblad-Toh.
    p. cm. -- (Cold Spring Harbor monograph series ; 44)
  Includes bibliographical references.
  ISBN 0-87969-742-3 (hardcover : alk. paper)
 1. Dogs--Genetics. 2. Dogs--Breeding. I. Ostrander, Elaine A.
II. Giger, Urs. III. Lindblad-Toh, Kerstin. IV. Series.
  SF427.2.D64 2005
  636.7'082--dc22

2005019391

10 9 8 7 6 5 4 3 2 1

All Cold Spring Harbor Laboratory Press publications may be ordered directly from Cold Spring Harbor Laboratory Press, 500 Sunnyside Boulevard, Woodbury, New York 11797-2924. Phone: 1-800-843-4388 in Continental U.S. and Canada. All other locations: (516) 422-4100. FAX: (516) 422-4097. E-mail: cshpress@cshl.edu. For a complete catalog of Cold Spring Harbor Laboratory Press publications, visit our World Wide Web Site at http://www.cshlpress.com/

We dedicate this book to Donald F. Patterson, DVM, DSc, whose early investigations of the genetic basis of canine health defined the field. His curiosity, tenacity, and scientific rigor have projected canine science to its enviable position in the comparative genetics world today. We also salute dog breeders and owners as well as veterinarians everywhere. Their devotion and compassion for their pets both humbles and inspires us.

Contents

CANINE ORIGINS

GENOMICS AND GENOME ORGANIZATION

GENETIC DISEASE AND GENE MAPPING

Preface

Since the domestic dog became the most important animal companion centuries ago, there has existed in humans a desire to cross-breed dogs with unique characteristics and to preserve the most desirable traits in the form of breeds. The resulting phenotypic variation observed across breeds is far greater than that observed for any other animal species, hence the dog and its genome have attracted enormous scientific and public interest. As geneticist, molecular biologist, and veterinary clinician, we have led many of the successful collaborations in canine genomics, clinical diagnostics, and molecular technology that culminated in 2004 with the completion of the canine genome sequence. We, hence, embraced the opportunity offered by Cold Spring Harbor Laboratory Press to present the recent developments and current molecular genetic understanding of the domestic dog and its genome at this exciting time.

We are most indebted to our many colleagues whose expertise and enormous efforts made their chapter contributions invaluable to this book. As we embarked on this project, we recognized how much progress has recently been made in this rapidly evolving field; we apologize that we had to select topics and could not acknowledge various collaborative and other research groups worldwide who made outstanding contributions to the better understanding of the genetics of the dog. All these achievements would not have been possible without the tremendous efforts of countless students, research and sequence production specialists, and postdoctoral fellows in our own and other laboratories. A very special thanks goes to all who did the actual hard work.

The book would never have gotten off the ground without the encouragement and support of the Cold Spring Harbor Laboratory Press staff. We especially acknowledge Alex Gann, Editorial Director, whose idea this was, and John Inglis, Excutive Director of CSHL Press, for his support of this project. We are also grateful to the editorial development office, including Jan Argentine, the Director of Editorial Development, and Inez Sialiano, Project Coordinator, the latter of whom spent countless hours with us revising and organizing the chapters. We thank also the production team including Denise Weiss, Production Manager, and Pat Barker, Production Editor, and Danny deBruin, Desktop Editor, all of whom helped us bring this daunting task to a timely completion.

A host of public and private agencies worldwide have supported scientific research summarized in these chapters, and we are grateful to all, especially the National Institutes of Health, American Cancer Society, Guide Dogs for the Blind, Seeing Eye, Morris Animal Foundation, American Kennel Club Canine Health Foundation, Burroughs Wellcome Foundation, Wellcome Trust, many breed clubs, and a host of others for their generous financial support. We thank also the American Kennel Club and the Kennel Club (United Kingdom) for their enthusiastic support for canine genetics and health during the last 15 years.

Additionally, we thank those who have inspired us. The dog owners and breeders who remind us daily why we started the "dog genome project" 15 years ago have offered unparalleled support for everything we do. Their enthusiasm and willingness to provide samples and information about their dogs propel us forward. They have taught us so much about this fascinating species, and we are constantly grateful for their commitment to the dog and to us.

Finally, we thank our families, who have remained ever patient while we continue on this journey to understand the species Lord Byron praised as possessing "Beauty without Vanity, Strength without Insolence, Courage without Ferocity, and all the Virtues of Man, without his Vices."

ELAINE A. OSTRANDER
KERSTIN LINDBLAD-TOH
URS GIGER

Foreword

All knowledge, the totality of all questions and answers, is contained in the dog.

FRANZ KAFKA, *Investigations of the dog*

B ECAUSE IT IS CLEAR THAT we are at a time of unprecedented advancement and accomplishment in dog genetics, the optimal time for a Cold Spring Harbor Laboratory (CSHL) volume about the dog is now. Through rigorous selection of phenotypic traits, dogs have been divided into more than 350 genetically closed populations, known as breeds. In the U.S., the American Kennel Club (AKC) recognizes 154 distinct breeds of dogs, each a genetic isolate defined by a breed standard that specifies every conceivable aspect of morphology from coat color to body size, leg length to head shape, and a range of behaviors (AKC 1998). Thus, locked within the wolf genome was the capacity for an extraordinary amount of genetic variation, and hence phenotypic variation, that humans have manipulated over the last several centuries to create breeds of dogs that are hunters, guardians, guide dogs, and companions.

The restricted gene flow associated with the formation of most modern breeds derives from several sources. In some cases, a limited number of founders were responsible for defining a breed. In other cases, bottlenecks associated with shifts in breed popularity, frequently used sires, and catastrophic world events generated breeds with limited gene pools. For an animal to be a registered member of a breed, both its parents must have been registered members of the same breed. Thus, at this time, the gene pool is closed for most breeds, and unless they arise as new mutations, no new alleles will enter the breeding stock.

The outcome of such restrictions is a comparatively high incidence of genetic disease in purebred dogs, which is unfortunate for both owners and breeders, as well as for the dogs themselves. For biologists, however, this great experiment in selection and natural variation offers a unique opportunity to understand the genetics of mammalian morphology, disease susceptibility, and behavior. Toward that end, for the last decade the canine research community has focused on (1) developing the resources needed to navigate the canine genome; (2) application of those resources to gene hunting for diseases important for both human and companion animal health; and finally, (3) unraveling the evolutionary relationship of the dog genome with that of other species and applying that knowledge to understanding the genetics of morphology, behavior, and disease susceptibility.

We began our own work several years ago, focused exclusively on development of mapping resources. As is evident from the chapters presented in this book, however, canine genetics has entered a period of unprecedented growth and discovery. No longer limited by molecular resources, we are now hindered only by our ability to accurately phenotype and collect families and populations featuring traits of interest. It is worth taking a few moments, however, to summarize some of the key advancements that have defined our journey to date.

HISTORICAL OVERVIEW

Our studies on the dog genome began 10 years ago with the construction of meiotic linkage and radiation hybrid maps (Mellersh et al. 1997; Priat et al. 1998), eventually reaching a density of one marker per megabase (Mb) in 2003 (Guyon et al. 2003). The characterization of the dog karyotype proved very challenging, as it is composed of 38 autosomes, most of which are small and acrocentric, plus the X and Y (Breen et al. 1999c; Yang et al. 1999). The DAPI-banded karyotype of the dog was not finalized until 1999 (Breen et al. 1999a) and then only through the use of reciprocal chromosome paints (Breen et al. 1999b). The integration of the maps was only achievable through the use of a common set of BACs, which were placed on all three maps (Breen et al. 2004). The recent development of a 10,000 gene map of the dog produced by ourselves and collaborators is a major step forward in understanding the human/dog comparative map (C. Hitte et al., in prep.). Certainly, however, the availability of a 1.5× survey sequence of the dog (Kirkness et al. 2003) and the production of a high-quality 7.5× draft sequence (K. Lindblad-Toh et al., in prep.) have to be counted as the crowning achievements.

The identification of loci and/or genes implicated in diseases such as canine renal cancer, narcolepsy, several forms of retinal atrophy, hematologic disorders, and metabolic diseases such as copper toxicosis and ceroid lipofucinosis, were direct consequences of these efforts, and many of these findings are reviewed in the following chapters or in other recent reviews (Galibert et al. 1998; Ostrander et al. 2000; Patterson 2000; Sutter and Ostrander 2004). These successes were important not only for the advancement of animal health, but perhaps more importantly, because they alerted the greater scientific community to the specific advantages offered by the dog model. In this regard, the often-cited identification of the hypocretin 2 receptor as the causative gene for inherited narcolepsy in the Doberman remains the best example (Lin et al. 1999), as it was subsequently found that the ligand of the receptor, when mutated (Peyron et al. 2000), was responsible for the human form of the disease. Thus, findings from genetic analysis of dog pedigrees opened an entirely new area of research for the study of human sleep disorders. A second excellent example is provided by the findings not only that the MURR1 gene is associated with copper toxicosis in the Bedlington terriers (Yuzbasiyan-Gurkan et al. 1997; van de Sluis et al. 2002), but also that mutations in the human MURR1 gene cause accelerated disease in people with Wilson's disease (Stuehler et al. 2004).

Parallel to the development of the various genomic resources and the cloning of several disease genes has been spectacular progress made in gene therapy using the canine model. Hemophilia type B (Herzog et al. 1999), Duchenne muscular dystrophy (Bartlett et al. 2000), congenital stationary night blindness (Acland et al. 2001), and mucopolysaccharidosis type VII (Ponder et al. 2002) are all monogenic diseases that are stably prevented through the use of gene therapy in affected lines of dogs. This remarkable achievement suggests to even the harshest critics that gene therapy may indeed become a vital piece of the arsenal for treating inherited diseases.

Why has it taken the medical community so long to recognize the dog system as a viable model for medical research? Certainly its value for developing protocols for marrow transplant is well understood, as evidenced by recognition with a Nobel Prize (Thomas 1994). But it is only recently that the value of mapping studies within dog pedigrees and populations has been recognized as a way to overcome the problem of locus heterogeneity that has plagued so many human studies. Canine studies of retinal disease (Acland et al. 1994, 1998, 1999; Zeiss et al. 2000; Sidjanin et al. 2002) and epilepsy (Lohi et al. 2005) demonstrate this point most clearly. It is only by analysis of simple populations, or breeds, where one rather than several disease genes plays a role, that we are likely to unravel the complexities associated with disorders like cancer and heart disease.

The availability of a high-quality 7.5× draft sequence of the dog (K. Lindblad-Toh et al., in prep.), together with an evolving understanding of the canine/human genome comparative maps (C. Hitte et al., in prep.), has leveled the playing field for those doing canine genetics. Libraries, maps, and markers are no longer rate-limiting, and as in human genetics, the acquisition of large, well phenotyped pedigrees or case/control materials is now the key step for any successful mapping study.

DOG GENETICS IN THE COMING DECADE

How will dog disease genes be found in the future? Certainly there remains a place, as in human genetics, for linkage analysis of extraordinary families using sets of dense, highly polymorphic markers. However, the availability of 2.6 million single-nucleotide polymorphisms (SNPs) in the public database (http://www.broad.mit.edu/mammals/dog/snp/), and an understanding of how linkage disequilibrium in the dog genome is organized (Sutter et al. 2004; K. Lindblad-Toh et al., in prep.), suggest the possibility of whole-genome association studies using modest numbers of SNPs, perhaps even fewer than 10,000, compared to the 500,000 that are hypothesized to be needed for human studies (Kruglyak 1999). The available SNPs were derived by the comparison of the boxer reference sequence to a modest number of sequence reads from nine selected breeds (K. Lindblad-Toh et al., in prep.), as well as the 1.5× survey sequence of the poodle (Kirkness et al. 2003). Analysis of data, as well as targeted data from five sampled regions, suggests that a single SNP map is likely to be useful for mapping studies in most dog breeds (Sutter et al. 2004; K. Lindblad-Toh et al., in prep.).

What will the focus of our efforts be? Certainly, studies aimed at disease gene mapping and cloning will continue. For diseases like cancer, diabetes, heart disorders, and neuromuscular diseases, this may be the only way to proceed, as dense high-risk human pedigrees are increasingly difficult to obtain. Similarly, analyses of large collections of small human pedigrees of defined phenotype increasingly fail at finding genes associated with complex disorders. Thus, we predict that scientists will turn increasingly to the canine model, as they have with narcolepsy, retinal disease, epilepsy, and osteosarcoma.

The greatest challenge, however, is faced by investigators seeking to clone genes controlling morphology and behavior. Some progress has been made in this arena by Lark and colleagues (Chase et al. 2002, K. Chase et al., in prep.), and by extension, QTLs involved in hip dysplasia have been reported for the Portuguese water dog (Chase et al. 2004). However, causative genes remain to be identified. More disappointingly, our understanding of canine behavioral genetics remains nearly as rudimentary as it was when we began our work nearly a dozen years ago.

THE ROAD LESS TRAVELED

We are thus only partway through what promises to be an exciting journey. The editors have selected chapters by sets of authors that have defined the field. We begin with a discussion of the history of the breed clubs, and then describe our current understanding of domestication from the wolf. The latter sets the stage for a chapter aimed at understanding the evolutionary relationship between the modern breeds.

We consider next the development of resources; each was an important milestone in unraveling the relationship between the human and dog genomes. Indeed, comparative genomics is a continuous thread running through many of the chapters, as it has proven indispensable in the mapping and cloning of several disease genes. The status of mapping and cloning experiments for several canine diseases follows. We then discuss the current status of gene therapy in the dog. Finally, we have included chapters presenting ideas on what is known about the genetics of biological changes. The first successful mapping studies in that field may very well be those that define the genetics of fox domestication (Trut et al. 2004).

For many years, the dog model was viewed as somehow "exotic." Although perhaps not yet quite mainstream, the dog can no longer be considered a model organism that resides on the fringes. Indeed, as the chapters in this book demonstrate, the dog is now set to take its rightful place as a valued system for genetic studies along with the mouse, rat, and several insect species. We hope you come away from this experience, as we have, not only with a sense of accomplishment for what the community has done, but also with a sense of awe and marvel for the extraordinary species the modern domestic dog has become.

E. Ostrander
F. Galibert

REFERENCES

Acland G.M., Blanton S.H., Hershfield B., and Aguirre G.D. 1994. XLPRA: A canine retinal degeneration inherited as an X-linked trait. *Am. J. Med. Genet.* **52:** 27–33.

Acland G.M., Ray K., Mellersh C.S., Langston A.A., Rine J., Ostrander E.A., and Aguirre G.D. 1999. A novel retinal degeneration locus identified by linkage and comparative mapping of canine early retinal degeneration. *Genomics* **59:** 134–142.

Acland G.M., Ray K., Mellersh C.S., Gu W., Langston A.A., Rine J., Ostrander E.A., and Aguirre G.D. 1998. Linkage analysis and comparative mapping of canine progressive rod-cone degeneration (prcd) establishes potential locus homology with retinitis pigmentosa (RP17) in humans. *Proc. Natl. Acad. Sci.* **95:** 3048–3053.

Acland G.M., Aguirre G.D., Ray J., Zhang Q., Aleman T.S., Cideciyan A.V., Pearce-Kelling S.E., Anand V., Zeng Y., Maguire A.M., et al. 2001. Gene therapy restores vision in a canine model of childhood blindness. *Nat. Genet.* **28:** 92–95.

American Kennel Club (AKC). 1998. *The complete dog book.* Howell Book House, New York.

Bartlett R.J., Stockinger S., Denis M.M., Bartlett W.T., Inverardi L., Le T.T., thi Man N., Morris G.E., Bogan D.J., Metcalf-Bogan J., and Kornegay J.N. 2000. In vivo targeted repair of a point mutation in the canine dystrophin gene by a chimeric RNA/DNA oligonucleotide. *Nat. Biotechnol.* **18:** 615–622.

Breen M., Bullerdiek J., and Langford C.F. 1999a. The DAPI banded karyotype of the domestic dog (*Canis familiaris*) generated using chromosome-specific paint probes. *Chromosome Res.* **7:** 401–406.

Breen M., Thomas R., Binns M.M., Carter N.P., and Langford C.F. 1999b. Reciprocal chromosome painting reveals detailed regions of conserved synteny between the karyotypes of the domestic dog (*Canis familiaris*) and human. *Genomics* **61:** 145–155.

Breen M., Langford C.F., Carter N.P., Holmes N.G., Dickens H.F., Thomas R., Suter N., Ryder E.J., Pope M., and Binns M.M. 1999c. FISH mapping and identification of canine chromosomes. *J. Hered.* **90:** 27–30.

Breen M., Hitte C., Lorentzen T.D., Thomas R., Cadieu E., Sabacan L., Scott A., Evanno G., Parker H.G., Kirkness E., et al. 2004. An integrated 4249 marker FISH/RH map of the canine genome. *BMC Genomics* **5:** 65.

Chase K., Lawler D.F., Adler F.R., Ostrander E.A., and Lark K.G. 2004. Bilaterally asymmetric effects of quantitative trait loci (QTLs): QTLs that affect laxity in the right versus left coxofemoral (hip) joints of the dog (*Canis familiaris*). *Am. J. Med. Genet. A* **124:** 239–247.

Chase K., Carrier D.R., Adler F.R., Jarvik T., Ostrander E.A., Lorentzen T.D., and Lark K.G. 2002. Genetic basis for systems of skeletal quantitative traits: Principal component analysis of the canid skeleton. *Proc. Natl. Acad. Sci.* **99:** 9930–9935.

Galibert F., Andre C., Cheron A., Chuat J.C., Hitte C., Jiang Z., Jouquand S., Priat C., Renier C., and Vignaux F. 1998. The importance of the canine model in medical genetics. *Bull. Acad. Natl. Med.* **182:** 811–821.

Guyon R., Lorentzen T.D., Hitte C., Kim L., Cadieu E., Parker H.G., Quignon P., Lowe J.K., Renier C., Gelfenbeyn B., et al. 2003. A 1-Mb resolution radiation hybrid map of the canine genome. *Proc. Natl. Acad. Sci.* **100:** 5296–5301.

Herzog R.W., Yang E.Y., Couto L.B., Hagstrom J.N., Elwell D., Fields P.A., Burton M., Bellinger D.A., Read M.S., Brinkhous K.M., et al. 1999. Long-term correction of canine hemophilia B by gene transfer of blood coagulation factor IX mediated by adeno-associated viral vector. *Nat. Med.* **5:** 56–63.

Kirkness E.F., Bafna V., Halpern A.L., Levy S., Remington K., Rusch D.B., Delcher A.L., Pop M., Wang W., Fraser C.M., and Venter J.C. 2003. The dog genome: Survey sequencing and comparative analysis. *Science* **301:** 1898–1903.

Kruglyak L. 1999. Prospects for whole-genome linkage disequilibrium mapping of common disease genes. *Nat. Genet.* **22:** 139–144.

Lin L., Faraco J., Li R., Kadotani H., Rogers W., Lin X., Qiu X., de Jong P.J., Nishino S., and Mignot E. 1999. The sleep disorder canine narcolepsy is caused by a mutation in the hypocretin (orexin) receptor 2 gene. *Cell* **98:** 365–376.

Lohi H., Young E.J., Fitzmaurice S.N., Rusbridge C., Chan E.M., Vervoort M., Turnbull J., Zhao X.C., Ianzano L., Paterson A.D., et al. 2005. Expanded repeat in canine epilepsy. *Science* **307:** 81.

Mellersh C.S., Langston A.A., Acland G.M., Fleming M.A., Ray K., Wiegand N.A., Francisco L.V., Gibbs M., Aguirre G.D., and Ostrander E.A. 1997. A linkage map of the canine genome. *Genomics* **46:** 326–336.

Ostrander E.A., Galibert F., and Patterson D.F. 2000. Canine genetics comes of age. *Trends Genet.* **16:** 117–124.

Patterson D. 2000. Companion animal medicine in the age of medical genetics. *J. Vet. Internal. Med.* **14:** 1–9.

Peyron C., Faraco J., Rogers W., Ripley B., Overeem S., Charnay Y., Nevsimalova S., Aldrich M., Reynolds D., Albin R., et al. 2000. A mutation in a case of early onset narcolepsy and a generalized absence of hypocretin peptides in human narcoleptic brains. *Nat. Med.* **6:** 991–997.

Ponder K.P., Melniczek J.R., Xu L., Weil M.A., O'Malley T.M., O'Donnell P.A., Knox V.W., Aguirre G.D., Mazrier H., Ellinwood N.M., et al. 2002. Therapeutic neonatal hepatic gene therapy in mucopolysaccharidosis VII dogs. *Proc. Natl. Acad. Sci.* **99:** 13102–13107.

Priat C., Hitte C., Vignaux F., Renier C., Jiang Z., Jouquand S., Cheron A., Andre C., and Galibert F. 1998. A whole-genome radiation hybrid map of the dog genome. *Genomics* **54:** 361–378.

Sidjanin D.J., Lowe J.K., McElwee J.L., Milne B.S., Phippen T.M., Sargan D.R., Aguirre G.D., Acland G.M., and Ostrander E.A. 2002. Canine CNGB3 mutations establish cone degeneration as orthologous to the human achromatopsia locus ACHM3. *Hum. Mol. Genet.* **11:** 1823–1833.

Stuehler B., Reichert J., Stremmel W., and Schaefer M. 2004. Analysis of the human homologue of the canine copper toxicosis gene MURR1 in Wilson disease patients. *J. Mol. Med.* **82:** 629–634.

Sutter N.B. and Ostrander E.A. 2004. Dog star rising: The canine genetic system. *Nat. Rev. Genet.* **5:** 900–910.

Sutter N.B., Eberle M.A., Parker H.G., Pullar B.J., Kirkness E.F., Kruglyak L., and Ostrander E.A. 2004. Extensive and breed-specific linkage disequilibrium in *Canis familiaris. Genome Res.* **14:** 2388–2396.

Thomas E.D. 1994. The Nobel Lectures in Immunology. The Nobel Prize for Physiology or Medicine, 1990. Bone marrow transplantation—Past, present and future. *Scand. J. Immunol.* **39:** 339–345.

Trut L.N., Pliusnina I.Z., and Os'kina I.N. 2004. An experiment on fox domestication and debatable issues of evolution of the dog. *Genetika* **40:** 794–807.

van De Sluis B., Rothuizen J., Pearson P.L., van Oost B.A., and Wijmenga C. 2002. Identification of a new copper metabolism gene by positional cloning in a purebred dog population. *Hum. Mol. Genet.* **11:** 165–173.

Yang F., O'Brien P.C., Milne B.S., Graphodatsky A.S., Solanky N., Trifonov V., Rens W., Sargan D., and Ferguson-Smith M.A. 1999. A complete comparative chromosome map for the dog, red fox, and human and its integration with canine genetic maps. *Genomics* **62:** 189–202.

Yuzbasiyan-Gurkan V., Blanton S.H., Cao V., Ferguson P., Li J., Venta P.J., and Brewer G.J. 1997. Linkage of a microsatellite marker to the canine copper toxicosis locus in Bedlington terriers. *Am. J. Vet. Res.* **58:** 23–27.

Zeiss C.J., Ray K., Acland G.M., and Aguirre G.D. 2000. Mapping of X-linked progressive retinal atrophy (XLPRA), the canine homolog of retinitis pigmentosa 3 (RP3). *Hum. Mol. Genet.* **9:** 531–537.

1

Canine History and Breed Clubs

Jessica A. Moody, Leigh Anne Clark, and Keith E. Murphy
Department of Veterinary Pathobiology
College of Veterinary Medicine
Texas A&M University
College Station, Texas 77843-4467

He is your friend, your partner, your defender, your dog. You are his life, his love, his leader. He will be yours, faithful and true, to the last beat of his heart. You owe it to him to be worthy of such devotion.

UNKNOWN

THE DOMESTIC DOG HAS BEEN HUMANKIND'S best friend for 140 centuries. The bond between dog and human is stronger than any other human–animal relationship and has pragmatic and emotional roots. The pragmatic side is that the dog has served as shepherd, guide, hunter, protector, and as a model organism for biomedical research. The emotional side, however, is equally important, and the dog's major role has been as companion. When one tries to understand the importance of a dog as companion, it is easiest to start with the simple fact that dogs are fun to look at and touch. These visual and tactile interactions are among the greatest pleasures of owning a dog. When one delves deeper into why the companionship of a dog matters so much, however, it is far more difficult to explain the range of emotions the friendship of a dog elicits. It is considerably easier to explain the value of the dog as a service, or working, companion.

The details of canine domestication are the subject of multiple studies (see Chapter 7), but the value of the dog to human populations since that event is indisputable. Throughout the ages, the dog has been integrated into various aspects of society to meet certain needs. These needs are constantly increasing and being revised. Recent events,

The Dog and Its Genome ©2006 Cold Spring Harbor Laboratory Press 0-87969-742-3

such as terrorist attacks, natural disasters, and war have highlighted the working dog's multiple roles. These highly visible activities, however, do not tell the full story regarding the importance of the dog to society. For example, the dog is still the most widely used non-rodent mammal in biomedical science. In the pharmaceutical industry, the dog has long served as a model because it has greater physiologic homology with the human than we have with the mouse or rat. The dog is also an alternative species to evaluate efficacy of potential therapeutic agents. Furthermore, due to recent advances in pharmacogenomics, toxicogenomics, and the sequencing of the canine genome, it is now possible to gain even greater benefit from medical research using the dog as a model. It is also important to recognize that the dog has been a model organism for studies pertaining to social science and psychology (for review, see Scott and Fuller 1965).

The desire to develop dogs with particular physical traits, behavioral characteristics, or unique skills was the driving force behind selective breeding practices that have been ongoing for centuries. Today there are more than 400 breeds of dogs, many of which arose from the desire to create a breed that meets a certain phenotypic standard (e.g., temperament, function). The goal of many breedings was to create dogs physically suited for specific purposes: the long-bodied dachshunds to hunt badgers underground; the tiny Chihuahuas to be lap dogs; the heavy-coated komondors to protect flocks; and the web-footed Newfoundlands to be powerful swimmers. As a result, dogs exhibit a vast array of phenotypic traits, with varying heights, weights, bone structures, and hair coats. In fact, the extent of morphological and behavioral variation observed in the dog is greater than in any other mammalian species.

Dogs possess a wide range of behavioral characteristics, with many of these behaviors unique to specific breeds. For example, the herding instincts of the Shetland sheepdog are not likely to be observed in the weimaraner, just as the pointing instincts of the weimaraner are not likely to be observed in the Shetland sheepdog. An experimental cross between two breeds with distinct behavioral differences (the intense and excitable Border collie and the easygoing and affectionate Newfoundland) resulted in puppies exhibiting a blending of these traits (Coren 2004). However, the next generation possessed unique combinations of those behaviors exhibited by their grandparents (Coren 2004). Efforts to exploit the instincts and behavioral tendencies of various breeds are ongoing in all sectors (i.e., military, medical research, foundations producing assistance dogs) in order to produce dogs even better suited physically, emotionally, and mentally for particular service activities.

Dogs serve humans in multiple capacities. They are indispensable as companions and as working dogs. Their contributions throughout the centuries, their changing roles, their unique abilities, and their importance to society are described herein.

MILITARY WORKING DOGS

This soldier, I realized, must have had friends at home and in his regiment, yet he lay there deserted by all except his dog.... I had looked on, unmoved, at battles which decided the future of nations. Tearless, I had given orders which brought death to thousands. Yet here I was stirred—profoundly stirred—stirred to tears. And by what? By the grief of one dog.

NAPOLEON BONAPARTE

Fossil and artistic evidence indicates that partnerships between humans and dogs have existed for 140,000 years (Coren 1994; Lemish 1996). Rock art depicting men hunting with what appear to be canids (quadrupeds with curly tails) suggests that, in the beginning, the dog was used primarily to aid in obtaining food (Lemish 1996; Thurston 1996). With the emergence of agricultural communities, dogs became a more integral part of society and were used to guard flocks, homes, and their masters (Thurston 1996). When conflicts between societies led to wars, the dog was utilized as a fighting tool and became an important part of military forces (Lemish 1996). In this role, the dog was a powerful and frightening weapon. For example, dogs wore spiked armor or carried lances and pots of burning resin attached to their bodies to wreak havoc on horses and infantry (Emert 1985c; Coren 1994).

The use of dogs in war was not indiscriminate. That is, specific breeds were often selected for battle. Ancient warring populations used an early version of the mastiff known as the Molossian dog, which weighed nearly 280 pounds and was highly aggressive (Coren 1994). The first organized war-dog units were established by the Romans (Emert 1985c). They selected intelligent, strong dogs that were territorial and exhibited the most aggressive tendencies (Coren 1994). To increase aggressiveness, the Romans kept dogs chained within small areas (Coren 1994). Attila the Hun used the Molossian dog and also the Talbot, an ancestor of the bloodhound, in his conquest of Europe (Coren 1994; Lemish 1996). In the 16th century, Spanish conquistadors used the swift greyhound and the savage Alaunt, an extinct ancestor of the bulldog and mastiff, to conquer the New World (Thurston 1996).

In World War I, more than 75,000 mixed and purebred dogs were used mainly by the military forces of Germany, France, and Belgium (Going 1944; Coren 1994). Dogs were selected for temperament and physical traits such as medium stature, dark coat, and superior olfactory and visual acuity (Lemish 1996). The German shepherd dog (GSD), bulldog, and Airedale terrier were among those breeds used in the war, as were various retrievers and sheepdogs (Lemish 1996). Dogs were used mainly as messengers and as sentries for protection of factories, railroads, and power plants (Stauffer 1943). Messenger dogs were a reliable and rapid means of communication between commanders and soldiers in trenches (Lemish 1996). The Red Cross also used dogs that worked primarily at night to carry medical supplies to the wounded and to lead medical personnel to them (Lemish 1996). Large-breed dogs could negotiate the terrain better than ambulances and were used to pull the wounded to aid stations (Lemish 1996). Interestingly, the U. S. was the only country to enter World War I without trained dogs, but quickly recognized their value and borrowed dogs from the French and British (Lemish 1996).

World War II saw increased numbers of working dogs, with more than 200,000 in service (Coren 1994). However, at the start of World War II, the U.S. had not yet established a military working dog (MWD) program (Lemish 1996). Frustrated dog fanciers lobbied government officials to adopt a war-dog program and in January, 1942, started a civilian agency, Dogs for Defense, which recruited and trained dogs for the military (Lemish 1996). Success with these dogs prompted the establishment of an extensive, although rudimentary, military dog program (Lemish 1996). During the course of the war, dog training and recruiting methods were refined (Lemish 1996). Specifically, instructions for handlers, technical manuals, facilities, and training and care films advanced the program (Lemish 1996).

During World War II, the U.S. did not have a breeding program and relied on dogs donated by breeders and pets contributed voluntarily (Lemish 1996). For admission into the program, dogs had to be 1–5 years of age, greater than 20 inches at the shoulder, and at least 50 pounds (Going 1944). Dogs were required to be in good physical condition, aggressive and alert, and tolerant of loud noises (Going 1944). Although originally 32 breeds and crosses were deemed suitable for training, the U.S. Army later accepted only five breeds: the GSD, Belgian sheepdog, Doberman pinscher, collie, and giant schnauzer (Emert 1985c). Unlike other branches of the military, the Marine Corps officially used the Doberman pinscher after the breed club pledged free dogs for the duration of the war (Putney 2001).

Dogs were used primarily as sentries, although technological advances such as new planes, bombs, and land mines broadened their utility in World War II (Thurston 1996). Dogs could warn of approaching aircraft and were trained to jump with troops from planes for nighttime missions in enemy territories (Coren 1994; Thurston 1996). Many countries trained suicide dogs to carry explosives into tanks and camps (Coren 1994).

In 1958, all training and care of MWDs was assumed by the Air Force and relocated to Lackland Air Force Base in San Antonio, Texas (Emert 1985c; Thurston 1996). Although the majority of dogs for the Vietnam War were trained at Lackland, some training occurred in Vietnam and at Fort Benning in Georgia, an area with similar terrain to southeast Asia (Emert 1985c). Even with increased recognition as to the value of MWDs, the U.S. did not have enough dogs to meet the needs presented by the Vietnam War. Therefore, it began purchasing GSDs from West Germany (Lemish 1996). Approximately 4,000 dogs served in the Vietnam War (Thurston 1996).

The heat and thick jungles of Vietnam and advances in post-World War II warfare required altering the roles of war dogs. Dogs were used mainly as scouts, sentries, and mine and tunnel detectors (Emert 1985c). Water dogs were trained to find the enemy hiding in swamps (Emert 1985c). In many instances, dogs provided companionship, and this eased the emotional toll endured by soldiers (Thurston 1996). War dogs were credited with saving at least 10,000 lives in the Vietnam War. Tragically, and unlike in previous wars after which dogs were returned to the U.S. and demilitarized, in the Vietnam War the government considered dogs to be surplus equipment and handlers were forced to leave them behind. The emotions that handlers and military personnel feel regarding MWDs is evidenced by the existence of memorials for war dogs found worldwide and in the videos and publications that are created to increase public awareness.

In 2004, the U.S. had an estimated 2,400 MWDs. Presently, between 500 and 600 dogs are purchased each year from vendors in Germany, Holland, and the Czech Republic. Dogs must be in excellent physical condition and are selected for aggressive tendencies and detection abilities. Preference is not given to specific breeds, although of the dogs purchased in 2004, 60% were GSDs and 40% were Belgian Malinois. These breeds offer intelligence and loyalty, have superior stamina, are versatile with regard to handling various environmental stresses, and possess superior olfactory abilities. MWDs are trained in an intensive 120-day program in one or more of four areas: patrol, explosives detection, narcotic detection, and mine detection.

The Air Force has piloted several breeding programs in the U.S. for MWDs. The most notable was the Canine Biosensor or Super Dog program of the 1960s. This produced almost 1,000 GSDs in 5 years. The goal of the program was to enhance the performance of MWDs by stimulating neurological development at an early age, but it was terminated primarily because of downsizing of military canine requirements. An ongoing breeding program was established in 2002 for the Belgian Malinois and Labrador retriever, selected for dual-purpose and detection work, respectively. Establishment of the breeding program in the midst of the global war on terrorism has been difficult because of the increased demand for MWDs.

Numbers of MWDs currently in service in Afghanistan and Iraq are classified. In general, there are between 1,000 and 2,000 dogs deployed at approximately 200 locations around the world. The U.S. spends seven to eight million dollars annually for the purchasing, training, and medical care of dogs. MWDs are in high demand by all services and are critical to current U.S. military efforts. Greater numbers will likely be needed in the future as new working tasks are developed for them.

LAW ENFORCEMENT DOGS

> *If these dogs only knew what a difference they make. Certainly, there's nothing that can replace the precision of a dog's nose—and absolutely nothing that can replace a dog's heart.*
>
> BOB SESSIONS, 9/11 RESCUE WORKER,
> FEDERAL EMERGENCY MANAGEMENT AGENCY

Law enforcement dogs have been used in Europe since the 13th century, and official police working dog programs were established there by the 20th century (Emert 1985b). The U.S. first launched a police dog program in 1907, although it failed due to lack of trainers, dogs, and funding (Emert 1985b). In 1956, a successful program was founded in Baltimore, Maryland and led to the development of canine units across the country (Emert 1985b).

Dogs assist law enforcement officials in a variety of tasks: protection, apprehension of criminals, building searches, missing person searches, explosives and narcotic detection, arson investigations, and confiscation of illegal imports. Most basic police dog training includes exercises in obedience, agility, attack methods, building searches, and tracking (Emert 1985b). Advanced training using odor signatures is necessary for dogs working in detection. Depending on the area of detection, dogs learn to alert either passively (e.g., sit, lie down, bark) or aggressively (e.g., paw at, dig). To maintain high performance levels, police dogs receive weekly training and an annual skills test (Emert 1985b).

The GSD, Belgian Malinois, Labrador retriever, and golden retriever are most commonly used in police work (Emert 1985b). Dogs must have a strong desire to please their handlers and are selected for intelligence, strength, aggression, loyalty, agility, and trainability (ARDA 1991). Breeds that excel at detection work must also possess a strong hunt-and-retrieve drive, endurance, and stable temperament. Dogs are donated to or purchased by law enforcement agencies from breeders or professional training facilities (Emert 1985b). Some agencies buy trained dogs from German vendors who have established breeding colonies specifically for police work (Emert 1985b).

The superior sense of smell in the dog (1,000–10,000 times more sensitive than the human) is the reason for its widespread use in detection work (Coren 2004). Dogs have a significantly larger region of sensory epithelium and devote 40% more of the brain to processing smells than do humans (Coren 2004; Olender et al. 2004). The canine olfactory repertoire is estimated to consist of 1,300 genes, with roughly 18% pseudogenes, in comparison to 60% of the 1,100 human olfactory receptor genes (Quignon et al. 2003; Olender et al. 2004, see Chapter 13).

This enhanced olfactory acuity allows dogs to recognize biological scents and pheromones and to use these odor signatures to distinguish between individuals (Coren 2004). Dogs have been used since the mid-1800s to assist in missing person searches (Jones et al. 2004). Search and Rescue (SAR) dogs use three methods to locate an individual: tracking, trailing, and air scenting; each one suited for different situations. Tracking uses ground scents, from the individual and disturbed terrain (e.g., broken vegetation), that are associated with the exact footsteps of the missing person (Coren 2004). These dogs work on a leash and must start with a scent article and point of origin (ARDA 1991). Trailing relies on the scent of skin cells that are shed and linger on or near the tracks (Coren 2004). Dogs work on or off leash and require a scent article, but no point of origin (Jones et al. 2004). Trailing dogs can perform even when others have contaminated the tracks, but both tracking and trailing dogs work best when other persons are not present (Jones et al. 2004). The bloodhound excels at tracking and trailing, and a recent study suggests that its olfactory acuity is far superior to that of the GSD (Harvey and Harvey 2003). These data also demonstrate that the sensitivity and discriminatory abilities of the bloodhound increase with the maturation of the neurological system (after ~1 year of age) (Harvey and Harvey 2003).

Tracking and trailing are ground-oriented methods that do not utilize airborne scents (ARDA 1991). Dogs trained in air scenting detect odors carried by the wind without following tracks and do not require a scent article or point of origin (ARDA 1991). This method allows dogs to work off leash and at the same time as other rescuers (Jones et al. 2004). Air scenting SAR dogs are capable of locating human cadavers and forensic evidence such as teeth, bones, and blood when trained to detect odors associated with the five stages of decomposition (Lasseter et al. 2003). "Cadaver dogs" can efficiently search large areas of terrain and identify scattered remains located above ground, in burial sites, and in water, thereby reducing the time, cost, and manpower necessary for investigations (Komar 1999; Lasseter et al. 2003). SAR dogs trained in air scenting are well suited for disaster situations in which the identity and number of victims are unknown. Approximately 250–300 SAR dogs were used to locate victims at the World Trade Center and Pentagon after the September 11th attacks (Otto et al. 2004). Although some groups exclusively use the GSD, various breeds that possess a coat suitable for all types of weather, proven scenting ability, and lack of inbred game instinct make excellent SAR dogs (ARDA 1991).

Explosives and narcotic detection dogs are widely used to aid in the confiscation of illegal goods at U.S. ports of entry and in public areas. Dogs are capable of identifying odor signatures associated with explosives and drugs including marijuana, cocaine, and methamphetamines. Dogs are able to distinguish between different scents to such a degree that it is possible to detect explosives and narcotics masked with other strong smells (Coren 2004). Upon detection of narcotics, dogs may be trained to passively or aggressively alert; however, explosive detection dogs are exclusively trained to passively alert. The demand for explosive detection dogs has risen drastically post-9/11.

Annually, arsonists destroy 75,000 buildings in the U.S. in order to defraud insurance companies and/or destroy evidence (Coren 2004). Mechanical devices are incapable of distinguishing between accelerants and compounds naturally produced by fire (Weisbord and Kachanoff 2000). Dogs, however, can identify accelerants such as gasoline, lighter fluid, paint thinner, alcohol, acetone, and kerosene for up to 18 days after fires (Weisbord and Kachanoff 2000; Coren 2004). In the U.S., the first arson detection dog was trained in 1986, and approximately 200 dogs were active in 2000. The short coat, gentle temperament, and olfactory sensitivity of the Labrador retriever make it the breed of choice in arson detection (Weisbord and Kachanoff 2000). Upon identifying an accelerant, dogs passively alert to minimize disruption of evidence, and samples are collected and sent to a laboratory for gas chromatographic analysis (Kurz et al. 1994). The demand for arson dogs is increasing among law enforcement and insurance companies because they reduce the time and money necessary for an investigation.

A group of beagles, owned by the U.S. Department of Agriculture (USDA) and known as the Beagle Brigade, is trained to protect American agricultural interests. More than 60 beagle teams inspect cargo in international airports and at border crossings to detect prohibited agricultural products that may harbor pests and/or diseases (www.usda.gov/news/releases/1995/03/0232). The program began in 1984 and utilizes beagles because of the superior olfactory acuity, good-natured personality, and small stature of this breed. Private owners, breeders, and rescue groups provide beagles between 9 months and 3 years of age to the USDA. In an 8–12-week training program, these dogs learn to passively alert to illegal products through a food reward system (www.usda.gov/news/releases/1995/03/0232). Beagle Brigade dogs are 90% accurate in detection after 2 years of service and have prevented roughly 75,000 agricultural goods from entry (www.usda.gov/news/releases/1995/03/0232).

SERVICE DOGS

It was glorious—just the dog and a leather strap linking me to life.

MORRIS FRANK

A service animal is any guide dog, signal dog, or other animal individually trained to provide assistance to an individual with a disability, as defined by the Americans with Disabilities Act. This law ensures that service animals are permitted into restaurants, theaters, modes of transportation, or any privately owned business that serves the public. The most well-known service dogs are guide dogs for the blind and hearing-assistance dogs. In recent years, dogs have been trained to assist people with other medical disabilities such as epilepsy, diabetes, paralysis, autism, and numerous psychological disorders.

A Chinese scroll dating to the 13th century depicts a dog leading a man with a cane, indicating that dogs have been used to guide the blind for hundreds of years (Fishman 2003). After observing schools in Germany training dogs for blind war veterans, Dorothy Harrison Eustis recruited Morris Frank, a blind American, to train with a GSD called Buddy and to become an ambassador for the use of guide dogs in the U.S. (Fishman 2003). Frank and Buddy traveled across the country to demonstrate that guide dogs offer independence, confidence, security, and companionship for blind people (Emert 1985e). In 1929, Frank and Eustis helped start the first guide dog school in America, The Seeing Eye, Inc., and the overwhelming demand for dogs eventually led to the establishment of fourteen other schools (Fishman 2003).

Dogs that are to be trained as guides are placed in homes as puppies in order to be socialized, taught basic commands, and exposed to various environments and situations (Emert 1985e). Studies have shown that puppies placed in homes at 8 weeks of age have a higher success rate in the training program than do those placed at a later age (Coren 2004). At The Seeing Eye, Inc., dogs return to the school at 18 months of age and, in a 4-month course, learn to work in a harness, stop at curbs, avoid dangerous situations, and even to disobey commands that could lead to harm (www.seeingeye.org). In the final phase, the dog and its new handler train together.

Selection of dogs for service activities requires the simultaneous consideration of aptitude, temperament, physical traits, and health. In 1941, The Seeing Eye, Inc. established its own breeding colonies for GSDs, Labrador retrievers, and golden retrievers (The Seeing Eye, Inc. 2005). To produce superior guide dogs, it maintains a comprehensive database of the colonies for use in studies of hip dysplasia, temperament, size, etc., and collaborates with other research groups to advance the health of the general canine population. Helmink et al. (2001, 2003) used these colonies to explore different breeding strategies for size, and they suggest that using a selection index may be the best method to obtain dogs within optimal height and weight ranges for handling. Many groups have conducted studies to show that behavioral characteristics (e.g., temperament, motivation, sharpness) are lowly to highly heritable and that certain traits can be selected for or against (Goddard and Beilharz 1985; MacKenzie et al.1985; Ruefenacht 2002). In the 1940s, Clarence Pfaffenberger of Guide Dogs for the Blind, Inc. bred for intelligence and personality by selecting for these characteristics and raised the percentage of dogs graduating from the program from 9 to 90 (Coren 2004).

Much like guide dogs for the blind, hearing-assistance dogs offer independence and security for deaf individuals, but they also alleviate feelings of isolation and loneliness (Fogle and Radcliffe 1983). In the 1970s, the American Humane Association in Denver started the first formal training program to produce hearing-assistance dogs (Coren 2002). Dogs learn to alert to sounds common in home and work environments. These sounds include the ringing telephone, doorbell, baby's cry, smoke alarm, incoming E-mail, and presence of other people. Interestingly, a survey of hearing-dog owners and prospective owners cited companionship and love as the most rewarding and anticipated aspect of owning a service dog (Hart et al. 1995). Most training programs obtain dogs from shelters and select for size (small to medium build), friendliness, and high energy (Emert 1985a).

People suffering from muscular dystrophy, multiple sclerosis, spinal cord injuries, or other conditions that result in ambulatory motor impairment may also benefit from service dogs. The utility of mobility-assistance dogs was first recognized in the U.S. in 1976 (Weisbord and Kachanoff 2000). Such dogs are specifically trained to fulfill the needs of their wheelchair-bound handlers, and some can perform almost 100 tasks (Allen and Blascovich 1996). Dogs assist with everyday chores including opening doors, picking up items, showering, getting dressed, washing/drying clothes, and grocery shopping. They are capable of pulling owners up from a sitting or lying position, removing them from dangerous situations, and calling for help in an emergency (Allen and Blascovich 1996). The temperament and intelligence of the Labrador retriever make it a popular breed for mobility assistance. Allen and Blascovich (1996) conducted the first study to evaluate the value of service dogs for people with ambulatory disabilities. They found that service dogs improved self-esteem, psychological well-being, and work/school attendance, and reduced the amount of assistance needed from others. Besides increasing independence, the latter is significant because it eases the financial burden of employing aides and relieves responsibilities of family and friends (Allen and Blascovich 1996).

In recent years, service dogs have proven able to provide assistance and, in some cases even treatment, to people suffering from seizure disorders. Dogs are trained to respond at the onset of the seizure (seizure-assist) or to detect the seizure prior to its onset (seizure-alert). These dogs remove the seizure victim from dangerous surroundings, provide support for the duration of the episode, and call for outside help if necessary (Weisbord and Kachanoff 2000). Benefits to owners of a seizure dog include a reduction in anxiety levels, increased self-esteem, and a feeling of control and predictability, all of which enhance the quality of life (Strong et al. 1999).

Strong et al. (1999) studied six epileptic patients with trained seizure-alert dogs and found that all dogs alerted 15–45 minutes prior to onset and were consistent in warning times. In this study, dogs were trained specifically for each individual using a Pavlovian associative technique in which they were conditioned to associate the seizure with a positive reward. It is believed that seizure-alert dogs respond to pre-seizure variants in the nervous system (e.g., small movements), but alerts from dogs that have been out of sight suggest the possibility of a scent or auditory signal (Weisbord and Kachanoff 2000; Dalziel et al. 2003). Interestingly, the Strong et al. study also found that each subject experienced a reduction in seizure frequency with the trained dogs. As of 2003, there were 15 organizations offering seizure-alert dogs, each with its own training methods and source of dogs (varying from pure breeds to rescued mongrels) (Dalziel et al. 2003).

DOGS AND HUMAN MEDICINE

The dog may have saved her owner's life by prompting her to seek treatment when the lesion was still at a curable stage.

DRS. HYWELL WILLIAMS AND ANDRES PEMBROKE

The concept of using dogs in medicine dates back to ancient times when healing dogs, known as cynotherapists, were believed to diagnose illness and relieve suffering (Thurston 1996). Although cynotherapists may have only been healing psychosomatic disorders, it is not inconceivable that the superior olfactory capability of the dog could aid in diagnosis by detecting metabolic changes that are associated with disease (Thurston 1996). Anecdotal reports over the past two decades have suggested this to be true; the best-known case was reported in *The Lancet* in 1989. A patient became concerned about a mole on her leg when her dog repeatedly sniffed at it and eventually tried to bite it off (Williams and Pembroke 1989). When the lesion was excised, histology showed it to be a malignant melanoma (Williams and Pembroke 1989). Other case reports state that dogs have detected breast and lung cancers, as well as malignancies of other internal organs (Willis et al. 2004).

To date, only two peer-reviewed studies that offer scientific data for cancer detection by dogs have been published (Pickel et al. 2004; Willis et al. 2004). The first of these, Willis et al. (2004) was a double-blind, proof-of-principle study that explored the use of dogs in the diagnosis of bladder cancer using odor signatures detectable in urine. Six dogs of various breeds were trained by operant conditioning to alert to one cancerous urine sample placed among six controls (Willis et al. 2004). The results were unambiguous: The dogs alerted to the urine from patients with bladder cancer an average of 41% of the time, which is highly statistically significant when compared to 14% expected by chance alone (Willis et al. 2004).

Inspired by the 1989 letter to *The Lancet*, a retired police dog handler conducted studies to determine whether his former bomb detection dog, a standard schnauzer, could identify malignant melanoma (Pickel et al. 2004). In these studies, which were expanded to include a golden retriever, the dogs were trained to retrieve a PVC tube containing malignant melanoma. In preliminary tests, the dogs had to identify the cancerous sample in two situations: (1) one melanoma sample and nine distractor stimuli placed in wooden boxes and (2) one melanoma sample and nine or ten distractor stimuli planted underneath bandages on healthy patients (Pickel et al. 2004). Adhesive bandages, gauze, rolls of tape, or other items expected to be present in a medical environment were used as the distractors (Pickel et al. 2004). Both dogs alerted to the melanoma in these situations with 100% accuracy (Pickel et al. 2004). The dogs were then tested using actual patients having a possible malignant melanoma with the suspect area bandaged, as well as several non-suspect areas (Pickel et al. 2004). The schnauzer alerted on 6 out of 7 patients with subsequently confirmed cancer and the golden retriever on 3 out of 4; the probabilities of these results occurring by chance alone were $10^{-7.1}$ and $10^{-3.5}$, respectively (Pickel et al. 2004).

Interestingly, in both the Willis et al. (2004) and Pickel et al. (2004) studies, dogs repeatedly alerted to control samples, prompting further testing, which revealed that the samples were in fact cancerous. Further research is necessary to determine what chemical compound(s) produced by the cancers is detectable by dogs. Nonetheless, if dogs are capable of recognizing odor signatures associated with cancers, the implications for human medicine are far-reaching. Dogs have the potential to supplement existing methods of diagnoses by providing early and accurate detection. Studies evaluating the use of detection dogs in the diagnosis of lung and prostate cancers are in progress, and it is likely that similar research efforts will follow.

Although the use of dogs as a diagnostic tool in medicine is still in its infancy, they have been used for decades as therapeutic tools in the treatment of patients with various disorders. Sigmund Freud, who was fond of dogs and usually had one in his office during psychotherapy sessions, observed that many patients were more open to the discussion of painful issues when a dog was present (Coren 2002). In the 1960s, Dr. Boris Levinson, a pediatric psychiatrist, observed the same phenomenon and presented the first scientific data on the subject (Coren 2002). Levinson concluded that pets may help people overcome difficulties in expressing themselves and meeting emotional needs and should be used in nursing homes, hospitals, schools, and prisons (Levinson 1970).

In 1977 at a psychiatric ward at Ohio State University, Drs. Sam and Elizabeth Corsen started the first pet-assisted therapy program, and by 2000 there were more than 1000 programs across the country (Coren 2002). As Levinson predicted, dogs have proven to be therapeutic in a number of situations. For example, canine programs implemented in correctional institutions allow inmates to acquire and practice job and interpersonal skills, learn to set and achieve goals, and receive unconditional love from the animals in their care. Many such programs adopt rescue dogs from shelters and train them as assistance dogs, providing a valuable service to the community. Working with the animals and other people in a positive manner aids in rehabilitation, while allowing the inmates to contribute to society. Successful programs have also been established in nursing homes and hospitals, in which dogs are permanent residents or frequent visitors. Interactions such as petting and playing with therapy dogs decrease loneliness, fear, and antisocial behavior in nursing home residents and long-term hospital patients. Anecdotal evidence suggests that dogs are capable of eliciting responses from people that humans are unable to generate (Burch 1996). For example, an elderly Alzheimer's patient spoke more clearly when visiting with a dog than she had previously spoken in conversations with people (Burch 1996). In each of the aforementioned settings, dogs provide the residents with a target for their attention, companionship activity, and a sense of responsibility. This results in increased morale, confidence, communication, and other positive socializing effects (Edney 1995).

In addition to psychological benefits, there is evidence for positive physiological effects from interactions with animals. Studies have shown that canine companionship can decrease blood pressure and increase survival rates after illness. Schuelke et al. (1991) evaluated blood pressure levels in normo- and hypertensive patients while they were petting a dog. The authors report a significant decrease in systolic and diastolic pressures in both groups. Friedmann et al. (1980) studied patients for 1 year after hospitalization for myocardial infarction or angina pectoris and found that ownership of a dog was correlated with increased survival rates. This finding is not a result of the physical activity associated with having a dog, because owners of other types of pets also had better survival rates (Friedmann et al. 1980).

SUMMARY

We give dogs time we can spare, space we can spare and love we can spare.
And in return, dogs give us their all. It's the best deal man has ever made.

M. FACKLAM

Although this brief report cannot possibly do justice to and address all the many contributions of the dog, it does serve to summarize the main categories of activities carried out by working dogs. A constant throughout the evolution of working dogs is that the number and types of activities and tasks performed by the dog are always changing. This happens because we are always identifying new tasks for which the dog is better suited than man or machine. For example, current efforts are directed toward training of dogs for use by the housing, insurance, and pest control industries. Specifically, the growth of particular fungi inside houses has become a major health issue. Often, the detection of such organisms is not accomplished until serious damage to the house and its inhabitants has occurred. The consequences of this include extensive medical treatment for respiratory diseases and destruction of the house. Early detection of fungal growth may prevent such drastic outcomes. Hence, dogs are being trained to detect fungi within the walls of houses. Another area of interest to the housing industry and homeowners is the detection of termites, which cause billions of dollars in damage worldwide to private and public structures. Dogs are vastly superior in detecting the presence of these pests and are increasingly employed for this.

Although only mentioned here, the dog's role in biomedical research, although not technically a "working" role, cannot be overlooked. The dog, due to its unique genetic structure and relative ease with which multigenerational pedigrees can be established, is an ideal model organism for study of human hereditary diseases. This was a prime reason for sequencing of the canine genome, work fully supported by the National Institutes of Health.

The International Working Dog Breeding Association annually brings together researchers from law enforcement agencies, the military, service dog training groups, and others to discuss issues involving the selective breeding, rearing, and assessment of dogs in working occupations. A major goal of those who train and care for working dogs is to improve breeding strategies so that dogs with desirable characteristics can be produced in greater numbers and comprise a greater percentage of all offspring. For example, this is the focus of breeding programs under the auspices of the U.S. Department of Defense (DoD). This is especially important for the DoD because it relies on purchase of dogs to maintain its population of MWDs. Obviously, if dogs could be produced in the U.S. and in a more directed fashion (i.e., breeding for specific traits), the efficiency of training would be markedly increased and the cost of this dramatically decreased to the DoD, and, in turn, the taxpaying public. An intriguing aspect pertaining to this is the possibility of cloning dogs. Although attempts to clone the dog have not been successful, it seems likely to be accomplished in the near future. One wonders if it may be practical one day to clone superior working dogs in order to produce dogs with genetic predisposition to function as their progenitors.

Finally, more than one-third of homes in the U.S. have at least one dog. The aid and comfort dogs provide through service activities and simply by just being there are impossible to estimate. However, one thing is clear, the human and dog are inextricably linked in a profound and ever-changing relationship with one constant: mutual devotion. It is only fair, then, that we provide dogs with medical surveillance, and superior treatment regimens, and continue to conduct research designed to eliminate diseases and improve the quality of life for this noble animal.

ACKNOWLEDGMENTS

We thank Walter F. Burghardt, Jr, D.V.M., Ph.D. of Lackland Air Force Base, San Antonio, Texas for his assistance in preparation of this manuscript. We also thank members of our laboratory for critical reading of this manuscript.

REFERENCES

Allen K. and Blascovich J. 1996. The value of service dogs for people with severe ambulatory disabilities. *J. Am. Med. Assoc.* **275:** 1001–1006.

American Rescue Dog Association (ARDA). 1991. *Search and rescue dogs: Training methods.* Howell Book House, New York.

Burch M.R. 1996. *Volunteering with your pet: How to get involved in animal-assisted therapy with any kind of pet.* Howell Book House, New York.

Coren S. 1994. *The intelligence of dogs.* Bantam Books, New York.

———. 2002. *The pawprints of history: Dogs and the course of human events.* Free Press, New York.

———. 2004. *How dogs think: Understanding the canine mind.* Free Press, New York.

Dalziel D.J., Uthman B.M., McGorray S.P., and Reep R.L. 2003. Seizure-alert dogs: A review and preliminary study. *Seizure* **12:** 115–120.

Edney A.T.B. 1995. Companion animals and human health: An overview. *J. R. Soc. Med.* **88:** 704P–708P.

Emert P.R. 1985a. *Hearing-ear dogs* (ed. Dr. H. Schroeder). Crestwood House, New York.

———. 1985b. *Law enforcement dogs* (ed. Dr. H. Schroeder). Crestwood House, New York.

———. 1985c. *Military dogs* (ed. Dr. H. Schroeder). Crestwood House, New York.

———. 1985d. *Search and rescue dogs* (ed. Dr. H. Schroeder). Crestwood House, New York.

———. 1985e. *Guide dogs* (ed. Dr. H. Schroeder). Crestwood House, New York.

Fishman G.A. 2003. When your eyes have a wet nose: The evolution of the use of guide dogs and establishing the seeing eye. *Surv. Ophthalmol.* **48:** 452–458.

Fogle B. and Radcliffe A. 1983. Hearing dogs for the deaf. *Practitioner* **277:** 1051–1053.

Friedmann E., Katcher A.H., Lynch J.J., and Thomas S.A. 1980. Animal companions and one-year survival of patients after discharge from a coronary care unit. *Public Health Rep.* **95:** 307–312.

Goddard M.E. and Beilharz R.G. 1985. A multivariate analysis of the genetics of fearfulness in potential guide dogs. *Behav. Genet.* **15:** 69–89.

Going C.G. 1944. *Dogs at war.* Macmillan, New York.

Hart L.A., Zasloff R.L., and Benfatto A.M. 1995. The pleasures and problems of hearing dog ownership. *Psychol. Rep.* **77:** 969–970.

Harvey L.M. and Harvey J.W. 2003. Reliability of bloodhounds in criminal investigations. *J. Forensic Sci.* **48:** 811–816.

Helmink S.K., Shanks R.D., and Leighton E.A. 2003. Investigation of breeding strategies to increase the probability that German shepherd dog and Labrador retriever dog guides would attain optimum size. *J. Anim. Sci.* **81:** 2950–2958.

Helmink S.K., Rodriguez-Zas S.L., Shanks R.D., and Leighton E.A. 2001. Estimated genetic parameters for growth traits of German shepherd dog and Labrador retriever dog guides. *J. Anim. Sci.* **79:** 1450–1456.

Jones K.E., Dashfield K., Downend A.B., and Otto C.M. 2004. Search-and-rescue dogs: An overview for veterinarians. *J. Am. Vet. Med. Assoc.* **225:** 854–860.

Komar D. 1999. The use of cadaver dogs in locating scattered, scavenged human remains: Preliminary field test results. *J. Forensic Sci.* **44:** 405–408.

Kurz M.E., Billard M., Rettig M., Augustiniak J., Lange J., Larsen M., Warrick R., Mohns T., Bora R., Broadus K., et al. 1994. Evaluation of canines for accelerant detection at fire scenes. *J. Forensic Sci.* **39:** 1528–1536.

Lasseter A.E., Jacobi K.P., Farley R., and Hensel L. 2003. Cadaver dog and handler team capabilities in the recovery of buried human remains in the southeastern United States. *J. Forensic Sci.* **48:** 617–621.

Lemish M.G. 1996. *War dogs: Canines in combat.* Brassey's, Washington.

Levinson B.M. 1970. Pets, child development, and mental illness. *J. Am. Vet. Med. Assoc.* **157:** 1759–1766.

Mackenzie S.A., Oltenacu E.A.B., and Leighton E. 1985. Heritability estimate for temperament scores in German shepherd dogs and its genetic correlation with hip dysplasia. *Behav. Genet.* **15:** 475–482.

Olender T., Fuchs T., Linhart C., Shamir R., Adams M., Kalush F., Khen M., and Lancet D. 2004. The canine olfactory subgenome. *Genomics* **83:** 361–372.

Otto C.M., Downend A.B., Serpell J.A., Ziemer L.S., and Saunders H.M. September 2004. Medical and behavioral surveillance of dogs deployed to the World Trade Center and the Pentagon from October 2001 to June 2002. *J. Am. Vet. Med. Assoc.* **225:** 861–867.

Pickel D., Manucy G.P., Walker D.B., Hall S.B., and Walker J.C. 2004. Evidence for canine olfactory detection of melanoma. *Appl. Anim. Behav. Sci.* **89:** 107–116.

Putney W.W. 2001. *Always faithful: A memoir of the Marine dogs of WWII.* Free Press, New York.

Quignon P., Kirkness E., Cadieu E., Touleimat N., Guyon R., Renier C., Hitte C., Andre C., Fraser C., and Galibert F. 2003. Comparison of the canine and human olfactory receptor gene repertoires. *Genome Biol.* **4:** R80.0–R80.9.

Ruefenacht S., Gebhardt-Henrich S., Miyake C., and Gaillard C. 2002. A behavior test on German Shepherd dogs: Heritability of seven different traits. *Appl. Anim. Behav. Sci.* **79:** 113–132.

Schuelke S.T., Trask B., Wallace C., Baun M.M., Bergstrom N., and McCabe B. 1991. Physiological effects of the use of a companion animal dog as a cue to relaxation in diagnosed hypertensives. *Latham Lett.* **13:** 14–17.

Scott J.P. and Fuller J.L. 1965. *Genetics and the social behavior of the dog.* The University of Chicago Press, Chicago, Illinois.

Stauffer A.P. 1943. *The war dog program*. Historical Section, Office of the Quartermaster General, Washington D.C.

Strong V., Brown S.W., and Walker R. 1999. Seizure-alert dogs—Fact or fiction? *Seizure* **8:** 62–65.

Thurston M.E. 1996. *The history of the canine race: Our 15,000-year love affair with dogs.* Andrews and McMeel, Kansas City, Missouri.

Weisbord M. and Kachanoff K. 2000. *Dogs with jobs: Working dogs around the world.* McArthur & Company, Toronto.

Williams H. and Pembroke A. 1989. Sniffer dogs in the melanoma clinc? *Lancet* **1:** 734.

Willis C.M., Church S.M., Guest C.M., Cook W.A., McCarthy N., Bransbury A.J., Church M.R.T., and Church J.C.T. September 2004. Olfactory detection of human bladder cancer by dogs: Proof of principle study. *Br. Med. J.* **329:** 712.

WWW RESOURCES

www.seeingeye.org/AboutUs.asp?sc=fq The Seeing Eye, Inc., accessed on January 15, 2005. Morristown, New Jersey.

www.usda.gov/news/releases/1995/03/0232 USDA's beagle brigade: Protecting American agriculture. United States Department of Agriculture (USDA), accessed on January 15, 2005.

2

The Kennel Club and the Early History of Dog Shows and Breed Clubs

J. Sampson
The Kennel Club, London
W1J 8AB, United Kingdom

M.M. Binns
The Royal Veterinary College, London
NW1 0TU, United Kingdom

A RECENT BOOK ABOUT DIFFERENT pedigree dog breeds (Morris 2001) described some 1000 breeds around the world. This chapter aims in part to explain how this wealth of variety has come about as a result of the enthusiasm of dog breeders for the organization of showing, trialing, and classifying the species they are so fascinated with.

CLASSIFICATION OF BREEDS

Animal breeds have been defined as "a group of animals that has been selected by man to possess a uniform appearance that is inheritable and distinguishes it from other groups of animals within the same species." The accurate classification of dog breeds based on their interrelationship presents particular difficulties. It is clear that certain features, such as size, are unhelpful, since breeds such as the mastiff and the pug are anatomically similar, though of vastly different size. Indeed, for each of the toy breeds there exists a full-size equivalent. The division into function is also not likely to group the most closely related breeds, as a wide variety of different types have historically been used for similar purposes.

The first list of recognized breeds printed in English is that published in 1486 of Dame Juliana Berners, a treatise on hunting in the *Boke of*

St Albans. "Thyse ben the names of houndes, fyrste there is the Grehoun, a Bastard, a Mengrell, a Mastiff, a Lemor, a Spanyel, Raches, Kennettys, Teroures, Butchers' Houndes, Myddyng dogges, Tryndel-taylles, and Prikherid currys, and small ladyes'poppees." Although the list only describes a small number of varieties, several names are recognizable today, and, even at that time, some importance was attached to the different breeds.

A formal classification of dogs, *De Canibus Britancicus,* was published in 1570 by Dr. Caius (the second founder of Gonville and Caius College Cambridge), with an English translation following in 1576. All known breeds of dog were divided into sections, and the origins of their names (etymology) were detailed. An indication of the functional nature of breeds at this time is apparent from the initial classification of his first group into "Venatici," used for hunting beasts, and "Acupatorii," used for hunting fowl. The Venatici included Harriers, Terrars, Bloodhounds, Gazehounds, Grehounds, Lyemmer, Tumblers, and Stealers, whereas the Acupatorii only comprised Setters and Spaniells.

The next significant grouping of dog breeds was undertaken by the father of modern classification schemes, Carl von Linne, better known as Linnaeus (1707–1778). The list assembled by Linnaeus comprised 35 breeds, nearly doubling the earlier efforts of Dr. Caius. The Latin names assigned by Linnaeus to the different breeds have disappeared, and current breeds are all categorized as *Canis familiaris,* with no attempt to include scientific names for the vastly increased numbers of modern breeds.

A contemporary of Linnaeus, Buffon (1707–1788), produced a genealogical tree of all the known races of dogs (Fig. 1), and he believed that the basic breeds derived from climatic influences, whereas the varieties that followed were generated by interbreeding these initial breeds. Buffon believed that all dog breeds derived from the shepherd dog, which he regarded as the most sensible of all breeds. The *Encyclopaedia Britannica* (Dublin Edition) of 1791 described five main classes of dogs, Sheepdogs, Hounds, Spaniels, Greyhounds, and Bulldogs, based on the work of Buffon.

The next major breakthrough was made by Baron Cuvier (1769–1832), who divided dogs into three classes, Matins, Spaniels, and Dogues, each of which was divided into further sections. His classification was based on the head and jaw shape, together with other morphological features, rather than being based on the dog's behavior. As such, his classification had a strong influence on future schemes.

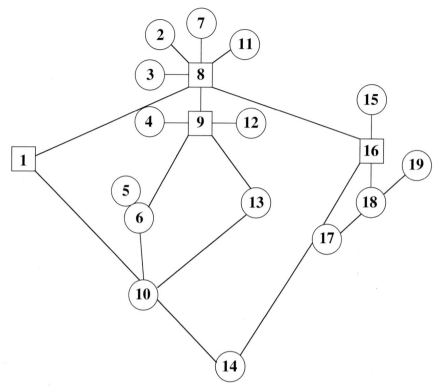

Figure 1. Buffon's (1707–1788) genealogical table of the different races of dogs. (1) Bulldog, (2) Iceland hound, (3) wolf dog, (4) terrier, (5) large spaniel, (6) small spaniel, (7) Lapland dog, (8) shepherd dog, (9) hound, (10) small water dog, (11) Siberian hound, (12) harrier, (13) water dog, (14) mastiff, (15) large Danish dog, (16) Irish greyhound, (17) English greyhound, (18) large greyhound, and (19) great hound mongrel.

THE FIRST DOG SHOWS AND THE ESTABLISHMENT OF THE KENNEL CLUB

The first dog show held in Britain is thought to have been an exhibition of spaniels held at the Zoological Gardens, London, in 1843. Victorian society was fascinated by exhibitions of all kinds, a thirst that was fueled by the Great Exhibition of 1851. It is therefore no surprise that during this period there was an increase in enthusiasm for competition between dogs, either in the show ring or in the field. The first dog show ever held, organized by a local sportsman and a local gunsmith, took place in the Town Hall, Newcastle-on-Tyne, on June 28 and 29, 1859, where exhibits were confined to pointers and setters. There were 60 dogs entered, about 15,000 supporters attended, and three judges were appointed for each breed. Messrs. J. Jobling, T. Robson, and J. H. Walsh judged the pointer classes and found an overall winner in Mr. R. Brailsford's liver and white dog by Lord Derby's Bang out of his Dora. Setters were judged by Messrs. F. Foulger, R. Brailsford, and J. H. Walsh, who found a winner in Mr. Jobling's Gordon setter, Dandy. The winner of the pointer classes was owned by one of the setter judges, and the winner of the setter classes was owned by one of the pointer judges!

The second dog show quickly followed in November of the same year and was organized in Birmingham. The Newcastle show was not repeated, but the Birmingham Dog Show became an annual event. Manchester Dog Show was first held at Belle Vue in 1861 and, like Birmingham, became an annual event. Both are still major championship dog shows in the present U.K. calendar. In the period between 1859 and 1873, when the Kennel Club was established, 50 different dog shows were held.

Dog shows preceded field trials by six years. The first trial of dogs in the field took place on Tuesday, the 18th of April, 1865, at Southill in Bedfordshire, over the estate of Mr. Samuel Whitbread, M.P. Sixteen pointers and setters entered the trial. According to a report in *The Field*, "hares and partridge were very plentiful, the fields being large, and a great portion of the ground under wheat. The day was not a good scenting one, being hot and close, with scarcely a breath of air." The trial was pronounced by *The Field's* leading article to have been a success, although it "occasioned a good deal of correspondence." Interestingly, with one or two exceptions, the dogs that ran in this first field trial were also exhibited at the Islington Dog Show held in June of the same year. Today we see a far greater separation between show dogs and field trial dogs, with few individual dogs taking part in both activities. Although a little slower to take off, when compared to dog shows, field trials increased in popularity, and in total, 19 trials were held between 1865 and 1873.

These early sporting activities for the dog were not without their problems. To quote from *The Kennel Club: A History and Record of Its Work*, by E. W. Jaquet (1905), "In the interval of fourteen years between the date of the first dog show and the foundation of the Kennel Club, many irregularities—not to say scandals—had arisen, and the need for legislation and guidance became an absolute necessity. This necessity arises in all forms of sport as soon as they become popular, and as in most respects the aims and objectives of the dog owner are similar to those which prevail among owners and breeders of bloodstock it was felt that a body with power to enforce their decisions should be brought into existence to do for dogs what the Jockey Club had done in connection with equine affairs." So it was that Mr. S. E. Shirley, of Ettington, decided to establish the Kennel Club in April, 1873. Again to quote Jaquet, "It is solely owing to the influence of the Club that today dog breeding and dog showing are pursuits which can be indulged by gentlewomen, and dog shows held under Kennel Club Rules are patronised by every class of the community, including their Majesties the King and Queen—a wonderful testimony to the acumen and foresight of Mr. Shirley."

S. E. Shirley was chairman of the Kennel Club from 1873 to 1899, and then its president from 1899 until his death in 1904. Shirley was an aristocrat and extremely wealthy, and his family was of Saxon descent. He was born in 1844 and educated at Eton and Christ Church, Oxford. Interested in all country pursuits, he had a pack of beagles at Eton, he rode to hounds and was an excellent shot, and he had a serious interest in breeding dogs and working his gun dogs. Shirley was not only a man involved in all aspects of dogs, he was an M.P., in government from 1868 to 1880, a Justice of the Peace, and also High Sherriff and Deputy Lieutenant for County Monaghan. It was Shirley who, on April 4, 1873, met with 12 colleagues at 4 Albert Mansions, Victoria Street, London, to establish the Kennel Club.

One of the earliest undertakings of the newly formed Kennel Club was the compilation of a studbook. In preparing the studbook, Mr. Shirley consulted the then-editor of *The Field*, who recommended one Mr. Frank C. S. Pearce as the book's editor. Duly appointed, Mr. Pearce began producing the very first *Kennel Club Stud Book* for distribution at the Birmingham Dog Show held on December 1–4, 1874. This first studbook covered the years from the first dog show in 1859 to 1874 and was de-signed to cover a blank in the history of the canine world. Its production was no mean feat; the final book contains the pedigrees of 4,027 dogs, and Pearce claims to have applied for pedigrees to at least 3,500 people during its production. These 4,027 dogs represented 40 different breeds divided into two groups: Sporting Dogs and Non-Sporting Dogs. In ad-dition, this first studbook contained the rules of the newly formed Ken-nel Club, a list of club members (numbering 51), a code of rules for the guidance of dog shows, a code of rules for the guidance of field trials for sporting dogs, together with a suggested scale of points for the judging of such trials, and, finally, a list of winners at all of the dog shows and field trials held between 1859 and 1874. *The Kennel Club Stud Book* has been published annually since.

1880 was one of the most eventful years in the annals of the Kennel Club because it was in this year that new rules came into place that contained the important enactments regarding registrations. That new rules governing registration were required is clearly demonstrated in Shirley's own words: "Does it tend to improvement or give much reliable information to breeders or purchasers to find a pointer called Wag, another Roy; mastiffs under the name of Turk and Nero; fox terrier Pincher; two Bedlingtons, side by side in the show, both 'Highly commended' and both being called Piper, one being a well-known dog, the other not; quantities of Shots, Bobs, Bangs, Jets, Nettles, Vics etc. most of them insufficiently described and none of them the well-known dogs of the same name." The new registration rules brought much-needed method to registration and ensured that each registered dog's name was unique. This same year also saw considerable debate over the precise definition of breeder and a discussion of a system of prefixes that could be used to identify dogs bred by the same breeder.

By 1905 the number of breeds had risen to 76. Dog breeds continued to be classified into two groups up until 1920, when 84 different breeds were recognized by the Kennel Club. In 1921 the Sporting breeds was subdivided into Sporting and Terriers, and in 1922 the Gundog group was added to the Sporting breeds; in the same year the Non-Sporting breeds were divided into Non-Sporting and Toy groups. This system of classifying breeds remained in place until 1962 when the Hound group was added to the Sporting breeds, the Non-Sporting breeds remaining as either the Non-Sporting group or the Toy group. 1968 saw the next major change to the breed classification scheme with the Sporting breeds being classified into one of three groups (Hound, Gundog, and Terrier) and the Non-Sporting breeds being classified into three groups (Utility, Working, and Toy). The final change to the breed classification system came in 2000, when the Working group was split to give a Working group and a Pastoral group. Thus, today the Sporting breeds are classified into Hound, Gundog, and Terrier groups and the Non-Sporting breeds into Utility, Working, Pastoral, and Toy groups. In 1875, the Kennel Club recognized 40 different breeds and registered some 719 new dogs; in 2003 it recognized 201 different breeds and registered 245,894 new dogs.

On September 17, 1884, 18 months after the establishment of the Kennel Club in London, twelve representatives gathered in the offices of the Philadelphia Kennel Club, in response to a meeting call issued by J.M. Taylor and E. Smith. The outcome of this meeting was the establishment of what we now know as the American Kennel Club (AKC). The next meeting of the group was held on October 22, 1884, when a constitution and by-laws were adopted and Major James M. Taylor became the AKC's first president. One of the first tasks of this newly formed club was to produce a reliable studbook, which of course was the main priority at the formation of the Kennel Club in the U.K. Since 1878 Dr. N. Rowe had been compiling three volumes of *The National American Kennel Club Stud Book*, which he donated to the American Kennel Club, a gift that was acknowledged in the publication of the fourth volume of *The American Kennel Club Stud Book* in 1887.

From these very early beginnings, the AKC has grown into an organization controlling a vast "canine empire." For example, in 1998 almost 2 million dogs competed in 15,000 member, licensed, and sanctioned events. In the same year, the AKC registered more than 1.2 million dogs, representing 555,000 litters.

The Fédération Cynologique Internationale (FCI) is an umbrella organization for many of the other National Kennel Clubs that have been founded since the formation of the Kennel Club and the American Kennel Club, and it was created on May 22, 1911, with the aim to promote and protect cynology and purebred dogs by any means it considers necessary. The founding nations of the FCI were as follows:

- **Germany** (Kartell für das Deutsche Hundewesen und Die Delegierten Kommission)
- **Austria** (Osterreichischer Kynologenverband)
- **Belgium** (Société Royale Saint-Hubert)
- **France** (Société Centrale Canine de France)
- **Netherlands** (Raad van Beheer op Kynologisch Gebied in Nederland)

The Fédération disappeared due to the first World War, and in 1921, the Société Centrale Canine de France and the Société Royale Saint-Hubert recreated it. New articles of association were adopted on April 10, 1921, and on March 5, 1968, the FCI got the legal personality by decree. The FCI is the World Canine Organisation. It includes 80 members and contract partners (one member per country) that each issue their own pedigrees and train their own judges.

CONTEXTUAL HISTORY

The formation of the Kennel Club and the establishment and interest in dog breed clubs should be set in context of what was happening in relation to other domestic species in Great Britain at the time. It is clear from a study of Charles Darwin's *The Variation of Animals and Plants under Domestication*, originally published in 1868, that a wide variety of breeds were recognized and valued at the time he was writing. There was intense debate at this time as to whether the huge phenotypic diversity seen in domestic dog breeds could have arisen by the selection of individuals with a common origin in the gray wolf, or whether other canidae had also contributed extensively to the gene pool in different domestic dog breeds. He concludes that "it is highly probable that the domestic dogs of the world are descended from two well-defined species of wolf (viz. *C. lupus* and *C. latrans*) and from two or three other doubtful species (namely, the European, Indian, and North African wolves); from at least one or two South American canine species; from several races or species of jackal; and perhaps from one or more extinct species." Darwin based his conclusions, which turned out to be erroneous, on detailed studies on the wide range of dog breeds with which he was familiar, including greyhounds, mastiffs, Scottish deerhounds, Irish wolf dogs, bloodhounds, bulldogs, pointers, shepherds, setters (Irish and English), King Charles spaniels, Eskimo dogs, foxhounds, retrievers, pugs, Newfoundlands, and English otterhounds.

In many cases, the development of breeds grew out of attempts to improve the quality of animals through selective breeding. Robert Bakewell (1725–1795) is regarded as the first person to have made strenuous efforts at animal improvement in the U.K. through the use of selective breeding. Bakewell was the first person to begin breeding specific individuals together. Previously, livestock of both sexes were kept together in the fields, breeding at random, resulting in progeny with random characteristics. Bakewell separated the male from female, and allowed only specific matings to take place. Furthermore, by inbreeding his livestock, he fixed and exaggerated those traits he felt to be desirable.

He was also well aware of the value of progeny testing and leased out his young bulls and rams to test their value as breeding animals. Some of the impact of assessing breeding value has subsequently been diluted by the influence that shows have had in identifying show champions as individuals, rather than in stressing the way in which it breeds and produces top quality offspring.

Toward the end of Bakewell's life, the first studbook for Thoroughbred horses was published. The Thoroughbred horse breed was generated in the U.K. in the early 1700s, from a mixture of "Arabian" horses and local breeds. In 1791, the first volume of the studbook was published, in which the early history of the breed was recapitulated. There was great interest in pedigrees from an early stage, with selective breeding for athletic ability advanced through the popularity of champion horses at stud.

Although other horse breeds, including the Suffolk Punch and the Clydesdale, were well recognized by the 1750s, it wasn't until the 1880s, more than a hundred years after Bakewell's initial breeding work, that studbooks were established for these breeds. In England at this period there was also a rapid expansion in the number of cattle, pig, and sheep breed clubs, i.e., at exactly the same time that there was a burgeoning interest in dog breed clubs and studbooks.

The Victorians were clearly fascinated by the ideas of breed purity and genetic improvement. Indeed, there was widespread concern about the concept of degeneration, the progressive ill health in succeeding generations of a family, and the need to actively reverse this trend. This in turn probably lay behind early ideas of eugenics and interests in physiognomy that also advanced in parallel with the ideas of breed purity in dogs and other species.

THE DEVELOPMENT OF THE GOLDEN RETRIEVER BREED

One of the best-documented histories of the development of a new breed of dogs from this period of intense activity is that of the golden retriever. Retrievers were not seen as specific breeds until the latter part of the 19th century, and their development owes much to concurrent improvements in guns. Prior to this date, retrievers were in a state of flux and were mated for their ability to work, rather than their overall appearance. Thus, breeds like the golden retriever are relatively recent arrivals on the dog scene, certainly when compared to setters and pointers; the English setter's work has been described in historical references since the 14th century.

In the days of muzzle-loading guns, not many birds would have been shot in a day, probably no more than 10 per day, and thus landowners could easily retrieve the shot birds using their pointers and setters. By the 1880s the steel-barreled breech-loading gun, with its single trigger, had arrived on the scene, which meant that birds could now be shot in flight, and led to the introduction of the French partridge into Britain. French partridge were particularly good for the driven shoot where birds were driven over standing guns. Further gun improvements, like the introduction of the break-and-load shotgun, led to the replacement of partridge by pheasants, because their higher and faster flight provided better sport for the guns. All this meant that many more birds were shot in a day and, because they were shot in flight, they came down over very large areas; there was now a requirement for dogs to retrieve the birds, hence the development of the retriever breeds.

The development of the golden retriever breed owes everything to a gentleman called Sir Dudley Coutts Majoribanks, who later became the first Lord Tweedmouth. He and his family lived in London, but like many of his compatriots of the time who moved in the highest echelons of Victorian society, he devoted most of his life and energy to sporting pursuits in the Highlands of Scotland. In 1854 he purchased the Guisachan Estate, 20,000 acres of prime Scottish land near Loch Ness, and set about creating a new sporting estate par excellence. At the time, the most common retriever was something called a wavy-coated retriever, probably the forerunner of today's flat-coated retriever. Like Labrador retrievers, these wavy coats came in black, yellow, and chocolate varieties, but the black color was greatly favored. Marjoribanks decided that he would prefer to develop a new strain of yellow retriever, which was later to become the golden retriever. The key dog in the heritage of the golden retriever is a dog purchased by Majoribanks called Nous, which is Gaelic for wisdom.

There is some debate as to how this dog came into his ownership, but the vast proportion of evidence suggests that Nous was a yellow wavy-coated retriever that Majoribanks purchased in Brighton in the 1860s, and was part of a litter bred by Lord Chichester on his Stanmer Estate, just outside Brighton. What happened then is documented in detail in the studbook that was maintained by Majoribanks. This original studbook is now kept in the Kennel Club library in the U.K. for all to examine. Its contents show the meticulous planning that went into Majoribank's breeding program from 1868 to his last litter in 1889–1890, before his death in 1894.

In 1868 Nous was mated to a Tweed water spaniel called Belle, giving rise to four yellow puppies called Ada, Cowslip, Crocus, and Primrose (see Fig. 2). Tweed water spaniels are now extinct as a breed, but they were quite common in that late 19th century, particularly in the Borders between Scotland and England. They were described at the time as being similar to a small retriever, liver-colored and curly-coated. Cowslip was eventually mated to a second Tweed water spaniel to produce a litter containing a yellow bitch puppy called Topsy. In a second mating, Cowslip was mated to a red setter (Sampson) to produce a male puppy called Jack. Topsy was then mated to a black wavy-coated retriever (Sambo). Zoe, a bitch from this litter, was then mated to Jack, and the subsequent litter contained a yellow bitch called Gill and a yellow male called Nous II. Gill was then mated to a black Labrador called Tracer, and a puppy from this litter, called Queenie, was mated to Nous II, her uncle. A sandy-colored bloodhound was also used in the crosses.

This represents the foundation of Majoribank's golden retriever line; furthermore, he gave progeny from earlier matings, like Ada,

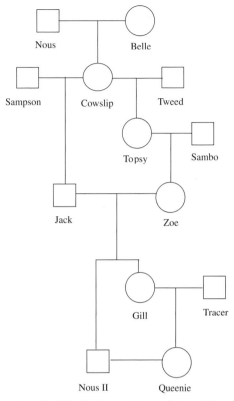

Figure 2. The crosses involved in the foundation of the golden retriever breed.

Crocus, and Primrose, to fellow breeders who used them to further develop the breed. The origins of this popular breed therefore lie in Scotland and the Border country, with all golden retrievers tracing back to Nous II and Lord Tweedmouth's methodical line-breeding between 1868 and 1890.

SUMMARY

The holding of dog shows and the subsequent establishment of the kennel club in Britain in 1873, the American Kennel Club in 1884, and the Fédération Cynologique Internationale in 1911 have had an enormous influence on the development of pedigree dog breeding and showing worldwide. The different kennel clubs around the world continue to have an important role in leading initiatives to ensure the health and welfare of man's best friends and the continued enjoyment they bring to dog owners and breeders.

REFERENCES

Darwin C. 1868. *The variation of animals and plants under domestication: Volume 1.* Johns Hopkins Press Ltd., London, 1998.

Jaquet E. 1905. *The Kennel Club: A history and record of its work.* The Kennel Gazette, London.

SUGGESTED READINGS

Foss V. 2005. *The Golden Retriever fifth book of champions, show champions, field trial champions and obedience champions 1999–2004.* NB Colour Print Ltd., Lancashire, United Kingdom.

Hubbard C.L.B. 1948. *Dogs in Britain: A description of all native breeds and most foreign breeds in Britain.* Macmillan and Company, London.

Leighton R. 1916. *The new book of the dog. Vols. I, II.* Cassell and Company, London.

Morris D. 2001. *Dogs: The ultimate dictionary of over 1000 dog breeds.* Ebury Press, London.

Pawson R. 1957. *Robert Bakewell pioneer livestock breeder.* Crosby Lockwood and Son, London.

Vesey-Fitzgerald B. 1948. *The book of the dog.* Nicholson and Watson, London and Brussels.

———.1957. *The domestic dog.* Routledge and Kegan Paul, London.

3

American Breed Clubs and Health Initiatives

Jerold S. Bell
Department of Clinical Sciences
Tufts Cummings School of Veterinary Medicine
North Grafton, Massachusetts 01536-1895

THE FORMATION OF DOG CLUBS and official studbook registries did not occur in America until the late 1800s. All-breed clubs were developed to organize conformation shows or field trials. Single-breed clubs were formed to develop a breed and to establish an American breed standard. The establishment and publication of pedigree studbooks provided legitimacy to breeding records and separated mixed-breed from purebred stock.

Many breeds were imported from Europe and other areas of the world. American breed standards initially mirrored those of the originating country. However, based on the function of the breed and American preferences, many breed standards changed from those of their country of origin.

Some American breeds were developed on the basis of a working phenotype that brought together founders from unrelated breeds. Other breeds were developed by inbreeding on a limited number of related founder dogs, and then expanding the population. There are also breeds or varieties that developed as offshoots of other pure breeds.

ESTABLISHMENT OF AMERICAN DOG CLUBS AND STUDBOOKS

In the U.S., dogs were bred as working dogs, for sport, or for companionship. The first American studbook of pedigrees, which included 327 sporting dogs, was published in 1876 in *The American Kennel & Sporting Field*. *The Field Dog Stud Book* continues to be published for field trail and hunting dogs. On January 26, 1876, the National American Kennel

Club (NAKC) was formed in St. Louis, Missouri, with the purpose of running field trials and bench (conformation) shows. Their first stud-book of 1,416 dogs was published in 1879. It included pointers, English, Gordon, and Irish setters; and Clumber, cocker, Irish water, and Sussex spaniels. The first volume was followed by Volume 2 in 1885 and Volume 3 in 1886, with a combined total of 5,397 dogs (The American Kennel Club 1985, 1998).

The American Kennel Club (AKC) was formed on September 17, 1884. Incorporating and taking over responsibility from the *NAKC Stud Book*, they published their first studbook, numbered Volume 4, in 1887, and continue publishing monthly. The AKC currently registers dogs from 151 breeds divided into seven groups: Sporting, Hound, Working, Terrier, Toy, Non-Sporting, and Herding. The Miscellaneous class is a provisional group that currently includes four breeds undergoing evaluation for full acceptance into the AKC studbook. The AKC maintains a foundation stock service for 48 other non-AKC breeds (The American Kennel Club 1998, 2004).

The Westminster Kennel Club ran its first all-breed dog show in New York City on May 8–10, 1877. This was an annual event that planted the seeds for the organization of specialty dog clubs. The American Fox Terrier Club formed in 1885 and was the first specialty breed club to join the AKC in 1888 (The American Fox Terrier Club 1924). The American Spaniel Club was formed February 15, 1881, and joined the AKC in 1889. One of the earliest acts of The American Spaniel Club was to draw up standards separating cocker spaniels from field spaniels, although they were interbred for years. The standards separated the two by weight, height, and length. Although they were judged separately in shows, the AKC continued to register them all as cocker spaniels in the studbook. Several motions to recombine the two "breeds" failed within the American Spaniel Club over the next 15 years, and in 1905 the AKC separated their listings in the studbook (The American Spaniel Club 1980).

The primary activity of the AKC is to act as a registry for purebred dogs. They maintain pedigree, breeding, and competition records for the member breeds. The individual parent breed clubs create their own breed standard, club rules, and codes of ethics.

In 2004, the AKC registered 958,270 dogs (The American Kennel Club 2005). This was down by over one-third from the peak of 1,528,392 dogs registered in 1992. The Labrador retriever maintained its hold as the number-one dog in America; a title it has held since 1991 when it replaced the cocker spaniel. There were three times as many Labrador

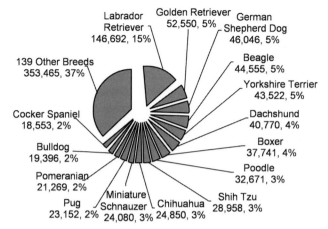

Figure 1. Top 15 Breeds in AKC Registration for 2004. Listed as breed, number of dogs registered, and percentage of total AKC registrations.

retrievers registered in 2004 as golden retrievers, the number-two breed in registrations. The top 15 breeds in AKC registrations represent almost two-thirds of all dogs of the 154 breeds registered by the AKC (Fig. 1). The breed with the least number of registrations in 2004 was the English foxhound, with 17 dogs registered.

The decline in AKC registrations since 1992 does not reflect a decrease in the number of purebred dogs in the U.S. What has changed is that commercial kennels which produce puppies for pet stores have created their own pedigree registries and have stopped registering their breeding stock and puppies with the AKC. In part, this was a financial decision to forgo payment for AKC registration. However, this change also occurred because of disciplinary actions by the AKC through inspections of facilities, breeding records, and DNA verification of parentage.

The United Kennel Club (UKC) was formed in the U.S. on February 10, 1898, in response to what it saw as the elite show dog registry of the AKC. They touted their studbook as a dog registry for breeders. The UKC publishes *Bloodlines* and the *Dog & Pet Stock Journal*. The UKC is noted for promoting the development of American breeds, which include the coon hounds, American Eskimo dog, American pit bull terrier, and toy fox terrier. They register breeds in eight groups: Guardian (Flock Guards & Mastiffs), Scent Hound, Sight Hound, Gun Dog, Northern Breeds, Herding, Terrier, and Companion Dog (United Kennel Club 1997). The UKC currently registers 308 breeds, with over 250,000 individual dog registrations per year.

DEVELOPMENT OF BREED STANDARDS

The development of breed standards by a parent club places selective pressure toward a specific phenotype. This acts to narrow the acceptable breeding population. Strict breed standards, competitions, closed studbooks, popular sires, and genetic disease control can all act to manipulate the gene pools of dog breeds. Depending on these selective pressures, gene frequencies can significantly change from generation to generation. What is considered natural selection is rarely the dominant force in dog breeding. The frequency of individual genes (whether of a beneficial, detrimental, or null effect) can increase, decrease, or even be lost to the breeding population due to selective pressure.

Breed-related genetic disorders develop based on the increased frequency of disease-causing genes within breeds. Sometimes this occurs due to the founder effect from popular sires. Disorders are unintentionally selected because of their linkage to desirable traits established by the club in their breed standards.

CLUB EXAMPLE: DALMATIAN CLUB OF AMERICA

The Dalmatian Club of America breed standard allows blue eyes (Fig. 2) (American Kennel Club 1998). This is a disqualification in all other international dalmatian breed standards. It has been found that dalmatians with blue eyes have a $1.7\times$ greater incidence of deafness at 50.9%, than brown-eyed dalmatians at 29.9% (Strain 2004). Motions have been made by Dalmatian Club of America members to disallow blue eyes, but these have not been successful. Another dalmatian phenotypic marker is a solid ear patch. All dalmatian standards disqualify dogs with a patched ear, although it is shown that dogs with an ear patch have a significantly lower incidence of deafness.

All dalmatians are homozygous for a defective autosomal recessive gene affecting purine metabolism and causing hyperuricosuria. In 1976, a dalmatian $\times$ pointer backcross family was established by Dr. Robert Schaible of Indiana University to reintroduce the normal purine-metabolism gene into the American dalmatian gene pool. Based on a request by the Dalmatian Club of America, the AKC agreed to register a male and female heterozygous carrier 5th-generation of the normal gene (96.87% dalmatian, 3.13% pointer).

Ensuing controversy within the dalmatian club resulted in a request to the AKC that the offspring of these dogs not be registered. Although their descendants could not be AKC registered, the breeding of this line

Figure 2. Blue-eyed dalmatian. Photograph courtesy of Barbara Augello.

of purine-metabolism heterozygous, and homozygous normal dalmatians, has continued for many generations. Of note is that it has never been possible to reproduce the larger, well-defined spots desired by dalmation breeders. It is suggested that selection for the dalmatian-specific spotting pattern may be tightly linked to the defective purine metabolism gene (Safra et al. 2004).

Other breeds have high frequencies of homozygous recessive phenotypes, and it must be considered whether this is due to random drift, the founders effect, or selective pressure. In the American rough and smooth-coated collie, the incidence of choroidal hypoplasia is 66.7%, based on 19,592 Canine Eye Registry Foundation (CERF) examinations by board-certified veterinary ophthalmologists during the period 1991–1999 (American College of Veterinary Ophthalmologists 1999). It is possible that the high incidence for this disorder is due to the breed standard recommendation to select for an extreme dolichocephalic skull and almond-shaped eye (The American Kennel Club 1998).

The Shiloh shepherd is a rare breed established from linebreeding on four female German shepherd dogs. Breeders have selected for a large size (100–140 lbs.), and a long body. A new inherited condition has been identified in 10-week to 18-month-old dogs in this breed, characterized by vertebral degenerative joint disease causing spinal cord compression.

It is suggested that this condition has arisen because of selection for this extreme body type (McDonnell et al. 2003).

These examples raise the question whether some breed clubs may need to choose a less desirable breed standard in order to increase the overall health of the breed. The identification of linked markers and gene mutations, and the adequacy of the response to selection with these tools, may provide answers.

AMERICAN BREED CLUB HEALTH INITIATIVES

Several health initiatives and breeder education programs have been originated by American dog clubs. These include single-breed and all-breed health foundations. Many breed clubs recognized their role in maintaining the health of their breed by supporting research. Some breed clubs established test breeding programs and open genetic registries to assist the breed and their breeders with managing genetic disorders.

In 1973, the American Miniature Schnauzer Club adopted a voluntary resolution that all breeding stock be screened for congenital cataracts, a simple autosomal recessive disorder. They recommended that breeding stock that were affected or that produced affected offspring be retired from breeding. In 1975, the club published the first list of test-bred normal miniature schnauzers for congenital cataracts. The test-breeding program included a "Sale of Carrier Puppies" contract that was approved by the AKC to refuse registration of those dogs unless documentation of sterilization was provided (American Miniature Schnauzer Club 1975).

In 1975, the United States Kerry Blue Terrier Club instituted a test breeding program for the autosomal recessive genetic disorder hereditary cerebellar cortical and extrapyramidal nuclear abiotrophy. The club published pedigrees of affected dogs and obligate carriers of the defective gene identified through test breeding (United States Kerry Blue Terrier Club 1975). The Irish Setter Genetic Registry, Inc., was established in 1979 to document and publish the pedigrees and results of test matings for early-onset progressive retinal atrophy (The American Irish Setter Foundation 1979).

American dog clubs have historically provided substantial funding for health and genetic initiatives. In December, 1952, the American Kennel Club appropriated $15,000 to the Cornell University College of Veterinary Medicine for research in dog health. In prior years, the AKC had sporadically funded research at Cornell, the University of Pennsylvania, and the University of Missouri. In 1961, the AKC funded a $5,000 fellowship at the Animal Medical Center (AMC) in New York City to study comparative radiology. In 1963 they funded a $5,000 fellowship at the AMC in canine cardiology, and a $4,000 symposium on canine hip dysplasia for the American Veterinary Medical Association. Between 1952 and 1984, the AKC appropriated $2,313,161 on canine health research and education (The American Kennel Club 1985).

The Morris Animal Foundation, founded in 1948 by veterinarian Mark L. Morris, originally funded research for dogs and cats. It now funds research for all companion animals and wildlife. Since its inception, it has funded over $32 million for more than 1,100 research studies; over $4 million in research was funded in 2004.

In 1995, the AKC established the Canine Health Foundation (CHF), whose mission is to develop significant resources for basic and applied health programs with emphasis on canine genetics (www.akcchf.org). Through 2004, the AKC has donated over $11 million to the CHF, which has been supplemented by contributions from breed clubs and individuals. The CHF has funded more than 240 grants totaling over $12 million.

GENETIC REGISTRIES

The first all-breed genetic health registry was established by The Orthopedic Foundation for Animals (OFA) in 1966. The industrialist John M. Olin established this foundation, based on a meeting in 1964 with representatives of the Golden Retriever Club of America, the German Shepherd Dog Club of America, and veterinarians. The OFA originally maintained a closed registry for hip dysplasia, listing only normal testing dogs. It now maintains a Web-based (www.ofa.org) semi-open registry for over 17 hereditary disorders. This registry allows owners to list normal as well as carrier and affected dogs. Epidemiological data from all dogs are available to researchers studying these disorders.

The Institute for Genetic Disease Control (GDC) was formed in 1990 to establish open health registries, to track both the phenotype and the genotype of genetic disorders in dogs. An open registry lists both normal and abnormal individuals. The GDC was originally established with the Bernese Mountain Dog Club of America to track elbow dysplasia and other musculoskeletal disorders. The Poodle Club of America added sebaceous adenitis and eye disease registries. In 1993, the Cairn Terrier Club of America established open registries for Legg-Perthes, craniomandibular osteopathy, and patellar luxation. In 1994, the Bernese Mountain Dog Club of America added an open tumor registry (Institute for Genetic Disease Control in Animals 1997). The GDC registries eventually covered more than 20 diseases. Breeders and owners could order KinReports[TM] that list phenotypic and genotypic data on a dog, its parents, littermates, half-siblings, and offspring. This breadth of pedigree information is instrumental to manage disorders with complex inheritance (Bell 2001b).

In July 2002, the OFA and GDC agreed to merge all but the tumor and eye registries into the OFA database. In 2002, CERF began forwarding all normal eye test results to the OFA database. Some genetic disease testing companies also forward test results to the OFA database. For example, the Australian Cattle Dog Club of America contracted with the company Optigen to forward all test results for prcd-PRA to the OFA for an open database. The Briard Club of America has arranged for all congenital stationary night blindness (CSNB) test results to be forwarded to the CERF registry.

The OFA Web site now allows on-line searches of their database for any dog, and generates full-family pedigree reports at no charge for all reported disorders. The OFA registries continue to be semi-closed, requiring owners to check a box on applications to allow the listing of abnormal test results. Breeding guidelines are offered for the proper use of the registry information (Keller 2003).

In 2001, the AKC CHF, in collaboration with the OFA, established the Canine Health Information Center (CHIC). This program was formulated to aid breeders, owners, and breed clubs to foster health testing and health consciousness. The AKC parent breed club must enroll in the CHIC program, by indicating which testable disorders should be screened for in their breed. Dogs receive CHIC certification if they complete the required testing, regardless of whether they test normal or abnormal for those conditions. With the increasing number of testable genetic conditions, it is less likely that dogs will be genetically normal for all of them. The program is based on the concept that conscientious breeders will health-test their dogs, and select mates based on producing quality, healthy offspring. The CHIC Web site (www.caninehealthinfo.org) is linked to the OFA database and allows owners and breeders to view results of all dogs and their close relatives.

CLUB EXAMPLE: PORTUGUESE WATER DOG CLUB OF AMERICA

Portuguese water dogs were first imported into America in the 1960s, and the Portuguese Water Dog Club of America (PWDCA) was founded in 1972. In late 1987, the PWDCA contacted the author about a genetic disease control program for the storage disease GM1-gangliosidosis. While a blood enzyme screening test was being developed for the breed, the PWDCA called for a six-month breeding moratorium. This was effective, as evidenced by a 24.3% drop in AKC registrations in 1988 from 1987 (333 versus 414 dogs). When the blood screening test was standardized in 1988, breeders began screening their breeding stock and 5- to 7-week-old puppies. Once again, the determination of the Portuguese water dog breeders was evident, as there were more puppies tested during 1989 (475) and 1990 (659) than AKC registered (406 and 601, respectively).

The PWDCA club newsletter counseled Portuguese water dog breeders to use the "breed and replace" method of managing the defective gene causing GM1-gangliosidosis, so as to not limit the genetic diversity of the population (Bell 1988). This method recommends breeding quality carrier dogs to normal testing mates and replacing the carrier parent with a normal testing offspring. However, the emotional response of a breeder receiving a carrier test result is to not breed the dog, regardless of its qualities. Consequently, many breeders simply eliminated their carrier-testing breeding stock. The frequency of carrier-testing dogs fell from 16.1% in 1988 to 8.8% in 1989 and to 6.2% in 1990. A direct mutation test was eventually developed, and the defective gene is now rare in the Portuguese water dog population (Wang et al. 2000).

During this period of time, Portuguese water dog breeders recognized that the founders of the defective gene causing GM1-gangliosidosis all originated from the Portuguese Algarbiorum line, while it was not present in the Portuguese de Alvalade line. The de Alvalade line soon became the major influence in the American Portuguese water dog population. Over time, it became apparent that de Alvalade dogs, including several influential affected dogs imported from Portugal, carried the autosomal recessive gene for late-onset prcd-PRA. This significantly increased the carrier frequency of the PRA gene to 35%. This defective gene was not present in the Algarbiorum line.

This example shows why it is important to provide genetic counseling information when instituting a genetic disease testing program (Bell 2001a). Due to their breeding choices, this breed was left with a defective recessive gene for a different disorder at a much higher frequency than the original one.

The PWDCA and its breeders have been proactive in promoting the health and well-being of their breed. The club has twelve separate health-related committees. In order to advertise dogs in their breed newsletter *The Courier*, owners must list the health test results of their dogs, or state that the health tests have not been performed. By not restricting the ability to advertise, but establishing recommended guidelines, the club has created an environment of health consciousness. The club maintains a long-standing voluntary open health registry for all medical conditions of Portuguese water dogs. This registry is open to all owners, not just members of the PWDCA. The club has supported and sponsored a wide variety of health research, including GM1-gangliosidosis, PRA, Addison's disease, juvenile cardiomyopathy, hip dysplasia, and the Georgie project of musculoskeletal variation.

BREED CLUB EFFORTS CONCERNING GENETIC DIVERSITY

With the increased recognition of genetic disorders, some American breed clubs have questioned whether established breeding practices can allow a healthy and diverse gene pool for pure breeds. There are clubs that have proposed codes of conduct that restrict the allowable level of inbreeding for litters to be registered. This effort is based on a fallacious premise that genetic diversity depends on the types of matings that take place.

Consider a theoretical situation where four dogs represent a breed; two A-line dogs and two B-line dogs. Whether inbreeding takes place (the two A dogs to each other and the two B dogs to each other) or outbreeding takes place (only A to B matings), the frequencies of the A and B genes remain the same in the resulting population. The notion that outbreeding is required to maintain genetic diversity is based on the premise that all representatives of the breed must maintain the genetic diversity of the breed within their own genes. Genetic diversity is a variable of the population that requires the maintenance of diverse pedigree lines.

Problems with genetic diversity occur due to selection of breeding stock, with the popular-sire effect being the major factor affecting diversity in dog breeds. If a popular stud dog contributes a disproportionate number of genes to the gene pool, it skews the gene pool in his direction. This can increase deleterious genes due to the founders effect. It also limits breed diversity because other unrelated quality males are displaced from breeding by the popular sire (Bell 2003).

AMERICAN BREED EXAMPLE: THE CHINOOK

Some breed clubs whose breeds have small gene pools or recent bottlenecks have concerns about whether enough genetic diversity exists to maintain a healthy gene pool. The Chinook breed was established in America in the early 1900s and was used for the Antarctic explorations of Arthur Treadwell Walden and Admiral Richard E. Byrd. The breed was based on a working phenotype, whose founders included huskies, mastiffs, St. Bernards, German shepherd dogs, Belgian sheepdogs, and

Eskimo dogs. After the expeditions were over, the breed proliferated, and dogs were sold as pets. In 1931, President Hoover declared a Chinook, Paugus, and his young owner, Lawrence Orne, to be America's most typical "boy and his dog." The breed was consolidated under the Perry Greene kennel in Maine in the 1940s, where it experienced growth and then decline. In 1965, the *Guinness Book of World Records* recorded the Chinook as the "Rarest Dog in the World," with only 125 living. By 1981, only 11 breedable Chinook dogs remained.

The modern-day Chinook population grew from these 11 dogs. Further limiting the genetic variation in the breed, 3 of the modern founders were full siblings (Tia, Barrow, and Fairbanks) and three more (Tavi, Juna, and Honey) were half-siblings to them (Fig. 3). All 11 dogs traced back to only 4 common ancestors within one generation. Based on a perceived limitation of genetic diversity, a Chinook cross-breeding program was established using working dogs of Alaskan malamute, Siberian husky, and other breed backgrounds.

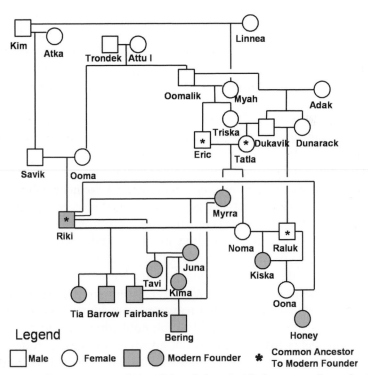

Figure 3. Modern ancestry of the Chinook breed. All dogs were from the Perry Greene kennel.

In 1997, the Chinook Owners Association and Chinooks Worldwide Inc. contacted the author to conduct a breed health survey and a small-scale genetic diversity study. For the genetic diversity study, 10 dogs were selected that represented the most diverse pedigree backgrounds in the breed. These included 8 purebred Chinooks and 2 third-generation cross dogs (88.5% Chinook, 12.5% other breeds). The eight-generation Wright's inbreeding coefficient for the purebred Chinooks ranged from 25.1% to 47.3%, and was 18.1% and 27.0% for the crosses. Six type-I canine polymorphic markers were analyzed in the study, none of which was known to contain disease-causing alleles in the breed. The average heterozygosity at all six loci was 47.9% for the purebred Chinooks, with an average of 3.17 alleles present at each locus. This compared with 54.0% heterozygosity at these loci across all breeds. The results of the two Chinook crosses were not significantly different from the purebred Chinooks (Bell and Johnson 1998).

The Chinook breed health survey showed several common hereditary conditions, including cryptorchidism, epilepsy, hip dysplasia, and atopy. The purebred Chinook population did not show decreased litter size, increased stillbirths, decreased longevity, or other signs of significant restricted diversity or inbreeding depression. The Chinook crosses did not show significant differences from the purebred population regarding health issues (Bell 1998).

These studies suggest that the selective pressure of a working phenotype and the diverse origins of the Chinook breed's founders may have allowed considerable health and diversity to remain despite a pedigree bottleneck and high consanguinity. This diversity was also evident in the health survey, based on the purebred Chinook variation of height, weight, color, ear carriage, and eye color (Fig. 4).

Problems with inbreeding depression arise at the gene locus level. If there is no diversity (nonvariable gene pairs for a breed), but the homozygote is not detrimental, there is no effect on breed health. Each breed has different founding events, breeding histories, and selective pressures. These results may not be the same for other breeds with small gene pools.

Today over 400 purebred Chinooks are registered with the United Kennel Club. Between 2003 and 2005, the Chinook Owners Association, through the United Kennel Club, registered four fourth and fifth generation Chinook crosses as purebred Chinook dogs. These dogs were selected through a stringent process of health testing, sled-pulling ability, temperament testing, and breed phenotype. The Chinook is presently enrolled in the AKC Foundation Stock Service program, a precursor to AKC recognition. The AKC, in consultation with the breed club, will have

Figure 4. Phenotypic diversity of Chinook dogs. Photograph courtesy of Nancy Bartol.

to decide whether to allow the UKC-registered former Chinook crosses into the AKC database.

Other American breed clubs have attempted to increase their genetic diversity by searching for and importing dogs from their country of origin. The Basenji Club of America African Stock Project supported an expedition in 1987 and 1988 to the Democratic Republic of the Congo (formerly Zaire) to increase the limited gene pool of the Basenjis. From these two trips a total of 19 Basenjis were imported to America. After petitioning the AKC, the studbook was reopened in 1990. Thirteen dogs and a previous import were AKC-registered as foundation stock. Similar efforts are under way by American Shar-Pei breeders through trips to China.

American breed clubs have been proactive in seeking guidance and progress concerning genetic health and diversity issues in their dogs. Great strides are being made investigating the genomic basis of disorders with both simple and complex inheritance (see Chapter 15). It is important not only to pass this information on to the breeders, but also to counsel them on the proper use of genetic tests, with guidelines for maintaining a healthy and diverse gene pool.

SUMMARY

American breed clubs were developed to promote the working, showing, and breeding of purebred dogs. Breed standards were adopted, and pedigree studbooks were established to maintain breed quality. Breed-related genetic disease and genetic diversity issues can arise in purebred dogs. American breed clubs have become proactive in funding genomics research, promoting genetic testing, and establishing health registries. Breeders must be counseled on the proper use of genetic tests and breeding management, or these positive tools could have negative consequences on the breed's gene pool.

REFERENCES

American College of Veterinary Ophthalmologists. 1999. *Ocular disorders presumed to be inherited in purebred dogs*, 3rd edition. American College of Veterinary Ophthalmologists, Meridian, Idaho.

The American Fox Terrier Club. 1924. *Club book of the American Fox Terrier Club 1885–1924*. The American Fox Terrier Club, Boston, Massachusetts.

The American Irish Setter Foundation. 1979. *Recommended procedures for test-mating Irish Setters for progressive retinal atrophy*. The American Irish Setter Foundation, Coraopolis, Pennyslvania.

The American Kennel Club. 1985. *The American Kennel Club 1884–1984: A source book by the American Kennel Club*. Howell Book House, New York.

———. 1998. *The complete dog book*, 19th edition revised. Howell Book House, New York.

———. 2004. American Kennel Club—Breeds. (accessed 10 Jan 2005) www.akc.org/breeds/complete_breed_list.cfm.

———. 2005. 2004 Dog registration statistics. *AKC Gazette* **122:** 10–11.

American Miniature Schnauzer Club. 1975. *The American Miniature Schnauzer Club list of dogs and bitches who have or are in the process of being test bred for congenital cataracts*. American Miniature Schnauzer Club, Bryan, Texas.

The American Spaniel Club. 1980. *A century of spaniels, vol. 1: The American Spaniel Club 1881–1981*. Hamilton I Newell, Amherst, Massachusetts.

Bell J.S. 1988. The blood analysis for GM_1-gangliosidosis. *The Courier* (newsletter of the Portuguese Water Dog Club of America). **XVI:** 41–42.

Bell J.S. 1998. Tufts 1997–1998 Chinook breed health survey. *Chinook News*. Fall, 1998.

———. 2001a. The effects of genetic testing: Constructive or destructive? *AKC Gazette* **118:** 24–25.

———. 2001b. Breeding strategies for the management of genetic disorders. *AKC Gazette* **118:** 24–25.

———. 2003. Popular sire syndrome and concerns of genetic diversity. In *Proceedings of the AKC Canine Health Foundation National Parent Club Canine Health Conference*, St. Louis, Missouri.

Bell J.S. and Johnson G.S. 1998. Chinook genetic diversity study. *Chinook News*. Fall, 1998.

Institute for Genetic Disease Control in Animals. 1997. *The time is NOW*. Monograph of the Institute for Genetic Disease Control in Animals, Davis, California.

Keller G. 2003. *The use of health databases and selective breeding—A guide for dog and cat breeders and owners*. Orthopedic Foundation for Animals, Columbia, Missouri.

McDonnell J.J., Knowles K.E., deLahunta A., Bell J.S., Lowrie C.T., and Todhunter R.J. 2003. Thoracolumber spinal cord compression due to vertebral process degenerative joint disease in a family of Shiloh Shepherd dogs. *J. Vet. Intern. Med.* **17:** 530–537.

Safra N., Schaible R.H., Ling G.V., and Bannasch D.L. 2004. Mapping Dalmatian hyperuricosuria using a historical interbreed backcross. In *Proceedings of the 2nd International Conference Adv. in Canine and Feline Genomics*, Utrecht, The Netherlands. Poster #45:101.

Strain G. 2004. Deafness prevalence and pigmentation and gender associations in dog breeds at risk. *Vet. J.* **167:** 23–32.

The United Kennel Club. 1997. *United Kennel Club, Inc. 1898–1997: The first 100 years.* The United Kennel Club, Kalamazoo, Michigan.

United States Kerry Blue Terrier Club. 1975. *Progressive neuronal abiotrophy—A genetically inherited disease in Kerry Blue Terriers.* United States Kerry Blue Terrier Club, Santa Rosa, California.

Wang Z.H., Zeng B., Shibuya H., Johnson G.S., Alroy J., Pastores G.M., Raghavan S., and Kolodny E.H. 2000. Isolation and characterization of the normal canine beta-galactosidase gene and its mutation in a dog model of GM1-gangliosidosis. *J. Inherit. Metab. Dis.* **23:** 593–606.

4

Morphological Variation in the Dog

Amy Young and Danika Bannasch

Department of Population Health and Reproduction
School of Veterinary Medicine
University of California, Davis
Davis, California 95616

FROM THE DIMINUTIVE CHIHUAHUA to the giant Irish wolfhound, the diversity of canine sizes and shapes has fascinated humans for thousands of years. This chapter aims to catalog the wide array of morphological variation in the purebred dog in a qualitative manner. The physical characteristics that define a breed can be summarized into a handful of categories that allow judges, breeders, and veterinarians to recognize hundreds of different breeds. These categories are based on morphology-altering mutations that have been fixed within breeds. The diversity of morphological variation that exists in domestic dogs makes them an excellent model for understanding human morphological variation. Human congenital malformations are important medical problems that are studied in great detail, particularly craniofacial changes, limb formation, and height, all of which vary in dog breeds. Virtually any and all combinations of traits have been created and fixed within particular breeds. In addition, although not addressed in this chapter, coat patterns and colors abound in the domestic dog. These colors and patterns, combined with variation in morphology and behavior, provide the framework for the greatest diversity recognized within any single species.

HISTORY OF MORPHOLOGICAL TYPES

It is widely thought that dogs were domesticated from wolves. However, the location, timing, and number of domestication events remain controversial. Bones of wolves, dating from the Middle Pleistocene epoch (~500,000 years ago), have been found in association with early hominid sites (Olsen 1985). Changes in skeletal anatomy, an indication of domestication, are not evident in these remains, suggesting that humans and wolves had not yet developed a dependent association. Once wolves were domesticated, various changes in anatomical features occurred. The earliest physical evidence of a domesticated dog, a small mandible, dates from the late Paleolithic, approximately 14,000 years ago (Clutton-Brock and Jewell 1993). Analysis of mitochondrial DNA implies that the origin of the dog is considerably more ancient than this fossil suggests. The actual timing of dog domestication, currently estimated to be between 15,000 and 150,000 years before the present, remains unclear, since it is possible that early dogs closely resembled wolves for a significant length of time (Vila et al. 1997; Savolainen et al. 2000).

The first evidence for potentially distinctive breeds of dogs dates to 3,000 years ago and is principally associated with ancient Egypt and Mesopotamia. Early Egyptian art illustrated two types of dogs: one that was slender, erect-eared, and curly-tailed; and another that was shorter, with a heavier muzzle, drop ears, and a curved or saber tail. It appears that both types were used primarily for hunting. Tomb carvings have been discovered from several archaeological sites that also suggest the presence of heavier mastiff-type dogs with massive muzzles, long tails, and drop ears. The function of the early mastiff type is unclear (Brewer et al. 2001).

In an effort to understand and identify the origins and relationships of modern breeds, dogs are frequently grouped by numerous characteristics, including body type, behavior, and function. Aside from known historical information regarding certain groups, attempts to categorize dog types remain largely speculative. In Bonnie Wilcox's *The Atlas of Dog Breeds of the World*, dogs are categorized on the basis of morphological similarities resulting from their original functional development into flock guard, mastiff, scent hound, gun dog, terrier, northern, herding, and southern dog (Wilcox and Walkowicz 1995). Although not defined by Wilcox, small companion dogs comprise an additional category. These groups are presented here as plausible divisions of canine types.

Early farmers developed the flock guard type of dog to protect their valuable flocks from carnivores and raids from neighboring villages. Their jobs required not only mass, but agility and power as well. They are generally tall, muscular, and large boned, with a long tail that is carried low, a broad, flat head, and small, drop ears. These dogs stayed with their flocks year round, often in mountainous regions where food was scarce. As a result, they generally have ample coats of moderate length to protect against the elements.

It is likely that these flock guard dogs are the stem type from which mastiff-type dogs were developed. Mastiffs were primarily used to guard homes and property. They were adapted to warfare and combat and used as fighting dogs against bears, lions, and men. Bred to be intimidating and aggressive, the first mastiff types are described as "leviathans," giant dogs with huge paws, drop, pendant-shaped ears, and a bulky skull. Mastiffs generally have short coats and low-carried tails. This group marks the first appearance of the brachycephalic head type, which is generally thought to have functioned to help a dog maintain its grip on an opponent.

It is possible that the mastiffs are the stem type from which many other types of dogs were developed. The scent hounds were likely selected from the mastiffs for their sensitive nose and tenacity, and early types were used to hunt wolves and wild boar. Scent hounds are medium-sized, heavily built, with heavy heads, hanging, long ears, and loose, heavy skin. Some scent hounds have lighter bodies, enabling them to be faster and better suited for mounted hunting. Conversely, heavier-bodied scent hounds were selected for methodical tracking. Some were bred with shorter legs for work in heavier cover or for following on foot. Many scent hounds are shorthaired, although wirehaired varieties are not uncommon.

On the basis of morphology, it is easy to divide certain dog breeds into the aforementioned categories. The gun dog group, however, comprises dogs of slightly more varied morphologies. Gun dogs originally accompanied hunters who used nets, falcons, and bows. Following the invention of firearms, they were used for finding and retrieving game. Gun dogs can be divided into four groups: pointers, land retrievers, water dogs, and setters or spaniels. The pointer subgroup is closely related to the scent hounds. Reflecting this ancestry, they have a basic hound body and head type. Originally bred to find and point to the game, many originated as heavy-boned and methodical, but were eventually modified to be lighter and faster. They are mostly shorthaired, although there are wirehaired and longhaired varieties. In contrast, the land retrievers are more mastiff-like. In order to retrieve game, they were required to be tough dogs that could handle cold water as well as heavy vegetation. As a result, they are heavy-bodied and shorter than pointers, with a heavier, more bracoid head. Also required to work in the water, water dogs are used most frequently for waterfowl shooting and are often used on boats. They generally have a shaggy coat with hair that is curly or corded to insulate them from cold water. Setters or spaniels were developed to crouch and creep in on the scent of birds, allowing hunters to approach more closely before sending the game into flight. They have a long, fringed tail, feathering on the legs and belly, and flat, drop ears.

Morphologically and behaviorally very distinct from the gun dogs, the terriers are another type of dog that was used for hunting. They are small to medium-sized dogs originally bred for vermin hunting. Their small size facilitated the frequent requirement of going underground after their quarry. Terriers are lean and long-headed, with a square jaw, pricked or semi-pricked ears, tails that are carried straight up, and a rough, wiry coat.

Terriers, along with many other types of dogs, have been bred to retain few distinctly wolf-like morphological characteristics. The Northern breeds, however, have kept the small, pricked ears, wedge-shaped head, and powerful, lean body reminiscent of wolves. Mainly developed for hunting and draft work, their physical characteristics reflect the power and stamina needed to cross difficult terrain over long distances. These northern dogs also endure harsh weather conditions. As a result, they have short plush coats that protect them from the cold weather.

Early northern dogs are the likely progenitors of the herding breeds, which have pricked or semi-pricked ears and a tail with a curve at the end. Bred for flock gathering, herding dogs are long-bodied, enabling them to make sudden turns. Many were originally bred to work with reindeer, but were later switched to sheep. In addition, some varieties of herding dogs were chosen to work with cattle and were bred to have short legs to avoid injuries from kicks.

In sharp contrast to the northern dogs, the southern dogs, which include the pariah and sighthound, were bred for hunting in arid, treeless areas by sighting and running down prey such as gazelle and antelope. Frequently represented in art, they are tall and slender, with a wedge-shaped, sharply angled head and a short coat to withstand the heat. They have large prick ears and a long, thin tail. Having been selected for light bone, deep chests, and narrow heads to minimize wind resistance and increase speed, they are best known as racing dogs.

Whereas the dogs in the aforementioned groups were bred to perform a specific function, pet dogs were "simply any that caught one's fancy." Early forms appear to have been too small to be guarding or herding dogs, so it is assumed that their primary purpose was to serve as companions. Pet dogs are mentioned in literature of the classical period in the writings of Aristotle, Pliny, and Strabo. From pictures of pet dogs from Pompeii, they appear to be similar in type to terriers or Maltese, a small dog with long, white hair, drop ears, and a curled tail (Brewer et al. 2001).

The division of dog types into these categories is one way to group them based on their original purpose and, to some extent, their geographical development. Most dog breeds fit neatly into one of these groups. However, many modern breeds, in which there are numerous variations on these "original phenotypes," can be included in one or more of these categories based on their ancestry. The ability to group dog breeds into these ancestral types provides insight into the functions of the extreme morphological variations seen in dogs today.

HEAD TYPE

Skull shape, along with body size, contributes to the most obvious differences between dog breeds. The three main classifications of dog head types are brachycephalic, mesaticephalic, and dolichocephalic. A short muzzle with a relatively short maxilla and a broad back skull is characteristic of the brachycephalic head type (Figs. 1a, 2d, e, and i). The mesaticephalic head type is the intermediate head type of medium proportions. It is the most common head type in purebred dogs and is represented by the beagle in Figure 1c, the skull of a dachshund in Figure 2k, and the skull of a Great Dane in Figure 2b. The dolichocephalic head type is long and narrow and is represented by the collie in Figure 1e, a collie skull in Figure 2g, and a Sealyham terrier skull in Figure 2j. In addition to the shape of the skull, the size can vary with the overall body size of the dog, producing the tiny head of the Chihuahua (Fig. 2a) and the massive head of the Great Dane (Fig. 2b).

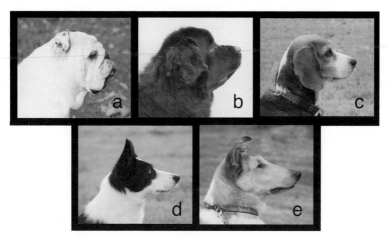

Figure 1. Canine head types. (a) Bulldog (brachycephalic). (b) Newfoundland (mesaticephalic). (c) Beagle (mesaticephalic). (d) Border collie (mesaticephalic). (e) Collie (dolichocephalic).

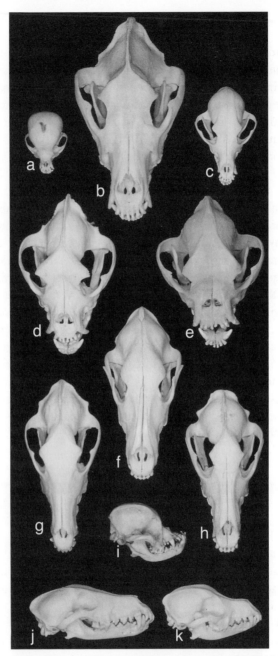

Figure 2. Canine skull types. (**a**) Chihuahua (mesaticephalic). (**b**) Great Dane (mesaticephalic). (**c**) Yorkshire terrier (mesaticephalic). (**d**) Boxer (brachycephalic). (**e**) Bulldog (brachycephalic). (**f**) Basset hound (mesaticephalic). (**g**) Collie (dolichocephalic). (**h**) Doberman pinscher (mesaticephalic). (**i**) Pekinese (brachycephalic). (**j**) Sealyham terrier (dolichocephalic). (**k**) Dachshund (mesaticephalic). (Modified, with permission, from Done et al. 1986).

The brachycephalic head type is characteristic of many breeds, including the Pekinese, pug, Boston terrier, and boxer. The bulldog is perhaps the most easily recognized of the brachycephalic breeds (Figs. 1a and 2e). The standards for breeds with brachycephalic heads all allow or require prognathism, or undershot bites, and include a short muzzle and wide-set, round eyes. The rounder skull shape is similar to a human or baby head and may explain why this head type was selected in the smaller breeds. In the larger breeds, it is associated with dogs bred for fighting, with the idea that the shorter muzzle and wider back skull are more powerful for biting. Unfortunately, normal respiration is compromised in dogs with brachycephalic head types, particularly in warm weather. A resulting condition, brachycephalic airway syndrome, which can include a hypoplastic trachea, everted laryngeal saccules, elongated soft palate, and stenotic nares, sometimes requires surgery to improve airflow in affected animals (Nelson and Couto 2003). Crosses performed between dogs with brachycephalic and mesaticephalic heads result in dogs with an intermediate head type, indicating partial dominance of the phenotype (Stockard et al. 1941).

The opposite extreme of the brachycephalic head type, the dolichocephalic head type, is represented by the borzoi, greyhound, German shepherd, and collie (Fig. 1e). The dolichocephalic breeds tend to have more almond-shaped eyes rather than the round eyes of the brachycephalic breeds. It has been observed that retinal ganglion cell distribution is highly correlated with nose length in dogs (McGreevy et al. 2004). This may partially explain the advantage of a long nose in these breeds that were originally bred to hunt by sight. Although they represent a contrast to the brachycephalic type, dolichocephalic breeds also encounter health problems associated with skull shape. For example, mycotic rhinitis and nasal tumors are more common in dolichocephalic breeds than in breeds of the other head types (Hayes 1982; Reif et al. 1998).

There is a much finer distinction between dolichocephalic and mesaticephalic head types (compare Fig. 1e to 1b, 1c, and 1d; and Fig. 2g to 2f and 2h, and 2j to 2k) than there is between brachycephalic and mesaticephalic head types (compare Fig. 1a to 1b, 1c, and 1d; and Fig. 2d and 2e to 2f and 2h and 2i to 2k). The Newfoundland in Figure 1b and the Border collie in Figure 1d have mesaticephalic head types but are intermediate between the extremes. As a consequence of this diversity, skull measurements can distinguish the three head types and have been used to categorize breeds (Evans 1993).

Whether dolichocephalic, mesaticephalic, or brachycephalic, head shape within a breed remains under strong selection. Breeders desire more extreme versions of head types because they tend to win in the conformation ring. As such, dramatic changes in skull morphology have occurred in the last 50 to 100 years in many breeds. An extreme example occurs in the brachycephalic breeds, where even wider heads with shorter muzzles have been produced. As a result of these larger head sizes, most of these breeds require caesarian section delivery of puppies (Feldman and Nelson 2004), which was not an option 50–100 years ago. Purebred dog breeds have closed studbooks (i.e., closed breeding pools), so the changes in head shape are not due to the introduction of new genetic material. These modifications may be explained by the occurrence of new mutations. Fondon and Garner (2004) argue that this rapid and continuous morphological evolution must occur through gene-associated tandem repeat length differences, since they have a high mutation rate. Wayne has suggested, since the dog undergoes much more dramatic skull shape changes postnatally than other canids (Wayne 1986), that the genes involved with the timing of early growth patterns cause dramatic changes in adult dogs (Wayne 2001).

To date, little is known about the molecular basis of skull shape variation in the dog. In a study performed in the Portuguese water dog, QTLs were identified for skull shape and size in this strictly mesaticephalic breed (Chase et al. 2002). In another study, a candidate gene, MSX2, which has an important function in normal head and face morphogenesis in people, was sequenced in dog breeds with varying skull morphologies, and no sequence differences were identified (Haworth et al. 2001a). This same group identified a missense mutation in a limited number of samples that seemed to be associated with the brachycephalic head type in the Treacher Collins syndrome gene (TCOF1) (Haworth et al. 2001c). Mutations in this gene are responsible for Treacher Collins syndrome in humans, which is characterized by craniofacial defects. Another group recently took a more generalized approach to understanding the basis for the morphological differences in dogs by looking at a set of genes involved in development to identify repeat polymorphisms that might be correlated with morphological differences. In the bullterrier breed, which is characterized by an oval or egg-shaped head, they identified a mutation in the Runx-4 gene that was associated with skull shape changes (Fondon and Garner 2004).

EAR MORPHOLOGY

Along with differences in skull shape and size, dog ears vary in length and carriage. Shorter, stiffer cartilage within the pinnae leads to an upright or prick ear similar to that of a wolf (Fig. 3a), whereas a long pendulous

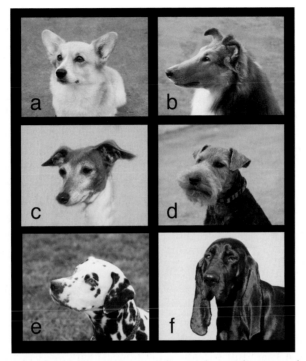

Figure 3. Canine ear morphology. From left to right: prick ears to drop ears. (**a**) Pembroke Welsh corgi. (**b**) Collie. (**c**) Italian greyhound. (**d**) Welsh terrier. (**e**) Dalmatian. (**f**) Black and tan coonhound.

ear with more flexible cartilage leads to a drop ear conformation (Fig. 3e, f). There are also ear conformations intermediate to completely erect and completely dropped (Fig. 3b, c, d). Generally, herding dogs have prick ears and hunting dogs have drop ears. An extreme drop ear is seen in the scent hounds (Fig. 3f), which were selected for longer ears because of the belief that they would trap more scent around the dog's nose when it was working.

Although breed standards are usually very specific about the desirable ear type, some ear types do not breed true. Tipped ears, like those seen in the collie and Shetland sheepdog, can be difficult for breeders to produce (Fig. 3b). When dogs with tipped ears are bred to each other, puppies with completely erect ears can be produced as well as puppies with folded ears (that drop to the side rather than forward) (Fig. 3c). Similarly, the inheritance of drop ears is not without complications. Some breeds that have been selected for a natural drop ear can have individual offspring with folded ears as well. As can be seen in the cases of tipped and drop ears, the mode of inheritance of ear carriage is complicated, because the size of the pinnae, as well as the thickness of the cartilage, are both involved (Burns 1952). To the casual observer, this may be further complicated in some breeds by surgical modifications intended to give the ears a distinctive pricked appearance.

The molecular basis of the variation in ear carriage and size has not been determined in dogs. However, known molecular variations relating to ear type in other species may indicate genetic regions that influence canine ear type. In mice, mutations in BMP5 cause a short-ear phenotype (Kingsley et al. 1992). In addition, a drop ear has been associated with the selection for tame behavior in the silver fox, leading to the hypothesis that a drop ear may have a pleiotrophic association with a behavioral change (Wayne 2001). In keeping with this hypothesis, it is interesting to note that the dog breeds with prick ear conformation, such as the spitz and Nordic types, tend to be the breeds with sharper temperaments, whereas those with drop ears tend to have a more docile temperament.

COAT TYPE

Among and within dog breeds there are numerous variations not only in ear and head type, but in coat type as well, with respect to thickness, texture, and length. Notably, coat color is also highly significant in many breeds. Since canine coat color is worthy of a lengthy discussion, and many works have been published that extensively explain the inheritance of coat colors in dogs, the topic is not presented in this chapter. *The Genetics of the Dog* by Ruvinsky and Sampson (2001), and *The Inheritance of Coat Color in Dogs* by Little (1957), both address coat color in depth.

Canine coat types can be divided into four categories based on length. The first, a "normal coat," is characteristic of wild canids, as well as some domestic breeds, such as German shepherd dogs, malamutes, and Siberian huskies (Fig. 4a). The short-hair coat is seen in breeds such as the boxer and dalmatian (Fig. 4b). The medium-length coat can be found in the golden retriever and the Nova Scotia duck tolling retriever (Fig. 4c). Last, the long-hair coat is characteristic of breeds such as the Afghan hound and the bearded collie (Fig. 4d). Additional coat-type variations include curly hair, as seen in the puli (Fig. 4e), and wire hair, as seen in the Scottish deerhound (Fig. 4f).

Based on numerous crosses between dogs possessing coats of varying lengths, it is understood that long hair is recessive to short hair. This mode of inheritance is illustrated in several crosses, including basset

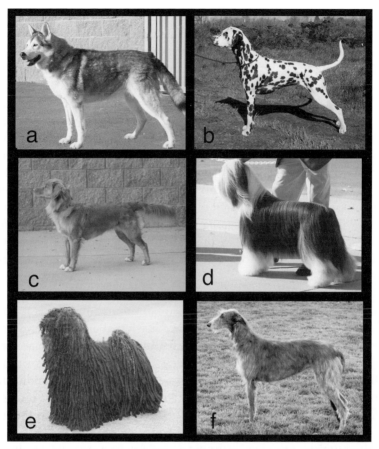

Figure 4. Canine coat types. (**a**) "Normal"coat (husky cross). (**b**) Shorthair coat (dalmatian). (**c**) Medium length coat (Nova Scotia duck tolling retriever). (**d**) Longhair coat (bearded collie). (**e**) Curly coat (puli). (**f**) Wirehair coat (Scottish deerhound).

hound × saluki, Great Dane × collie, St. Bernard × French bulldog, and bloodhound × Newfoundland, in which the resulting F_1 progeny all had short hair (Stockard et al. 1941; Whitney 1947). The inheritance of coat length has also been explored within breeds. Crosses between longhaired and shorthaired St. Bernards supported the dominance of short hair over long (Crawford and Loomis 1978).

Another coat type, wirehair, which is often associated with breeds in the terrier group, is dominant to short hair. Wiry coats offer protection from the elements and require minimal grooming (Wilcox and Walkowicz 1995). Crosses between the wirehaired and shorthaired varieties of the Brussels griffon breed demonstrated that the wirehaired coat is dominant to a shorthaired coat (Little 1934). Whitney (1947) provided further evidence for the dominance of the wirehaired coat in crosses between wirehaired Airedales and shorthaired beagles. The "beard," or bristly hairs on the muzzle and chin in wirehaired dogs, is especially evident in F_1 crosses between wirehaired and shorthaired breeds (Burns 1952).

In addition to wirehair, a curly coat, characteristic of breeds such as poodles and puli, is also dominant to a short coat. This is evidenced by the curly-coated offspring that were reported in a cross between a curly-coated retriever and a pointer. Further evidence is seen in poodle crosses, which usually have curly coats similar to their poodle parent (Burns 1952). Curly coats are often waterproof and quick-drying, both important characteristics for dogs that retrieve game from the water (Wilcox 1995). The degree of curl in a dog's coat can be described as a function of the shape of the hair follicles. Hair follicles that are more oval in shape, as opposed to round, result in tighter curling (Whitney 1947).

Hairlessness in dogs can be caused by two different mutations, one dominant and one recessive. The autosomal dominant form is found in Mexican hairless dogs and is homozygous lethal in utero (Robinson 1990). Heterozygotes are generally hairless and frequently have abnormal dentition (Goto et al. 1987). Hairlessness is also inherited as an autosomal dominant in the Chinese crested breed. Although the gene has not yet been identified, a LOD score of 3.93 between the hairless locus and a microsatellite on canine chromosome 17 has been reported as the result of a genome scan (O'Brien 2004).

Alternatively, a recessive gene causes hairlessness in American hairless terriers. This type of hairlessness first appeared in litters of American rat terriers. Sponenberg et al. (1988) noted that hairless × hairless matings produced only hairless offspring. As opposed to the Mexican hairless, there appear to be no other morphologic abnormalities associated with hairlessness in the recessive form.

As seen in the above examples, coat type is influenced by genetics, but other factors, such as environment, also play a role. For example, breeds that originated in hot, arid climates generally have shorter, thinner coats than do northern breeds. Seasonal variation in coat is also stimulated by temperature and climate. In addition, some coat variations are due to variations in the implantation angle of the hair follicle. Many breeds show an implantation angle of 30° to 40°. Airedales and Scottish terriers generally have implantation angles closer to 45°. Longhaired dachshunds, cocker spaniels, and Irish setters have an implantation angle <30°. Generally, it appears that longer-haired dogs have a lower implantation angle than do shorter-haired dogs. In almost all breeds, the angle of the hair tends to slant toward the tail (Evans 1993).

TAIL MORPHOLOGY

Dogs have been bred to have tails that vary in appearance as much as their coats, ears, and head types. Variations in tail morphology can be caused by differences in caudal vertebrae number, shape, and size, as well as the articulation between vertebrae. The average number of caudal vertebrae in the dog is 20, although the number can vary from 2 to 23 (Evans 1993). For example, tail kinks occur in some breeds due to fusion of two caudal vertebrae. Additionally, a screw tail, seen in the Boston terrier and the bulldog, has variable expression in the number and direction of kinks and the number of vertebrae (2–6). Other factors may also influence tail morphology. Curled tails that occur naturally in some breeds are not due to boney changes in the caudal vertebrae. Therefore, ligamentous or muscular attachments must be responsible for the curly tail trait in these breeds.

The mode of inheritance is known for some tail types. The curly tail is inherited as a dominant trait in crosses between greyhounds and German shepherd dogs and between collies and chow chows. Conversely, a screw tail seems to be inherited as a recessive trait (Whitney 1947). An additional tail type, the bobtail, is a naturally short tail that is inherited as a dominant trait. The bobtail is characteristic of some Pembroke Welsh corgis. A missense mutation in the T-box transcription factor T locus has been shown to cause this phenotype (Haworth et al. 2001b).

BODY SIZE AND TYPE

Perhaps the most obvious differences among breeds are related to body size and type. The Pomeranian and the Great Dane represent two size extremes among dog breeds. The Pomeranian in Figure 5b weighs approximately 5 pounds, whereas the Great Dane (Fig. 5e) weighs in excess of 120 pounds, more than a 20-fold difference. The Pomeranian measures approximately 9 inches in height, and the Great Dane stands over 30 inches. Such size extremes are unparalleled among other domesticated species.

As noted in Figure 5, leg type is one of the most variable traits in dogs. Stockard et al. (1941) categorize leg type as long and slender; long and strong; short and strong; short, straight, and stocky; or short and

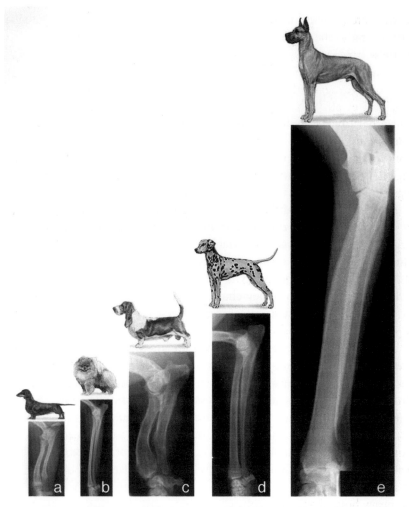

Figure 5. Radiographic appearance of the distal forelimb (radius and ulna) in various canine breeds. (**a**) Miniature dachshund (weight = 9 lbs). (**b**) Pomeranian (weight = 5 lbs). (**c**) Basset hound (weight = 53 lbs). (**d**) Dalmatian (weight = 68 lbs). (**e**) Great Dane (weight = 123 lbs). (Photographs © American Kennel Club.)

twisted. The latter is most notable in the basset hound (Fig. 5c) and dachshund breeds (Fig. 5a). The dwarfism phenotype of short, bent legs, or achondroplasia, is due to a change in the growth of the long bones. In dogs of "normal" leg type, such as the dalmatian in Figure 5d, the long bones grow in length and width until the ends of the bones are completely ossified. In achondroplastic breeds, the ends of the bones ossify and fuse earlier. The continued growth of the bone causes an increase in diameter, resulting in bones that are short and thick (Willis 1989). Interestingly, the achondroplastic growth is limited to the extremities in these breeds; the skull and vertebral column are unaffected (Stockard et al. 1941).

To investigate the mode of inheritance for leg length and type, Stockard et al. (1941) performed many crosses between basset hounds and other breeds of various heights and types. In crosses between basset hounds and German shepherds, the F_1 hybrids all had short legs. However, heterozygotes were not as short or as crooked as homozygotes, leading them to conclude that short legs are incompletely dominant to long legs.

The dominance of short legs to long was further evidenced in crosses of basset hounds to salukis. The F_1 hybrids all had short legs, but the degree of achondroplasia was less than that seen in the basset hound × German shepherd hybrids. The basset hound × saluki hybrids also showed a strong tendency toward the slender bone type of the saluki. Achondroplastic growth was seen in both bone types in the F_2 generation, and backcrosses confirmed that the saluki bone constitution is dominant to the thicker basset hound bone type.

To test the influence of bone type on the degree of achondroplasia, Stockard crossed a basset hound to an English bulldog, a breed that has short, sturdy, straight legs. Offspring in the F_1 generation were all achondroplastic, and were as short, or shorter, than their basset hound parent. It was also evident that the bulldog bone constitution was dominant over the basset hound bone type. The combination of achondroplasia and bulldog-type bone resulted in a greater degree of achondroplasia than was seen in the basset hound × saluki crosses. From these crosses, Stockard et al. (1941) determined that achondroplastic factors have a greater effect on heavy bone than on light bone. As such, short-legged dogs have a better chance of having straight legs if they are also light-boned.

The dwarfism seen in basset hounds and dachshunds is a desirable characteristic in those breeds. They seldom have orthopedic problems associated with their achondroplasia. Rarely, they can have elbow dysplasia from the asynchronous growth of the radius and the ulna, which can affect the conformation of the elbow joint. Notably, however, achondroplastic breeds are more susceptible to intervertebral disk disease than are non-achondroplastic breeds (Verheijen and Bouw 1982).

Achondroplasia appears undesirably in other breeds such as beagles, Alaskan malamutes, Norwegian elkhounds, and Great Pyrenees. In these breeds, there are more frequent orthopedic problems associated with the dwarfism than in the breeds fixed for the trait. Disproportionate dwarfism appears to be inherited as an autosomal recessive trait in these breeds, but the underlying molecular mechanism remains unidentified (Sande and Bingel 1983).

Achondroplasia in humans is most often caused by mutations in the fibroblast growth factor receptor-3 (FGFR3) gene. Martinez et al. (2000) obtained genomic sequence from dachshund, basset hound, bulldog, and German shepherd dogs in the region of the FGFR3 gene that contains the most common human achondroplasia mutation. No mutations were found in the four dog breeds, indicating that the responsible mutation is either located in another part of the FGFR3 gene, or that it resides in a different gene. There are many other genes that cause dwarfism in humans as well as in other mammals. Several of these genes would be good candidates for canine achondroplasia.

In addition to height, weight in dogs is also genetically controlled to some degree. It has been suggested that it is likely to be a polygenic character, and it should be noted that environment is an influential factor (Willis 1989). In a study of German shepherd 60-day weights, Reuterwall and Ryman (1973) found a 44% heritability among males, and a 35% heritability among females. These heritabilities are sufficient to suggest that it would be effective to select for an increase or decrease in weight.

In a later study on growth traits in guide dogs, Helmink et al. (2001) discerned heritabilities for height and weight and investigated the correlation between the two. They estimated mature weight heritability for German shepherds to be 0.57 ± 0.07, whereas mature weight heritability for Labrador retrievers was 0.44 ± 0.07. The heritability of mature height is reported as 0.35 ± 0.08 for German shepherds and 0.46 ± 0.08 for Labrador retrievers. In their analyses, they discovered a direct positive genetic correlation between height and weight. This implies that selection to increase or decrease one trait would result in changes in the same direction in the other trait.

In many animals, overall body size is mainly determined by growth hormone (GH) and insulin-like growth factors (IGFs). In dogs of different breeds, body size appears to be correlated with circulating concentrations of IGF-I (Eigenmann et al. 1984). In a study of Great Dane and miniature poodle pups, changes were observed in levels of GH and IGF-I during early growth. In addition, a QTL associated with overall skeleton size in Portuguese water dogs is closely linked to IGF-I (Chase et al. 2002). To investigate the differences in adult body size between medium and giant dog breeds, Favier et al. (2001) measured plasma concentrations of GH, IGF-I, and IGF-II in Great Danes and beagles. They found that GH levels in Great Danes were significantly higher than in beagles and that IGF-I and IGF-II levels did not show a significant difference. They propose that the differences in body size between medium-sized and giant dog breeds are mainly due to differences in the release of GH.

As evidenced by the extreme differences in body size, as well as the many variations in head, ear, tail, and coat types, domestication has resulted in a marked morphological divergence from the ancestral wolf form. In fact, the dog exhibits the richest array of morphological diversity in the animal world. Within the past decade, scientists have elucidated the probable evolutionary history of the dog, but the genes that influence morphology remain largely unidentified. Further advances in canine genetics, in particular the sequencing of the dog genome, should provide the catalyst to understanding the molecular basis of these fascinating morphological traits.

SUMMARY

The diversity of canine sizes and shapes has fascinated humans for thousands of years. The physical characteristics that define a breed are based on morphology-altering mutations that have been fixed within breeds. Dog breeds vary in head type, ear length and carriage, coat type, tail type, and body size and shape. Although the modes of inheritance for many of these traits are known, most of the underlying molecular causes remain unidentified. Advances in canine genetics, in particular the sequencing of the dog genome, should lead to a better understanding of these varied morphological traits.

ACKNOWLEDGMENTS

We thank the owners of dogs that were photographed for the figures, the Radiology Department at the Veterinary Medical Teaching Hospital at the University of California, Davis, for the radiographs in Figure 5, and the American Kennel Club for the breed drawings in Figure 5.

REFERENCES

Brewer D., Clark T., and Phillips A. 2001. *Dogs in antiquity: Anubis to Cerberus: The origins of the domestic dog.* Aris & Phillips, Warminster, United Kingdom.

Burns M. 1952. *The genetics of the dog.* Commonwealth Agricultural Bureaux, Farnham Royal, Slough, Bucks, United Kingdom.

Chase K., Carrier D.R., Adler F.R., Jarvik T., Ostrander E.A., Lorentzen T.D., and Lark K.G. 2002. Genetic basis for systems of skeletal quantitative traits: Principal component analysis of the canid skeleton. *Proc. Natl. Acad. Sci.* **99:** 9930–9935.

Clutton-Brock J. and Jewell P. 1993. Origin and domestication of the dog. In *Miller's anatomy of the dog* (ed. Howard E. Evans), pp. 21–31. W.B. Saunders, Philadelphia, Pennsylvania.

Crawford R.D. and Loomis G. 1978. Inheritance of short coat and long coat in St. Bernard dogs. *J. Hered.* **69:** 266–267.

Done S.H., Goody P.C., Evans S.A., and Stickland N.C. 1986. *Color atlas of veterinary anatomy: The dog and cat.* Mosby-Wolfe, London.

Eigenmann J.E., Patterson D.F., Zapf J., and Froesch E.R. 1984. Insulin-like growth factor I in the dog: A study in different dog breeds and in dogs with growth hormone elevation. *Acta Endocrinol.* **105:** 294–301.

Evans H.E. 1993. *Miller's anatomy of the dog.* W.B. Saunders, Philadelphia, Pennsylvania.

Favier R.P., Mol J.A., Kooistra H.S., and Rijnberk A. 2001. Large body size in the dog is associated with transient GH excess at a young age. *J. Endocrinol.* **170:** 479–484.

Feldman E.C. and Nelson R.W. 2004. *Canine and feline endocrinology and reproduction,* 3rd edition. Saunders, St Louis, Missouri.

Fondon J.W., III, and Garner H.R. 2004. Molecular origins of rapid and continuous morphological evolution. *Proc. Natl. Acad. Sci.* **101:** 18058–18063.

Goto N., Imamura K., Miura Y., Ogawa T., and Hamada H. 1987. The Mexican hairless dog, its morphology and inheritance. *Jikken Dobutsu* **36:** 87–90.

Haworth K., Breen M., Binns M., Hopkinson D.A., and Edwards Y.H. 2001a. The canine homeobox gene MSX2: Sequence, chromosome assignment and genetic analysis in dogs of different breeds. *Anim. Genet.* **32:** 32–36.

Haworth K., Putt W., Cattanach B., Breen M., Binns M., Lingaas F., and Edwards Y.H. 2001b. Canine homolog of the T-box transcription factor T; failure of the protein to bind to its DNA target leads to a short-tail phenotype. *Mamm. Genome* **12:** 212–218.

Haworth K.E., Islam I., Breen M., Putt W., Makrinou E., Binns M., Hopkinson D., and Edwards Y. 2001c. Canine TCOF1; cloning, chromosome assignment and genetic analysis in dogs with different head types. *Mamm. Genome* **12:** 622–629.

Hayes H.M., Wilson G.P., and Fraumeni J.F. 1982. Carcinoma of the nasal cavity and paranasal sinuses in dogs: Descriptive epidemiology. *Cornell Vet.* **72:** 168–179.

Helmink S.K., Rodriguez-Zas S.L., Shanks R.D., and Leighton E.A. 2001. Estimated genetic parameters for growth traits of German shepherd dog and Labrador retriever dog guides. *J. Anim. Sci.* **79:** 1450–1456.

Kingsley D.M., Bland A.E., Grubber J.M., Marker P.C., Russell L.B., Copeland N.G., and Jenkins N.A. 1992. The mouse short ear skeletal morphogenesis locus is associated with defects in a bone morphogenetic member of the TGF beta superfamily. *Cell* **71:** 399–410.

Little C. 1934. Inheritance in Toy Griffons. *J. Hered.* **25:** 198–200.

———. 1957. *The inheritance of coat color in dogs.* Comstock Pub. Associates, Ithaca, New York.

Martinez S., Valdes J., and Alonso R.A. 2000. Achondroplastic dog breeds have no mutations in the transmembrane domain of the FGFR-3 gene. *Can. J. Vet. Res.* **64:** 243–245.

McGreevy P., Grassi T.D., and Harman A.M. 2004. A strong correlation exists between the distribution of retinal ganglion cells and nose length in the dog. *Brain Behav. Evol.* **63:** 13–22.

Nelson R. and Couto C. 2003. *Small animal internal medicine.* Mosby, St Louis, Missouri.

O'Brien D.P., Taylor J.F., Johnson G.S., and Johnson G.C. 2004. Mapping of multiple system atrophy and hairlessness in Chinese Cresteds and Kerry Blue Terriers. In *2nd*

International Conference Advances in canine and feline genomics, Utrecht University, The Netherlands.

Olsen S.J. 1985. *Origins of the domestic dog: The fossil record.* University of Arizona Press, Tucson.

Reif J.S., Bruns C., and Lower K.S. 1998. Cancer of the nasal cavity and paranasal sinuses and exposure to environmental tobacco smoke in pet dogs. *Am. J. Epidemiol.* **147:** 488–492.

Reuterwall C. and Ryman N. 1973. An estimate of the magnitude of additive genetic variation of some mental characters in Alsatian dogs. *Hereditas* **73:** 277–284.

Robinson R. 1990. *Genetics for dog breeders.* Pergamon Press, Oxford, United Kingdom.

Ruvinsky A. and Sampson J. 2001. The genetics of the dog. CABI Publishing, Wallingford, United Kingdom.

Sande R.D. and Bingel S.A. 1983. Animal models of dwarfism. *Vet. Clin. North Am. Small Anim. Pract.* **13:** 71–89.

Savolainen P., Arvestad L., and Lundeberg J. 2000. A novel method for forensic DNA investigations: Repeat-type sequence analysis of tandemly .repeated mtDNA in domestic dogs. *J. Forensic Sci.* **45:** 990–999.

Sponenberg D.P., Scott E., and Scott W. 1988. American hairless terriers: A recessive gene causing hairlessness in dogs. *J. Hered.* **79:** 69.

Stockard C.R., Anderson O.D., James W.T., and Wistar Institute of Anatomy and Biology. 1941. *The genetic and endocrinic basis for differences in form and behavior, as elucidated by studies of contrasted pure-line dog breeds and their hybrids.* The Wistar Institute of Anatomy and Biology, Philadelphia, Pennyslvania.

Verheijen J. and Bouw J. 1982. Canine intervertebral disc disease: A review of etiologic and predisposing factors. *Vet. Q.* **4:** 125–134.

Vila C., Savolainen P., Maldonado J.E., Amorim I.R., Rice J.E., Honeycutt R.L., Crandall K.A., Lundeberg J., and Wayne R.K. 1997. Multiple and ancient origins of the domestic dog. *Science* **276:** 1687–1689.

Wayne R.K. 1986. Cranial morphology of domestic and wild canids: The influence of development on morphological change. *Evolution* **4:** 243–261.

——— 2001. Consequences of domestication: Morphological diversity of the dog. In *the genetics of the dog* (ed. A. Ruvinsky and J. Sampson). CABI Publishing, Wallingford, United Kingdom.

Whitney L. 1947. Heredity in dogs. In *How to breed dogs.* Howell Book House, New York, pp. 268–320.

Wilcox B. and Walkowicz C. 1995. *The atlas of dog breeds of the world.* T.F.H. Publications, Neptune City, New Jersey.

Willis M.B. 1989. *Genetics of the dog.* Howell Book House, New York.

5

Genetic Analysis of the Canid Skeleton: Morphological Loci in the Portuguese Water Dog Population

Karl G. Lark, Kevin Chase,
David R. Carrier, and Fred R. Adler
Department of Biology
University of Utah
Salt Lake City, Utah 84112

SCIENCE HAS OFTEN BEEN LIKENED TO A VOYAGE of discovery. During my more than 50 years as a research scientist, that has never been more evident than in our journey into the quantitative genetics of dog morphology. The death of our first Portuguese water dog, Georgie, brought our laboratory into contact with Karen Miller, a dog breeder intensely interested in dog genetics. This happenstance, coupled with the fortuitous population structure of this breed, led into a collaboration between our laboratory and dog owners/breeders that revealed a world of unexpected treasures, all the more beautiful because of their simplicity. (K.G.L.)

In what follows, we demonstrate that dog populations with good pedigrees will allow estimation of relatedness within the population. This, coupled with accurate phenotypic data from owners, allowed us to look at heritabilities of phenotypes and the relation of heritable traits to the founders of the population. The availability of such data led to the genetic analysis of polygenic phenotypes. Grouping correlated traits using principal component (PC) analysis allowed us to look for underlying biological structure. Eventually, with the use of genetic markers, it became possible to look for quantitative trait loci (QTLs) regulating individual traits such as diseases, as well as those that regulate combinations of traits representing a higher level of genetic control. Our results are a prelude to much more that will come from the study of the dog.

The Dog and Its Genome ©2006 Cold Spring Harbor Laboratory Press 0-87969-742-3

DEVELOPING THE SYSTEM

Feasibility

Canine isolates with accurate pedigree information can provide an excellent compromise in the trade-off between the breadth of phenotypes or phenotypic variation available and the statistical power available. Combined with molecular markers and a genetic map, they can become valuable tools in the study of polygenic phenotypes (Williams et al. 1997; Slate et al. 1999).

On the basis of pedigree information, the Portuguese water dog (PWD) population appeared useful: It arose from relatively few founders, consanguinities ranged from ~0 to 0.6, and owners were eager to cooperate. Comparing consanguinities based on pedigree information to relatedness based on SSR alleles demonstrated the accuracy of the pedigrees (Chase et al. 1999).

We used simulations to relate trait heritability to the probability of identifying QTLs of different orders of magnitude (Chase et al. 1999). Subsequently, we sent tape measures to about 500 owners and asked them to measure 12 different morphological parameters (e.g., height at shoulder, length of tail, length of body, height at hock, distance between eyes). More than 300 responded, demonstrating the high level of cooperation that we could expect from owners. Of the 12 traits, 9 provided sufficient heritability in the simulation studies to allow QTL identification. Since these measurements provided a "least accurate assay," we felt confident in proceeding to invest in X-rays of the population.

Characterizing the Population

The population derives from three major founding events involving two different kennels, Algarbiorum and Alvalade (Molinari 1993). From the pedigrees, it is possible to calculate the consanguinity between dogs in the current population and each of the founders. Using these consanguinities as a data matrix, we analyzed the relatedness to founders using PC analysis. The first two PCs account for 66% of the variation in consanguinities. From the loadings we have identified the founders involved in these PCs. PC1 represents the variation introduced by the two major kennels, Algarbiorum and Alvalade, involved with founding the breed. PC2 represents variation involving contributions from early and late founders from the Alvalade kennel. When these two PCs are graphed (Fig. 1), the resulting distribution presents a "genetic topography" that displays the relation of each dog to the founders from which it is

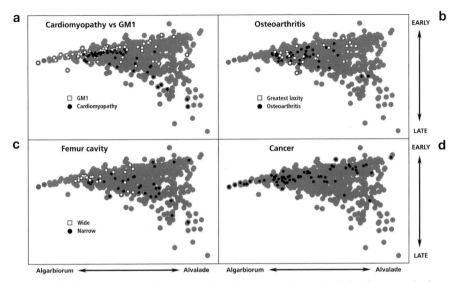

Figure 1. Genetic topography of specific phenotypes. In each of the four panels the entire PWD breeding population has been presented (*gray circles*) such that the first principal component of relationship to the founders (*x* axis) is graphed against the second (*y* axis). The *x* axis distinguishes between the Algarbiorum (negative values) and Alvalade (positive values) founding kennels. The *y* axis distinguishes between the early and late Alvalade founders. Specific phenotypes are indicated.

descended. This distribution is useful when combined with phenotypic data as a first estimate of the complexity of a phenotype. Thus, two recessive diseases, GM1 (Warren and Alroy 2000) and juvenile cardiomyopathy (Alroy et al. 2000) that can be traced to the founders Tabu and Taro (Algarbiorum and Alvalade, respectively) represent quite different distributions reflecting their origins (Fig. 1a). In contrast, both Addison's disease and seizures present two phenotype distributions that are indistinguishable from the distribution of the breeding population (data not shown), suggesting greater genetic or phenotypic complexity and possible multiple genetic origins for the phenotype.

This form of representation becomes useful when dealing with polygenic phenotypes. Two examples are the internal diameter of the femur and the relation of hip joint laxity to osteoarthritis: The difference between femur phenotypes (Fig. 1b) suggested that a major gene for width was segregating between the two founding kennels (*x* axis), which subsequently proved to be the case. Early measurements of hip laxity and osteoarthritis in the breed led to distributions suggesting that hip laxity might not be genetically related to the development of osteoarthritis (Fig. 1c). Subsequently, we identified two loci affecting laxity (Chase et al.

2004), but not osteoarthritis, and another locus that regulates osteoarthritis, but not laxity (Chase et al. 2005a). The major haplotype responsible for laxity traces back to the Algarbiorum kennel, whereas the relatively rare haplotype for osteoarthritis can be traced to an ancestor grouped with the Alvalades in Figure 1c. Genetic topography also suggests that cancer genes are segregating in the PWD population (Fig. 1d). A significant absence of cancers associated with the later Alvalade founders (lower right quadrant) suggests either an absence of cancer genes or genetic protection against the disease.

THE GENETIC BASIS OF MORPHOLOGY

Collection of data is an ongoing process. From owners throughout North America, we receive blood for DNA genotyping, as well as 5 X-rays (2 of the skull, 1 each from pelvis, fore-, and hind leg). These provide some 70 different skeletal metrics (http://www.georgieproject.com; Chase et al. 2002).

More than 500 SSR markers have been used for genotyping, and 330 dogs were analyzed using the complete set of metrics (limbs, pelvis, and skull). Pelvis and limb metrics have been measured on an additional 120 animals, and skull metrics from these animals will shortly be available. We investigate the association between a marker and a phenotype using an allele-sharing technique described by Chase et al. (2002). The idea behind the test is that pairs of dogs which share the same QTL alleles will be phenotypically more similar than pairs of dogs which do not share QTL alleles. We are using a mixed model to estimate the QTL means (Kennedy et al. 1992), employing the best linear unbiased estimate (BLUE) of the genotypic means (Lynch and Walsh 1998), and estimating the additive genetic variation (pedigree effects) using the polygenic function of SOLAR (Almasy and Blangero 1998). In simulation studies we have found that the pair-wise test gives greater power to detect additive QTLs than the QTL characterization by estimation of means (Lynch and Walsh 1998), but that BLUE estimates of the genotypic means were much better than ordinary least-square estimates. We also found that the jackknife estimates of the standard errors were more precise than sub-sampling estimates. A detailed description of the methods used for identifying and characterizing QTLs is contained in Chase et al. (2004).

We have identified and continue to identify QTLs for skeletal morphology. Figure 2 presents a map of work in progress locating 44 putative QTLs for heritable skeletal phenotypes on 21 different autosomes and the X chromosome. Of these, 14 are significant or highly significant, 30 are suggestive.

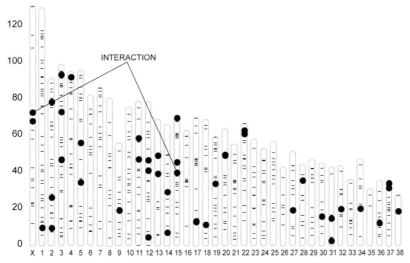

Figure 2. Physical map of SSR markers used in QTL identification. The physical positions for all the markers used in the genome scan to identify QTLs are indicated (−). The y axis is the physical position in Mb. Markers which show a significant association with a phenotype are indicated with circles. An interaction between the X chromosome and chromosome 15 is indicated.

Four major results relevant to the genetic architecture of the canid skeleton are presented here: (1) Principal component analysis dissects trait variation into independent groupings of functional skeletal complexes for which QTLs can be identified. These include components of both size and shape and represent trade-offs, compromises between extremes of force and speed. (2) Unlinked QTLs affect either the right or the left side of the animal, regulating variation in hip joint laxity associated with bilateral asymmetry. (3) An interaction between an autosomal locus and one on the X chromosome results in size sexual dimorphism. (4) Although QTLs identified for hip laxity do not regulate osteoarthritis, a QTL has been identified that regulates ∼16% of the osteoarthritis variation.

Heritable Principal Components

In Chase et al. (2002), we described PCs that characterized both size and shape of PWDs and, based on an analysis of 233 dogs, identified QTLs that regulated variation of the PCs. We used PC analysis of skeletal variation to reveal systems of traits defining skeletal structures. This analysis classified phenotypic variation into independent components that could be used to dissect genetic networks regulating complex biological systems. We showed that unlinked QTLs associated with these PCs individually promoted either correlations or inverse correlations within structures (e.g., within the skull, *or* among the limb bones) and, most strikingly, inverse correlations between structures (e.g., skull vs. limb bones). These latter were consistent with regulatory loci (single or closely linked genes) that inhibit growth of some bones while enhancing growth of others. These systems of traits exemplified the skeletal differences between such divergent breeds as greyhounds and pit bulls.

Since then, we have identified many more such QTLs (Fig. 2). In 2002, we described a QTL for PC1 (size) that was closely linked to canine *IGF-1* as well as syntenic with a mouse locus, *IGF-1sl2,* that regulates serum levels of IGF-1. (In poodles, size is correlated with serum levels of IGF-1.) Currently, we have identified seven QTLs for size located on six linkage groups. Additional PCs describe variations in shape, independent aspects of skeletal variation involving trade-offs between functional complexes of skeletal traits. Figure 3 presents radiographs illustrating two such trade-offs: the length versus width of limb bones and the shape (length vs. width) of the pelvis. Trade-off phenotypes associated with particular genotypes are presented in Figure 4. The trade-off between skull and postcranial body involves, among other loci, one that regulates both the length of the snout and the width of the humerus (Fig. 4A). Other loci in Figure 4 include a locus regulating the length and width of the tibia (Fig. 4B) and the length of the pelvis versus the width of the tibia (Fig. 4C). It is of interest to note that a major locus, regulating the trade-off between length and width of the

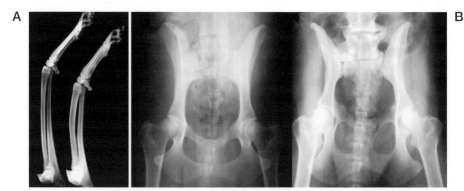

Figure 3. Radiographs illustrating shape "trade-offs" of two different principal components. (**A**) Two size-matched foreleg X-rays from dogs extreme for PC3 (limb length vs. limb width). (**B**) Two size-matched hip X-rays from dogs extreme for PC4 (ilium length vs. ischium length).

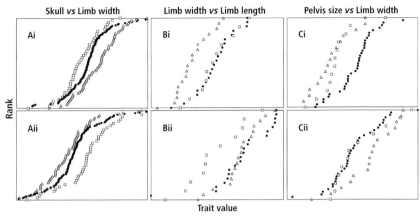

Figure 4. Marker genotypes associated with QTLs characterizing three different shape principal components. For each trait pedigree effects were removed using jack-knife techniques. Effects of PCs unrelated to the QTL (e.g., size) also were removed (see Chase et al. 2004). In each graph, dogs are ranked on the *y* axis in order of increasing phenotype (*x* axis) resulting in a cumulative distribution of phenotypes for each genotype. Each column compares phenotypes for selected genotypes for two different traits: (Ai) Snout length; (Aii) humerus width; (Bi) tibia width (I.D.); (Bii) tibia length; (Ci) tibia width (I.D.); (Cii) oscoxa. The graphs contain selected haplotypes: 2 majority haplotypes in *A*; 2 of 3 in *B*; and 2 of 4 in *C*. (A) *FH2356* [*ΔBB•BD DD*]; (B) *G02610* [*ΔAA•AB BB*]; (C) *FH3324* [*ΔDD•DG GG*].

limb bones, produces a marked effect on the width of limb bones, including the femur, as predicted by genetic topography (Fig. 1b).

The examples in Figure 4 focus on genotypes involving specific haplotypes that illustrate different aspects of genome architecture:

1. Trade-off haplotypes can (a) increase facial length while decreasing humerus width *or* decrease facial length while increasing humerus width; (b) give rise to a long, thin, tibia *or* a short, thick tibia; (c) give rise to a longer pelvis and thin tibia or a shorter pelvis coupled with thicker legs. All of these trade-offs characterize the difference between the extremes found in greyhounds versus pit bulls and are characteristic of compromises between force and speed.

2. Different modes of haplotype interaction are suggested in Figure 4: (A) additive; (B) dominant; and (C) synergistic.

3. The data in Figure 4B suggest that a haplotype may be dominant for one aspect of a trade-off and recessive for another: e.g., *A* dominant to *B* for tibia width, but recessive to *B* for tibia length. It seems probable that these haplotypes contain at least two relevant genes— one on *A* increases bone width, another on *B* increases bone length.

4. Variation in a particular trait may be partitioned between different PCs (independent functional complexes) and each aspect controlled by independent (unlinked) QTLs. Thus, variation in limb width is associated (inversely) with variation in skull length, limb length, or length of pelvis.

It is of interest that none of these QTLs appears to contain *Hox* genes. Although this is due, in part, to gaps in the marker distribution used to cover the genome, it also could represent the fixation of Hox phenotypes within the breed.

It should be noted that those functional trade-offs, represented by the PCs discussed above, also represent differences between puppies and adult dogs, and therefore the possibility arises that the genetic control of these trade-offs is involved in the developmental transition from juvenile to adult morphology, as suggested by Wayne (2001). Additional information on canine functional trade-offs is presented in Chapter 6, in which the PWD, *Canis familiaris*, is compared to the fox, *Vulpes vulpes*.

Size Sexual Dimorphism

Size can be estimated as the first principal component (PC1) of skeletal variation (Chase et al. 2002, 2005b). PC1 represents an averaging over each dog's normalized skeletal metrics and is significantly correlated with body mass (r = 0.7). Female PWDs are smaller than males. This is immediately apparent from the differences in the mean values of PC1 (first two lines of Table 1). A QTL on CFA 15, linked to IGF1 and associated with marker FH2017, regulates PC1. Alleles of FH2017 identify two major haplotypes (A and B) that account for >90% of FH2017 genotypes. Variation in PC1 associated with genotypes of FH2017 (Table 1, a) is consistent with an additive mode of inheritance such that genotypic means are ranked: BB(small)<AB<AA(large). However, if the population is further subdivided according to sex (Table 1, b and c) haplotype A (large) is dominant in males (AA = AB>BB); whereas haplotype B (small) is dominant in females (BB = AB<AA). This difference in itself provides a genetic basis for size dimorphism between the male and female populations. An interaction with a QTL associated with CHM on the X chromosome provides an additional basis for size sex dimorphism. Two major haplotypes, α and β, are associated with CHM. FH2017 AA female genotypes also *homozygous* for these CHM haplotypes (Table 1, d) are almost as large as FH2017 AA males (in fact, they do not differ

Table 1. Pedigree corrected PC1 (size) values for Portuguese water dogs subdivided according to sex.

Population	FH2017 Genotype	Count	Mean	S.E.
Male		164	2.52	0.28
Female		256	−1.54	0.26
Total[a]	AA	67	1.69	0.47
	AB	126	0.03	0.39
	BB	168	−0.91	0.35
Male[b]	AA	19	3.54	0.89
	AB	49	3.15	0.38
	BB	67	1.60	0.47
Female[c]	AA	48	0.96	0.53
	AB	77	−1.96	0.47
	BB	101	−2.57	0.41
Female CHM homozygotes[d]	AA	25	2.43	0.77
	AB	41	−2.33	0.55
	BB	63	−2.64	0.50
Female CHM heterozygotes[e]	AA	23	−0.63	0.57
	AB	32	−1.34	0.85
	BB	35	−1.98	0.69

[a]*FH2017* alleles (AB). [b,c]Sex and *FH2017* alleles. [d,e]Female *FH2017* alleles and CHM genotypes ([d]homozygotes[$\alpha\alpha$ or $\beta\beta$]; [e]heterozygotes[$\alpha\beta$]).

For each subpopulation the number of dogs (count), PC mean, and standard error are shown.

significantly in size from large males). In contrast, FH2017 AA female genotypes *heterozygous* for these CHM haplotypes (Table 1, e) are smaller than small males but somewhat larger than the smallest female genotypes. Overall, these effects are responsible for ∼50% of the size sexual dimorphism in the breed.

The data suggest the hypothesis diagrammed in Figure 5: The FH2017 QTL contains two distinct genes associated with two different haplotypes: *FH2017* haplotype A acts in both males and females to up-regulate size. *FH2017* haplotype B not only does *not* up-regulate size, but contains another gene, active in females only (silenced in males), that suppresses this up-regulator. Activation of the up-regulator requires interaction with a CHM-associated multimeric product, either $\alpha\alpha$ or $\beta\beta$. The *CHM* heterozygote data can be explained by assuming that a mixed multimeric product ($\alpha\beta$) is inactive (spoiler effect) (Kadouri et al. 1978). Our data require that this region of the X chromosome escapes inactivation—i.e., both copies of the X are active—a phenomenon observed for several other X-linked loci (Brown

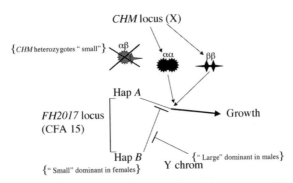

Figure 5. Proposed interactive network between three chromosomes: X (CHM), Y, and CFA 15 (see text).

and Greally 2003). Because all of the effects (inhibitions and up-regulation) are dominant, the system as a whole will respond rapidly to selection.

Sexual selection, acting on body size genes on the X and autosomal chromosomes, thus makes species larger, with both males and females becoming larger as a result of sexual selection on males. Sexual dimorphism evolves later as a result of natural selection for optimal body size, favoring genes that secondarily reduce the size of females. This scenario and our data are consistent with Rensch's rule, that the level of sexual size dimorphism most often is positively correlated with body size (Rensch 1950, 1960; Abouheif and Fairbairn 1997). Moreover, the species with the highest levels of sexual size dimorphism tend to have relatively large females, as expected given the proposed genetic mechanism. Reduction of female size relative to that of males, through an inhibition of genes that enhance growth, may commonly be associated with the evolution of sexual size dimorphism. Such a pattern of inverse dominance, in which small is dominant in females whereas large is dominant in males, should be relatively easily tested in other species where controlled mating is feasible and hybrid progeny can be examined.

Disease-related Polygenic Phenotypes

Two common skeletal pathologies in dogs are hip dysplasia and osteoarthritis. Both have been shown to be correlated with laxity in the coxofemoral (hip) joint. Laxity and osteoarthritis can be measured from pelvic radiographs. The Norberg angle is a heritable, quantitative measure of hip (coxofemoral) joint laxity in dogs—the larger the angle, the "tighter" the joint. A significant, nonheritable, directed asymmetry in laxity is observed in PWDs (Chase et al. 2004), represented by smaller Norberg angles (greater laxity) in the left hip than in the right. The basis for this might be the result of behavioral or physiological traits that are fixed in the breed. For example, left or right "footedness" (analogous to handedness in humans) might lead to greater stress on one joint than on the other; or organ asymmetry could produce unequal loading, hence differential stress, during running or jumping. Variation in the extent or type of nutrition or exercise might then be translated into variation in the extent of the asymmetry.

Two QTLs have been identified using the metrics of the left and right hips as separate data sets. Both are on CFA 1, separated by about 95 Mb. One QTL is associated significantly with variation in the left, but not the right, hip; whereas the other is associated significantly with variation in the right, but not the left, hip. These QTLs accounted for ~30% of the heritable variation observed in each hip. An interesting observation was that the right hips with the tightest phenotypes resulted from heterozygous genotypes involving synergism between haplotypes that produced phenotypically lax hip joints. Thus, matings between dogs with "excellent" hip joints would segregate offspring with lax hip joints.

Two possible explanations for the genetic discrimination between the right and left hips are:

1. The QTL associated with FH2524 can only express its phenotypic variation under conditions of greater laxity (low values of the Norberg angle),whereas the phenotypic variation derived from genotypes of the QTL associated with FH2598 can only be observed under conditions of low laxity (large Norberg angles).
2. The phenotypic variation is dependent on some form of bilateral asymmetry, and genotypes distinguish the right from the left side of the animal.

These two hypotheses can be distinguished on the basis of genetic variation within the extreme values of the population; however, a much larger database must be assembled before that will be possible.

Although the Norberg angle is correlated with osteoarthritis (OA), the Norberg angle QTLs on CFA 1 are not significantly associated with OA. Recently, we have identified a QTL for OA, associated with the SSR marker FH2320 on CFA 3 (Chase et al. 2005a). This QTL accounts for about 16% of the OA variation ($R^2 = 16.4\%$) and involves primarily cranial and caudal acetabular marginal osteophytes. This same QTL affects the shape of the pelvis through its effect on the ischial tuberosity, the width of which is segregating in the PWD population (Fig. 3). It also affects several other bone metric residuals including the length of the second vertebra above the sacrum, the inner diameter of the radius, as well as lever arms on the limb bones, e.g., two metrics of the heel, the calcaneus and inlever, as well as the pisiform or wrist.

We have shown that the OA phenotype cannot be due to the differences in the shape of the pelvis (illustrated in Fig. 3). A more acceptable hypothesis is that the QTL regulates normal skeletal growth of several bones, and the CFA haplotype responsible for OA contains a gene that has a propensity for promoting abnormal skeletal growth. A comparison of the canine genome with the mouse, rat, and human genomes suggests possible candidate genes close to the *FH2320* marker sequence: *IGF1R*, an IGF1 receptor (Abbot et al. 1992); or *DMN*, a cytoskeletal protein conferring resistance to mechanical stress (Mizuno et al. 2001).

SUMMARY: THE LESSONS LEARNED

In principle, purebred dog populations should represent ideal isolates with which to analyze complex polygenic phenotypes. With the sequencing of the dog genome, results from such analyses become applicable to other mammalian systems such as the human or mouse. We undertook the research described above to demonstrate the feasibility of this principle. The results obtained exceeded our expectations. The discovery of novel phenotypes has suggested new ways to explore genome architecture and subsequently focus on the evolution of the mammalian skeleton: (1) Functional trade-offs between suites of traits can result in rapid changes in shape in response to selection for function or shape. (2) Genetic regulation of bilateral asymmetry suggests an in-depth search for additional genes that distinguish between the right and left sides of the animal. (3) Interaction between unlinked loci leading to sexual size dimorphism suggests that similar interactions could maintain pools of selectable loci allowing response at the population level to environmental change. It seems likely that similar relationships will be found when other complex phenotypic systems are investigated in the dog.

ACKNOWLEDGMENTS

We thank Karen Miller, director of the Georgie Project, and all of the Portuguese water dog owners for making available the materials (blood and X-rays) used in this study. Ms. Deborah Broughton coordinated their contributions. This research was supported by grant GM-63056 from the National Institutes of Health and by gifts from the Judith Chiara Family

Trust, the Nestle Purina Company, and more than 100 Portuguese water dog owners. Tyler Jarvik, Dennis Lawler, Travis Lorentzen, Karen Miller, and Elaine Ostrander were co-authors on primary publications of research reviewed in this chapter.

REFERENCES

Abbott A.M., Bueno R., Pedrini M.T., Murray J.M., and Smith R.J. 1992. Insulin-like growth factor I receptor gene structure. *J. Biol. Chem.* **267:** 10759–10763.

Abouheif E.F. and Fairbairn D.J. 1997. A comparative analysis of allometry for sexual size dimorphism: Assessing Rensch's rule. *Am. Nat.* **149:** 540–562.

Almasy L. and Blangero J. 1998. Multipoint quantitative trait linkage analysis in general pedigrees. *Am. J. Hum. Genet.* **62:** 1198–1211.

Alroy J., Rush J.E., Freeman L., Amarendhra Kumar M.S., Karuri A., Chase K., and Sarkar S. 2000. Inherited infantile dilated cardiomyopathy in dogs: Genetic, clinical, biochemical, and morphologic findings. *Am. J. Med. Genet.* **95:** 57–66.

Brown C.J. and Greally J.M. 2003. A stain upon the silence: Genes escaping X inactivation. *Trends Genet.* **19:** 432–438.

Chase K., Adler F.R., Miller-Stebbings K., and Lark K.G. 1999. Teaching a new dog old tricks: Identifying quantitative trait loci [in dogs] using lessons from plants. *J. Hered.* **90:** 43–51.

Chase K., Carrier D.R., Adler F.R., Jarvik T., Ostrander E.A., Lorentzen T.D., and Lark K.G. 2002. Genetic basis for systems of skeletal quantitative traits: Principal component analysis of the canid skeleton. *Proc. Natl. Acad. Sci.* **99:** 9930–9935.

Chase K., Lawler D.F., Adler F.R., Ostrander E.A., and Lark K.G. 2004. Bilaterally asymmetric effects of quantitative trait loci (QTLs): QTLs that affect laxity in the right versus left coxofemoral (hip) joints of the dog (*Canis familiaris*). *Am. J. Med. Genet.* **124A:** 239–247.

Chase K., Lawler D.F., Carrier D.R., and Lark K.G. 2005a. Genetic regulation of osteoarthritis: A QTL regulating cranial and caudal acetabular osteophyte formation in the hip joint of the dog (*Canis familiaris*). *Am. J. Med. Genet.* **135A:** 334–335.

Chase K., Carrier D.R., Adler F.R., Ostrander E.A., and Lark K.G. 2005b. Interaction between the X chromosome and an autosome regulates size sexual dimorphism in Portuguese Water Dogs. *Genome Res.* (in press).

Kadouri A., Kunce J.J., and Lark K.G. 1978. Evidence for dominant mutations reducing HGPRT activity. *Nature* **274:** 256–259.

Kennedy B.W., Quinton M., and van Arrendonk J.A.M. 1992. Estimation of effects of single genes on quantitative traits. *J. Anim. Sci.* **70:** 2000-2012.

Lynch M. and Walsh B. 1998. *Genetics and analysis of quantitative traits.* Sinauer, Boston, Massachusetts, pp. 745–751.

Mizuno Y., Puca A., O'Brien K., Beggs A., and Kunkel L. 2001. Genomic organization and single-nucleotide polymorphism map of desmuslin, a novel intermediate filament protein on chromosome 15q26.3. *BMC Genet.* **2:** 8.

Molinari C. 1993. *The Portuguese water dog.* ELO-Publicidade, Portugal.

Rensch B. 1950. Die Abhangigkeit der relativen sexual differenz von der Korpergrosse. *Bonner Zoologische Beitraege* **1:** 58–69.

———. 1960. *Evolution above the species level.* Columbia University Press, New York, pp. 157–159.

Slate J., Pemberton J.M., and Visscher P.M. 1999. Power to detect QTL in a free-living polygynous population. *Heredity* **83:** 327–336.

Warren C.D. and Alroy J. 2000. Morphological, biochemical and molecular biology approaches for the diagnosis of lysosomal storage diseases. *J. Vet. Diagn. Invest.* **12:** 483–496.

Wayne R. 2001. Consequences of domestication: Morphological diversity of the dog. In *The genetics of the dog* (ed. A.R.J. Sampson), pp. 15–43. CABI Publishing, Wallingford, United Kingdom.

Williams J.T., Duggirala R., and Blangero J. 1997. Statistical properties of a variance components method for quantitative trait linkage analysis in nuclear families and extended pedigrees. *Genet. Epidemiol.* **14:** 1065–1070.

WWW RESOURCE

http://www.georgieproject.com, Georgie Project home page

6

Morphology and Behavior: Are They Coupled at the Genome Level?

Lyudmila N. Trut and Anastasia V. Kharlamova
Institute of Cytology and Genetics
Siberian Branch Russian Academy of Sciences
Novosibirsk, 630090, Russia

Anna V. Kukekova and Greg M. Acland
James A. Baker Institute for Animal Health
College of Veterinary Medicine
Cornell University, Ithaca, New York 14853

David R. Carrier, Kevin Chase, and Karl G. Lark
University of Utah, Department of Biology
Salt Lake City, Utah 84112

IN ALL ORGANISMS, BEHAVIOR AND MORPHOLOGY are inextricably coupled. Morphology constrains the range of behavioral patterns, and behavior regulates and modifies morphological function. Over centuries, this constraint has been recognized in the selective breeding of dogs to perform specific tasks. Here, we explore the genetic coupling of morphology and behavior in a fox model system: the heritable morphological variation that has accompanied selection for tame and aggressive behavioral traits in the silver fox.

In what follows, we demonstrate that similar functional trade-offs in morphology are found in both foxes and dogs and that this morphological variation may be concomitant with behavioral variation. Finally, we describe preliminary results from experiments designed to test whether these two phenotypes are related at the genetic level (thus co-selected) or have become associated serendipitously as a result of random genetic drift and founder effects.

The red fox, *Vulpes vulpes*, of which the silver fox is a subspecies, is the out-group for modern canids. Its phylogenetic lineage has been

The Dog and Its Genome ©2006 Cold Spring Harbor Laboratory Press 0-87969-742-3

separated from *Canis familiaris* for about 10 million years (Wayne 1993). *V. vulpes* has the largest natural geographic distribution of any terrestrial mammal other than humans. It occupies habitats ranging from mixed forest to alpine tundra, from sand dunes to urban outskirts (Nowak and Paradiso 1991; Sheldon 1992). Associated with this diversity of habitats is a wide range of anatomical configurations and a behavioral plasticity that makes the red fox the bane of chicken farmers, the favorite of hunting sportsmen, and a mythic symbol of crafty intelligence and creativity. In its various natural habitats, it displays a diversity of behaviors and morphologies similar to the spectrum encountered in different breeds of dogs.

Two silver fox populations, one tame and the other aggressive, have been established in Novosibirsk, Russia (Belyaev 1979; Belyaev et al. 1984, 1985; Trut 2001; Chapter 26). The inherited behavioral changes were accompanied by morphological variations that distinguished tame from aggressive animals (Trut 1999, 2001). These populations present a unique opportunity to analyze the genetic basis for interaction between behavior and functional morphology: How are behavioral and morphological phenotypes related at the genome level?

Here we present a quantitative description of the differences in morphology that characterize the different silver fox populations, together with preliminary findings on the co-segregation of behavioral and morphological characters. It seems reasonable to assume that information obtained from this system will be applicable to analysis of the behavior of *C. familiaris*.

Our experiments have been guided by hypotheses developed to explain the morphology of different breeds of dogs (Wayne 1986, 2001), and hypotheses to explain the behaviors and physiological parameters that characterize these fox populations (Trut 1999, 2001). Trut (2001) has suggested that a genetic hierarchy regulates behavior, physiology, and morphology such that "master" loci regulate both phenotypes by interacting with other loci particular to specific suites of traits. As a consequence, selection for behavior may already have fixed genetic loci responsible for behavior in the Tame and Aggressive populations while allowing many morphological parameters to continue to segregate. Data on dogs (Wayne 1986, 2001) and on these foxes led Trut (2001) to suggest that the highest hierarchical level of regulation might involve genetic loci that govern postnatal ontogeny, the stages of development from juvenile to adult forms.

BEHAVIORAL VARIATION OF FOX POPULATIONS

Five silver fox populations, described in Chapter 26, were used in this study: Tame, Aggressive, Unselected, Hybrid (Tame × Aggressive), and

Table 1. Behavior of parental F$_1$ and backcross (F$_1$ by Tame) progeny

Population	Sex	Count	Mean	S.E.
Aggressive	female	4	−2.8	0.25
Aggressive	male	5	−3.4	0.24
F$_1$	female	13	1.7	0.47
F$_1$	male	10	1.3	0.47
Backcross	female	68	2.8	0.14
Backcross	male	58	2.3	0.18
Tame	female	6	3.9	0.09
Tame	male	3	3.8	0.17

Means and standard errors of female and male fox behavioral scores are given together with the number of animals used (count). Aggressive and Tame animals were scored on separate scales: Aggressive from −1 (least aggressive) to −4 (most aggressive); Tame from +1 (least tame) to +4 (most tame). For a description of the assay criteria see Trut (1980); Chapter 26.

Backcross (Tame × Hybrid). Kukekova et al. (Chapter 26) have detailed the genetic basis for differences in behavioral phenotypes between the Tame and Aggressive populations. The results, summarized in Table 1, indicate a high additive heritability for behavior in both males and females.

MORPHOLOGICAL VARIATION BETWEEN THE FOX POPULATIONS

An analysis of externally measured metrics was undertaken on a total of 608 foxes: 414 Tame, 63 Aggressive, 31 Hybrid (these comprised all of the foxes from the three different populations), as well as 100 foxes from the large farm population (Unselected) similar to the population that furnished the founders of the present Tame and Aggressive populations some 40 years ago. Sixteen morphological traits were measured on each animal, including body mass, 8 length metrics (hip; mandible; radius; skull; snout; tibia; tail; and body from back of skull to base of tail), and 7 width metrics (minimum diameter of tibia; span of canine and molar teeth; mediolateral width of elbow and knee; hip; and maximum width of skull).

We used principal component analysis (Jackson 1991) to define independent components of size and shape variation (see Chapter 5). Principal components (PCs) were calculated after correcting the individual metrics for differences in age of the animals. Sexual dimorphism was observed for all metrics (see below). The four populations were combined in a common matrix, comprising the 16 metrics derived for all of the animals.

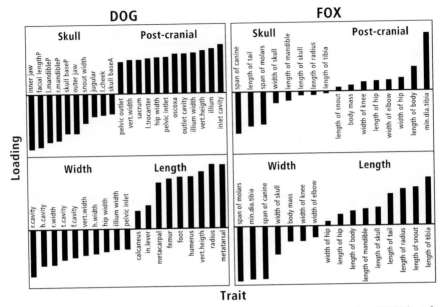

Figure 1. Principal Component (PC) loadings for Portuguese water dog (PWD) and silver fox populations. PWD values were based on radiographic data involving ~70 metrics (Carrier et al. 2005). Fox data were based on 16 external metrics using 608 foxes from the four populations, combined to form a single PC matrix (see text).

The magnitude of trait loadings and their relative signs (correlation or inverse correlation) describe the influence of different traits in determining size or shape and offer insight into the biological nature of each PC (Jackson 1991; Chase 2002). Thus, all of the trait loadings for PC1 (42% of the variation) were positively correlated, consistent with overall size. Loadings for the remaining PCs were inversely correlated, consistent with different parameters of shape (Fig. 1). PC2 (11% of the variation) is a generalized length-versus-width axis in which metrics of skeletal length are inversely correlated with metrics of skeletal width. Individual foxes that have relatively short limbs and snouts tended to have relatively wide snouts, skulls, and limb bones. In PC4 (7% of the variation), metrics of body size are inversely correlated with metrics of head size. Individuals with narrow limb bones, small pelvic girdles, low body mass, and short body lengths tended to have large heads in which the length of the snout is particularly pronounced. Sexual dimorphism was observed for the first, second, and fourth principal components (Table 2).

The same main shape components, representing trade-offs between bone lengths and widths as well as between skull and postcranial structures,

Table 2. Sexual dimorphism of principal components

External measurements (Fig. 2)				Backcross (F_1 by Tame)			
	PC mean				PC mean		
Principal component	male (180)	female (428)	Dimorphism p-value	Principal component	male (74)	female (57)	Dimorphism p-value
PC 1	3.889	−1.635	$<10^{-7}$	PC 1	−3.73	5.52	$<10^{-7}$
PC 2	−0.169	0.071	0.0076	PC 2	−0.26	−1.49	0.0009
PC 3	−0.009	0.004	0.863	PC 3	0.66	−0.44	0.0006
PC 4	−0.241	0.101	$<10^{-6}$	PC 4	0.72	−0.36	$<10^{-4}$
PC 5	0.021	−0.009	0.637	PC 5	0.44	−0.14	0.035
				behavior	2.80	2.25	0.003*

External metrics comparing populations (Fig. 2) and radiographic metrics of the backcross population (see below). Means of the male and female populations are given. The numbers of animals are in parentheses. Significance was determined by a t-test except for behavior in the backcross population where a non-parametric test (Mann-Whitney) was used.

are found in both foxes and dogs. The loadings for PC2 and PC4 represent shape variations similar to those found in Portuguese water dogs (Fig. 1) and are consistent with changes that would be expected to occur along a trade-off continuum between speed and force (Chase 2002; Pasi and Carrier 2003). Foxes are the out-group for other modern canids. The *V. vulpes* and *C. familiaris* lineages separated 7–10 million years ago (Wayne 1993). Our data suggest that the shape trade-offs we observe may have been present in the common ancestor from which they arose.

As proposed for Portuguese water dogs (Chase 2002), the patterns of variation in skeletal metrics observed in silver fox PCs (Fig. 1) may be related to the ontogenetic transition from the juvenile to the adult state. In mammals, newborns require anatomical specializations for nursing and competition with siblings for access to nipples (de Beer 1985; Frank et at. 1991; Emerson and Bramble 1993) that result in relative uniformity of newborn body shape. Adult behaviors associated with feeding, loco-motion, reproduction, and sociality require changes in the shape and pro-portion of the skeletal system during postnatal growth in most or all species of mammals (Gould 1977; Shea 1981; Carrier 1996). Genetic components that regulate the sets of inversely correlated characters in Figure 1 could account for much of this transformation in canids. For example, appropriate temporal activation of different genes could pro-duce the short-broad face and limbs of newborn foxes on the one hand, followed by the development of the relatively longer and narrower face and more gracile limbs of adult foxes on the other. This suggestion is consistent with analyses of skeletal dimensions in canids (Wayne 1986, 2001), which indicate that allometry among adults of different species is often nearly identical to the shapes encountered during the course of postnatal growth.

Accurate pedigree information, together with the large number of Tame foxes (404), allowed us to calculate heritabilities (Almasy and Blangero 1998) for the different principal components of morphology within the Tame population. Each of the first five PCs, which together account for more than 60% of the total morphological variation, has a large genetic component: Heritabilities ranged from 40% to 65%. It is interesting to note that behavioral heritability could not be detected within the Tame population using the assay criteria by which the population was selected. Thus, within this population, the major genetic components are fixed with respect to behavior, whereas morphological characters are still segregating. However, as documented in Table 1, behavioral differences between the Tame and Aggressive populations have a high heritability.

Because the PCs were derived from a matrix comprising all of the foxes (Tame, Aggressive, Unselected, and Hybrid), we were able to compare the PC means of each of these four populations (Fig. 2). Significant differences between the means of the Tame and Aggressive populations were observed for PCs 2 and 4, between the Aggressive and Unselected for PCs 3 and 4, and between the Tame and Unselected for PCs 2, 3, and 5. Because these populations have been maintained in identical environments (living space and nutrition), significant differ-

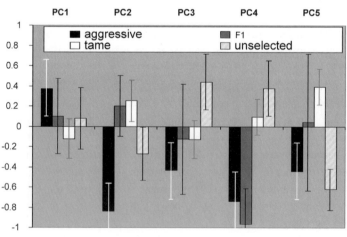

Figure 2. Means and standard errors of PC values for the four different fox populations: Tame, Aggressive, F₁ Hybrid, and Unselected. Values have been standardized to the same scale. See text and legend to Fig. 1.

ences between population phenotypes reflect differences in the genetic composition of the populations. Clearly, the populations are genetically different in various suites of traits that govern shape (PCs 2–5). The Hybrid population provided further insight into the genetic control of the different morphological PCs. Although this population was small, the mean values of this population differed significantly from the Aggressive population for PC2 and from the Aggressive population for PC4, suggesting that for these components, dominant/recessive modes of inheritance characterize many of the genes that regulate the morphological differences between the two populations (however, see Conclusions and Caveats, page 90).

A large variation in PC values was observed within each population. Figure 3 presents the cumulative distribution of PC values for the first three PCs. There is a continuous distribution of PC values for each population, as would be expected from quantitative traits. It can be seen that although the population means differ significantly between the three populations for PC2 and PC3, there is an overlap in the values for individual foxes. Such differences between the populations might be expected if PCs 2 and 3 represent trade-off continuums along which a population might move as a result of founder effects, selection, or drift.

CO-SEGREGATION OF BEHAVIOR AND MORPHOLOGY

Are the morphological changes the result of random drift, or are behavior and anatomy coordinated at a genetic level such that selection for behavior also selects for morphology appropriate to that behavior? (This might occur either during selection of founders or during the selective breeding that produced the Tame and Aggressive populations.) Coordinate selection might occur if both behavior and morphology were under the control of some common regulatory genes, or if genes for these two phenotypic classes were linked and therefore segregated together. To test this, we have used radiographic metrics (Fig. 4), similar to those used for the Portuguese water dog, to analyze morphological metrics of fox segregants produced by backcrossing Hybrid with Tame animals. We have already seen that behavior segregates in this backcross (Table 1). If behavior and morphology have some common genetic basis, we expect to find a significant correlation between the two phenotypes in the backcross population. However, since behavior is fixed within the Tame population, but morphology is not, we cannot expect this correlation to be complete.

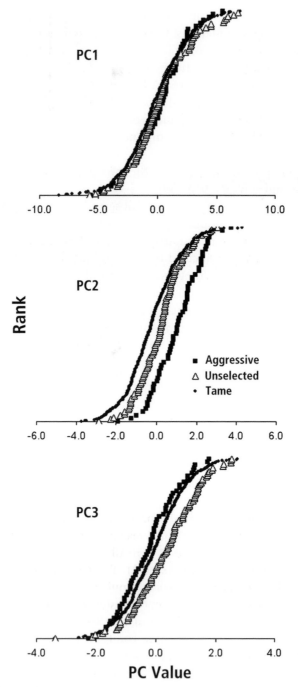

Figure 3. Cumulative distributions of PC values ranked on the *y*-axis according to increasing PC value on the *x*-axis. PC values based on external metrics are graphed for each subpopulation.

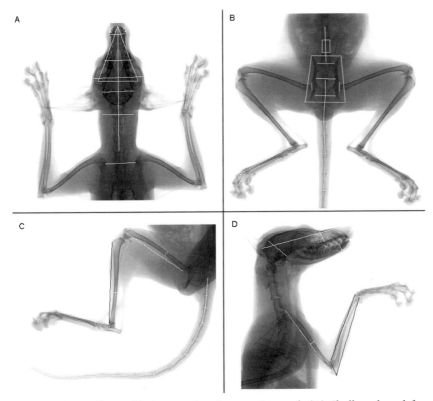

Figure 4. Fox radiographic images showing metrics used. (**A**) Skull neck and fore limbs (dorsal–ventral). (**B**) Pelvis hind limbs and tail (dorsal–ventral). (**C**) Hind limb and pelvis (lateral). (**D**) Skull and forelimb (lateral).

Our analysis of such a correlation is preliminary. At present we have only analyzed the data from 120 one-year-old animals from the backcross population (F_1 by Tame, Tables 1, 2). All of the first five PCs showed significant sexual dimorphism (Table 2). Therefore, we prepared separate PC matrices for the male and female populations. No significant correlation between behavior and morphology was found in the female backcross population. However, in the male population one PC was highly correlated with behavior (Spearman correlation $= -0.49$; p-value $= 0.00004$). The correlation of behavior with this PC is presented in Figure 5.

This PC accounts for 9% of the skeletal variation in the male backcross population. The loadings comprise a trade-off between length and width that involves the length and width of limb bones, skull width metrics, as well as the length of the neck and lumbar vertebrae.

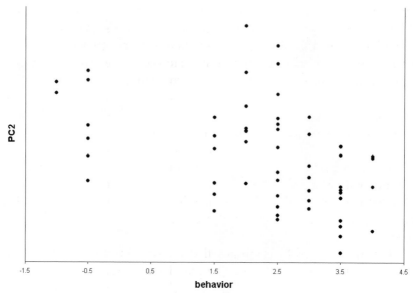

Figure 5. Scatter graph of behavior and a principal component of morphological variation in one-year-old backcross males. Details are presented in the text. Employment of different behavior assays with different scales (aggressive −4 to 0; tame 0 to +4) necessitated the use of non-parametric analysis (Spearman correlation). Values of PC2 range from negative (long and thin bones) to positive (short and thick bones).

CONCLUSIONS AND CAVEATS

Our data clearly support a genetic basis for the difference in behavior between the Aggressive and Tame populations. However, the behavioral scoring for the F_1 and backcross populations utilized two different scales for scoring: one appropriate to the Tame population (0 to +4) and one appropriate to the Aggressive population (0 to −4). This created some ambiguity for animals close to 0, a problem that is evident in the behavioral scoring of 8 "Aggressive" BC segregants seen in Figure 5. The current development of a new behavioral assay based on analysis of videos and a single scoring system that comprises the entire spectrum of behavior (A.V. Kukekova and N.V. Trut, in prep.) should solve this problem.

The Tame, Aggressive, and Unselected populations differ morphologically, and again, there is good reason to believe that the differences have a genetic basis. Heritability estimates in the Tame population demonstrate that morphological variation is segregating in this population. The smaller size of the Aggressive population precluded significant estimation of heritability. Principal component analysis of the fox populations (Aggressive, Tame, Unselected, and F_1) revealed shape differences that had been established for *C. familiaris* (Portuguese water dogs) indicating axes of variation consistent with the transition from the juvenile to the adult state (neotony). However, the large majority of animals were 2 years of age or older. In contrast, the 23 F_1 animals were all 1 year old. Thus, it is not clear at this time whether conclusions of dominance are warranted or whether the differences in means between F_1 and Tame or Aggressive animals reflect an effect of age.

Perhaps the most promising conclusion is the co-segregation of shape (PC2) and behavior in 1 year backcross males. Further analysis at 2 and 3 years will establish whether this is a transient stage in juvenile development or whether this persists in older animals. Similarly, it is too early to conclude that similar correlations will not be found in older females.

Finally, it should be cautioned that the two different selections used to produce the Tame and Aggressive populations could easily have led to different morphological correlates of behavior. The two populations, Tame and Aggressive, may not represent opposite extremes of the same behavioral continuum. Thus, in the backcross we may eventually see segregation of Tame from "Not Tame" and Aggressive from "Not Aggressive" phenotypes.

The next stage of this research will be to identify quantitative trait loci (QTLs) for behavior and morphology and to see whether some of these regulate both types of phenotypic variation. For this research, the dog genome is invaluable, since it is already providing the information needed as a source of SSR markers and eventually will allow comparison of genetic regions between these two canine lineages that have been separated for 10 million years.

SUMMARY

Starting from a farm population of silver foxes (*V. vulpes*), Belyaev, Trut, and collaborators used behavioral criteria to select Tame and Aggressive fox populations. They reported a genetic basis for the behavioral differences between the two populations and that morphological differences between the populations also appeared during the behavioral selection. We have characterized these morphological differences and used genetic crosses between the Aggressive and Tame populations to investigate the possibility of a common genetic basis for morphological and behavioral variation.

Behavioral differences between Aggressive and Tame parents, F_1 hybrids, and F_1 by Tame backcross animals confirmed the genetic basis for behavioral differences between the two populations. Using PC analysis, we characterized the morphological variation in terms of size and various components of shape. Dogs and foxes have similar components that define variation in shape (e.g., independent components of variation that represent trade-offs in metrics of length versus width, or of cranial versus post-cranial morphology). Aggressive and Tame populations differ significantly in several of these shape components. In preliminary results, behavior and one morphological shape component were significantly correlated in backcross segregants, suggesting a common genetic basis resulting in co-segregation of behavior with morphology.

ACKNOWLEDGMENTS

This research was supported by National Science Foundation grant IBN-0212141 to D.R.C.; National Institutes of Health grant GM63056 to K.G.L.; National Institutes of Health National Institute of Mental Health grant MH069688-01 to A.V.K.; NATO LST.CLG.979216 to L.N.T., G.M.A., and A.V.K.; and grants from the Russian Fund of Basic Research (Scientific School grant #2303.2003.4) and from the Program of the Basic Research of the Presidium of Russian Academy of Science to L.N.T.

REFERENCES

Almasy L. and Blangero J. 1998. Multipoint quantitative trait linkage analysis in general pedigrees. *Am. J. Hum. Genet.* **62:** 1198–1211.

Belyaev D.K. 1979. Destabilizing selection as a factor in domestication. *J. Hered.* **70:** 301–308.

Belyaev D.K., Plyusnina I.Z., and Trut L.N. 1984/85. Domestication in the silver fox (*Vulpes vulpes*): Changes in physiological boundaries of the sensitive period of primary socialization. *Appl. Anim. Behav. Sci.* **13:** 359–370.

Carrier D.R. 1996. Invited perspective: Ontogenetic limits on locomoter performance. *Physiol. Zool.* **69:** 467–488.

Carrier D.R., Chase K., and Lark K.G. 2005. Genetics of canid skeletal variation: Size and shape of the pelvis. *Gen. Res.* (in press).

Chase K., Carrier D.R., Adler F.R., Jarvik T., Ostrander E.A., Lorentzen T.D., and Lark K.G. 2002. Genetic basis for systems of skeletal quantitative traits: Principal component analysis of the canid skeleton. *Proc. Natl. Acad. Sci.* **99:** 9930–9935.

de Beer G.R. 1985. *The development of the vertebrate skull.* University of Chicago Press, Illinois.

Emerson S.B. and Bramble D.M. 1993. Scaling, allometry, and skull design. In *The skull* (ed. J. Hanken and B.K. Hall), pp. 384–421. University of Chicago Press, Illinois.

Frank L.G., Glickman S.E., and Licht P. 1991. Fatal sibling aggression, precocial development, and androgens in neonatal spotted hyenas. *Science* **252:** 702–704.

Gould S.J. 1977. *Ontogeny and phylogeny*. Belknap Press of Harvard University Press, Cambridge, Massachusetts.

Jackson J.E. 1991. *A user's guide to principal components*, p. 10. Wiley, New York.

Nes N.N., Einarsson E.J., Lohi O., and Jorgensen G. 1988. *Beautiful fur animals—and their colour genetics*, p. 271. SCIENTIFUR, Tjele, Denmark.

Nowak R.M. and Paradiso J.L. 1991. *Walker's mammals of the world*. Johns Hopkins University Press, Baltimore, Maryland.

Pasi B.M. and Carrier D.R. 2003. Functional tradeoffs in the limb muscles of dogs selected for running versus fighting. *J. Evol. Biol.* **16:** 324–332.

Shea B.T. 1981. Relative growth of the limbs and trunk in the African apes. *Am. J. Phys. Anthro.* **56:** 179–201.

Sheldon J.W. 1992. *Wild dogs. The natural history of the nondomestic canidae*. Academic Press, San Diego, California.

Trut L.N. 1980. The genetics and phenogenetics of domestic behaviour. Problems in general genetics. In *Problems in general genetics* (ed. Y.P. Althukov), pp. 123–137. Proceedings of the XIV International Congress of Genetics, Mir, Moscow.

Trut L.N. 1999. Early canid domestication: The farm fox experiment. *Amer. Sci.* **87:** 160–169.

———. 2001. Experimental studies of early canid domestication. In *The genetics of the dog* (ed. A. Ruvinsky and J. Sampson), pp. 15–43. CABI, Wallingford, United Kingdom.

Wayne R.K. 1986. Developmental constraints on limb growth in domestic and some wild canids. *J. Zool.* **210:** 381–399.

———. 1993. Molecular evolution of the dog family. *Trends Genet.* **9:** 218–224.

———. 2001. Consequences of domestication: Morphological diversity of the dog. In *The genetics of the dog* (ed. A. Ruvinsky and J. Sampson), pp. 15–43. CABI, Wallingford, United Kingdom.

7

From Wild Wolf to Domestic Dog

Jennifer A. Leonard and Carles Vilà
Department of Evolutionary Biology, Uppsala University
SE-752 36 Uppsala, Sweden

Robert K. Wayne
Department of Ecology and Evolutionary Biology
University of California, Los Angeles, California 90095-1606

THE DOG (*CANIS FAMILIARIS*) WAS THE FIRST SPECIES to be domesticated. This event was a crucial step in the history of humankind and it occurred more than 15,000 years ago when humans were generally nomadic hunter–gatherers (Price and Gebauer 1995). Dogs were domesticated at least several thousand years before any other plant or animal species, and the few ancient remains found so far come from Europe, North America, and the Near East, suggesting they rapidly spread throughout the world after initial domestication events. As a result of the scarce and highly fragmented archaeological evidence, little is known about the specific location, conditions, or causes of domestication. Knowledge of the pattern and process of domestication is essential to understanding human civilization at the end of the Stone Age and the transition from hunter–gatherer to agrarian societies.

Domestic dogs are morphologically diverse, and differences in cranial and skeletal proportions among dog breeds exceed that among wild canids (Wayne 1986a,b). Domestic dogs are also behaviorally diverse and have behavioral patterns that are distinct from those of their wild ancestors (Coppinger and Coppinger 2001; Hare et al. 2002; Miklósi et al. 2003). Despite this dramatic diversity in phenotype, dogs have diverged very recently from their wild progenitor, the gray wolf (*Canis lupus*), and consequently, the two species have very similar genomes. Understanding the small subset of genes that have changed during domestication will provide insights into how rapid diversification occurs in domestic and wild species, as well as a more precise understanding of the process of domestication. Moreover, the wide variety of behavior, phenotype, and disease susceptibility that are partitioned in different breeds, in combination with a fully sequenced genome, make dogs a particularly important model organism for the study of traits affecting human health. More human disease syndromes have been described in dogs than any in other domestic mammal (Ostrander and Kruglyak 2000), and disease genes (see Jonasdottir et al. 2000) and genes influencing phenotypic traits (see Chase et al. 2002) are being characterized. However, a major limitation of these efforts is the lack of understanding of how breeds were formed and their evolutionary relationships (Sutter and Ostrander 2004).

Dogs were domesticated from wild gray wolves (Seal et al. 1970; Sarich 1977; Ferrell et al. 1978; Wayne and O'Brien 1987; Wayne et al. 1989; Vilà et al. 1997, 1999a). This is fortunate, because the gray wolf is one of the few wild progenitors of a domestic species that is still widespread and available for study, allowing the comparison of domestic and wild lineages. Wild gray wolves are distributed over much of Eurasia and North America. However, because wolves are highly mobile, and high rates of genetic exchange (gene flow) characterize wild populations, few genetic subdivisions have developed among them. Consequently, identifying the particular location of domestication events is very difficult.

New genetic tools are available to address questions about the origin, location, and timing of domestication events. Previous research has used mitochondrial DNA (mtDNA) sequences of modern dogs to reconstruct the history of domestication (Okumura et al. 1996; Tsuda et al. 1997; Vilà et al. 1997, 1999a; Savolainen et al. 2002), and recently, mtDNA sequences have been analyzed from archaeological remains, providing a direct historical record of sequence evolution (Leonard et al. 2002). mtDNA has proved to be a useful tool for the study of dog domestication, but limitations, such as maternal inheritance, can result in a biased view of its evolutionary history. However, these limitations can be surmounted through the study of paternally inherited Y chromosome markers (Olivier and Lust 1998; Olivier et al. 1999; Sundqvist et al. 2001) and rapidly evolving biparentally inherited markers, such as autosomal microsatellite loci (Parker et al. 2004) or X chromosome microsatellites (see Vilà et al. 2002). In general, most microsatellites are assumed to be selectively neutral, and hence their evolution reflects mutation and genetic drift. Additional insights about the process of domestication will come from the study of genes under selection, such as those from the major histocompatibility complex (MHC; Seddon and Ellegren 2002) or disease-related genes (Neff et al. 2004). In addition to the many DNA-based tools now available, other genetic tools such as RNA expression patterns measured by microarray or real-time PCR analyses have been used to study the origin and diversification of dogs (Saetre et al. 2004). Finally, phenotypic analyses such as those based on morphological and behavioral data also may lead to important insights when combined with molecular genetic approaches (see Chase et al. 2002).

LOCATION OF DOMESTICATION

mtDNA diversity in wild populations of gray wolves has been extensively surveyed (Vilà et al. 1999b; Flagstad et al. 2003; Sharma et al. 2004; Leonard et al. 2005). Although wolves have high levels of genetic diversity, it is not partitioned into distinct populations, and only the following population groupings have been identified: North America, the Indian subcontinent, the Himalayan region, and the rest of Eurasia. Wolves in the Indian subcontinent and Himalayan regions contain divergent mtDNA haplotypes not similar to any found in dogs (Sharma et al. 2004). Consequently, a comparison of the haplotypes found in modern domestic dogs with those in wild wolves can only narrow the location of domestication to a large geographic region such as Eurasia, exclusive of the Indian subcontinent and the Himalayas (Vilà et al. 1997; Sharma et al. 2004). Better geographic resolution may be possible through the future analysis of DNA from archaeological specimens and highly variable nuclear markers.

Many of the dogs that have been genetically surveyed were purebred dogs of European origin. This bias and the ancient archaeological record of dogs in North America suggest that dogs could conceivably have been domesticated separately in the New and Old Worlds. The earliest archaeological remains of dogs are about 14,000 years old and were found in the Old World, in the Near East and eastern Europe (Nobis 1979; Dayan 1994; Clutton-Brock 1995, 1999; Sablin and Khlopachev 2002), and coincide closely in time with the earliest archaeological evidence of humans in the New World (Jablonski 2002). The earliest archaeological dog remains in the New World came several thousand years later at about 9,000 years before present (ybp) (Jennings 1957; Olsen 1985). Gray wolves, the wild ancestors of domestic dogs, are also present in North America and thus were candidates for domestication. To test for a separate domestication in America, Vilà et al. (1999a) surveyed the genetic variation in the Mexican hairless dog. This breed has a long history in America. It is represented in ancient pottery more than 2000 years old and, due to a pathological reduction in the size and number of the premolars, has been identified in the archaeological record well before contact with Europeans. However, all of the mtDNA haplotypes identified in the modern Mexican hairless dogs were closely related or identical to haplotypes identified in Eurasian dogs that, in turn, were closely related to Eurasian wolves. This suggests that Old and New World dogs derived from the domestication of Old World wolves. Alternatively, extensive interbreeding between Mexican hairless dogs and dogs that arrived with Europeans might have caused the loss of unique American haplotypes.

To distinguish between these alternatives, we examined mtDNA sequences from New World archaeological remains of dogs (Leonard et al. 2002). Thirty-seven pre-Columbian remains from Mexico, Bolivia, and Peru were analyzed for variability in mtDNA control region sequences, as well as 11 samples from Alaska deposited before the first arrival of European explorers. Due to the degraded nature of ancient DNA, not all of the specimens that were analyzed yielded results. Twenty-four of the 48 specimens (13 from Latin America and 11 from Alaska) yielded DNA sequences that could be analyzed. All the ancient sequences were assigned to two clades that contained modern Eurasian dogs (Fig. 1), and all except one sequence were assigned to the clade containing the sequences of about 80% of all the modern dogs analyzed so far, most of them of European origin (clade I, Fig. 1). The results of this study strongly support the hypothesis that humans brought dogs domesticated in the Old World with them when they first colonized North America. No evidence was found to support a separate domestication event in America, or backcrossing between domestic dogs and American wolves (although some researchers suggest that it may have happened later; Koop et al. 2000).

The general location of domestication has been estimated through an extensive analysis of mtDNA sequence diversity of extant dogs and gray wolves worldwide (Savolainen et al. 2002). These authors found that East Asian dogs have considerably more diversity than dog populations elsewhere, suggesting an origin there. This analysis assumes that the current Asian dog population is a direct descendant of ancient Asian dogs. However, the genetic composition of dog populations may have fluctuated with societal changes and human demography over the millennia that have passed since dogs were first domesticated. For example, Leonard et al. (2002) found a sequence clade in pre-Columbian American dogs absent from modern dogs, suggesting that current levels of diversity may not be an accurate guide to the past. Furthermore, the East Asian sample used by Savolainen et al. contained a large proportion of mongrels, whereas dogs from the rest of the world were mostly represented by purebred dogs. Most breeds originate with only a few founders and are highly inbred. As a consequence, the East Asian sample may have higher

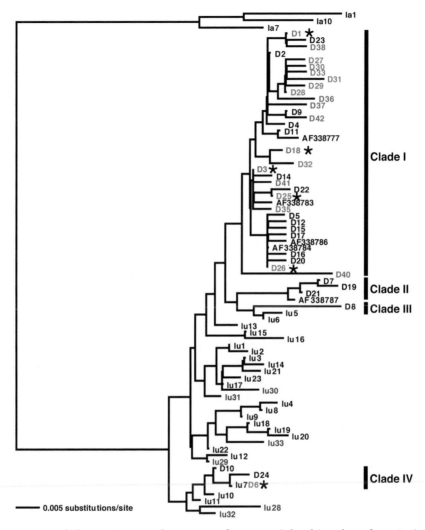

Figure 1. Phylogenetic tree of sequences from pre-Columbian dogs from Latin America (*red*), ancient Alaskan dogs (*green*), modern published dogs (*prefix D* for sequences from Vilà et al. 1997 and GenBank accession numbers), Eurasian wolves (*black lu*, from Vilà et al. 1999a), America wolves (*blue lu*, from Vilà et al. 1999a), and coyotes (*la*, from Vilà et al. 1999a) as an outgroup. Dog clades (I–IV) are indicated as in Vilà et al. (1997). Haplotypes shared between modern and ancient dogs are marked with an asterisk. (Reprinted from Leonard 2002.)

diversity. Nonetheless, the Savolainen et al. study provides the strongest evidence thus far for a specific geographic locale where dogs were first domesticated. Similar analyses have clearly defined an East Asian origin of the dingo about 5,000 years ago (Savolainen et al. 2004). Future research utilizing nuclear markers, ancient remains, and more extensive samples of purebred and aboriginal dogs may better establish the center(s) of dog domestication.

TIMING OF DOMESTICATION

The timing of dog domestication is particularly interesting because they are the first domesticate. It seems clear from archaeological remains that dogs were domesticated at least by 14,000–15,000 years ago (Nobis 1979; Dayan 1994; Clutton-Brock 1995; Sablin and Khlopachev 2002), when humans lived primarily as nomadic hunter–gatherers, without metal tools, in an Ice Age World that was cooler than our world (Dawson 1992). The timing of domestication is important to understanding the environmental and cultural conditions that promoted domestication.

Vilà et al. (1997) demonstrated that most modern dogs fall into four main mtDNA control region sequence clades within the diversity of gray wolves. Sequences from about 80% of the dogs analyzed, including some ancient breeds such as the Australian dingo and the New Guinea singing dog, formed the largest and most diverse of the dog clades (clade I, Fig. 1). To estimate the time of domestication independent of the archaeological record, a molecular clock was assumed to calculate the age of the common ancestor of clade I. The molecular clock was calibrated considering that wolf and coyote (*Canis latrans*) lineages diverged at least one million years ago (by that time, both species already existed in the fossil record). Because these two species have 6.5% sequence divergence in the control region and clade I dogs differ by a maximum of about 1%, their inferred divergence time was about 135,000 ybp. Although the confidence interval around this date is uncertain, these data suggest that dogs were domesticated more than 15,000 years ago.

More recently, Savolainen et al. (2002) attempted to date the time of domestication using a more extensive sample of mitochondrial control region sequences and a revised molecular clock. Instead of assuming that clade I had an origin from a single matriline, they suggested that several subclades can be defined within clade I. If each subclade represented a different founding matriline, then the time to the common ancestor of each subclade provides a range of possible origin dates. This process yielded dates about 15,000 years ago, a result which is more consistent with the fossil record. However, Savolainen et al. did not state the criteria used to define or choose specific subclades for dating, and consequently, there is uncertainty in these dates as well. Depending on how the subclades are defined, a variety of dates could be inferred. Furthermore, an East Asian origin and divergence date close to 15,000 years ago fail to explain the existence of morphologically differentiated dogs in

Central/Eastern Europe and West Asia by that time (Nobis 1979; Dayan 1994; Clutton-Brock 1995, 1999; Sablin and Khlopachev 2002), and the presumed spread of dogs to the New World from a source in Beringia by about 12,000 years ago (Leonard et al. 2002). Alternatively, Savolainen et al. assumed a single origin of clade I and deduced an origin of about 40,000 ybp. Clearly, further DNA analyses of modern and ancient Eurasian dogs and wolves using multiple independent markers are needed to provide independent tests of the ancient or more recent domestication hypotheses. Additionally, parallel archaeological research on remains of wolf- and dog-like canids from the late Pleistocene could provide critical support for these theories of domestication.

NUMBER OF FOUNDERS

The first molecular analysis of the origin of dogs used mitochondrial control region sequences to construct a phylogeny of domestic dog and gray wolf haplotypes (Vilà et al. 1997). Four clades of dog sequences were evident in this phylogeny, suggesting that at least four female wolf lineages founded or were subsequently introduced into domestic dogs (Fig. 1). A more extensive analysis by Savolainen et al. (2002) suggested six or more lineages. However, the number of actual founding lineages is uncertain because many of the initial matrilines may have been lost through stochastic processes (Avise 2000), and others may have been introduced through backcrossing between dogs and female gray wolves. The primary conclusion to be derived from these mtDNA data is that a limited number of founding events explain the genetic diversity of dogs. This result is consistent with genetic studies of other domesticated animals (cattle, sheep, goats, pigs; Bruford et al. 2003). In contrast, horses have incorporated numerous matrilines from their wild progenitors, suggesting repeated domestication (or backcrossing) events over a large geographic area (Vilà et al. 2001). However, the limited number of founding events for dogs does not necessarily imply that the number of founders was small.

The major histocompatibility complex (MHC) encodes proteins that are essential to the normal functioning of the immune system and typically show a high level of genetic diversity that is maintained by balancing selection (Hughes and Yeager 1998). One consequence of this form of selection is that MHC alleles are maintained for longer periods of time than expected under neutrality. Recently, a large number of alleles have been described for the MHC in dogs, wolves, and coyotes (Kennedy et al. 1999, 2000, 2001, 2002; Seddon and Ellegren 2002, 2004). Seddon and Ellegren (2002) observed that a large number of alleles are shared

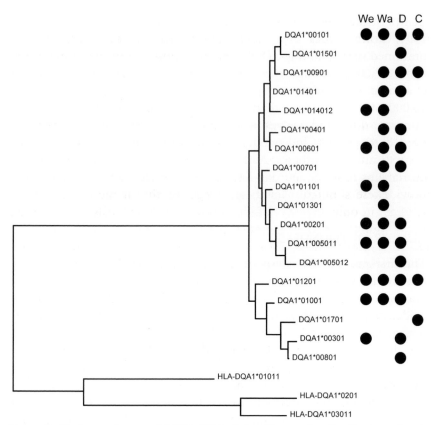

Figure 2. Phylogenetic tree of MHC *DQA* alleles. The species distribution of each allele is indicated by filled circles. (We) European gray wolf, (Wa) North American gray wolf, (D) dog, (C) coyote. (Based on Seddon and Ellegren 2002.)

between the different species (Fig. 2), which suggests that they may have existed before the species divergence (over one million years ago for gray wolves and coyotes). Hence, these alleles can provide evolutionary information over a longer time frame than mtDNA and have been used to provide estimates of effective population sizes in humans (Ayala et al. 1994) and to calculate founding population sizes for other species (Vincek et al. 1997). Predictive models based on selection and drift can be used to estimate the minimum numbers of wolf founders under a variety of demographic conditions (Vilà et al. 2005).

Vilà et al.(2005) collated information from 42 MHC *DRB* alleles for dogs, almost all of them differing by at least one nonsynonymous change. Considering the rate of nonsynonymous substitution estimated for MHC class II loci (Klein et al. 1993), the authors concluded that almost the totality of the alleles likely originated before the time of domestication. Consequently, at least 21 founders are required to explain the large MHC diversity observed in dogs. However, this number represents a minimum estimate, since it assumes that all founders are heterozygous for different alleles and are equally successful producing offspring and that no alleles are removed from the population by drift. To obtain more realistic estimates of the number of founders, Vilà et al. (2005) used models that varied the number of founders, the number of founding populations, the rate of population growth after domestication, the diversity in the ancestral population, and the strength of selection acting on the MHC locus. These simulations strongly suggested that if the domestication occurred in only one population, hundreds of individuals were involved in the founding of the domestic dog (Fig. 3). Obviously, this is not a realistic possibility at the time of the domestication of the dog, because humans were still hunter–gatherers, living in small groups, and they lacked the resources to support a large population of domestic animals. Since several lineages of mtDNA had been identified in dogs (Vilà et al. 1997, 1999; Savolainen et al. 2002), one possibility was that the domestication would have taken place in different populations. If this was the case, drift would lead to the fixation of different alleles in each population, and a lower number of founders would be enough to explain the level of diversity found today. Indeed, the simulations show that the estimated number of founders decreased as the number of populations involved increased (Fig. 3). If six populations were involved (this number is more

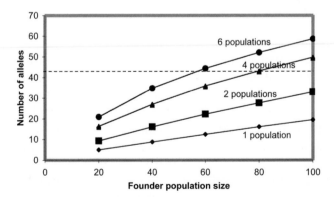

Figure 3. Number of MHC alleles that would be expected in modern dogs depending on the number of wolves involved in the founding event (divided into 1–6 populations of equal size). Each dot represents the average of 1000 simulations. The dashed line indicates the number of alleles described so far in modern dogs. (Based on Vilà et al. 2005.)

than suggested by mtDNA studies [Vilà et al. 1997, 1999; Savolainen et al. 2002] and more than suggested for most livestock species [Bruford et al. 2003]), about 60 founders would be required, which is an underestimate, considering that the model assumes that all individuals are equally likely to reproduce, and consequently, the real number of founders would likely be several times larger (Frankham 1995).

In conclusion, the simulations showed that the number of wolves involved in the domestication process was larger than deduced from mtDNA alone. Second, since it seems improbable that such a large founding population could have been kept separated from wild wolves, hybridization between wolves and dogs was likely an important influence on the diversity of the dog MHC. Presumably, extensive backcrossing with male wolves early in the history of dog domestication may have enriched nuclear diversity to a greater extent than maternally inherited mtDNA. Future research utilizing male-specific markers on the Y chromosome may help clarify the situation.

SELECTION ON BEHAVIOR

Initially, docility was likely the object of intense artificial selection in dogs. However, because behavioral changes leave no archaeological record, only inferences from experiments and observation with modern dogs provide clues to the behavioral changes that occurred in the first domestic dogs. The most extensive experiment in domestication involved selection on silver foxes (*Vulpes vulpes*) in Russian fur farms (Trut 2001). A part of the population was selected for tameness and continued selective breeding. Since the initiation of the experiment in the early 1940s, remarkably tame foxes that can be handled as pets and desire human attention have been produced. In addition, traits common in domestic dogs have appeared in these foxes as a consequence of selection on linked traits or through drift. These traits include floppy ears, spotted coats, and a white spot on the tail (see Chapter 26). Thus, morphological traits typical of dogs may have arisen as an accidental by-product of selection for docile behavior, which may have intensified as humans and dogs started living in close proximity.

On an evolutionary scale, the divergence of dogs from gray wolves, ranging from 15,000 to over 100,000 ybp, is relatively recent. The mutation rate of most functional genes is on the order of 10^{-5} mutations per gamete per generation (Hartl and Clark 1997), and consequently, few new mutations in functional genes have likely occurred. Despite this recent history, dogs are clearly behaviorally and morphologically divergent from gray wolves. Changes in gene regulation rather than mutation in struc-

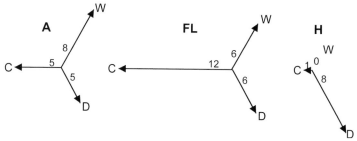

Figure 4. Representation of relative expression differences in dogs (D), wolves (W), and coyotes (C) in three regions of the brain: (A) amygadala, (FL) frontal cortex, (H) hypothalamus. The branch lengths correspond to the relative expression differences, and the number of species-specific genes is indicated next to each branch. (Based on Saetre et al. 2004.)

tural genes may account for some of theses behavioral differences. Saetre et al. (2004) surveyed the expression pattern of 7762 genes in three different regions in the brains of domestic dogs and two of their wild relatives, gray wolves and coyotes. They found that the pattern of gene expression in the hypothalamus of domestic dogs was very different from that found in both gray wolves and coyotes (Fig. 4). In contrast, patterns of gene expression in other regions of the dog brain (amygdala and frontal cortex) were less differentiated from those in their wild relatives. This is particularly interesting because the hypothalamus controls many of the behavioral responses essential to survival and forms a crucial node to link exploratory behavior and specific emotional, endocrinological, and autonomic responses of dogs. The hypothalamus is highly conserved throughout mammals, and small injuries to it can be fatal. The results of Saetre et al. (2004) suggest that behavioral selection on dogs may have affected this central part of the brain, having a cascade of effects on the behavior of the dogs compared to wolves and coyotes.

In addition to genetic studies, behavioral studies have also attempted to estimate the magnitude of behavioral selection that has been applied to domestic dogs. Hare et al. (2002) assessed comprehension of cues in untrained dogs, captive gray wolves, and chimpanzees. The tests revealed that dogs were able to understand cues given by the experimenter more easily than were wolves and chimpanzees. These results support early and strong selection on dogs to understand human cues, although they could also represent a by-product of selection for tameness (Hare et al. 2005). Miklósi et al. (2003) suggested that a key difference between dog and wolf behavior "is the dog's ability to look at the human's face" and this may have allowed the development of complex forms of communication between humans and dogs.

SELECTION ON MORPHOLOGY

The massive amount of morphological variation among domestic dog breeds has aroused interest in its genetic basis. Dogs may represent the ideal model organism to study the association between phenotype and genotype. The recent origin of domestic dogs from gray wolves suggests that morphological differences do not generally reflect a large number of nonsynonymous mutations in genes of large phenotypic effect (see Ostrander and Kruglyak 2000). The process by which large amounts of morphological variation can be rapidly generated is of interest to understanding the processes leading to rapid divergence and speciation (Wayne 1986a,b). Patterns of inheritance indicate that single gene mutations may be responsible for some specific morphological traits in dogs, such as achondroplasia or brachycephaly (Stockard 1941; Hutt 1979) and hairlessness (Cordy-Collins 1994). However, in order for mutations in structural genes to account for a substantial proportion of the phenotypic variation among domestic dogs, a higher mutation rate than commonly assumed for these genes seems necessary. Dogs have a shorter generation time and high reproductive rate relative to wolves (two estrus periods per year vs. a single heat in wolves; Mech and Seal 1987), and effective population sizes are larger. About 200,000 wolves live today worldwide (Boitani 2003), whereas the number of dogs may be over 400 million (Coppinger and Coppinger 2001). However, given mutation rates of 10^{-5} per gamete per generation for many structural genes (Hartl and Clark 1997), it seems dubious that structural gene mutation alone can explain the unique genetic diversity of dogs.

Candidate gene surveys may allow the identification of genes with significant effects on dog morphology. For example, the homeobox gene *MSX2* is expressed during human craniofacial development, and mutations in this gene cause abnormalities (Haworth et al. 2001a). Haworth et al. (2001a) sequenced *MSX2* in dogs from 10 different breeds with very different skull and face shape and found that it was highly conserved across dog breeds. In this case, the candidate gene approach failed to identify a gene with selectively important variation. Consequently, Haworth et al. (2001b) studied another candidate gene, *TCOF1*, implicated in human craniofacial development. After constructing and screening a cDNA library, cloning the gene and then sequencing it in a panel of dog DNA, variation was identified at four nucleotide positions in introns and five in exons. On the basis of the location of the intron variable sites, Haworth et al. determined that they were unlikely to have any impact on mRNA splicing and, thus, the structure of the expressed product. Of the five variable sites located in exons, two were silent and three resulted in amino acid substitutions. One of these amino acid substitutions was significantly associated with short, broad head shapes in dogs, suggesting that candidate gene approaches can be used to find genes that influence phenotypic variation in dogs.

Much of the morphological variation in domestic dogs, as well as other species, is not in discrete characters (such as haired vs. hairless); rather, it is manifest as continuously varying quantitative traits (such as size). Chase et al. (2002) examined the genetics of quantitative trait loci (QTLs) influencing skeletal dimensions in Portuguese water dogs. A principal component analysis revealed four main axes that accounted for 61% of the morphological variation. These components were correlated to more than 500 genetic markers, and 9 were found to be significantly associated with one of the four principal components. Some of these have been mapped and were linked to genes that affect skeletal morphology in other species. The combination of measurements represented by each principal component changed in a correlated fashion during development, suggesting that they may be responding to a common set of regulatory genes. This idea fits well with earlier morphological studies showing a strong ontogenetic association between skeletal measurements of size and shape (Wayne 1986a,b).

Nonclassical genetic variation may also be important in dogs. Given their high mutation rates, simple tandem repeats embedded in genes may be the source of new variation in recently developed lines (King 1994; Kashi et al. 1997). To test this hypothesis, Fondon and Garner (2004) made three-dimensional high-resolution models of dog skulls from each of 20 breeds, and seven mongrels. In representatives of 92 different breeds, they also sequenced 37 repeat-containing regions from 17 genes known or thought to be involved in craniofacial development. They examined the number and structure of repeats (perfect vs. imperfect) and found that dogs in general had more perfect repeats than humans. This result suggests that tandem repeat loci are changing in length faster in dogs than in humans. Furthermore, Fondon and Garner found that the size and the ratio of lengths of two tandem repeats in the *Runx-2* gene correlated with the degree of dorsoventral nose bend (clinorhynchy) and midface length across a variety of breeds. Tandem repeat length mutations could potentially provide a mechanism for rapid phenotypic change in dogs, but more detailed studies associating repeat change with specific phenotypic traits are needed (Pennisi 2004).

ORIGIN OF MODERN BREEDS

An understanding of the origin and evolution of modern breeds is important to models of genetic disease and theories about the genetic basis of morphological and behavioral variation (Wayne and Ostrander 1999). However, reconstructing breed relationships has proved to be difficult. Studies based on mtDNA found no correlation between mtDNA haplotype and breed (Vilà et al. 1997, 1999a; Savolainen et al. 2002). A single breed often had many different mtDNA haplotypes derived from different sequence clades. For example, Vilà et al. (1999a) screened 19 Mexican hairless dogs and found seven different mtDNA haplotypes that fall into three of the four main clades of dogs (Fig. 5). In addition, a single mtDNA haplotype could be found across a wide variety of breeds (Table 1). For example, haplotype D3 from Vilà et al. (1997) was identified in chow chow, Norwegian elkhound, Mexican hairless, Siberian husky, papillon, poodle, rottweiler, English setter, Icelandic sheepdog, springer spaniel, Japanese spitz, Border terrier, fox terrier, and whippet. This lack of differentiation between breeds for mtDNA markers is surprising, given the morphological uniformity within breed compared to the large differences between breeds, and suggests that modern breeds have a recent origin from a well mixed and genetically diverse dog population. However, this conclusion conflicts with historical evidence suggesting an ancient and restricted geographic origin of some dog breeds. For example, archaeological evidence from ancient Egypt suggests that several types of morphologically differentiated dogs similar to mastiffs and greyhounds existed there 4000 years ago, and Romans may have been the first people to develop dog breeds in Europe (Clutton-Brock 1999). Additionally, pottery from the Colima culture (250 B.C.–450 A.D., western Mexico) clearly represents Mexican hairless dogs (Cordy-Collins 1994). This breed also was found by the Spanish conquistadors on their arrival in Mexico in the 16th century (Valadez 1995). Furthermore, dogs depicted in European paintings during that time can be easily recognized today as spaniels, mastiffs, hounds, pointers, etc.

Analysis of microsatellite loci proved to be more informative about the genetic structure of breeds and correctly assigned dogs to breed of origin, suggesting that breeds are clearly differentiated (Koskinen 2003; Parker et al. 2004). Parker et al. (2004) typed 96 microsatellite loci in 85 breeds, each represented by 5 dogs. An analysis of molecular variance showed that variation between breeds accounted for about 27% of the genetic diversity observed in all dogs. Similarly, Parker et al. also assessed genetic diversity within and between breeds with nuclear single nucleotide polymorphisms (SNPs). In a survey of 19,867 bp of sequence

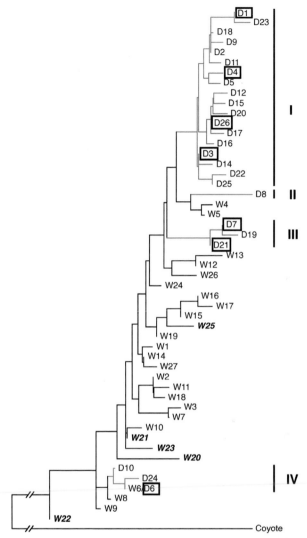

Figure 5. Neighbor-joining relationship tree of wolf (W) and dog (D) mtDNA control region sequences (261 bp in length; Vilà et al. 1997). Dog haplotypes are grouped in four clades, I–IV. Boxes indicate haplotypes found in the 19 Mexican hairless dogs (Vilà et al. 1999a). Bold characters indicate haplotypes found in New World wolves (W20–W25). (Based on Vilà et al. 1999b.)

from 120 dogs, the authors identified 75 SNPs, and 14 of these were breed-specific in the small number of individuals screened per breed. These results suggest that many breeds have been isolated since their origination, in stark contrast to the maternally inherited mtDNA results. This

Table 1. Mitochondrial DNA haplotypes found in dog breeds and their distribution in the four groups of dog haplotypes of Figure 5

	I	II	III	IV
Chow chow (n = 3)	D1, D2, D3			
Border collie (n = 3)	D1, D5			
Wirehaired dachshund (n = 3)	D5			D10
Australian dingo (n = 4)	D18			
Norwegian elkhound (n = 9)	D3	D8		
German shepherd (n = 8)	D4, D5		D7, D19	D6
Afghan hound (n = 3)				D6
Siberian husky (n = 3)	D3, D18		D7	
Jämthund (n = 3)		D8	D7	
Flat-coated retriever (n = 3)	D4			D10
Golden retriever (n = 6)	D4, D15			D6, D24
Labrador retriever (n = 6)	D4, D12			
Samoyed (n = 3)	D1, D4, D5			
Giant schnauzer (n =3)	D4		D7	
English setter (n = 4)	D3, D5			
Irish setter (n = 3)	D1, D9			
Mexican hairless (n = 19)	D1, D3, D4, D26		D7, D21	D6

n = sample size. Based on Vilà et al. (1999a).

disparity implies that breeds were founded by a larger diversity of females than males. Alternatively, it is also possible that the history of modern breeds has been marked by higher rates of genetic exchange of females than males.

Nuclear microsatellite variation within breeds has also been used to reconstruct their phylogenetic relationships. Parker et al. (2004) found that many of the most recognizable and common breeds of European origin are so closely related that branching patterns were difficult or impossible to elucidate. This suggests a recent origin for all of these breeds, and that they share a common recent ancestry from a diverse population of dogs. This result is consistent with historical evidence which suggests that most European breeds originated in the past few hundred years (Ash 1927; American Kennel Club 1997).

The Fédération Cynologique Internationale (FCI) has organized breeds into ten different groups based on function and, to some extent, supposed origin. However, the genetic analysis of Parker et al. (2004) defined only four fundamental genetic subdivisions in dogs (see Chapter 9). The most primitive and divergent groupings included a wide diversity of breeds ranging in geographic origin from Africa (basenji) to Asia (chow chow and Akita) to the Arctic (Siberian husky and Alaskan malamute). The wide geographic area represented by these primitive breeds suggests that dogs spread over a large geographic area soon after domestication. Three other breed clusters can be defined by the genetic analysis: (1) European breeds that diverged in the 1800s (see above); (2) herding dogs, including the collie and sheepdogs and similar to FCI group 1 (sheepdogs and cattle dogs); and (3) mastiff-like dogs, including the bulldog and boxer. With the exception of the second group, none of these four clusters has been previously recognized, and this suggests that there have been four primary radiations in the history of dog domestication. The earliest radiation involved the formation of breeds over a wide geographic area that remains highly divergent and genetically distinct today. Subsequent radiations occurred recently in Europe and include dogs representing a wide diversity in form and function. In contrast, mastiff-like dogs and sheepdogs form separate and more phenotypically limited groups. Such genealogical relationships need to be considered in studies attempting to explain patterns of genetic disease or the origin of breed-defining traits.

FUTURE RESEARCH

Although much has been learned about dog domestication from mtDNA analysis, it has been of limited use in understanding the origin, timing, and genealogical relationship of dog breeds. The current evidence suggests an origin in East Asia 15,000–40,000 years ago (Savolainen et al. 2002). However, the presence of differentiated dogs in western Russia 15,000 years ago (Sablin and Khlopachev 2002) casts some doubt on the more recent age or suggests a separate origination. Similarly, the mtDNA evidence does not indicate well the size or number of original founding populations. Additional research is needed on highly polymorphic nuclear genes such as those of the MHC complex (Vilà et al. 2005) and Y chromosome sequences. Moreover, genetic typing needs to include a wide variety of populations, such as the native dogs of Asia, Africa, and America, rather than primarily purebred dogs. The genetic diversity present in native and feral dogs has clearly been shown by the study of street dogs in a region of Bali, where the diversity observed surpassed that in American Kennel Club breeds (Irion et al. 2005). Such native populations may contain unique diversity that provides insights into the pattern and process of domestication and will document the vast reservoir of genetic diversity that could be tapped to rescue breeds with disease problems. Finally, characterization of DNA from ancient dog remains from throughout the past distribution of dogs needs to be undertaken. Such remains provide the only direct evidence of the genetic prehistory of dogs and the founding variation preserved in modern breeds (Leonard et al. 2002).

Some progress has been made in characterizing genes and the genetic changes that influence morphological and behavioral diversity in dogs (Chase et al. 2002). However, specific genes or regulatory elements accounting for the substantial variation in size and shape of dog breeds have not yet been identified. "Selective sweep" mapping (Kohn et al. 2000, 2003; Luikart et al. 2003), as well as more traditional association and candidate gene studies, may allow the identification of genes critical to breed differences. Furthermore, surveys of gene expression in the brain and other organs may narrow the search for functional genes to a more limited subset (Saetre et al. 2004). Similarly, comparative studies of differences in expression patterns in tame and wild foxes may help identify key changes associated with the evolution of docility.

With the publication of the 1.5x genome sequence of a poodle (Kirkness et al. 2003) and the recent completion of the 7.8x genome sequence of a boxer (http://www.ncbi.nih.gov/Genbank) (Chapter 11), a critical resource is now available for studies of dog domestication. Access to the complete dog genome will facilitate development of microsatellite and SNP markers, and potentially new sources of phylogenetic information such as bimorphic short interspersed elements (SINEs) (Kirkness et al. 2003) and functional genes and regulatory elements. The latter two gene classes are exposed to selection, in contrast to phenotypically neutral changes in microsatellite size or mtDNA control region sequences. Hence, analysis of functional genes and regulatory elements may provide a specific record of the events that have transformed the phenotype of wolves into the diverse assemblage of breeds existing today. Indeed, Darwin's extensive use of the domestic dog as a metaphor for the process of evolution by natural selection will only be complete if the underlying genetic changes that transformed wolf into dog are identified and analyzed in an evolutionary framework.

ACKNOWLEDGMENTS

J.A.L. is supported by a grant from The Swedish Research Council. Financial support to C.V. was obtained from the Swedish Research Council for Environment, Agricultural Sciences and Spatial Planning. Ancient DNA research was supported by a National Science Foundation grant to R.K.W. and Blaire Van Valkenburgh (OPP-9617068).

REFERENCES

American Kennel Club. 1997. *The complete dog book; the histories and standards of breeds admitted to AKC registration, and the feeding, training, care, breeding, and health of pure-bred dogs, 19th. edition.* Doubleday, New York.

Ash E.C. 1927. *Dogs: Their history and development.* E. Benn Limited, London.

Avise J.C. 2000. *Phylogeography: The history and formation of species.* Harvard University Press, Cambridge, Massachusetts.

Ayala F.J., Escalante A., O'Huigin C., and Klein J. 1994. Molecular genetics of speciation and human origins. *Proc. Natl. Acad. Sci.* **91:** 6787–6794.

Boitani L. 2003. Wolf conservation and recovery. In *Wolves. Behavior, ecology, and conservation* (ed. L.D. Mech and L. Boitani), pp. 317–344. University of Chicago Press, Chicago, Illinois.

Bruford M.W., Bradley D.G., and Luikart G. 2003. DNA markers reveal the complexity of livestock domestication. *Nat. Rev. Genet.* **4:** 900–910.

Chase K., Carrier D.R., Adler F.R., Jarvik T., Ostrander E.A., Lorentzen T.D., and Lark K.G. 2002. Genetic basis for systems of skeletal quantitative traits: Principal component analysis of the canid skeleton. *Proc. Natl. Acad. Sci.* **99:** 9930–9935.

Clutton-Brock J. 1995. Origins of the dog: Domestication and early history. In *The domestic dog, its evolution, behaviour and interactions with people* (ed. J. Serpell), pp. 7–20. Cambridge University Press, United Kingdom.

———. 1999. *A natural history of domesticated mammals, 2nd edition.* Cambridge University Press, United Kingdom.

Coppinger R. and Coppinger L. 2001 *Dogs.* The University of Chicago Press, Illinois.

Cordy-Collins A. 1994. An unshaggy dog history. *Nat. Hist.* **2:** 34–40.

Dawson A.G. 1992. *Ice age earth, late Quaternary geology and climate.* Routledge, New York.

Dayan T. 1994. Early domesticated dogs of the Near East. *J. Archaeol. Sci.* **21:** 663–640.

Ferrell R.E., Morizot D.C., Horn J., and Carley C.J. 1978. Biochemical markers in species endangered by introgression: The red wolf. *Biochem. Genet.* **18:** 39–49.

Flagstad Ø., Walker C.W., Vilà C., Sundqvist A.-K., Fernholm B., Hufthammer A.K., Wiig Ø., Koyola I., and Ellegren H. 2003. Two centuries of the Scandinavian wolf population: Patterns of genetic variability and migration during an era of dramatic decline. *Mol. Ecol.* **12:** 869–880.

Fondon J.W., III, and Garner H.R. 2004. Molecular origins of rapid and continuous morphological evolution. *Proc. Natl. Acad. Sci.* **101:** 18058–18063.

Frankham R. 1995. Effective population size/adult population size ratios in wildlife: A review. *Genet. Res.* **66:** 95–107.

Hare B., Brown M., Williamson C., and Tomasello M. 2002. The domestication of social cognition in dogs. *Science* **298:** 1634–1636.

Hare B., Plyusnina I., Ignacio N., Schepina O., Stepika A., Wrangham R., and Trut L. 2005. Social cognitive evolution in captive foxes is a correlated by-product of experimental domestication. *Curr. Biol.* **15:** 226–230.

Hartl D.L. and Clark A.G. 1997. *Principles of population genetics, 3rd edition.* Sinauer, Sunderland, Massachusetts.

Haworth K., Breen M., Binns M., Hopkinson D.A., and Edwards Y.H. 2001a. The canine homeobox gene *MSX2*: Sequence, chromosome assignment and genetic analysis in dogs of different breeds. *Anim. Genet.* **32:** 32–36.

Haworth K.E., Islam I., Breen M., Putt W., Makrinou E., Binns M., Hopkinson D., and Edwards Y. 2001b. Canine *TCOF1*; cloning, chromosome assignment and genetic analysis in dogs with different head types. *Mamm. Genome* **12:** 622–629.

Hughes A.L. and Yeager M. 1998. Natural selection at major histocompatibility complex loci of vertebrates. *Annu. Rev. Genet.* **32:** 415–435.

Hutt F.B. 1979. *Genetics for dog breeders*. W.H. Freeman, San Francisco.

Irion D.N., Schaffer A.L., Grant S., Wilton A.N., and Pedersen N.C. 2005. Genetic variation analysis of the Bali street dog using microsatellites. *BMC Genet.* **6:** 6.

Jablonski N.G., ed. 2002. *The first Americans: The Pleistocene colonization of the New World. Memoirs of the California Academy of Sciences.* In *The 4th Wattis Foundation Endowment Symposium held October 29, 1999, no. 27* University of California Press, Berkeley.

Jennings J.D. 1957. *Danger cave*. University of Utah Press, Salt Lake City.

Jonasdottir T.J., Mellersh C.S., Moe L., Heggebo R., Gamlem H., Ostrander E.A., and Lingaas F. 2000. Genetic mapping of a naturally occuring heredity renal cancer syndrome in dogs. *Proc. Natl. Acad. Sci.* **97:** 4132–4137.

Kashi Y., King D.G., and Soller M 1997. Simple sequence repeats as a source of quantitative genetic variation.*Trends Genet.* **13:** 74–78.

Kennedy L.J., Carter S.D., Barnes A., Bell S., Bennett D., Ollier B., and Thomson W. 1999. DLA-DRB1 polymorphisms in dogs defined by sequence-specific oligonucleotide probes (SSOP). *Tissue Antigens* **53:** 184–189.

Kennedy L.J., Angles J.M., Barnes A., Carter S.D., Francino O., Gerlach J.A., Happ G.M., Ollier W.E., Thomson W., and Wagner J.L. 2001. Nomenclature for factors of the dog major histocompatibility system (DLA), 2000: Second report of the ISAG DLA Nomenclature Committee. *Anim. Genet.* **32:** 193–199.

Kennedy L.J., Altet L., Angles J.M., Barnes A., Carter S.D., Francino O., Gerlach J.A., Happ G.M., Ollier W.E., Polvi A., Thomson W., and Wagner J.L. 2000. Nomenclature for factors of the dog major histocompatibility system (DLA), 1998: First report of the ISAG DLA Nomenclature Committee. *Anim. Genet.* **31:** 52–61.

Kennedy L.J., Barnes A., Happ G.M., Quinnell R.J., Bennett D., Angles J.M., Day M.J., Carmichael N., Innes J.F., Isherwood D., Carter S.D., Thomson W., and Ollier W.E. 2002. Extensive interbreed, but minimal intrabreed, variation of DLA class II alleles and haplotypes in dogs. *Tissue Antigens* **59:** 194–204.

King D.G. 1994. Triplet repeat DNA as a highly mutable regulatory mechanism. *Science* **263:** 595–596.

Kirkness E.F., Bafna V., Halpern A.L., Levy S., Remington K., Rudch D.B., Delcher A.L., Pop M., Wang W., Fraser C.M., and Venter J.C. 2003. The dog genome: Survey sequencing and comparative analysis. *Science* **301:** 1898–1903.

Klein J., Satta Y., O'Huigin C., and Takahata N. 1993. The molecular descent of the major histocompatibility complex. *Annu. Rev. Immunol.* **11:** 269–295.

Kohn M.H., Pelz H.J., and Wayne R.K. 2000. Natural selection mapping of the warfarin-resistance gene. *Proc. Natl. Acad. Sci.* **97:** 7911–7915.

———. 2003. Locus-specific genetic differentiation at Rw among warfarin-resistant rat (*Rattus norvegicus*) populations. *Genetics* **164:** 1055–1070.

Koop B.F., Burbidge M., Byun A., Rink U., and Crockford S.J. 2000. Ancient DNA evidence of a separate origin for North American indigenous dogs. In *Dogs through time: An archaeological perspective* (ed. S.J. Crockford), pp. 271–285. British Archaeological Reports, Oxford, United Kingdom.

Koskinen M.T. 2003. Individual assignment using microsatellite DNA reveals unambiguous breed identification in the domestic dog. *Anim. Genet.* **34:** 297–301.

Leonard J.A. 2002. *"The use of ancient DNA to directly monitor genetic changes in vertebrate populations over millennium time scales."* Ph.D. thesis. University of California, Los Angeles.

Leonard J.A., Vilà C., and Wayne R.K. 2005. Legacy lost: Genetic variability and population size of extirpated US grey wolves (*Canis lupus*). *Mol. Ecol.* **14:** 9–17.

Leonard J.A., Wayne R.K., Wheeler J., Valadez R, Guillen S., and Vilà C. 2002. Ancient DNA evidence for Old World origin of New World dogs. *Science* **298:** 1613–1616.

Luikart G., England P.R., Tallmon D., Jordan S., and Taberlet P. 2003. The power and promise of population genomics: From genotyping to genome typing. *Nat. Rev. Genet.* **4:** 981–994.

Mech L.D. and Seal U.S. 1987. Premature reproductive activity in wild wolves. *J. Mammal.* **68:** 871–873.

Miklósi Á., Kubinyi E., Topál J., Gácsi M., Virányi Z., and Csányi V. 2003. A simple reason for a big difference: Wolves do not look back at humans, but dogs do. *Curr. Biol.* **13:** 763–766.

Neff M.W., Robertson K.R., Wong A.K., Safra N., Broman K.W., Slatkin M., Mealey K.L., and Pedersen N.C. 2004. Breed distribution and history of canine *mdr1-1Δ*, a pharmacogenetic mutation that marks the emergence of breeds from the collie lineage. *Proc. Natl. Acad. Sci.* **101:** 11725–11730.

Nobis G. 1979. Der älteste Haushund lebte vor 14,000 Jahren. *Umschau* **19:** 610.

Okumura N., Ishiguro N., Nakano M., Matsui A., and Sahara M. 1996. Intra- and interbreed genetic variations of mitochondrial DNA major non-coding regions in Japanese native dog breeds (*Canis familiaris*). *Anim. Genet.* **27:** 397–405.

Olivier M. and Lust G. 1998. Two DNA sequences specific for the canine Y chromosome. *Anim. Genet.* **29:** 146–149.

Olivier M., Breen M., Binns M.M., and Lust G. 1999. Localization and characterization of nucleotide sequences from the canine Y chromosome. *Chromosome Res.* **7:** 223–233.

Olsen S.J. 1985. *Origins of the domestic dog: The fossil record.* University of Arizona Press, Tucson.

Ostrander E.A. and Kruglyak L. 2000. Unleashing the canine genome. *Genome Res.* **10:** 1271–1274.

Parker H.G., Kim L.V., Sutter N.B., Carlson S., Lorentzen T.D., Malek T.B., Johnson G.S., DeFrance H.B., Ostrander E.A., and Kruglyak L. 2004. Genetic structure of the purebred domestic dog. *Science* **304:** 1160–1166.

Pennisi E. 2004. A ruff theory of evolution: Gene stutters drive dog shape. *Science* **306:** 2172.

Price T.D. and Gebauer A.B., eds. 1995. *Last hunters-first farmers. New perspectives on the prehistoric transition to agriculture.* School of American Research Press, Santa Fe, New Mexico.

Sablin M.V. and Khlopachev G.A. 2002. The earliest Ice Age dogs: Evidence from Eliseevivhi I. *Curr. Anthropol.* **43:** 795–799.

Saetre P., Lindberg J., Leonard J.A., Olsson K., Pettersson U., Ellegren H., Bergström T.F., Vilà C., and Jazin E. 2004. From wild wolf to domestic dog: Gene expression changes in the brain. *Mol. Brain Res.* **126:** 198–206.

Sarich V.M. 1977. Albumin phylogenetics. In *Albumin structure, function and uses* (ed. V.M. Rosenoer et al.), pp. 85–111. Pergamon Press, New York.

Savolainen P., Leitner T., Wilton A.N., Matisoo-Smith E., and Lundeberg J. 2004. A detailed picture of the origin of the Australian dingo, obtained from the study of mitochondrial DNA. *Proc. Natl. Acad. Sci.* **101:** 12387–12390.

Savolainen P., Zhang Y.P., Luo J., Lundeberg J., and Leitner T. 2002. Genetic evidence for an East Asian origin of domestic dogs. *Science* **298:** 1610–1613.

Seal U.S., Phillips N.I., and Erickson A.W. 1970. Carnivora systematics: Immunological relationships of bear serum albumins. *Comp. Biochem. Physiol.* **32:** 33–48.

Seddon J.M. and Ellegren H. 2002. MHC class II genes in European wolves: A comparison with dogs. *Immunogenetics* **54:** 490–500.

———. 2004. A temporal analysis shows major histocompatibility complex loci in the Scandinavian wolf population are consistent with neutral evolution. *Proc. R. Soc. Lond. B Biol. Sci.* **271:** 2283–2291.

Sharma D.K., Maldonado J.E., Jhala Y.V., and Fleischer R.C. 2004. Ancient wolf lineages in India. *Proc. R. Soc. Lond. B Biol. Sci.* (suppl.) **271:** S1–S4.

Stockard C.R. 1941. *The genetic and endocrinic basis for differences in form and behavior, as elucidated by studies of contrasted pure-line dog breeds and their hybrids.* The Wistar Institute of Anatomy and Biology, Philadelphia, Pennsylvania.

Sundqvist A.-K., Ellegren H., Olivier M., and Vilà C. 2001. Y chromosome haplotyping in Scandinavian wolves (*Canis lupus*) based on microsatellite markers. *Mol. Ecol.* **10:** 1959–1966.

Sutter N.B. and Ostrander E.A. 2004. Dog star rising: The canine genetic system. *Nat. Rev. Genet.* **5:** 900–910.

Trutt L.N. 2001. Experimental studies of early canid domestication. In *The genetics of the dog* (ed. A. Ruvinsky and J. Sampson), pp. 15–41. CABI Publishing, New York.

Tsuda K., Kikkawa Y., Yonekawa H., and Tanabe Y. 1997. Extensive interbreeding occurred among multiple matriarchal ancestors during the domestication of dogs: Evidence from inter- and intraspecies polymorphisms in the D-loop region of mitochondrial DNA between dogs and wolves. *Genes Genet. Syst.* **72:** 229–238.

Valadez R. 1995. *El perro mexicano.* Instituto de Investigaciones Antropológicas, Universidad Nacional Autónoma de México, Mexico.

Vilà C., Maldonado J., and Wayne R.K. 1999a. Phylogenetic relationships, evolution and genetic diversity of the domestic dog. *J. Hered.* **90:** 71–77.

Vilà C., Seddon J., and Ellegren H. 2005. Genes of domestic mammals augmented by backcrossing with wild ancestors. *Trends Genet.* **21:** 214–218.

Vilà C., Leonard J.A., Götherström A., Marklund S., Sandberg K., Lidén K., Wayne R.K., and Ellegren H. 2001. Widespread origins of domestic horse lineages. *Science* **291:** 474–477.

Vilà C., Amorim I.R., Leonard J.A., Posada D., Castroviejo J., Petrucci-Fonseca F., Crandall K.A., Ellegren H., and Wayne R.K. 1999b. Mitochondrial DNA phylogeography and population history of the grey wolf *Canis lupus*. *Mol. Ecol.* **8:** 2089–2103.

Vilà C., Savolainen P., Maldonado J.E., Amorim I.R., Rice J.E., Honeycutt R.L., Crandall K.A., Lundeberg J., and Wayne R.K. 1997. Multiple and ancient origins of the domestic dog. *Science* **276:** 1687–1689.

Vilà C., Sundqvist A.-K., Flagstad Ø., Seddon J., Björnerfeldt S., Kojola I., Casulli A., Sand H., Wabakken P., and Ellegren H. 2002. Rescue of a severely bottlenecked wolf (*Canis lupus*) population by a single immigrant. *Proc. R. Soc. Lond. B Biol. Sci.* **270:** 91–97.

Vincek V., O'Huigin C., Satta Y., Takahata Y., Boag P.T., Grant P.R., Grant B.R., and Klein J. 1997. How large was the founding population of Darwin's finches? *Proc. R. Soc. Lond. B Biol. Sci.* **264:** 111–118.

Wayne R.K. 1986a. Cranial morphology of domestic and wild canids: The influence of development on morphological change. *Evolution* **4**: 243–261.

———. 1986b. Limb morphology of domestic and wild canids: The influence of development on morphologic change. *J. Morphol.* **187**: 301–319.

Wayne R.K. and O'Brien S.J. 1987. Allozyme divergence within the Canidae. *Syst. Zool.* **36**: 339–355.

Wayne R.K. and Ostrander E.A. 1999. Origin, genetic diversity, and genome structure of the domestic dog. *Bioessays* **21**: 247–257.

Wayne R.K., Benveniste R.E., Janczewski D.N., and O'Brien S.J. 1989. Molecular and biochemical evolution of the Carnivora. In *Carnivore behavior, ecology and evolution* (ed. J.L. Gittleman), pp. 465–494. Cornell University Press, Ithaca, New York.

8

mtDNA Studies of the Origin of Dogs

Peter Savolainen
Department of Biotechnology
Albanova University Centre
KTH-Royal Institute of Technology
SE-10691 Stockholm, Sweden

POPULATION GENETIC STUDIES OF DOGS of the last few years, based on sequence analysis of mitochondrial DNA (mtDNA), has meant a great leap forward in the unraveling of the origin and early history of the domestic dog. Based on the archaeological record, which was earlier the principal source of information, it was not possible to say anything more precise than that the dog probably originated from the wolf some 15,000 years ago. Such basic questions as where this happened, and whether it happened just once or several times at different places, could not be answered. In studies of human history, population genetic analyses of mtDNA have been of considerable importance, establishing the time and place for the first origin of modern humans and monitoring the major migrations (Cavalli-Sforza and Feldman 2003). In this chapter, we show how, similarly, studies of mtDNA in dogs, starting on a large scale in 1996, have begun to create an increasingly sharper picture of the origin of dogs. Most prominently, two studies from 2002 (Leonard et al. 2002; Savolainen et al. 2002) have given the first solid evidence for the geographic origin of dogs, indicating a single origin in East Asia. The results also show the potential for further, more detailed studies, which should help to map the routes and time points for the first spread of dogs around the world, and to describe the development of morphological variants that eventually resulted in today's dog breeds.

The Dog and Its Genome ©2006 Cold Spring Harbor Laboratory Press 0-87969-742-3

The domestic dog is perhaps the most fascinating of the domestic animals. According to the available knowledge, it was the first domestic animal, probably domesticated by mobile hunter–gatherers rather than by settled farmers (Clutton-Brock 1995). It is the morphologically most diverse mammalian species, with an extreme range in size and shape, with the close to hundredfold difference in weight between the Chihuahua and the Great Dane as the perhaps most obvious example (Wayne 1986a,b). Dogs and humans share a number of social and behavioral signals facilitating communication, many of which are also shared with the wolf and undoubtedly have had significance in the domestication of the wolf (Houpt and Willis 2001). This gives the dog a behavior that appeals to us more than that of other domestic animals, explaining its status as man's best friend.

Taking this into consideration, surprisingly little is known about the first origin of the dog and about how the immense morphological variation, and eventually the hundreds of dog breeds, developed. Numerous texts have handled these topics, and there have been many theories about how, where, and when the dog originated, but there have been very few facts to build on, making reading about this subject an often frustrating experience. However, as shown in this chapter, genetic analysis based on sequence analysis of mtDNA has implied a large step forward for the unraveling of the origin and early history of the domestic dog and promises to contribute more information in future studies.

The question of which animal is the wild ancestor of the domestic dog has been relatively uncontroversial. An origin primarily from the wolf has been the main theory from the start of evolutionary theory-building (Clutton-Brock 1995). It has been hypothesized that other canids, such as the jackal or the coyote, could also have contributed to the forming of the dog, explaining some of the morphological and behavioral variation among domestic dogs. However, numerous studies of physical and behavioral characteristics (Scott and Fuller 1965; Wayne 1986a,b; Houpt and Willis 2001), as well as of molecular genetic markers, both mtDNA (Wayne et al 1992) and nuclear markers (Wayne and O'Brien 1987; Garcia-Moreno et al. 1996), have built an increasingly stronger case for the wolf origin. Although not disproving some degree of contribution from other species, the total collection of evidence strongly favors the wolf as the only ancestor of the dog.

However, although the origin from the wolf has been relatively well supported, the other basic questions, where and when and how many times the wolf was domesticated into the domestic dog, have remained to be answered. Before the molecular tools began to be used, the only source of information was the archaeological record. Archaeological finds, probably originating from dogs, have been found at Mesolithic sites in Europe, Asia, and America, the earliest going back to at least 14,000 years before present (ybp), indicating that the dog was the first domestic animal (Clutton-Brock 1995). However, except for this, there have been virtually no clues to a more detailed picture of the origin of the dog.

The earliest finds believed to be from domestic dogs are from central and eastern Europe and the Middle East: a single jaw from 14,000 ybp in Germany (Nobis 1979), two skulls from 13,000–17,000 ybp in western Russia (Sablin and Khlopachev 2002), and an assemblage of small canids from 12,000 ybp in the Middle East (Tchernov and Valla 1997). The earliest finds from East Asia and North America are younger, from 7,500 (Li 1990) and 8,500 ybp (Clutton-Brock and Noe-Nygaard 1990), respectively. The archaeological record therefore indicates that the dog originated in Southwest Asia or Europe approximately 15,000 ybp.

There are several uncertainties with the archaeological evidence, however. It is problematic to interpret remains of canids, since it is difficult to discriminate between species; for example, between small wolves and domestic dogs (Olsen 1985), and the dating of the finds may be obscured by, for example, disturbed layers. Most importantly, archaeological excavations have not been evenly performed in different parts of the world. Thus, the fact that the oldest remains of dogs have been found in the relatively thoroughly excavated regions of Europe and the Middle East could be attributed to a greater chance of finding remains there than in less explored regions.

The morphology and size of early archaeological dog remains suggest an origin from one of the small South Asian wolf subspecies rather than from the large North Eurasian and North American wolves. Furthermore, one osteological feature of the jaw, diagnostic for dogs, is also found among Chinese wolves but rarely in other wolves, indicating an origin from East Asia (Olsen 1985). Based on the archaeological evidence, there have been two main theories for the origin of the domestic dog (Clutton-Brock 1995). The Middle East has been proposed, based on the earliest finds of dogs in Europe and the Middle East, and the small size of the local species of wolf. An origin from several different wolf populations has been suggested, based on the widespread occurrence of early finds of dog remains, for example, in both the New and the Old World by 8,500 ybp. This would also to some extent offer an explanation for the extreme morphological variation among dog breeds. There is also very little known about the early history of different dog breeds, and even for the large number of breeds which have been developed in the last few hundred years, the exact origin is mostly unknown (Clutton-Brock 1995).

THE FIRST MAJOR POPULATION GENETIC STUDIES

The unclear picture given by the archaeological and historical data prompted population genetic studies when the techniques for large-scale DNA-sequencing studies became available. The first three major studies of mtDNA from dogs and wolves were published in 1996 and 1997. In one study (Vila et al. 1997), 261 bp of the mtDNA control region (CR) was analyzed in 140 domestic dogs representing 67 breeds, mostly European but also a few from Asia, Africa, and America, and 162 wolves from throughout Europe, Asia, and North America. In two other studies (Okumura et al. 1996; Tsuda et al. 1997), approximately 970 bp and 670 bp, respectively, of the CR was analyzed in a total of 128 dogs, representing mainly Japanese breeds, and 19 wolves. These studies showed that, in phylogenetic analyses, the domestic dog mtDNA sequences were distributed into several distinct groups (clades) interspersed by wolf sequences (Figs. 1 and 2). These clades were in a later study (Savolainen et al. 2002) given the names clades A–F, which we will use herein. The separation into distinct clades interspersed by wolf sequences showed that the dog originates from several female wolf lines, at least as many as the number of clades. Importantly, these studies of a total of 268 domestic dogs showed that all dog mtDNA sequences were closer to those of wolves than to those of any other canid. Thus, there is no indication of genetic contribution from any other canid than the wolf to the domestic dog population, on the female side.

Comparing the three studies, there were clear similarities between the genetic variation found in the European and the Japanese breeds. The three most frequently represented clades, clades A, B, and C, which comprised >95% of the individuals, were found at approximately similar frequencies in all three studies, approximately 65–75% for clade A, 15–20% for clade B, and 5–10% for clade C. This showed that dogs across the Eurasian continent to a large degree share a common origin. However, of the individual sequence types forming the three clades, >50% were specific to either European or Japanese breeds, showing regional differences possibly caused by founder effects during migrations of dogs, or from development of novel mtDNA types by mutations. The three other clades were found only regionally in small numbers of

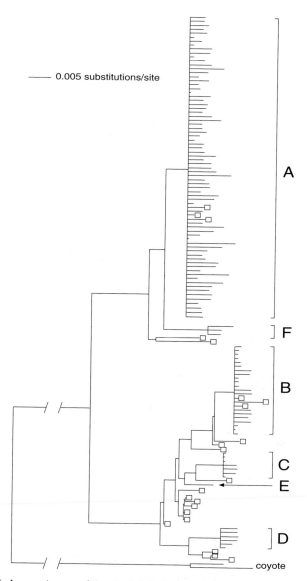

Figure 1. Phylogenetic tree of dog (unlabeled) and wolf (*open squares*) mtDNA types (Savolainen et al. 2002). Six clades (A–F) of dog mtDNA types are indicated. Branch lengths are according to the indicated scale; the branch leading to the outgroup (coyote) has been reduced by 50%. Clades A–D and F were imploded because of high levels of homoplasy. The relationships between mtDNA types within the clades are instead shown as minimum-spanning networks in Figs. 2 and 3. (Reprinted, with permission, from Savolainen et al. 2002.)

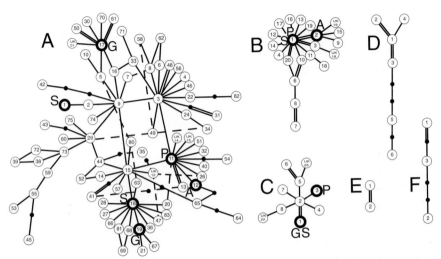

Figure 2. Minimum-spanning networks showing genetic relationships among dog mtDNA types of clades A–F based on 573-bp sequence of the mtDNA CR. mtDNA types (*white circles*) are separated by one mutational step, single strokes represent substitutions, and double strokes represent indels. Black dots are hypothetical intermediates. The largely random distribution of breeds and morphological types of dogs among clades A, B, and C is exemplified by marking of the mtDNA types found in German shepherd (mtDNA types A17, A19, and C1, marked "G"), Pekingese (mtDNA types A11, B1, and C3, marked "P"), Samoyed (mtDNA types A1, A18, B1, and C1, marked "S") and Afghan dog (mtDNA types A12 and B2, marked "A"), and by highlighting of the respective mtDNA types by bold circles. (Modified, with permission, from Angleby and Savolainen 2005.)

individuals, clade D only in Scandinavian breeds (Vila et al. 1997), and clades E and F only in dogs of East Asian or Siberian breeds (Okumura et al. 1996; Tsuda et al. 1997).

There was generally very limited correlation between mtDNA type and breed, or morphologic type, of dog. Most of the well-represented breeds had several different mtDNA types, often belonging to two or three different clades. Although there were clear similarities between the European and Japanese data sets, indicating a common history for the two dog populations, the geographic location of the domestication of the wolf, or any detailed conclusions about the history of specific dog breeds, could not be determined from these data sets.

The time of the origin of the dog was also estimated, based on the largest genetic distance of the sequences in clade A and the mutation rate of the analyzed region (Vila et al. 1997). The mutation rate was estimated from the sequence difference between wolves and coyotes and the divergence time between wolves and coyotes, assumed to be at least 1 million years, as based on the fossil record. The age of clade A was thus estimated at 135,000 years. Based on the monophyly of the dogs in clade A and the lack of resolution between individual mtDNA types, it was assumed that this clade had an origin from a single wolf mtDNA type. Under this assumption, the domestic dog would have existed much longer than the 15,000 years suggested by the archaeological record. Furthermore, it would also imply that clades B–F, which have a much shorter divergence time than clade A, would have originated much later, suggesting several instances of crossbreeding between dogs and wolves through history, or possibly some instances of later separate domestications of wolves. However, clade A does not necessarily originate from a single wolf mtDNA type. As shown below, later analyses of a larger part of the CR shows signs of substructure in clade A, indicative of several different origins from wolf mtDNA types for clade A, and three wolf mtDNA types were later found to fall into clade A, which is therefore not monophyletic for dog.

To conclude, the first major studies of the mtDNA variation among domestic dogs showed conclusively that dogs, at least on the female side, originate from the wolf. They also showed that the domestic dog mtDNA types originated from at least six wolf mtDNA types. Thus, at least six female wolves, and possibly many more individual female wolves with identical mtDNA types, were founders of the domestic dog population. A very early time for the first origin of the dog has also been suggested. However, a definitive dating of the origin of the dog, and an estimation of the number of origins and subsequent crossbreedings, and their locations, were not possible based on the available data sets.

THE NEXT STEP

The greatest step forward so far in the study of the origin of dogs was taken in 2002 with the publication of two articles, one study of extant dogs from throughout the world (Savolainen et al. 2002) and one of archaeological remains of pre-European American dogs (Leonard et al. 2002). The study by Leonard et al. showed that domestic dogs on all continents have a common origin from one gene pool, and the study by Savolainen et al. suggested that this gene pool originated just once, somewhere in East Asia.

There are principally two approaches available to elucidate the place of domestication of the wolf, using mtDNA analysis. The most straightforward approach would be to study the mtDNA sequences of the domestic dog population of today and compare them to sequences from a worldwide sample of wolves. The wolves having the most similar, or even identical, sequences compared to the dogs would then be presumed to be ancestors of the wolf population from which the domestic dogs descend, and the geographical region in which these wolves live would be presumed to be the geographical origin of the dog. However, wolves are extremely mobile, resulting in little geographic partitioning of mtDNA types. In a study of the worldwide mtDNA sequence variation among wolves (Vila et al. 1999b), it was, for example, found that a Bulgarian and a Saudi Arabian wolf had identical mtDNA types, and that a Mongolian and another Saudi Arabian wolf shared another type. Thus, the finding of a few wolf mtDNA types closely related to domestic dog types is not in itself a sufficient basis for the determination of the location of domestication. Furthermore, the wolf populations have in large parts of their original geographic distribution experienced severe reductions in population size, or total extinction. Thus, there is a possibility that wolves have become extinct in the region where wolves became dogs, rendering such studies impossible.

The other approach would be to make an intraspecific study of the phylogeographic mtDNA variation among today's domestic dogs; i.e., to make a sequence analysis of today's domestic dog population, making sure to obtain a representative sample from all parts of the world, and comparing the genetic variation between geographic regions. The region showing the largest genetic variation would then be assumed to be the region in which the domestic dog originated, based on the assumptions that only part of the genetic variation in the region of origin was spread to the other parts of the world, and that any subsequent mixing of populations through migration would not have been thorough enough to conceal these differences.

The second approach was used in the study by Savolainen et al. (2002). Using published sequences and 526 more samples collected from dogs from most parts of the world, a total of 654 dogs and 38 Eurasian wolves were studied. This resulted in the first comprehensive picture of the genetic variation among domestic dogs worldwide. Importantly, it gave a good representation of dogs in Europe, Southwest Asia, and East Asia, the three regions that have been the strongest candidates for the geographic origin of the domestic dog (Table 1). The dogs represented a geographic region based on either of two criteria: that they were of a breed of known geographic origin or that they had been sampled in a geographic region with little import of foreign dogs. A flaw in the sample collection was that the American continent was represented only by Arctic breeds. This was because the original Indian American dog populations are thought to have been largely obliterated or mixed with

Table 1. Number and proportion of individuals, and number of mtDNA types and unique mtDNA types, for the phylogenetic clades A–F and for all six clades (Total) in different populations

	Clade A		Clade B		Clade C		Clade D		Clade E		Clade F		Total	
Region	number of individuals (proportion, %)	number of mtDNA types (unique types)	number of individuals (proportion, %)	number of mtDNA types (unique types)	number of individuals (proportion, %)	number of mtDNA types (unique types)	number of individuals (proportion, %)	number of mtDNA types (unique types)	number of individuals (proportion, %)	number of mtDNA types (unique types)	number of individuals (proportion, %)	number of mtDNA types (unique types)	number of individuals	number of mtDNA types (unique types)
Africa	30 (85.7)	14 (5)	4 (11.0)	1 (0)	1 (2.9)	1 (0)	0	0	0	0	0	0	35	16 (5)
America	25 (100)	5 (1)	0	0	0	0	0	0	0	0	0	0	25	5 (1)
Europe	140 (67.6)	20 (9)	36 (18.0)	4 (1)	12 (5.9)	3 (0)	19 (9.1)	3 (3)	0	0	0	0	207	30 (13)
East Asia	192 (73.8)	44 (30)	39 (15.0)	10 (7)	24 (9.2)	4 (1)	0	0	3 (1.1)	1 (1)	2 (0.8)	2 (2)	260	61 (41)
SW Asia	51 (56.7)	16 (4)	32 (35.0)	4 (2)	5 (5.4)	2 (0)	2 (2.2)	1 (1)	0	0	0	0	90	23 (7)
Siberia	17 (70.8)	9 (1)	2 (8.3)	1 (0)	4 (17.0)	2 (1)	0	0	0	0	1 (4.1)	1 (1)	24	13 (3)
India	11 (84.6)	4 (0)	1 (7.7)	1 (0)	1 (7.7)	1 (0)	0	0	0	0	0	0	13	6 (0)
Total	466 (71.3)	71	114 (17.4)	13	47 (7.2)	5	21 (3.2)	4	3 (0.5)	1	3 (0.5)	3	654	97

mtDNA types are defined by substitutions only, disregarding indels.

European dogs in connection with the arrival of Europeans, which would make the interpretation of American dog sequences complicated. However, in an earlier study of mtDNA in 19 Mexican hairless, a Mexican breed believed to have remained essentially isolated, the mtDNA types all belonged to clades A, B, and C, indicating a common origin of these dogs with those of the Old World (Vila et al. 1999a). Furthermore, the study of archaeological remains of pre-European American domestic dogs (Leonard et al. 2002) gives important information about the American dog population (see below).

A phylogenetic analysis of the samples distributed the mtDNA sequences into the six clades earlier found among Japanese and European breeds, thus not adding any more major phylogenetic groups in the dog mtDNA phylogenetic tree (Fig. 1). The phylogenetic tree has relatively low bootstrap values for some of the phylogenetic groups, with values <50% for clades A and B. However, the earlier Japanese studies of a larger region gave bootstrap values >50% for all clades, and analysis of the whole mitochondrial genome gives very strong support for all groups (>98%; P. Savolainen et al., unpubl.). Thus, the general structure of the mtDNA sequence variation among dogs seems to be well represented by analysis of the 582 bp of the CR. Still, the low bootstrap values are a sign of a large degree of homoplasy. This homoplasy, and the fact that most of the dog mtDNA types are separated from their closest neighbors by a single mutation, implies that although the relations between the mtDNA molecules at large genetic distances can be relatively truthfully depicted by analysis of the 582-bp region, it is not sufficient for obtaining a phylogenetic resolution of dog mtDNA types within the clades. Therefore, the branchings in the six dog clades were imploded, and the relations between the mtDNA types within the clades were instead displayed as minimum-spanning networks (Fig. 2). The homoplasy is apparent in the network of phylogenetic clade A, which has a complicated pattern with many alternative pathways. However, as shown below, the analyzed region contains enough information to allow a number of valuable conclusions about the history of domestic dogs.

GEOGRAPHIC ORIGIN OF THE DOG

The sample collection in Savolainen et al. (2002), with representation from most parts of the Old World, allowed for the first time a phylogeographic approach to be made in the study of the history of domestic dogs. Comparing the genetic variation between different parts of the world, the similarities in the distribution of sequences belonging to the three major clades is striking (Table 1). All the three major clades A–C are present in all parts of the world except Arctic America, and at similar frequencies. Thus, these three clades, comprising >95% of the dog mtDNA sequences, constitute a common source for a very large proportion of the mtDNA genetic variation in all domestic dog populations (clades D, E, and F were found only regionally and are discussed on page 136, under the heading Possible Crossbreedings with Wolf). The absence of clades B and C in Arctic America can probably be attributed to a genetic bottleneck in the forming of this dog population; probably this population was introduced to America by the ancestors of today's Inuits rather than by the ancestors of the Indians. However, information about the native American dog population is given in two other studies. As mentioned above, the Mexican hairless, which is believed to have remained largely isolated from European dogs, has mtDNA types belonging to all the three major clades, A, B, and C (Vila et al. 1999a), and the study of archaeological remains of pre-European American domestic dogs (Leonard et al. 2002) showed that among 13 samples from Mexico, Peru, and Bolivia, from 800 ybp or earlier, 12 had mtDNA types clustering with clade A, and one sample had a type clustering with clade B. Thus, both clades A and B, and possibly clade C, according to the data on the Mexican hairless, were present among the native American dogs, showing that the dog populations of the New World have a common origin with the dogs of the Old World. Perhaps more importantly, none of the American samples had sequences that did not fall into the same mtDNA clades as the Old World samples. Thus, in the available mtDNA data there is no sign of a separate domestication of dogs in America, or of crossbreeding with the wolf.

An important observation by Savolainen et al. (2002) was that clades A, B, and C had similar frequencies in all regions (Table 1). This suggests that, unless there has been a very effective gene flow along the Eurasian continent, the major present-day dog populations have had a common origin from a single gene pool containing clades A, B, and C. Had the three clades had different geographic origins, from separate domestications of wolf, the frequencies of the clades would be expected to be different in different regions. Had, for example, clade A originated in Europe and clade B in East Asia, we would expect to find a higher frequency of clade A in Europe than in East Asia and, vice versa, a highest frequency of clade B in East Asia, with intermediate frequencies in Southwest Asia. The only major deviation from the uniform frequencies of the three clades is found in the Southwest Asian sample, which has a higher frequency of clade B sequences and a lower frequency of clade A sequences. This could indicate that clade B originated in Southwest Asia separately from clades A and C. However, if this were the case, it would also be expected that the genetic variation among the sequences belonging to clade B would be higher in Southwest Asia than in other regions. On the contrary, it was considerably lower in Southwest Asia than in East Asia, and equal between Southwest Asia and Europe. This is further discussed below, but is clearly exemplified by the number of mtDNA types belonging to clade B, for the three regions (Table 1). A single first origin for the three clades was further indicated by the fact that there was no clear division of the main morphologic types of dogs (spitz, mastiff, greyhound) or of large and small breeds between the three main clades, except for a lack of greyhounds in clade C (Fig. 2). This clearly suggests that the extreme morphologic variation among dog breeds is not the result of different geographically distinct domestications of wolf. It is obviously impossible today to find out how much the dog populations have been mixed through migration during the thousands of years since the domestication of the wolf, and whether this could explain the universal presence, at similar frequencies, of clades A, B, and C. However, considering the large distance between the two island groups, the British Isles and Japan, situated outside opposite parts of the immense Eurasian continent, the relatively similar frequencies of clades A, B, and C, among breeds from the two regions (British Isles: 81.5%, 13.8%, and 4.6%; Japan: 62.5%, 18.8%, and 15.6%) is striking, and a strong indication of a single original gene pool containing the three major dog-mtDNA clades, A, B, and C (Savolainen et al. 2002).

To find the geographic origin of the dogs, a comparison of the genetic variation between geographic regions can be used, based on the assumption that if an ancestral population and a derived population (formed from a subset of the genetic types of the ancestral population) are compared, the number of mtDNA types and the nucleotide diversity are expected to be higher in the ancestral population. Comparing Europe, Southwest Asia, and East Asia, the three regions that have been the strongest candidates for geographic origin of the domestic dog, East Asia had a larger genetic variation for clades A and B as measured by a number of methods, whereas for clade C there were no significant differences (Savolainen et al. 2002). Thus, for clade A, the mean pair-wise sequence distance, which gives a rough measure of the amount of genetic variation, was 3.39 (S.D. = 0.13) substitutions in East Asia, 2.28 (S.D. = 0.23) in Southwest Asia, and 2.97 (S.D. = 0.08) in Europe. Furthermore, the number of mtDNA types was much larger in the East Asian sample than in the European and the Southwest Asian samples (Table 1). The difference between the East Asian and the European samples is striking, and, when corrected for sample size by resampling with replacement to correct for different sample size, there were 20.2 (S.D. = 2.4) mtDNA types in clade A among 51 East Asian dogs, which is significantly more than the 16 mtDNA types found among the 51 Southwest Asian dogs ($p<0.05$; 1000 replicates). It is notable that out of the 44 types found in East Asia, 30 were unique to this region. Thus, the number of mtDNA types unique to East Asia was larger than the total number of types in Europe (Table 1). In addition, compared to the other regions there were more mtDNA types in East Asia, when corrected for sample size, but at a significant level only compared to America. For clade B, the mean pair-wise sequence distance was larger for East Asia (0.93 substitutions, S.D. = 0.17) than for Europe (0.45, S.D. = 0.14) and Southwest Asia (0.36, S.D. = 0.11), the East Asian sample had significantly more mtDNA types than those of Europe and Southwest Asia, and a majority of all mtDNA types were unique to East Asia. The difference in number and distribution of mtDNA types in East Asia, Europe, and Southwest Asia is visually discernable in minimum-spanning networks of the three clades (Fig. 3). For clade A it is notable that East Asian mtDNA types were distributed throughout the network, whereas for Europe and Southwest Asia, parts of the network, largely the same in the two populations, were empty. For clade B, the larger number of mtDNA types is distinctive, whereas for clade C there are no large differences. A concern about these comparisons of genetic variation could be that in some samples, especially the European sample, many of the dogs belong to specific dog breeds that may have experienced inbreeding and thereby lost genetic variation compared to the more generalized dogs of other regions. However, also when only dogs of pure breeds are included in the analyses, the genetic variation is larger in East Asia than in other regions (Savolainen et al. 2002).

To conclude, in a comparison of the dog populations of East Asia, Europe, and Southwest Asia, there is a much higher genetic variation in East Asia for the two major dog mtDNA clades, clades A and B, which comprise >90% of all domestic dog mtDNA sequences. This suggests an origin somewhere in East Asia, and a subsequent spread to the rest of the world, for these clades. For clade C there is no significant difference between the regions, but the similar frequency of clade C among dog populations of the Old World suggests a common origin of clade C together with clades A and B. Thus, the available mtDNA data indicate that the domestic dog has a single geographic origin, somewhere in East Asia.

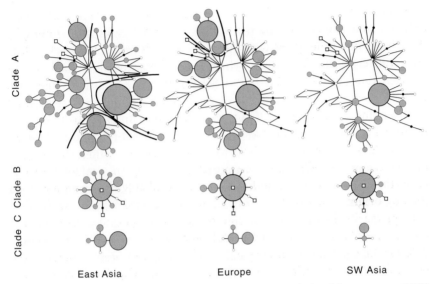

Figure 3. Minimum-spanning networks showing genetic relationships among mtDNA types of phylogenetic clades A, B, and C. mtDNA types (*circles*) are separated by one mutational step, ignoring indels. mtDNA types found in East Asia, Europe, and Southwest Asia are indicated in separate networks. The sizes of gray circles are proportional to mtDNA-type frequency in the respective populations. Small open circles denote mtDNA types not found in the regional population. Black dots are hypothetical intermediates. Open squares are wolf mtDNA types. Subclusters of clade A discussed in the main text, three in the East Asian and one in the European network, are marked by black lines. (Modified, with permission, from Savolainen et al. 2002.)

DATING OF THE ORIGIN OF DOGS

Although the present mtDNA data present a relatively strong case for an origin of clades A and B somewhere in East Asia, a dating of this origin, based on mtDNA data, is more problematic. The time of origin for each dog mtDNA clade can be estimated from the mean genetic distance of the sequences in each clade to the original wolf mtDNA type, and the mutation rate of the analyzed region. Under the assumption that the clade originates from a single wolf mtDNA type, the age of the clade will approximately correspond to the origin of the dogs originating from that female wolf. However, two problems with the available data render impossible a precise dating of the origin of the dog at present.

The first problem concerns the dating of the paleontological fix point needed to calibrate the mutation rate. The wild canid most closely related to the wolf is the coyote, and the time for the split into two species from their common ancestor is the best available calibrating point for the mutation rate of dog mtDNA. Counting the number of sequence differences between wolf and coyote in the analyzed mtDNA region, and dividing by the divergence time, gives the mutation rate. However, although a first appearance of wolves ~700,000 years ago and of coyotes ~1 million ybp (Kurtén 1968; Kurtén and Anderson 1980) indicates a date for the divergence between the two species of approximately 1 million ybp, the time of divergence has not been definitely established and an earlier date of up to ~2 million years cannot be ruled out (Nowak 1978, 1996; Kurtén and Anderson 1980). Therefore, without more exact paleontological evidence, a precise dating of the origin of the dog will not be possible. Nevertheless, it can give an indication of the probable range of time when the dog originated, and this information can be used together with other data, primarily the information from the archaeological record on the first appearance of dogs. To simplify the following argument, the 1-million-year date for the wolf–coyote divergence will be used, but bearing in mind the possibility that the split occurred up to 2 million ybp, in which case the age of the datings should be doubled.

The second problem is that the mutation rate of the analyzed 582-bp region does not give resolution between mtDNA types in the timescale needed to monitor the last 20,000 or 30,000 years. According to the calculations discussed above, the rate of mutation is approximately one mutation per 24,000 years in a lineage (Savolainen et al. 2002). This implies that if two wolves, having mtDNA types differing by a single mutation, were domesticated say 15,000 years ago, the dogs of today originating from those two lineages would have mtDNA types differing from each other only by the single original mutation or by just one or two additional mutations. If there were several wolves having mtDNA types differing by just a few mutations, it would not be possible to fully resolve the mtDNA types of the dogs originating from them. Thus, the period around 15,000 ybp, which is suggested to be the time for the origin of the domestic dog according to the archaeological record, cannot be fully studied using the 582-bp mtDNA region available at present. This implies that even though clade A is an almost completely continuous group of mtDNA types separated from each other by single mutational steps (Fig. 2), it does not necessarily originate from a single wolf mtDNA type. It is possible that clade A originally was a clade of several different wolf mtDNA types, and that several wolves having a number of these closely related mtDNA types belonging to clade A were domesticated.

In a domestication event with a subsequent population expansion, a starlike phylogeny, with the founder mtDNA type in the center and new mtDNA types distributed radially, would be expected. The networks of clades B and C are starlike, indicating an origin from a single wolf mtDNA type (Figs. 2 and 3). In contrast, clade A has a complicated pattern without an easily identifiable central node. A distance of up to eleven substitutional steps between mtDNA types would indicate that clade A is much older than clades B and C, and derives from an initial domestication of wolves. However, instead of a single central node, there are several subclusters with starlike shapes, suggesting that clade A may have originated from several wolf mtDNA types. However, as discussed above, the data set does not give the resolution necessary to determine the exact number of founding wolf mtDNA types in clade A. The approximate age of clade A, assuming a single origin from wolf and a subsequent population expansion, is estimated from the mean pair-wise distance between East Asian dog mtDNA sequences (3.39 substitutions, S.D. = 0.13) and the mutation rate to 41,000 ± 4,000 years. This calculation may be biased by the population history among the dogs and wolves. Alternatively, the maximum age of clade A can be estimated from the number of steps between the most distantly related mtDNA types, 11 substitutions apart, which corresponds to ~120,000 years. According to these calculations, clade A would have originated 40,000–120,000 ybp, and if it is supposed that it was formed in a domestication event from a single wolf mtDNA type, the domestic dog would have originated 40,000–120,000 ybp. If, instead, an origin of clade A from several different wolf mtDNA types is assumed, several reasonably defined subclusters can be found. To give an alternative dating of the domestication of the dog, three subclusters of clade A, possibly representing three origins of dog from wolf, marked by lines in Figure 3, can be studied. The mean genetic distances of the sequences belonging to these subclusters to their respective nodes (0.45, 0.65, and 1.07 substitutions with S.D. = 0.13, 0.09, 0.27, respectively) give estimates of 11,000 ± 4,000, 16,000 ± 3,000, and 26,000 ± 8,000 years for their ages, respectively. Thus, clade A has a substructure, suggesting that it could have been formed by several wolf mtDNA types, possibly ~15,000 ybp. Assuming single wolf mtDNA types as founders of clades B and C, the mean distances among East Asian sequences to the nodes (0.54 and 0.71 substitutions, S.D. = 0.08 and 0.10) give estimated ages of 13,000 ± 3,000 and 17,000 ± 3,000 years for clades B and C, respectively.

Thus, clade A was formed ~40,000–120,000 ybp but has a substructure indicating later population events, and clades B and C were formed ~15,000 ybp. Depending on how these data are interpreted, it can suggest a first origin of domestic dogs either ~40,000–120,000 ybp forming only clade A, after which clades B and C were introduced into the dog gene pool through crossbreeding between dogs and wolves ~15,000 ybp, but it can also suggest a single origin ~15,000 ybp involving all three clades, A, B, and C. Considering the very similar frequencies of clades A, B, and C in dog populations around the world, a simultaneous origin of all three clades seems most probable.

Looking at the other alternative, that clade A originated from an initial domestication ~40,000 ybp, the dog population originating from that domestication event would be expected to have spread to other parts of the world. An introduction of clades B and C, 15,000 ybp, into the already existing domestic dog gene pool, by regional crossbreedings with wolf, would require a very thorough mixing to have occurred, through migration to all parts of Eurasia, to level the frequencies of the three clades to the very similar levels now found throughout the world. Alternatively, if clade A originated from a domestication 40,000 ybp, these dogs could have remained isolated in the original geographic region, not spreading to other parts of the world. Clades B and C would then, 15,000 ybp, have been introduced into the domestic dog gene pool by regional crossbreedings with wolf, after which the dogs would have spread to other parts of the world. Although the amount of migration and trade 15,000 ybp is not well known, this scenario seems unlikely. It is also contradicted by the age of the oldest subcluster of clade A in Europe (as determined from the mean genetic distance from mtDNA types unique to the western part of the world to the nodal mtDNA type shared with East Asia; 0.39 substitutions, S.D. = 0.09), which is estimated to be only 9,000 ± 3,000 years old (Fig. 3).

A first origin of dog mtDNA clade A a long time before the origin of dog clades B and C, therefore, does not seem probable, considering the very similar frequencies of clades A, B, and C around the world. The total sum of circumstantial mtDNA evidence therefore indicates a single origin, in both place and time, for the three clades, ~15,000 ybp. Although this conclusion is a bit speculative, it can be concluded that the mtDNA data do not contradict an origin of the domestic dog ~15,000 ybp as suggested by the archaeological record. Thus, the best evidence for the time of the first origin of the domestic dog remains the archaeological record, which implies that ~15,000 ybp must be considered the best estimate for the time of origin of the domestic dog.

It should be noted that the mtDNA-dating calculations are based on a divergence time for the coyote and the wolf of 1 million years. If this divergence occurred as much as 2 million ybp, the calculations give a date for the origin of dogs of ~30,000 ybp. However, in the context of the archaeological record, an origin ~15,000 ybp seems to be a probable scenario. The archaeological record cannot define the number of geographic origins or their locations, but suggests the date at 9,000–17,000 ybp. The earliest finds are from Europe, and there is no certain evidence for domestic dogs in late Paleolithic China, but in the earliest Neolithic, finds are numerous, dating back to 7,500 ybp (Olsen 1985; Li 1990). Considering the relatively limited amount of archaeological work done in East Asia, the lack of late Paleolithic finds does not exclude a much earlier origin of domestic dogs in East Asia. A synthesis of available mtDNA and archaeological data therefore points to an origin of the domestic dog in East Asia ~15,000 ybp. In this event, clade A would have had several origins from wolf mtDNA types, at least around ten, and the first domestication of wolves would not have been an isolated event, but rather a common practice in the human population in question.

POSSIBLE CROSSBREEDINGS WITH WOLF

The origin of the three other phylogenetic clades, D, E, and F, remains to be solved. They could either represent late crossbreedings between male dogs and female wolves, or separate domestications of wolf, or they could have originated together with clades A, B, and C at a single domestication event. A problem with clades E and F is that they have so far been found in very few individuals, three for each clade, which makes population genetic analyses virtually impossible. Since they are found in East Asian dogs, an origin together with clades A, B, and C seems possible, but separate origins from wolf cannot be ruled out. For clade D the situation is very different. This clade is found only in Europe and Turkey, and it therefore seems that it has an origin outside East Asia, independent from that of the other clades. There is a clear geographic division within clade D: mtDNA types D1–D4, which constitute a separate subcluster, were all found in Scandinavian spitz breeds, a total of 18 out of 49 Scandinavian dogs, whereas D5 and D6, which are several mutations away from D1–D4, were found in two Turkish and one Spanish dog, respectively (Fig. 2). The distance between the subclusters D1–D4 and D5–D6, 3 substitutions, corresponds to ~36,000 years. It therefore seems probable that clade D originates from at least two separate origins from wolf. The divergence within subcluster D1–D4, a single substitution in 18 individuals, indicates a recent origin, a few thousand years at most.

TRACES OF MIGRATIONS

In two recent studies, some light has been shed on the earliest migrations of the dog. The Australian dingo is a wild canid, morphologically resembling

South Asian domestic dogs, which, according to the archaeological record, arrived in Australia between 3,500 and 12,000 ybp (Clutton-Brock 1995). However, the precise ancestry and time of arrival, and whether it originally was domesticated, or half-domesticated becoming feral, or a truly wild dog, is not known. In a study of 211 dingoes, sampled in all states of Australia, 582 bp of mtDNA was analyzed and compared to dog and wolf data, giving very distinct results (Savolainen et al. 2004). All dingo sequences clustered in dog clade A, forming a single internal cluster of mtDNA types unique to dingoes, around a central type, A29, found in both dingoes and East Asian dogs (Fig. 4). This indicates that the dingo population originates from a small population of domestic dogs coming from East Asia carrying type A29. The mean distance among the dingo sequences to

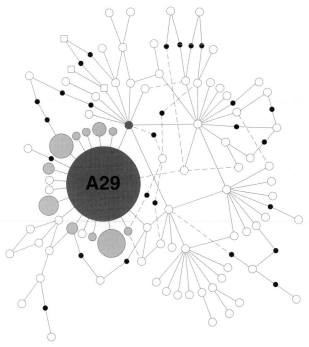

Figure 4. Minimum-spanning network of the main dog clade (clade A). mtDNA types (*circles*) and hypothetical intermediates (*solid dots*) are separated by one mutational step (substitutions, indels not shown). White circles are mtDNA types found in dogs only; light gray circles are types found in dingo only; dark gray circles are types found in both dingo and dog. Squares denote wolves. Areas of gray circles are proportional to frequencies among dingoes, but the area of A29 is reduced by 50%. (Reprinted, with permission, from Savolainen et al. 2004 [©National Academy of Sciences].)

A29 indicates their arrival to Australia ~5,000 ybp. In a study of European dog breeds (Angleby and Savolainen 2005), some patterns in the mtDNA variation were found between different regions of Europe. For example, a comparison of dog breeds from the northern and the southern parts of the European continent showed clear differences in the occurrence of mtDNA types. This suggests that dogs were introduced to the southern and northern parts of Europe by different routes, and indicates that, with a denser sampling of dogs, information about the early population history of the domestic dog may be obtained from further mtDNA studies.

SUMMARY

Studies of mtDNA variation have taken the knowledge about the origin of the domestic dog a large step forward. The universal presence of three groups of mtDNA types, phylogenetic clades A, B, and C, shows that all dogs originate from a common gene pool and indicates that they originated at a single domestication event. A larger genetic variation in East Asia than in other parts of the world suggests that the geographic region where this happened lies somewhere in East Asia. Two later instances of crossbreeding between dog and wolf seem to have occurred in Europe. The available mtDNA data do not give resolution enough to precisely determine the date for the origin of the dog, but the archaeological record indicates an origin ~15,000 ybp, a date which is not contradicted by the mtDNA data. No clear correlations between the breed or morphological type of dogs and their mtDNA type have been found so far. However, with an increasingly denser mapping of the mtDNA variation around the world, mtDNA studies have a potential to give information about the first migrations of dogs, and possibly also details about the development of morphological types and breeds. Together with studies of Y-chromosomal and autosomal haplotypes, which will no doubt be initiated within a few years, we may anticipate a detailed picture of the development of the domestic dog.

REFERENCES

Angleby H. and Savolainen P. 2005. A study of the forensic usefulness of the mitochondrial DNA variation among and within populations, breeds and types of domestic dogs. *Forensic Sci. Int.* (in press).

Cavalli-Sforza L.L. and Feldman M.W. 2003. The application of molecular genetic approaches to the study of human evolution. *Nat. Genet.* (suppl.) **33:** 266–275.

Clutton-Brock J. 1995. Origins of the dog: Domestication and early history. In *The domestic dog, its evolution, behaviour, and interactions with people* (ed. J. Serpell), pp. 7–20. Cambridge University Press, Cambridge, United Kingdom.

Clutton-Brock J. and Noe-Nygaard N. 1990. New osteological and C-isotope evidence on mesolithic dogs: Companions to hunters and fishers at Starr Carr, Seamer Carr and Kongemose. *J. Archaeol. Sci.* **17:** 643–653.

Garcia-Moreno J., Matocq M.D., Roy M.S., Geffen E., and Wayne R.K. 1996. Relationship and genetic purity of the endangered Mexican wolf based on analysis of microsatellite loci. *Conserv. Biol.* **10:** 376–389.

Houpt K.A. and Willis M.B. 2001. Genetics of behaviour. In *The genetics of the dog* (ed. A. Ruvinsky and J. Sampson), pp. 371–400. CABI Publishing, New York.

Kurtén B. 1968. *Pleistocene mammals of Europe.* Aldine, Chicago.

Kurtén B. and Anderson E. 1980. *Pleistocene mammals of North America.* Columbia University Press, New York.

Leonard J.A., Wayne R.K., Wheeler J., Valadez R., Guillen S., and Vila C. 2002. Ancient DNA evidence for Old World origin of New World dogs. *Science* **298:** 1613–1616.

Li Q. 1990. The history of dog breeding in China (in Chinese). *Chin. J. Cynology* **2:** 25–27.

Nobis G. 1979. Der älteste Haushund lebte vor 14,000 Jahren. *Umschau* **79:** 610.

Nowak R.M. 1978. Evolution and taxonomy of coyotes and related *Canis.* In *Coyotes: Biology, behaviour and management* (ed. M. Bekof), pp. 3–16. Academic Press, New York.

———. 1996. Another look at wolf taxonomy. In *Ecology and behavior of wolves in a changing world* (ed. L.N. Carbyn et al.), pp. 375–397. Canadian Circumpolar Institute, Edmonton.

Okumura N., Ishiguro N., Nakano M., Matsui A., and Sahara M. 1996. Intra- and interbreed genetic variations of mitochondrial DNA major non-coding regions in Japanese native dog breeds (*Canis familiaris*). *Anim. Genet.* **27:** 397–405.

Olsen S.J. 1985. *Origins of the domestic dog.* University of Arizona Press, Tucson, Arizona.

Sablin M.V. and Khlopachev G.A. 2002. The earliest ice age dogs: Evidence from Eliseevichi 1. *Curr. Anthropol.* **43:** 795–798.

Savolainen P., Zhang Y., Luo J., Lundeberg J., and Leitner T. 2002. Genetic evidence for an East Asian origin of the domestic dog. *Science* **298:** 1610–1613.

Savolainen P., Leitner T., Wilton A., Matisoo-Smith E., and Lundeberg J. 2004. A detailed picture of the origin of the Australian dingo, obtained from the study of mitochondrial DNA. *Proc. Natl. Acad. Sci.* **101:** 12387–12390.

Scott J.P. and Fuller J.L. 1965. *Genetics and the social behavior of the dog.* University of Chicago Press, Illinois.

Tcherncov E. and Valla F.R. 1997. Two new dogs, and other Natufian dogs, from the southern Levant. *J. Arch. Sci.* **24:** 65–95.

Tsuda K., Kikkawa Y., Yonekawa H., and Tanabe Y. 1997. Extensive interbreeding occurred among multiple matriarchal ancestors during the domestication of dogs: Evidence from inter- and intraspecies polymorphisms in the D-loop region of mitochondrial DNA between dogs and wolves. *Genes Genet. Syst.* **2:** 229–238.

Vila C., Maldonado, J., and Wayne R.K. 1999a. Phylogenetic relationships, evolution, and genetic diversity of the domestic dog. *J. Hered.* **90:** 71–77.

Vila C., Amorim I.R., Leonard J.A., Posada D., Castroviejo J., Petrucci-Fonseca F., Crandall K.A., Ellegren H., and Wayne R.K. 1999b. Mitochondrial DNA phylogeography and population history of the grey wolf *Canis lupus. Mol. Ecol.* **8:** 2089–2103.

Vila C., Savolainen P., Maldonado J.E., Amorim I.R., Rice J.E., Honeycutt R.L., Crandall K.A., Lundeberg J., and Wayne R.K. 1997. Multiple and ancient origins of the domestic dog. *Science* **276:** 1687–1689.

Wayne R.K. 1986a. Cranial morphology of domestic and wild canids: The influence of development on morphological change. *Evolution* **4:** 243–261.

———. 1986b. Limb morphology of domestic and wild canids: The influence of development on morphological change. *J. Morphol.* **187:** 301–319.

Wayne R.K. and O'Brien S.J. 1987. Allozyme divergence within the Canidae. *Syst. Biol.* **36:** 339–355.

Wayne R.K., Lehman N., Allard M.W., and Honeycutt R.L. 1992. Mitochondrial DNA variability of the gray wolf: Genetic consequences of population decline and habitat fragmentation. *Conserv. Biol.* **6:** 559–569.

9

Understanding Genetic Relationships among Purebred Dogs: The PhyDo Project

Heidi G. Parker, Nathan B. Sutter, and Elaine A. Ostrander
Cancer Genetics Branch
National Human Genome Research Institute
National Institutes of Health
Bethesda, Maryland 20892-8000

With the recent availability of the dog genome sequence (see Chapter 11), there is growing recognition of the domestic dog's importance as a system for identifying genes that control basic aspects of mammalian development. The growing popularity of the dog stems, in part, from its unique history and population structure. Centuries of intense selective breeding have produced more than 400 recognized breeds of dogs, whose members are characterized by dramatic differences in morphology, behavior, and disease susceptibility (Wayne and Ostrander 1999; Chase et al. 2002). Whereas the variation observed between breeds is optimal for studies of population structure, uniformity within breeds makes the dog ideal for identifying and mapping genes involved in complex traits, particularly those associated with disease susceptibility and progression (Ostrander and Giniger 1997; Galibert et al. 1998; Ostrander et al. 2000; Sutter and Ostrander 2004).

Key to our utilization of the canine system is a deeper understanding of how the 400 recognized breeds were created and how their morphological and behavioral features are selectively maintained. Ideally, such an understanding should derive not from historical lore, but from objective genetic analysis. In this chapter, we summarize recent findings about the population structure of the domestic dog, outline approaches for further elucidation of breed relationships, and describe how knowledge of

The Dog and Its Genome ©2006 Cold Spring Harbor Laboratory Press 0-87969-742-3

population structure is a vital component in the search for the genetic variants that give rise to breed-specific morphology, behavior, and disease susceptibility.

PUREBRED DOGS AND DISEASE

According to fossil records, dogs and humans have coexisted for about 17,000 years (Sablin and Khlopachev 2002). Studies of canine mitochondrial DNA suggest that the divergence of the dog from the gray wolf took place 15,000–100,000 years ago (Vila et al. 1997; Savolainen et al. 2002). For at least the past 3,000–4,000 years, dogs of variable size and shape have existed, but the breeds as we know them today, with narrowly defined morphologies, are largely a product of the "breed boom" that occurred in Europe during the 18th and 19th centuries (American Kennel Club 1998; Fogle 2000). That period saw the formation of recognized breed clubs, the advent of breed standards, and the development of breed barriers, which have resulted in the very diverse set of breeds in existence today (American Kennel Club 1998).

The "breed barrier rule" has been instrumental in defining the population structure of most modern breeds. Simply put, no dog may become a registered member of a breed unless both its sire and dam are members of the same breed (American Kennel Club 1998). As a result, gene flow between breeds is almost entirely prevented, since no dog may legitimately immigrate into a pure breed. At best, each breed can merely maintain the level of genetic diversity brought to the breed by the animals that founded it. Furthermore, restrictive breeding practices reduce the effective population size of each breed, making it possible for genetic drift to turn a new mutation or a rare allele into a common one within that breed. As we discuss, the health consequences of this can be quite striking.

In addition to a restricted gene pool, strong selection to produce animals of uniform appearance, and displaying similar behavior patterns, may have contributed to the high incidence of genetic disease in many breeds. Driving this selection by size, shape, and behavior is the "breed standard;" a formalized description of each breed, drafted by their respective breed clubs (American Kennel Club 1998). See for example, Figure 1. All members of a breed are judged by conformity to the standard, and the reproductive success of any single purebred dog is almost wholly determined by this evaluation. In addition, many breeds have experienced bottlenecks or changes in popularity that have affected the resulting population structure. The above factors, combined with world events

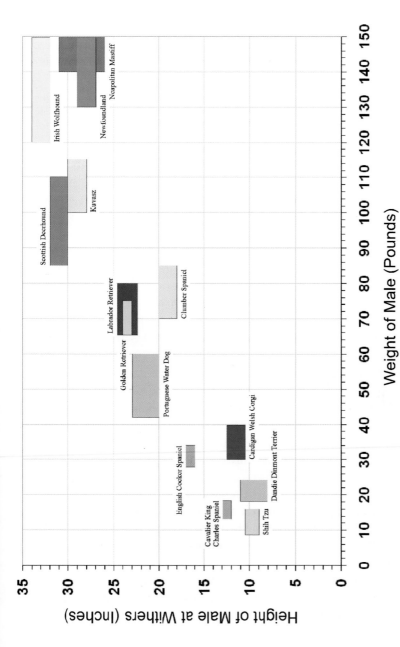

Figure 1. For many breeds a precise range of allowable heights and weights is specified for both males and females. Other features of body shape, length, head shape, etc., are also specified. Frequently breed-specific morphological features are absolutely required (e.g, the ridge of reversed hairs on a Rhodesian Ridgeback, or the coat curls of a curly-coated retriever). Failure to fit within these parameters leads to penalization in conformation competitions, which in turn severely reduces the nonconforming individuals' chance to reproduce.

limiting the number of founders for some breeds, together with the continued overuse of popular sires, i.e., dogs that perform well in shows, has led to allele distortions that contribute to the high frequency of genetic disease in domestic dogs (Patterson et al. 1982; Ostrander and Kruglyak 2000; Patterson 2000). The identification of susceptibility alleles for a myriad of diseases, many of which show specificity for one or a few breeds, is now a major focus of the canine genetics community.

HETEROZYGOSITY WITHIN AND BETWEEN BREEDS

Most U.S. and European breed clubs maintain extensive pedigree records chronicling the genealogy of each breed. However, such records only chronicle what has happened since each breed was established, and as such, provide little or no information about the creation of the breed and the events that took place prior to official recognition, which will have affected the amount of genetic variation in the population. We have therefore used molecular markers to determine levels of heterogeneity both within and between breeds. Several previous studies examined breed-to-breed differences in allele patterns using small sets of markers in studies that encompassed as few as 3 and as many as 28 breeds. In all cases, some variation in heterozygosity and allele frequency was noted. Deviation from Hardy-Weinberg equilibrium was also frequently observed (Zajc et al. 1997; Koskinen and Bredbacka 2000; Brouillette and Venta 2002; Irion et al. 2003).

Several studies have also addressed the issue of breed relatedness. Koskinen (2003) measured phylogenetic distances between five breeds and found that they were much larger than comparable distances between human subpopulations. In the same study, microsatellite allele frequencies were used to correctly assign individual dogs to their correct breed, indicating that distinctive patterns of genetic variation existed for each of the five breeds. These data support the expectation that there is considerably less variation within breeds than between breeds. Moreover, perhaps the genetic differentiation between breeds, if measured with sufficient resolution, could be used to identify nearly any dog's breed membership.

To address this challenge, we first undertook a detailed examination of heterozygosity between breeds. We discovered and genotyped 75 single-nucleotide polymorphisms (SNPs) by resequencing a set of BAC ends in at least two dogs from 60 breeds (Parker et al. 2004). We observed a greater than fourfold range of heterozygosity across the breeds (Fig. 2). The mean heterozygosity across all 120 dogs, without regard to breed

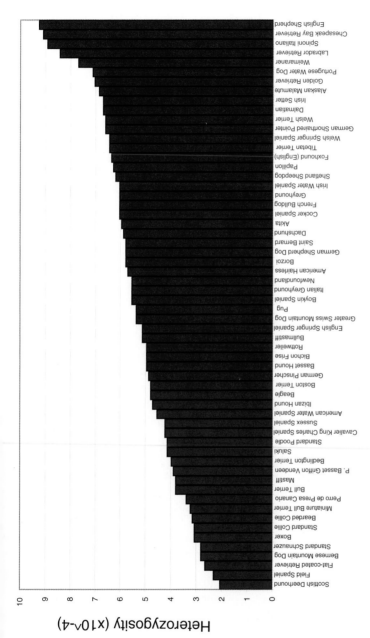

Figure 2. Heterozygosity of 60 breeds calculated from 75 SNPs. End sequence obtained from 60 BACs was resequenced in 120 dogs, 2 from each of the 60 breeds listed. SNPs were identified and genotyped by aligning sequence reads from all dogs and looking for single base changes (insertion deletion polymorphisms were also identified but not included in the heterozygosity calculation). 75 SNPs were identified and heterozygosity was calculated at each locus and averaged over all 75 SNPs. The average heterozygosity within a breed was 5.3×10^{-4} with a range of 2.1×10^{-4} to 9.3×10^{-4}.

membership, was very similar to that observed in humans, 8.0×10^{-4} (Sachidanandam et al. 2001; Venter et al. 2001). By comparison, the mean heterozygosity within individual breeds averaged 5.3×10^{-4}. Calculations of genetic variation showed that 27% of variation in domestic dogs is found between the breeds, in contrast to the 5–10% of variation observed between human populations (Parker et al. 2004). These data suggest that breeds are both well-isolated and highly structured populations. This view agrees with the historical evidence, which describes, for most breeds, founding events by small numbers of animals, other population bottlenecks, and often the use of small numbers of animals for breeding stock. All of these forces reduce heterozygosity within breeds but do not reduce levels of heterozygosity within the dog population as a whole. On the strength of these data, we next applied these data to understanding the hierarchical relationships between breeds.

UNDERSTANDING BREED RELATIONSHIPS

Assignment, phylogenetic analysis, and clustering are three approaches we have used to deconstruct the relationship between dog breeds (Parker et al. 2004). Initially, 96 microsatellite markers spanning all canine autosomes at approximately 30 Mb resolution were genotyped on DNA samples isolated from a panel of 85 dog breeds. Each breed was represented by five dogs, and all dogs were unrelated to one another at the grandparent level. Thus, multiple lineages within each breed were assayed.

Assignment methods were used first to test whether any given individual belonged to a defined "group." Excluding data from Belgian sheepdog and Belgian Tervuren, the above set of 96 microsatellite markers could be used to correctly assign 99% of dogs to their breed (Table 1). The method of assignment involved a "leave one out" analysis implemented by the *Doh* assignment calculator. Each dog's allele patterns are tested in turn against those of all remaining dogs in the data set. In only

Table 1. 414 dogs assigned to 85 breeds

	Correct assignment	Incorrect assignment
Number of individuals	405	9
Percent of data set	98%	2%
Percent assuming single Belgian breed	99%	1%

Excluding Belgian sheepdogs, just 4 dogs were assigned incorrectly, listed here as actual breed (breed assignment): beagle (Perro de Presa Canario), Chihuahua (cairn terrier), and two German shorthaired pointers (Kuvasz and standard poodle). Two Belgian sheepdogs were assigned as Belgian Tervurens and three Belgian Tervurens were assigned as Belgian sheepdogs.

4 of 414 dogs was the analysis method unable to correctly match an allele set with the appropriate breed. Incorrect assignment of the Belgian sheepdog as Tervuren and vice versa was not considered a failure of the method but, rather, a predicted outcome. These two breeds are considered coat-color variants of a single breed collectively known as the Belgian sheepdog outside of the U.S. and have only been recognized as two separate breeds within the U.S. since 1959 (Wilcox and Walkowicz 1995; American Kennel Club 1998). The assignment test therefore clearly demonstrates that established dog breeds are well-isolated populations with distinct patterns of microsatellite allele frequencies.

Phylogenetic analysis, although typically reserved for establishing relationships between species, can also be used to identify population groups within a species. Typically, because phylogenetic analysis deals with species, it does not allow for hybridization between the populations examined. However, based on calculations of genetic distance between individuals, a neighbor-joining tree can display groupings that represent subspecies or populations, with the assumption that individuals of the same population will have genotypes more similar to each other than to those of different populations (Cornuet et al. 1999). For example, Zhivotovsky et al. (2003) built trees describing human population divisions that were based on estimations of divergence time taken from microsatellite genotypes. Three of the previously mentioned microsatellite studies in dogs attempted phylogenetic analyses but with only limited success (Zajc et al. 1997; Koskinen and Bredbacka 2000; Irion et al. 2003). Koskinen (2003) used neighbor-joining trees to group dogs according to breed, but his studies only involved five breeds. Irion et al. (2003) were able to deduce the relationships between only two pairs of breeds out of 28 they studied and were not able to establish a significant tree structure.

In our experiments we also utilized phylogenetic methods to try to understand the relationships between breeds. When distance measures and tree-building methods were applied to microsatellite data from 85 breeds, representing all seven AKC-specified groups, we observed significant branching of 9 of the breeds (Fig. 3). These 9 breeds shared one striking feature: None were of European origin. In addition, 12 breeds showed significant pairing one to another, and a triplet of breeds also grouped together, suggesting very recent divergence. These included, for instance; the collie and Shetland sheepdog, which is a miniature version of the collie, and the three Asian lap dogs: shih tzu, Lhasa apso, and Pekingese. Interestingly, the majority of breeds on the canine phylogenetic tree appeared to stem from a single node without significant branch

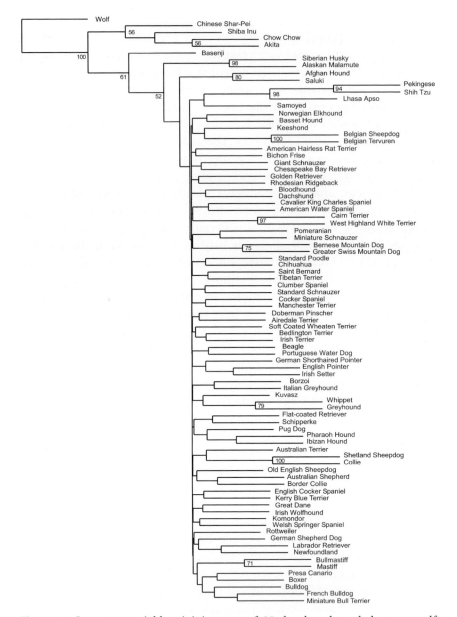

Figure 3. Consensus neighbor-joining tree of 85 dog breeds and the gray wolf. The trees that formed the consensus are based on the chord distance measure. 500 bootstrap replicates of the data were carried out, and the fraction of bootstraps supporting each branch is indicated at the corresponding node as a percentage. Only those branches supported in >50% of the replicates are labeled. Branch lengths are proportional to bootstrap values. (Adapted, with permission, from Parker et al. 2004 [©AAAS].)

structure. This flat topology, or "hedge" of branches, is indicative of hybridization between the breeds.

The final analysis method, clustering, is not restricted by the need for a training set of genotypes to designate a population. Populations are determined solely by the genotypes rather than by any subjective classification schema, such as breed identity or morphological or functional grouping (Pritchard et al. 2000). We used the computer program *structure* to implement a Bayesian model-based clustering algorithm that attempts to identify genetically distinct subpopulations based on patterns of allele frequencies (Pritchard and Rosenberg 1999; Pritchard et al. 2000). The approach has proven effective in defining human, chicken, and bacterial populations; identifying hybrids in chicken breeds; and elucidating the relationships between populations, both human and bacterial (Rosenberg et al. 2001, 2002; Falush et al. 2003). *Structure* was applied to the same set of genotypes as described above, derived from 96 microsatellites and 414 dogs representing 85 breeds. When asked to pick a maximum number of populations in which to assign the 414 individual dogs, the algorithm placed 335 dogs in 69 unique breed-specific clusters. Twenty dogs representing 4 breeds did not form perfect breed clusters. The last 59 dogs were placed into six clusters, each composed of 2 historically related breeds: the Bernese mountain dog and Greater Swiss mountain dog, collie and Shetland sheepdog, greyhound and whippet, Alaskan malamute and Siberian husky, mastiff and bullmastiff, and the Belgian sheepdog and Belgian Tervuren. The Bernese mountain dog and Greater Swiss mountain dog are two of four distinctive farm-type dogs of Swiss origin, both of which were saved from extinction and revitalized by the same breeder in the late 1800s (Fogle 2000). The greyhound/whippet and collie/Shetland sheepdog pairing features breeds distinguishable primarily by size. The whippet and Shetland sheepdogs appear to be miniaturized versions of the greyhound and collie, respectively. An additional pairing included the Alaskan malamute and Siberian husky, both of which are popular sled dogs of similar appearance and shared origin according to breed histories (Wilcox and Walkowicz 1995; Fogle 2000). The last pairing was unsurprising; the bullmastiff is reported to be 60% mastiff and 40% bulldog and was created by crossing the two breeds in the mid 1800s (Rogers and Brace 1995). All six of these closely related pairs were also identified by phylogenetic analysis, and one was predicted by assignment as well. It is worth noting that despite the close genetic relationships between these pairs of breeds, the individual breeds could nevertheless be readily distinguished for five of the pairs when the dogs from just the two breeds were analyzed apart from the other dogs.

As with the assignment method, clustering could not separate the two Belgian breeds. The inability of clustering algorithms to distinguish the two Belgian breeds, while clearly separating the first five pairs, supports, as breeders have recorded, a more distant relationship between the former than between the Belgian breeds.

We next asked whether this same approach could be used to determine more distant relationships between breeds. By minimizing the number of groups the dogs could be sorted into, we successfully deduced the higher-level relationships between most breeds. When the entire set of dogs is assigned membership in two, three, and finally four groups, four reproducible clusters were discerned. The first cluster repeats the finding of the phylogenetic analysis by grouping the Asian and sled dogs together and separating them from the rest of the breeds (Fig. 4). When added to the analysis, gray wolves from eight countries all group in the first cluster as well (Parker et al. 2004). The second cluster comprises guarding dogs, which mostly share a distinct mastiff-type morphology of broad heads and heavily muscled bodies. These include the mastiff, bullmastiff, bulldog, boxer, Bernese mountain dog, Greater Swiss mountain dog, and others. The third cluster places herding breeds such as the collie and Shetland sheepdog together with a subset of the sighthounds, including the greyhound. This supports reports that greyhounds were popular in Europe in the Middle Ages and may have contributed to the development of later European breeds (Rogers and Brace 1995; Wilcox and Walkowicz 1995). The remaining cluster comprises mostly modern breeds used in hunting, and includes gun dogs, hounds, and terriers.

We have described three different methods for analyzing a genomic data set to determine population structure. With each method we have arrived at complementary but distinct conclusions depicting different levels of relationship, from the nearly identical Belgian breeds identified by all three methods to the slightly amorphous breed groupings recognized only by cluster analysis. The early divergence of the Asian breeds on the phylogenetic tree and their association with the wolf in clustering analysis support the conclusions of mitochondrial DNA analysis that domestication first took place in Asia (Savolainen et al. 2002). Furthermore, a recent mutation analysis, described in further detail below, identified haplotypes that were identical by descent in a set of herding and sighthound breeds, supporting our clustering schema for these two breed groups (Neff et al. 2004). Additional studies are currently under way to refine our understanding of all these issues. The addition of breeds to fill out underrepresented groups may help to identify new breed clusters, particularly within the modern hunting set. In addition, by using more

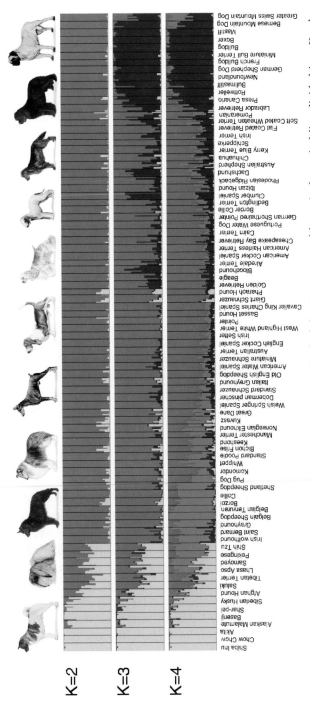

Figure 4. Population structure of 85 domestic dog breeds. Each individual dog is represented by a single vertical line divided into K colors where K is the number of clusters assumed. Each color represents one cluster, and the length of the colored segment shows the individual's estimated proportion of membership in that cluster. Thin black lines separate the breeds that are labeled below the figure. Pictures of dogs from representative breeds for each grouping are shown at the top. Results shown are averages over 15 *structure* runs at each value of K. (Reprinted, with permission, from Parker et al. 2004 [©AAAS].)

highly mutable tetranucleotide microsatellite markers (Francisco et al. 1996), or less mutable SNPs and insertion/deletion markers, we may be able to build phylogenies that, when compared to the dinucleotide microsatellite results presented here, will better define the temporal relationships between breeds.

One of the major hindrances in understanding the historical relationships between dog breeds is the existence of hybrid populations; i.e., those made up of crosses between two or more existing populations. This is a common occurrence in the development of new breeds as breeders attempt to combine the most desirable phenotypes from a variety of breeds. One way to navigate these difficulties and reduce the complexity of mixed genomes is to look at only one region of the genome at a time. This approach was utilized to look at a specific multi-drug resistance gene (MDR1) mutation and four microsatellite markers to which it is closely linked (Neff et al. 2004). The MDR1 mutation was found to segregate in nine breeds, seven herding breeds and two new variations of sighthound, with probable outcrosses to at least one of the herding breeds. Haplotype analysis of the markers revealed that the mutation was identical by descent in all nine breeds and was therefore acquired from a common ancestor. Each region of the genome, examined independently, could be expected to provide a distinct phylogeny depending on the function of genes in the region and the selective pressures applied to them (Templeton et al. 2000). Perhaps a number of independent regions analyzed separately and then combined would provide an overall look at the history of each breed.

LINKAGE DISEQUILIBRIUM AND HAPLOTYPE SHARING

Understanding canine population structure and selection will facilitate the cloning of disease genes of interest for both human and companion-animal health. The existence of well-defined populations assists the mapping effort by reducing the genetic complexity of the disorder, under the assumption that an isolated subpopulation will demonstrate less genetic heterogeneity than the entire population (Kruglyak 1999b; Sundin et al. 2000; Ophoff et al. 2002). Investigators have frequently gone to great lengths to identify small, isolated populations in order to map disease genes in humans. Particular examples include the use of isolated island populations to map genes for achromatopsia (Winick et al. 1999), studies of Amish to identify genes for mental illness (Egeland and Hostetter 1983), and analysis of Ashkenazi Jewish and Icelandic populations to better understand the penetrance of the *BRCA1* and *BRCA2* breast cancer susceptibility genes (Neuhausen 1999) and to identify prostate cancer genes (Friedrichsen et al. 2004). However, in human populations such rare isolated cohorts are unusual. In canine genetics, a multitude of well-isolated populations with restricted gene flow are readily available.

In a recent small-scale study, we measured the extent of linkage disequilibrium (LD) between SNPs in five dog breeds, each with a distinct population history (Sutter et al. 2004). In each breed, the value of D', a statistical measure of LD, decreases with greater distance between markers. In golden retrievers and Labrador retrievers, D' falls to half of its maximum at a distance between SNPs of approximately 400–750 kb. In Bernese mountain dog, Pekingese, and Akita, the distance at which D' is half of maximum is 2–3.5 Mb. In the U.S., each of these three breeds underwent population bottlenecks and was founded by relatively few animals. Both of these factors help to explain the very extensive LD observed. In comparison, the golden and Labrador retrievers are the two most popular U.S. breeds, and their populations are not thought to have experienced bottlenecks over the past century.

These results are interesting for two reasons. First, they demonstrate that there is at minimum a tenfold difference in extent of LD between dog breeds. This has strong implications for mapping strategies. In mapping common disease genes, it makes sense to consider carefully in what breed the study should be carried out, as the breed choice alone may increase or decrease the number of SNPs that need to be genotyped by tenfold.

Second, these LD values are 20–50 times more extensive than what is observed in studies of human LD, even for isolated human populations (Reich et al. 2001; Weiss and Clark 2002). It has been estimated that upward of 500,000 SNPs must be genotyped for a typical whole-genome association study in humans (Kruglyak 1999a; International HapMap Consortium 2003). This suggests that a similarly designed whole-genome association scan in dogs would require only about 10,000–15,000 SNPs. Thus, it may be more economical to tackle the mapping of common diseases like lymphoma, osteosarcoma, heart disease, deafness, and autoimmune disease in dogs rather than in humans.

Finally, we also examined the extent of haplotype sharing between loci in these five breeds. Approximately 8% of all haplotypes identified were found in all five breeds. Few private haplotypes were found, and the overall degree of haplotype sharing, measured as the proportion of a breed's chromosomes carrying haplotypes shared with another breed, ranged from 46% to 84%. In accordance with previous findings that separate the Asian breeds from those of European descent, the Akita was

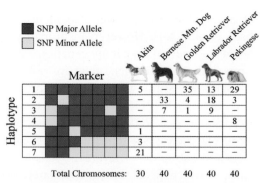

Figure 5. Haplotype sharing across breeds. Most chromosomes within breeds carry haplotypes shared with other breeds. Shown on the left are inferred haplotypes across a set of seven SNPs that span ~50 kb and are in linkage disequilibrium with one another. The number of chromosomes carrying the inferred haplotype within each breed is shown on the right. This locus is typical of the ten studied for haplotype diversity and sharing, in that some haplotypes are very common and are shared among most or all breeds assayed, whereas others are rare and possibly breed-specific.

found to share the fewest haplotypes with the other four breeds (Fig. 5). The largest number of shared haplotypes was found between the golden and Labrador retrievers, two breeds with possible shared ancestry dating back to the water dogs of Newfoundland (Fogle 2000). We observed that 80% of all the chromosomes carried an average of just 4.5 haplotypes for the sample set as a whole and only 2–3 haplotypes within each breed. These findings suggest that an examination of shared haplotypes across the genome, without the prior knowledge of specific gene associations, could be used to identify populations and establish relationships between breeds as well. They suggest also that because we observe low haplotype diversity and high haplotype sharing, a single universal SNP set can be used for whole-genome association studies in most dog breeds.

CONCLUSIONS AND FUTURE AIMS

The domestic dog has accompanied man on many journeys of discovery. Perhaps we have been together for so long that we have taken the phenotypic and behavioral diversity of the dog for granted, not considering what changes must have occurred to take a fierce predator and turn it into a docile bed warmer or a loyal family guardian. With the genomic advances that have taken place in the last decade and our new understanding of the relationships between the breeds, we are now in a position to address these changes directly and to begin identifying the genes responsible. These findings will have far-reaching implications from providing clues regarding possible genetic components of human behaviors to identifying genetic changes that may have led to speciation events in natural populations.

As we continue our analysis of genome structure and content (see Chapter 11), we once again face an exciting quest that will change the way we look at ourselves and our surroundings. Mapping disease susceptibility genes has been the major focus of the canine genetics community and will rightly remain a prominent focus in the future. For diligent gene hunters, however, the dog genome holds a treasure trove of information that will enlighten and inform us about the development of our own species. As we move from the realm of intellectual comfort, finding disease genes, to the search for genes important in appearance, and eventually behavior, we will be forced to think again about the degree to which we, or our genes, guide our destiny. It is somehow fitting, or at least comforting, that we take this difficult journey with our closest companion, as ever, at our side.

REFERENCES

American Kennel Club. 1998. *The complete dog book. Official publication of the American Kennel Club* (ed. J. Crowley and B. Adelman). Howell Book House, New York.

Brouillette J.A. and Venta P.J. 2002. Within-breed heterozygosity of canine single nucleotide polymorphisms identified by across-breed comparison. *Anim. Genet.* **33:** 464–467.

Chase K., Carrier D.R., Adler F.R., Jarvik T., Ostrander E.A., Lorentzen T.D., and Lark K.G. 2002. Genetic basis for systems of skeletal quantitative traits: Principal component analysis of the canid skeleton. *Proc. Natl. Acad. Sci.* **99:** 9930–9935.

Cornuet J.M., Piry S., Luikart G., Estoup A., and Solignac M. 1999. New methods employing multilocus genotypes to select or exclude populations as origins of individuals. *Genetics* **153:** 1989–2000.

Egeland J.A. and Hostetter A.M. 1983. Amish study, I: Affective disorders among the Amish, 1976–1980. *Am. J. Psychiatry* **140:** 56–61.

Falush D., Wirth T., Linz B., Pritchard J.K., Stephens M., Kidd M., Blaser M.J., Graham D.Y., Vacher S., Perez-Perez G.I., et al. 2003. Traces of human migrations in *Helicobacter pylori* populations. *Science* **299:** 1582–1585.

Fogle B. 2000. *The new encyclopedia of the dog.* Dorling Kindersley, New York.

Francisco L.V., Langston A.A., Mellersh C.S., Neal C.L., and Ostrander E.A. 1996. A class of highly polymorphic tetranucleotide repeats for canine genetic mapping. *Mamm. Genome* **7:** 359–362.

Friedrichsen D.M., Stanford J.L., Isaacs S.D., Janer M., Chang B.L., Deutsch K., Gillanders E., Kolb S., Wiley K.E., Badzioch M.D., et al. 2004. Identification of a prostate cancer susceptibility locus on chromosome 7q11-21 in Jewish families. *Proc. Natl. Acad. Sci.* **101:** 1939–1944.

Galibert F., Andre C., Cheron A., Chuat J.C., Hitte C., Jiang Z., Jouquand S., Priat C., Renier C., and Vignaux F. 1998. The importance of the canine model in medical genetics. *Bull. Acad. Natl. Med.* **182:** 811–821.

International HapMap Consortium. 2003. The International HapMap Project. *Nature* **426:** 789–796.

Irion D.N., Schaffer A.L., Famula T.R., Eggleston M.L., Hughes S.S., and Pedersen N.C. 2003. Analysis of genetic variation in 28 dog breed populations with 100 microsatellite markers. *J. Hered.* **94:** 81–87.

Koskinen M.T. 2003. Individual assignment using microsatellite DNA reveals unambiguous breed identification in the domestic dog. *Anim. Genet.* **34:** 297–301.

Koskinen M.T. and Bredbacka P. 2000. Assessment of the population structure of five Finnish dog breeds with microsatellites. *Anim. Genet.* **31:** 310–317.

Kruglyak L. 1999a. Prospects for whole-genome linkage disequilibrium mapping of common disease genes. *Nat. Genet.* **22:** 139–144.

———. 1999b. Genetic isolates: Separate but equal? *Proc. Natl. Acad. Sci.* **96:** 1170–1172.

Neff M.W., Robertson K.R., Wong A.K., Safra N., Broman K.W., Slatkin M., Mealey K.L., and Pedersen N.C. 2004. Breed distribution and history of canine *mdr1-1Δ*, a pharmacogenetic mutation that marks the emergence of breeds from the collie lineage. *Proc. Natl. Acad. Sci.* **101:** 11725–11730.

Neuhausen S.L. 1999. Ethnic differences in cancer risk resulting from genetic variation. *Cancer* **86:** 1755–1762.

Ophoff R.A., Escamilla M.A., Service S.K., Spesny M., Meshi D.B., Poon W., Molina J., Fournier E., Gallegos A., Mathews C., et al. 2002. Genomewide linkage disequilibrium mapping of severe bipolar disorder in a population isolate. *Am. J. Hum. Genet.* **71:** 565–574.

Ostrander E.A. and Giniger E. 1997. Semper fidelis: What man's best friend can teach us about human biology and disease. *Am. J. Hum. Genet.* **61:** 475–480.

Ostrander E.A. and Kruglyak L. 2000. Unleashing the canine genome. *Genome Res.* **10:** 1271–1274.

Ostrander E.A., Galibert F., and Patterson D.F. 2000. Canine genetics comes of age. *Trends Genet.* **16:** 117–123.

Parker H.G., Kim L.V., Sutter N.B., Carlson S., Lorentzen T.D., Malek T.B., Johnson G.S., DeFrance H.B., Ostrander E.A., and Kruglyak L. 2004. Genetic structure of the purebred domestic dog. *Science* **304:** 1160–1164.

Patterson D. 2000. Companion animal medicine in the age of medical genetics. *J. Vet. Intern. Med.* **14:** 1–9.

Patterson D.F., Haskins M.E., and Jezyk P.F. 1982. Models of human genetic disease in domestic animals. *Adv. Hum. Genet.* **12:** 263–339.

Pritchard J.K. and Rosenberg N.A. 1999. Use of unlinked genetic markers to detec population stratification in association studies. *Am. J. Hum. Genet.* **65:** 220–228.

Pritchard J.K., Stephens M., and Donnelly P. 2000. Inference of population structure using multilocus genotype data. *Genetics* **155:** 945–959.

Reich D.E., Cargill M., Bolk S., Ireland J., Sabeti P.C., Richter D.J., Lavery T., Kouyoumjian R., Farhadian S.F., Ward R., and Lander E.S. 2001. Linkage disequilibrium in the human genome. *Nature* **411:** 199–204.

Rogers C.A. and Brace A.H., eds. 1995. *The international encyclopedia of dogs.* Howell Book House, New York.

Rosenberg N.A., Pritchard J.K., Weber J.L., Cann H.M., Kidd K.K., Zhivotovsky L.A., and Feldman M.W. 2002. Genetic structure of human populations. *Science* **298:** 2381–2385.

Rosenberg N.A., Burke T., Elo K., Feldman M.W., Freidlin P.J., Groenen M.A., Hillel J., Maki-Tanila A., Tixier-Boichard M., Vignal A., et al. 2001. Empirical evaluation of genetic clustering methods using multilocus genotypes from 20 chicken breeds. *Genetics* **159:** 699–713.

Sablin M.V. and Khlopachev G.A. 2002. The earliest ice age dogs: Evidence from Eliseevichi 1. *Curr. Anthropol.* **43:** 795–798.

Sachidanandam R., Weissman D., Schmidt S.C., Kakol J.M., Stein L.D., Marth G., Sherry S., Mullikin J.C., Mortimore B.J., Willey D.L., et al. 2001. A map of human genome sequence variation containing 1.42 million single nucleotide polymorphisms. *Nature* **409:** 928–933.

Savolainen P., Zhang Y.P., Luo J., Lundeberg J., and Leitner T. 2002. Genetic evidence for an East Asian origin of domestic dogs. *Science* **298:** 1610–1613.

Sundin O.H., Yang J.M., Li Y., Zhu D., Hurd J.N., Mitchell T.N., Silva E.D., and Maumenee I.H. 2000. Genetic basis of total colourblindness among the Pingelapese islanders. *Nat. Genet.* **25:** 289–293.

Sutter N.B. and Ostrander E.A. 2004. Dog star rising: The canine genetic system. *Nat. Rev. Genet.* **5:** 900–910.

Sutter N.B., Eberle M.A., Parker H.G., Pullar B.J., Kirkness E.F., Kruglyak L., and Ostrander E.A. 2004. Extensive and breed-specific linkage disequilibrium in *Canis familiaris. Genome Res.* **14:** 2388–2396.

Templeton A.R., Weiss K.M., Nickerson D.A., Boerwinkle E., and Sing C.F. 2000. Cladistic structure within the human *lipoprotein lipase* gene and its implications for phenotypic association studies. *Genetics* **156:** 1259–1275.

Venter J.C., Adams M.D., Myers E.W., Li P.W., Mural R.J., Sutton G.G., Smith H.O., Yandell M., Evans C.A., Holt R.A., et al. 2001. The sequence of the human genome. *Science* **291:** 1304–1351.

Vila C., Savolainen P., Maldonado J.E., Amorim I.R., Rice J.E., Honeycutt R.L., Crandall K.A., Lundeberg J., and Wayne R.K. 1997. Multiple and ancient origins of the domestic dog [see comments]. *Science* **276:** 1687–1689.

Wayne R.K. and Ostrander E.A. 1999. Origin, genetic diversity, and genome structure of the domestic dog. *Bioessays* **21:** 247–257.

Weiss K.M. and Clark A.G. 2002. Linkage disequilibrium and the mapping of complex human traits. *Trends Genet.* **18:** 19–24.

Wilcox B. and Walkowicz C. 1995. *Atlas of dog breeds of the world.* T.F.H. Publications, Neptune City, New Jersey.

Winick J.D., Blundell M.L., Galke B.L., Salam A.A., Leal S.M., and Karayiorgou M. 1999. Homozygosity mapping of the Achromatopsia locus in the Pingelapese. *Am. J. Hum. Genet.* **64:** 1679–1685.

Zajc I., Mellersh C.S., and Sampson J. 1997. Variability of canine microsatellites within and between different dog breeds. *Mamm. Genome* **8:** 182–185.

Zhivotovsky L.A., Rosenberg N.A., and Feldman M.W. 2003. Features of evolution and expansion of modern humans, inferred from genomewide microsatellite markers. *Am. J. Hum. Genet.* **72:** 1171–1186.

10

Karyotype and Chromosomal Organization

Matthew Breen and Rachael Thomas
Department of Molecular Biomedical Sciences
College of Veterinary Medicine
North Carolina State University
Raleigh, North Carolina 27606

THE GENOME OF THE DOMESTIC DOG (*Canis familiaris*) is divided into
78 chromosomes, comprising 38 pairs of acrocentric autosomes and
2 metacentric sex chromosomes. Historically, the high degree of morphological similarity shared by many of the autosomes has presented significant challenges to cytogeneticists. Over the past decade, the development
of reagents and resources for molecular cytogenetic analyses of the canine
genome has played a vital role in the generation of comprehensive and
effective genome and comparative maps for the dog. This chapter provides
a summary of the molecular cytogenetic studies that have been conducted on the domestic dog, leading to the development of standardized
chromosome nomenclature, comparative cytogenetic maps, integrated
genome maps, and the anchoring of the canine genome assembly. The
application of molecular cytogenetic approaches to the investigation of
chromosome aberrations in canine cancers is also discussed.

THE KARYOTYPE OF THE DOG

The chromosome number of the dog, 2n = 78, was first determined from
studies of meiotic cells (Minouchi 1928) and later confirmed using cultured lymphocytes (Gustavsson 1964). The diploid karyotype comprises
38 pairs of acrocentric autosomes, a large sub-metacentric X chromosome,
and a small metacentric Y chromosome (Fig. 1). At approximately 137 Mb,
the largest autosome, dog chromosome 1, is equivalent in size to human

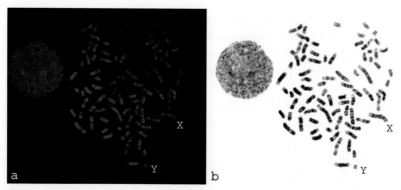

Figure 1. Metaphase chromosomes of the dog. (**a**) DAPI-stained metaphase preparation. (**b**) Inverted DAPI-stained metaphase preparation, revealing banding similar
to that of conventional GTG banding. The X and Y chromosomes are indicated.

chromosomes 9–12, and all but the four largest dog chromosomes are
smaller than human chromosome 19. Conventional Giemsa staining of
the chromosomes allows precise identification of only the sex chromosomes
(due to their size and morphology) and chromosome 1 (by virtue of its
size). Since the remaining autosomes gradually decrease in size, reliable
recognition of conventionally stained homologous pairs represents an
arduous task.

The development of chromosome banding techniques in the early
1970s provided the ability to identify homologous chromosome pairs and
thus presented the means to initiate the generation of a standard karyotype. The first GTG-banded karyotype of the dog (Selden et al. 1975)
was followed by a number of attempts over the subsequent 25 years to
produce reliable, complete karyotypes of the domestic dog using a variety
of conventional banding techniques such as GTG-banding (see, e.g.,
Manolache et al. 1976; Stone et al. 1991; Graphodatsky et al. 1995, 2000;
Reimann et al. 1996), R-banding (e.g., Howard-Peebles and Pryor 1980;
Mayr et al. 1986; Poulsen et al. 1990; Moreno-Millan et al. 1991), and also
fluorescence banding (Langford et al. 1996; Pienkowska and Switonski
1998; Breen et al. 1999a) and image analysis methodology (Christian
et al. 1998).

Despite the attempts of numerous investigators, it was evident that
conventional cytogenetic approaches, based on chromosome banding
patterns alone, would not permit the complete karyotype of the dog to
be associated with a standardized chromosome nomenclature. In an effort to
resolve this issue, the Committee for the Standardization of the Karyotype
of the Domestic Dog (CSKDD) was established, under the auspices of

the DogMap Workshop. Using conventional cytogenetics, the CSKDD was able to reach a consensus on the identification of dog chromosomes 1–21 (Switonski et al. 1996) and concluded that a complete standardized karyotype would require the use of molecular cytogenetic reagents, based on the application of fluorescence in situ hybridization (FISH) techniques. The use of high-resolution bivariate flow sorting of dog chromosomes facilitated the development of whole-chromosome paint probes (WCPP) for FISH analysis of the dog (Langford et al. 1996). The use of these WCPP by the CSKDD enabled all chromosomes to be identified conclusively and led to the development of a consensus chromosome nomenclature (Breen et al. 1998). The chromosome numbering recommended by the CSKDD was later endorsed by the International Society of Animal Genetics (ISAG) DogMap workshop held in Minneapolis, July 2000 (G. Dolf, pers. comm.).

With the development of sophisticated cooled charge coupled device (CCD) cameras and multicolor FISH analysis, an increasing number of laboratories began to adopt fluorochrome-based banding techniques (e.g., DAPI banding) to facilitate concurrent chromosome identification during FISH analyses. Sequential chromosome painting analysis of fluorescently labeled chromosome paint probes (Langford et al. 1996) onto single metaphases produced a series of complete DAPI-banded karyotypes of the dog in which the chromosome numbering followed that adopted by the CSKDD (Breen et al. 1999a). To facilitate the accurate assignment of FISH-mapped loci to dog chromosomes, a 460-band ideogram of the dog karyotype with five gray levels was also produced (Fig. 2) (Breen et al. 1999a). Independently, a second set of dog chromosome paint probes was generated (Yang et al. 1999) and a comparative chromosome painting approach (dog vs. red fox vs. human chromosomes) was applied to identify the dog chromosomes indirectly. The authors attempted to follow the established chromosome nomenclature and presented a DAPI banding pattern for the whole karyotype in the form of a black and white ideogram. However, it was indicated by Graphodatsky et al. (2000) that there were ten differences in chromosome nomenclature when compared to the paper of Breen et al. (1999a) and hence the recommendations of the CSKDD.

In conclusion, the standardization of the first 21 autosomes of the dog karyotype was achieved using conventional cytogenetics (Switonski et al. 1996), whereas reliable identification of the remaining 17 pairs of autosomes required the use of molecular cytogenetic reagents. The combined efforts of many researchers led to an internationally accepted numbering of all 38 autosomes of the dog (Switonski et al. 1996; Breen

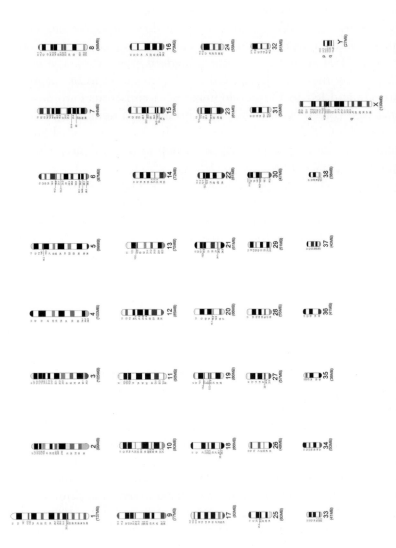

Figure 2. DAPI-banded ideogram of the domestic dog karyotype presented in five levels of gray shading. This is widely used to describe cytogenetic regions of the canine genome in both genomic mapping and clinical applications.

et al. 1998, 1999a), which has formed a common basis for subsequent cytogenetic investigations of the dog.

MOLECULAR CYTOGENETICS OF THE DOG

The application of molecular cytogenetics techniques, in combination with the increasing sophistication of image capture and analysis software, has enabled significant progress to be made in our understanding of the gross organization of the canine genome. The development of FISH reagents such as genome-wide sets of WCPPs allows the application of chromosome painting techniques for chromosome identification and comparative cytogenetics studies. The use of multicolor single locus probes (SLPs) has enabled the development of chromosome-specific panels of markers that have provided key genomic anchor points for successive iterations of the radiation-hybrid map and more recently for the canine genome assembly. The development of comparative genomic hybridization (CGH) analysis of canine cancers opened an important route for the investigation of chromosome aberrations associated with neoplasia, in which molecular cytogenetic studies of canine tumors have played a key role in advancing the recognition of this species as a model for human cancers.

FISH

FISH is a general term that describes any procedure in which the location of a nucleic acid probe is determined by the use of fluorochrome-based detection. In the most widely used form, DNA-FISH, the probe may range from a single small fragment of DNA to complex pools of DNA fragments representing whole genomes. For "direct" labeling, the probe is typically labeled with one of a variety of spectrally resolvable fluorochromes, either as conjugated nucleotides (e.g., fluorescein-isothiocyanate conjugated dUTP [FITC-dUTP], rhodamine-conjugated dUTP, Cyanine3-dUTP, Cyanine5-dUTP) or by chemical modification. Since the direct probe is itself fluorescent, the probe–target complex is immediately visible following hybridization to the template. For "indirect" labeling, the probe is labeled with a hapten, such as biotin-16-dUTP or digoxygenin-11-dUTP. These molecules are not fluorescent but have affinity for fluorochrome-conjugated antibodies, or reporter molecules. For indirectly labeled probes, visualization of the hybridization sites therefore requires that they first be detected with one of a variety of methods that attach fluorochromes to the hapten(s) in the probe–target complex. Since there are a limited number of haptens available but a wide variety of fluorochrome-based reagents, there has been an increasing trend toward the use of direct labeling techniques, although both approaches are widely used. The last decade has seen the introduction of numerous FISH-based techniques that have the potential to further improve the efficiency and accuracy of molecular cytogenetic studies in the dog such that the limitations of its challenging karyotype are now readily overcome.

Comparative Cytogenetics

Mammals have evolved with vastly different chromosome numbers and morphologies, such that different species may be defined by their gross karyotypic architecture. Despite this karyotypic diversity, all mammals are essentially differential arrangements of the same collection of ancestrally related genes, within genomes that have retained a remarkable degree of homology during many millions of years of evolution. The extent of this high level of structural conservation has resulted in large chromosomal regions being retained as intact, contiguous blocks of genetic material, which can be detected within the genomes of different species that share a common ancestor. These regions are referred to as evolutionarily conserved chromosome segments (ECCS). The genome-wide sets of whole-chromosome paint probes that have been generated for a range of mammals represent a tremendous resource with which to perform heterologous chromosome painting (Zoo-FISH) (see, e.g., Scherthan et al. 1994; Rettenberger et al. 1995; Chowdhary et al. 1996; Fronicke et al. 1996; Raudsepp et al. 1996) to determine the gross distribution of the ECCS that are shared between species. In this process, chromosome-specific paint probes from one species are hybridized to metaphase spreads of another species. The pattern of hybridization observed on the latter identifies ECCS that are the evolutionary equivalents of the chromosome of the former species from which the paint probe was derived. Where paint probes are available for both species of interest, this technique can be used in a reciprocal manner, providing positional data regarding the location of ECCS shared between the two species. For example, when a chromosome paint probe for dog chromosome 5 (CFA 5) is hybridized to human chromosomes (HSA), defined regions of HSA 1, 11, 16, and 17 are highlighted, indicating that CFA 5 is evolutionarily related to regions of these four human chromosomes (Breen et al. 1999b; Thomas et al. 1999; Yang et al. 1999). This process does not indicate which regions of CFA 5 are shared with HSA 1, 11, 16, and 17, and therefore, subsequent hybridization of chromosome paint probes representing HSA 1, 11, 16, and 17 onto dog chromosomes is required to permit identification of the chromosomal regions of CFA 5 that are shared with each of these four human chromosomes. Gene content, and frequently also gene order, are highly conserved within the corresponding ECCS of different species, and as a result, the identification of corresponding ECCS provides an effective means by which to extrapolate mapping data between their genomes, as well as increasing our understanding of genome structure and evolution. The availability of high-quality chromosome paints for the human and dog (Langford et al. 1996; Yang et al. 1999) allowed two complete reciprocal chromosome painting studies to be performed independently (Breen et al. 1999b; Yang et al. 1999), resulting in the identification of regions of conserved synteny between their karyotypes. These two reports were grossly comparable, although the study of Yang et al. (1999) identified a greater total number of ECCS. Whereas the study of Breen et al. (1999b) reported a number of small gaps in the comparative maps of human and dog, Yang et al. (1999) were able to suggest the chromosome homologies that filled these gaps. These data, based on a cytogenetic resolution in the order of 5 Mb, indicated that human and dog share up to 70 ECCS. The recent release of the 7.5X canine genome assembly suggests that there are 156 shared segments which are each >500 kb and a further 100 segments >3 Mb in size (K. Lindblad-Toh et al., in prep.), thus corresponding well with previous FISH estimates.

The Canidae as a family have played an important role in providing a better understanding of the evolution of karyotypes. Although gross similarities exist between the karyotypes of many species of the Carnivora, this does not extend to the Canidae, in which an extreme degree of karyotypic variation exists. Canids in general have very limited regions of similarity between their banded chromosomes and models of an ancestral carnivore karyotype based on banding homologies (Dutrillaux and Couturier 1983) and Zoo-FISH data (Fronicke et al. 1997). Comparative chromosome painting within the Canidae indicated that dog chromosomes evolved by extensive fissions and fusions from the ancestral carnivore karyotype and that the dog has one of the most diverged karyotypes within the mammals investigated thus far (Yang et al. 1999; Graphodatsky et al. 2000; Nie et al. 2003).

Comparative chromosome painting studies provide valuable information regarding the gross distribution of conserved segments in related species. By the nature of the technology, the resolution of this approach is limited by the capabilities of fluorescence microscopy, and so the boundaries of ECCS may be regarded as accurate only to within 5–10 Mb. In addition, such a low-resolution approach does not offer any indication of the gene order within each conserved segment. Refinement of the boundaries of ECCS and evaluation of the orientation and gene order within these regions is made possible using comparative FISH mapping and/or radiation-hybrid mapping of type-I (gene) markers. This approach was initially used in a study of dog chromosome 5 (Thomas et al. 1999) and formed the basis of subsequent genome-wide studies (Breen et al. 2001b, 2004; Guyon et al. 2003).

FISH Mapping of Single Loci

In routine FISH mapping of small genomic DNA fragments, the cytogenetic location of the probe may be determined using metaphase chromosome preparations. In most cases, the probe represents a cloned region of genomic DNA, ranging in size from a few kilobases (e.g., plasmid clones) to hundreds of kilobases (e.g., bacterial artificial chromosome [BAC] clones). Generally, the size of the resulting FISH signal corresponds to the length of the genomic DNA insert and the amount of probe used. Using multiple fluorochromes, it is possible to assign many loci simultaneously in a single hybridization reaction, according to the formula $n = (2^x) - 1$, where n is the number of unique color combinations that can be generated using x different, spectrally resolvable fluorochromes. With the use of five fluorochromes and a fluorescence microscope equipped with a series of high-quality filters, it is thus possible to generate up to 31 resolvable fluorescence signals. Figure 3a shows the simultaneous hybridization of 10 canine BAC clones to 10 different chromosomes in a metaphase preparation. The resolution of metaphase FISH is in the range of 2–4 Mb, and thus if two probes are separated by a smaller distance (Fig. 3b), higher-resolution FISH is required to resolve the signals and determine the precise order of the probes. This can be achieved by visualizing signals in interphase nuclei (Fig. 3c), where the limits of resolution are in the region of 50–100 kb, although chromosomes themselves cannot be distinguished as unique structures. In most instances, metaphase spreads and interphase nuclei exist on the same slide, such that probes can be ordered and chromosomally assigned in a single experiment.

Single-locus FISH probes have been used to assign numerous loci to the canine karyotype, including genes relating to traits such as coat color and deafness as well as diseases (see Table 11.1 in Breen et al. 2001a and also see, for example, van de Sluis et al. 2001; Rak et al. 2003; Tiret et al. 2003, including cancer; Thomas et al. 2003a). In addition, the

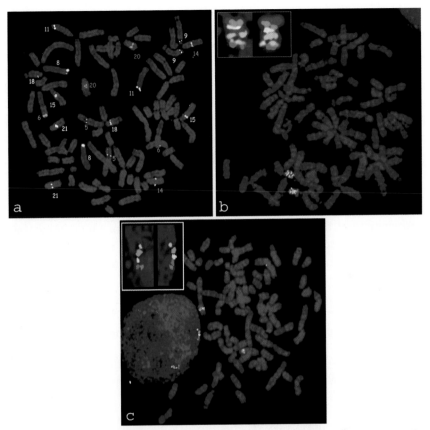

Figure 3. Multicolor single-locus probe (SLP) FISH of canine BAC clones. (**a**) Ten-color FISH of 10 canine BAC clones assigned to different chromosome in canine metaphase preparation. (**b**) Five-color FISH of canine BAC clones separated at sufficient distances to be resolved and ordered conclusively along dog chromosome (CFA) 38. Inset shows the two CFA 38 homologs correctly oriented and enlarged. (**c**) Five-color FISH of five canine BAC clones to a canine metaphase spread and an interphase nucleus. All five BAC clones map to the same cytogenetic band on CFA 10 and so are not resolvable by metaphase FISH, but resolve clearly by interphase FISH. Inset shows enlarged segments of the interphase nucleus to reveal the clone order within this small region of CFA 10.

canine karyotype has recently been "tiled" with a series of 1,250 evenly BAC-spaced clones (Thomas et al. 2005), generating a valuable resource that will be welcomed by cytogeneticists as a means to identify chromosome aberrations in canine tumors either by FISH or array-CGH (see below), and will also aid in the detection of candidate genes for other specific traits of interest.

Single-Locus Probes as Genomic Anchors

In the first integrated genome map of the dog (Breen et al. 2001b), a panel of 266 cosmid clones, each containing a polymorphic microsatellite marker, was used as chromosome-specific SLPs to define their cytogenetic location. Accordingly, these clones served as key cytogenetic points distributed across all chromosomes. Their use allowed the 1,800-marker integrated meiotic/radiation hybrid (RH) map of the canine genome (Breen et al. 2001b) and the subsequent 3,279-marker RH map of the dog (Guyon et al. 2003) to be firmly anchored to the karyotype. In the most recent integrated RH map (Breen et al. 2004), the number of markers was increased to 4,249. This map includes 1,760 BAC end sequences and 1,589 microsatellite markers and features a marker every 900 kb. At the cytogenetic level, this map reported the precise location of 1,000 canine BAC clones, which had been ordered along the length of each chromosome using stepwise multicolor SLP analysis (illustrated in Fig. 4). Of these 1,000 BAC clones, 804 were in common with the RH map. The use of a common set of markers for both the cytogenetic and the RH maps ensured that the integrated map was firmly anchored at multiple locations along the length of each chromosome, providing a very high level of confidence in marker order.

In a similar manner, molecular cytogenetics served a key role in anchoring the canine genome assembly (K. Lindblad-Toh, in prep.). With

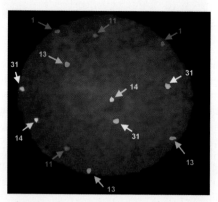

Figure 4. Application of multicolor SLP analysis to canine cancer research. Cytogenetic evaluation of canine lymphoma has indicated that CFA 1, 11, 13, 14, and 31 demonstrate recurrent copy number aberrations (Thomas et al. 2003). In this example, an interphase nucleus from a canine lymphoma has been probed with differentially labeled BAC clones representing each of these five chromosomes and indicate that in this cell, the copy number status of CFA 1, 11, and 14 are normal (2n = 2) but that the CFA 13 and 31 loci are both present as three copies.

an RH map (C. Hitte et al., in prep.) anchored and oriented correctly on the karyotype through the integrated map (Breen et al. 2004), it was possible to ensure that the genome assembly was correctly placed on the chromosomes. In addition, for regions of the genome where there was any doubt in genome location or orientation of supercontigs, molecular cytogenetics was able to resolve these issues and the data were incorporated into the second genome release (Canfam2.0).

MOLECULAR CYTOGENETIC EVALUATION OF CHROMOSOMAL ORGANIZATION IN CANINE CANCERS

Cytogenetics has thus far revealed clonal chromosome aberrations in almost 27,000 human neoplasms, representing 75 different types of cancer (see http://cgap.nci.nih.gov/Chromosomes/Mitelman). There is significant clinical value in the identification of recurring chromosome abnormalities in malignant cells, by cytogenetic or molecular means, for the diagnosis, subclassification, and prognosis of a neoplastic disorder. Knowledge of such aberrations has been shown to assist in the selection of the most appropriate therapeutic approach. In addition, the characterization of recurrent chromosome aberrations has played a major role in pinpointing the location of key cancer-associated genes. The application of molecular cytogenetics to the analysis of human neoplasia has revolutionized the way in which we interrogate tumor cells for cytogenetic changes, whether numerical or structural aberrations.

Humans and Dogs—Similar Genomes, Similar Cancers

Despite millions of years of divergent evolution, the genomes of humans and dogs remain highly comparable. In addition, since dogs are our pets, the canine genome is exposed to the same environmental influences as is our own genome. It is therefore not surprising that there are considerable pathophysiological similarities shared by many forms of human and canine cancer (Withrow and MacEwen 2001). These features, combined with the sophisticated genomic resources now available for the dog, have placed "Man's best friend" in a position of high visibility as a model system for cancer research. In addition, the demographic history and genetic structure of many purebred domestic dogs have resulted in breeds that are phenotypically distinct genetic isolates, with restricted genetic heterogeneity, and which are characterized by susceptibility to different malignancies (Priester and McKay 1980; Patterson 2000; Parker et al.

2004; Sutter and Ostrander 2004). For example, in canine lymphoma there is clear evidence of breed-related differences in lymphoma immunophenotypes (see Chapter 22). Such breed-associated predisposition suggests that selected breeds of dogs are inheriting "at risk" alleles for very few genes, perhaps even a single gene, with a profound effect.

The combination of pathophysiological and genetic similarity shared between humans and dogs also suggests that canine tumors may contain the natural variety of chromosome aberrations that are observed in human cancers, a feature that is not evident with induced rodent tumors. A detailed cytogenetic investigation of recurrent canine tumors may therefore reveal the locations of key genes involved in tumorigenesis, which have thus far remained undetectable in human cancers. Using molecular cytogenetics, it has already been shown that tumors in humans and dogs share evolutionarily related chromosome aberrations (M. Breen and J.F. Modiano, in prep.) and so further research to determine the diagnostic and prognostic significance of these shared aberrations will be an important goal for veterinary oncology.

Direct Evaluation of Tumor Karyotypes

With conventional cytogenetic techniques, the evaluation of complex canine tumor karyotypes is a daunting proposition, which is reflected by the relative paucity of published work in the field. Analysis of tumor genomes by direct molecular cytogenetic methods, however, allows a rapid assessment of the gross numerical and structural characteristics of chromosomes within individual tumor cells (see, e.g., Pinkel et al. 1988; Trask et al. 1988; Anastasi et al. 1990; Flactif et al. 1995; Werner et al. 1997). Chromosome painting analysis is widely used to unravel the complex nature of chromosome rearrangements and to determine chromosomal organization in the karyotypes of human cancers and syndromes (see, e.g., Blennow 2004; Langer et al. 2004). The availability of high-quality chromosome paints for application to canine tumor cytogenetics has now enabled this approach to be used in the investigation of karyotypic abnormalities in canine tumors, including mammary carcinoma (Tap et al. 1998), lymphoma (Thomas et al. 2001), and soft tissue sarcoma (Milne et al. 2004). The development of multiplex-FISH (M-FISH) (Speicher et al. 1996) and spectral karyotyping (Schrock et al. 1996) for human cytogenetics has led to many new cytogenetic discoveries due to their ability to increase the amount of genomic data that can be generated in a single experiment. To identify all 40 chromosome types of the dog (38 autosomes + X + Y) in a single FISH reaction with different colors, i.e., 40-color M-FISH, would require the use of six spectrally resolvable fluorochromes, which is beyond the scope of many laboratories. To overcome this limitation, Milne et al. (2004) used a robust 7-color system in their study of canine sarcomas and established karyotype coverage over a series of six different separate FISH reactions.

Comparative Genomic Hybridization—The Indirect Approach

CGH is a form of FISH analysis that provides a means by which to screen an entire tumor genome for regions of copy number aberrations (CNAs) in a single experiment (Kallioniemi et al. 1992, 1994). CGH has been used to identify CNAs in a range of human malignancies including cancers of the brain (Petersen et al. 2000; Wiltshire et al. 2000; Scheil et al. 2001), prostate (Cher et al. 1994), breast (Tirkkonen et al. 1998; Rodriguez et al. 2000), bladder (Obermann et al. 2003), thyroid (Wreesmann et al. 2002), skin (Bastian et al. 1998), and hematopoetic system (Allen et al. 2002; Berglund et al. 2002; Franke et al. 2002).

Chromosome-based CGH was developed recently for application to canine tumors (Dunn et al. 2000; Thomas et al. 2001) and has been used to demonstrate that recurrent chromosome aberrations are present in canine lymphoma (Thomas et al. 2003b). For example, this study identified that the most frequent chromosome aberration in a range of morphological types of canine lymphoma was gain of chromosome 13 (CFA 13). This aberration was apparent in almost 50% of the cases analyzed and was found in four of the six morphological subgroups of lymphoma identified, being absent in both lymphoplasmacytic and lymphoblastic lymphomas (see Table 1 in Thomas et al. 2003b). Comparative genome data indicated that CFA 13 is evolutionarily related to sites on two human (HSA) chromosomes, HSA 8q23-qtel and HSA 4pprox-qprox (Breen et al. 2004). The former of these human chromosome sites harbors the c-*myc* oncogene, and the latter the c-*KIT* oncogene. The canine orthologs of both these genes have been chromosomally assigned to CFA 13 (Thomas et al. 2003a). Activation of human c-*myc* has been shown to be a common event in intermediate- and high-grade NHL (Gaidano and Dalla-Favera 1993), and thus the recurrent gain of CFA 13 observed in this study suggests that it is crucial to examine the potential involvement of c-*MYC* and c-*KIT* in canine tumor development.

Despite the power of CGH analysis, the use of metaphase chromosomes as the target limits the ability to detect copy number changes to those that are generally greater than 5–10 Mb, and is unable to resolve closely spaced aberrations (Pinkel et al. 1998; Lichter et al. 2000). The recent development of microarray-based CGH analysis overcomes many of the limitations of conventional metaphase-based CGH and has been shown to provide increased resolution and information (see, e.g., Solinas-Toldo et al. 1997; Cai et al. 2002; Fiegler et al. 2003; Veltman et al. 2003). Array CGH has been used to describe a wide variety of DNA copy number changes in human cancers, providing new and valuable insight into the underlying gene involvement of tumors (Pinkel et al. 1998; Albertson et al. 2000; Zhao et al. 2002; Albertson and Pinkel 2003; Kraus et al. 2003; Lage et al. 2003; Veltman et al. 2003). Array-based CGH analysis has been developed for application to the dog, based on the use of degenerate-oligonucleotide-primed PCR products from canine BAC clones, each of which has been assigned to a precise cytogenetic location using FISH analysis (Thomas et al 2003c; Breen et al. 2004). The first canine BAC array was a small array of canine BAC clones containing a series of cancer genes and targeted clones, specifically for analysis of canine lymphomas (Thomas et al. 2003c). This array was followed by a genome-wide array of 1,250 clones with an average clone spacing of 2 Mb (Thomas et al. 2005). The power of this 2-Mb genome-wide array for a variety of applications derives from the source of the BAC clones. Over 1,000 of these clones have a defined position within the canine genome assembly, determined by FISH and/or in silico mapping. This integrated approach allows canine cytogenetic data to be immediately translated into canine DNA sequence data and thus offers the potential to progress from cytogenetic aberration to gene sequence with relative ease. The integration of BAC mapping data with the canine genome assembly also allows the generation of higher-resolution arrays, which may either be genome-wide, or comprise a targeted selection of additional clones for the construction of chromosome-specific microarrays. These will be particularly useful for higher-resolution studies of small genomic regions. We are currently exploiting the advantages of an integrated canine genome map in the generation of a more comprehensive dog array that will double current resolution limits to approximately 1 Mb.

SUMMARY

The genome of the dog is organized into 38 pairs of acrocentric auto-somes and a pair of metacentric sex chromosomes, thus comprising many more chromosomes than most other mammalian genomes. Using con-ventional cytogenetic techniques, it is very challenging to identify reliably many of the chromosomes comprising the dog karyotype. The development of molecular cytogenetic reagents (single-locus probes and chromosome paints) for the dog allowed an accurate chromosome nomenclature to be established, which was adopted by the scientific community. With a standardized nomenclature, the dog became a focus of interest for the development of comparative cytogenetic maps, and the process of genome mapping for the dog was accelerated. Integrated genome mapping for the dog relied on the use of FISH analysis to ensure that maps were tightly anchored to the karyotype. In a similar fashion, the canine genome assembly has also been anchored to the corresponding chromosomes to ensure the generation of a high-quality end product.

The high degree of pathophysiological similarity shared between many forms of human and canine malignancies has placed the dog as an ideal model system for cancer research. Molecular cytogenetic evaluation of canine tumor cells, using both direct and indirect approaches, has begun to reveal recurrent chromosome aberrations in a variety of tumor types. It is apparent that some of these cytogenetic aberrations are evolutionarily conserved between human and dog. As these studies progress, we will be able to determine whether the diagnostic and prognostic aberrations that are used to define human cancers are also of benefit to canine patients. Furthermore, the combination of high-quality molecular cytogenetic resources and the high level of breed predisposition to certain tumors may allow us to reveal cancer-associated genes in the canine genome that are not detectable in the corresponding human cancers. The study of chromosome aberrations in malignant cells ultimately has great potential in improving the health and welfare of both species.

ACKNOWLEDGMENTS

Molecular cytogenetics of canine cancer at North Carolina State University is supported by grants to M.B. from the American Kennel Club Canine Health Foundation (award nos. 2038, 2214, 2254, 249).

REFERENCES

Albertson D.G. and Pinkel D. 2003. Genomic microarrays in human genetic disease and cancer. *Hum. Mol. Genet.* **12:** R145–R152.

Albertson D.G., Ylstra B., Segraves R., Collins C., Dairkee S.H., Kowbel D., Kuo W.L., Gray J.W., and Pinkel D. 2000. Quantitative mapping of amplicon structure by array CGH identifies CYP24 as a candidate oncogene. *Nat. Genet.* **25:** 144–146.

Allen J.E., Hough R.E., Goepel J.R., Bottomley S., Wilson G.A., Alcock H.E., Baird M., Lorigan P.C., Vandenberghe E.A., Hancock B.W., and Hammond D.W. 2002. Identification of novel regions of amplification and deletion within mantle cell lymphoma DNA by comparative genomic hybridization. *Br. J. Haematol.* **116:** 291–298.

Anastasi J., Le Beau M.M., Vardiman J.W., and Westbrook C.A. 1990. Detection of numerical chromosomal abnormalities in neoplastic hematopoietic cells by in situ hybridization with a chromosome-specific probe. *Am. J. Pathol.* **136:** 131–139.

Bastian B.C., LeBoit P.E., Hamm H., Brocker E.B., and Pinkel D. 1998. Chromosomal gains and losses in primary cutaneous melanomas detected by comparative genomic hybridization. *Cancer Res.* **58:** 2170–2175.

Berglund M., Enblad G., Flordal E., Lui W.O., Backlin C., Thunberg U., Sundstrom C., Roos G., Allander S.V., Erlanson M., et al. 2002. Chromosomal imbalances in diffuse large B-cell lymphoma detected by comparative genomic hybridization. *Mod. Pathol.* **15:** 807–816.

Blennow E. 2004. Reverse painting highlights the origin of chromosome aberrations. *Chromosome Res.* **12:** 25–33.

Breen M., Bullerdiek J., and Langford C.F. 1999a. The DAPI banded karyotype of the domestic dog (*Canis familiaris*) generated using chromosome-specific paint probes. *Chromosome Res.* **7:** 401–406.

Breen M., Switonski M., and Binns M.M. 2001a. Cytogenetics and physical chromosome maps. *The genetics of the dog* (eds. A. Ruvinsky and J. Sampson), pp. 299–328. CABI Publishing, Oxon, United Kingdom.

Breen M., Thomas R., Binns M.M., Carter N.P., and Langford C.F. 1999b. Reciprocal chromosome painting reveals detailed regions of conserved synteny between the karyotypes of the domestic dog (*Canis familiaris*) and human. *Genomics* **61:** 145–155.

Breen M., Hitte C., Lorentzen T.D., Thomas R., Cadieu E., Sabacan L., Scott A., Evanno G., Parker H.G., Kirkness E.F., et al. 2004. An integrated 4249 marker FISH/RH map of the canine genome. *BMC Genomics* **5:** 65.

Breen M., Jouquand S., Renier C., Mellersh C.S., Hitte C., Holmes N.G., Cheron A., Suter N., Vignaux F., Bristow A.E., et al. 2001b. Chromosome-specific single-locus FISH probes allow anchorage of an 1800-marker integrated radiation-hybrid/linkage map of the domestic dog genome to all chromosomes. *Genome Res.* **11:** 1784–1795.

Breen M., Reimann N., Bosma A.A., Landon D., Zijlstra S., Bartnitzke S., Switonski M., Long S.E., de Haan N.A., Binns M.M., et al. 1998. Standardisation of the chromosome nos. 22-38 of the dog (*Canis familiaris*) with the use of chromosome painting probes. In *Proceedings of the 13th European Colloquium on Cytogenetics of Domestic Animals*, June 1–6. Hungarian Academy of Sciences, Budapest, Hungary.

Cai W.W., Mao J.H., Chow C.W., Damani S., Balmain A., and Bradley A. 2002. Genome-wide detection of chromosomal imbalances in tumors using BAC microarrays. *Nat. Biotechnol.* **20:** 393–396.

Cher M.L., MacGrogan D., Bookstein R., Brown J.A., Jenkins R.B., and Jensen R.H. 1994. Comparative genomic hybridization, allelic imbalance, and fluorescence in situ hybridization on chromosome 8 in prostate cancer. *Genes Chromosomes Cancer* **11:** 153–162.

Chowdhary B.P., Fronicke L., Gustavsson I., and Scherthan H. 1996. Comparative analysis of the cattle and human genomes: Detection of ZOO-FISH and gene mapping-based chromosomal homologies. *Mamm. Genome* **7:** 297–302.

Christian A., McNiel E., Robinson J., Drabek R., LaRue S., Waldren C., and Bedford J. 1998. A versatile image analysis approach for simultaneous chromosome identification and localization of FISH probes. *Cytogenet. Cell Genet.* **82:** 172–179.

Dunn K.A., Thomas R., Binns M.M., and Breen M. 2000. Comparative genomic hybridization (CGH) in dogs—Application to the study of a canine glial tumour cell line. *Vet. J.* **160**: 77–82.

Dutrillaux B. and Couturier J. 1983. The ancestral karyotype of Carnivora: Comparison with that of platyrrhine monkeys. *Cytogenet. Cell Genet.* **35**: 200–208.

Fiegler H., Carr P., Douglas E.J., Burford D.C., Hunt S., Scott C.E., Smith J., Vetrie D., Gorman P., Tomlinson I.P., and Carter N.P. 2003. DNA microarrays for comparative genomic hybridization based on DOP-PCR amplification of BAC and PAC clones. *Genes Chromosomes Cancer* **36**: 361–374.

Flactif M., Zandecki M., Lai J.L., Bernardi F., Obein V., Bauters F., and Facon T. 1995. Interphase fluorescence in situ hybridization (FISH) as a powerful tool for the detection of aneuploidy in multiple myeloma. *Leukemia* **9**: 2109–2114.

Franke S., Wlodarska I., Maes B., Vandenberghe P., Achten R., Hagemeijer A., and De Wolf-Peeters C. 2002. Comparative genomic hybridization pattern distinguishes T-cell/histiocyte-rich B-cell lymphoma from nodular lymphocyte predominance Hodgkin's lymphoma. *Am. J. Pathol.* **161**: 1861–1867.

Fronicke L., Chowdhary B.P., Scherthan H., and Gustavsson I. 1996. A comparative map of the porcine and human genomes demonstrates ZOO-FISH and gene mapping-based chromosomal homologies. *Mamm. Genome* **7**: 285–290.

Fronicke L., Muller-Navia J., Romanakis K., and Scherthan H. 1997. Chromosomal homeologies between human, harbor seal (*Phoca vitulina*) and the putative ancestral carnivore karyotype revealed by Zoo-FISH. *Chromosoma* **106**: 108–113.

Gaidano G. and Dalla-Favera R. 1993. Biologic and molecular characterization of non-Hodgkin's lymphoma. *Curr. Opin. Oncol.* **5**: 776–784.

Graphodatsky A.S., Beklemisheva V.R., and Dolf G. 1995. High-resolution GTG-banding patterns of dog and silver fox chromosomes: Description and comparative analysis. *Cytogenet. Cell Genet.* **69**: 226–231.

Graphodatsky A.S., Yang F., O'Brien P.C., Serdukova N., Milne B.S., Trifonov V., and Ferguson-Smith M.A. 2000. A comparative chromosome map of the Arctic fox, red fox and dog defined by chromosome painting and high resolution G-banding. *Chromosome Res.* **8**: 253–263.

Gustavsson I. 1964. The chromosomes of the dog. *Hereditas* **51**: 187–189.

Guyon R., Lorentzen T.D., Hitte C., Kim L., Cadieu E., Parker H.G., Quignon P., Lowe J.K., Renier C., Gelfenbeyn B., et al. 2003. A 1-Mb resolution radiation hybrid map of the canine genome. *Proc. Natl. Acad. Sci.* **100**: 5296–5301.

Howard-Peebles P.N. and Pryor J.C. 1980. The R-banding pattern of the canine karyotype. *J. Hered.* **71**: 361–362.

Kallioniemi O.P., Kallioniemi A., Piper J., Isola J., Waldman F.M., Gray J.W., and Pinkel D. 1994. Optimizing comparative genomic hybridization for analysis of DNA sequence copy number changes in solid tumors. *Genes Chromosomes Cancer* **10**: 231–243.

Kallioniemi A., Kallioniemi O.P., Sudar D., Rutovitz D., Gray J.W., Waldman F., and Pinkel D. 1992. Comparative genomic hybridization for molecular cytogenetic analysis of solid tumors. *Science* **258**: 818–821.

Kraus J., Pantel K., Pinkel D., Albertson D.G., and Speicher M.R. 2003. High-resolution genomic profiling of occult micrometastatic tumor cells. *Genes Chromosomes Cancer* **36**: 159–166.

Lage J.M., Leamon J.H., Pejovic T., Hamann S., Lacey M., Dillon D., Segraves R., Vossbrinck B., Gonzalez A., Pinkel D., et al. 2003. Whole genome analysis of genetic alterations in small DNA samples using hyperbranched strand displacement amplification and array-CGH. *Genome Res.* **13**: 294–307.

Langer S., Kraus J., Jentsch I., and Speicher M.R. 2004. Multicolor chromosome painting in diagnostic and research applications. *Chromosome Res.* **12:** 15–23.

Langford C.F., Fischer P.E., Binns M.M., Holmes N.G., and Carter N.P. 1996. Chromosome-specific paints from a high-resolution flow karyotype of the dog. *Chromosome Res.* **4:** 115–123.

Lichter P., Joos S., Bentz M., and Lampel S. 2000. Comparative genomic hybridization: Uses and limitations. *Semin. Hematol.* **37:** 348–357.

Manolache M., Ross W.M., and Schmid M. 1976. Banding analysis of the somatic chromosomes of the domestic dog (*Canis familiaris*). *Can. J. Genet. Cytol.* **18:** 513–518.

Mayr B., Krutzler J., Schleger W., and Auer H. 1986. A new type of Robertsonian translocation in the domestic dog. *J. Hered.* **77:** 127.

Milne B.S., Hoather T., O'Brien P.C., Yang F., Ferguson-Smith M.A., Dobson J., and Sargan D. 2004. Karyotype of canine soft tissue sarcomas: A multicolor, multi-species approach to canine chromosome painting. *Chromosome Res.* **12:** 825–835.

Minouchi O. 1928. The spermatogenesis of the dog, with special reference to meiosis. *Jpn. J. Zool.* **1:** 255–268.

Moreno-Millan M., Rodero A., Alonso F.J., and Sanz A. 1991. Contribution to the establishment of the R-banded karyotype in the dogs. *Genet. Sel. Evol.* (suppl. 1) **23:** 179–182.

Nie W., Wang J., Perelman P., Graphodatsky A.S., and Yang F. 2003. Comparative chromosome painting defines the karyotypic relationships among the domestic dog, Chinese raccoon dog and Japanese raccoon dog. *Chromosome Res.* **11:** 735–740.

Obermann E.C., Junker K., Stoehr R., Dietmaier W., Zaak D., Schubert J., Hofstaedter F., Knuechel R., and Hartmann A. 2003. Frequent genetic alterations in flat urothelial hyperplasias and concomitant papillary bladder cancer as detected by CGH, LOH, and FISH analyses. *J. Pathol.* **199:** 50–57.

Parker H., Kim L., Sutter N., Carlson S., Lorentzen T., Malek T., Johnson G., DeFrance H., Ostrander E.A., and Kruglyak L. 2004. Genetic structure of the purebred domestic dog. *Science* **304:** 1160–1164.

Patterson D.F. 2000. *Canine genetic disease information system: A computerized knowledge base of genetic diseases in the dog.* Mosby-Harcourt, St. Louis, Missouri.

Petersen I., Hidalgo A., Petersen S., Schluns K., Schewe C., Pacyna-Gengelbach M., Goeze A., Krebber B., Knosel T., Kaufmann O., et al. 2000. Chromosomal imbalances in brain metastases of solid tumors. *Brain Pathol.* **10:** 395–401.

Pienkowska A. and Switonski M. 1998. Chromosomal localization and activity of nucleolar organizer regions in the dog (*Canis familiaris*). *Genet. Sel. Evol.* **30:** 79–84.

Pinkel D., Landegent J., Collins C., Fuscoe J., Segraves R., Lucas J., and Gray J. 1988. Fluorescence in situ hybridization with human chromosome-specific libraries: Detection of trisomy 21 and translocations of chromosome 4. *Proc. Natl. Acad. Sci.* **85:** 9138–9142.

Pinkel D., Segraves R., Sudar D., Clark S., Poole I., Kowbel D., Collins C., Kuo W.L., Chen C., Zhai Y., et al. 1998. High resolution analysis of DNA copy number variation using comparative genomic hybridization to microarrays. *Nat. Genet.* **20:** 207–211.

Poulsen B.S., Shibasaki Y., Ikeuchi T., and Ronne M. 1990. Banding studies in *Canis familiaris.* I. Replication patterns in karyotypes from lymphocyte cultures. *Cytobios* **62:** 161–165.

Priester W.A. and McKay F.W. 1980. The occurrence of tumors in domestic animals. *Natl. Cancer Inst. Monogr.* **54:** 157–159.

Rak S.G., Drogemuller C., Leeb T., Quignon P., Andre C., Scott A., Breen M., and Distl O. 2003. Chromosomal assignment of 20 candidate genes for canine congenital sensorineural deafness by FISH and RH mapping. *Cytogenet. Genome Res.* **101:** 130–135.

Raudsepp T., Fronicke L., Scherthan H., Gustavsson I., and Chowdhary B.P. 1996. Zoo-FISH delineates conserved chromosomal segments in horse and man. *Chromosome Res.* **4:** 218–225.

Reimann N., Bartnitzke S., Bullerdiek J., Schmitz U., Rogalla P., Nolte I., and Ronne M. 1996. An extended nomenclature of the canine karyotype. *Cytogenet. Cell Genet.* **73:** 140–144.

Rettenberger G., Klett C., Zechner U., Bruch J., Just W., Vogel W., and Hameister H. 1995. ZOO-FISH analysis: Cat and human karyotypes closely resemble the putative ancestral mammalian karyotype. *Chromosome Res.* **3:** 479–486.

Rodriguez C., Causse A., Ursule E., and Theillet C. 2000. At least five regions of imbalance on 6q in breast tumors, combining losses and gains. *Genes Chromosomes Cancer* **27:** 76–84.

Scheil S., Bruderlein S., Eicker M., Herms J., Herold-Mende C., Steiner H.H., Barth T.F., and Moller P. 2001. Low frequency of chromosomal imbalances in anaplastic ependymomas as detected by comparative genomic hybridization. *Brain Pathol.* **11:** 133–143.

Scherthan H., Cremer T., Arnason U., Weier H.U., Lima-de-Faria A., and Fronicke L. 1994. Comparative chromosome painting discloses homologous segments in distantly related mammals. *Nat. Genet.* **6:** 342–347.

Schrock E., du Manoir S., Veldman T., Schoell B., Wienberg J., Ferguson-Smith M.A., Ning Y., Ledbetter D.H., Bar-Am I., Soenksen D., et al. 1996. Multicolor spectral karyotyping of human chromosomes. *Science* **273:** 494–497.

Selden J.R., Moorhead P.S., Oehlert M.L., and Patterson D.F. 1975. The Giemsa banding pattern of the canine karyotype. *Cytogenet. Cell Genet.* **15:** 380–387.

Solinas-Toldo S., Lampel S., Stilgenbauer S., Nickolenko J., Benner A., Dohner H., Cremer T., and Lichter P. 1997. Matrix-based comparative genomic hybridization: Biochips to screen for genomic imbalances. *Genes Chromosomes Cancer* **20:** 399–407.

Speicher M.R., Gwyn Ballard S., and Ward D.C. 1996. Karyotyping human chromosomes by combinatorial multi-fluor FISH. *Nat. Genet.* **12:** 368–375.

Stone D.M., Jacky P.B., and Prieur D.J. 1991. The giemsa banding pattern of canine chromosomes, using a cell synchronization technique. *Genome* **34:** 407–412.

Sutter N.B. and Ostrander E.A. 2004. Dog star rising: The canine genetic system. *Nat. Rev. Genet.* **5:** 900–910.

Switonski M., Reimann N., Bosma A.A., Long S., Bartnitzke S., Pienkowska A., Moreno-Milan M.M., and Fischer P. 1996. Report on the progress of standardization of the G-banded canine (*Canis familiaris*) karyotype. Committee for the standardized karyotype of the dog (*Canis familiaris*). *Chromosome Res.* **4:** 306–309.

Tap O.T., Rutteman G.R., Zijlstra C., de Haan N.A., and Bosma A.A. 1998. Analysis of chromosome aberrations in a mammary carcinoma cell line from a dog by using canine painting probes. *Cytogenet. Cell Genet.* **82:** 75–79.

Thomas R., Breen M., Langford C.F., and Binns M.M. 1999. Zoo-FISH analysis of dog chromosome 5: Identification of conserved synteny with human and cat chromosomes. *Cytogenet. Cell Genet.* **87:** 4–10.

Thomas R., Smith K.C., Gould R., Gower S.M., Binns M.M., and Breen M. 2001. Molecular cytogenetic analysis of a novel high-grade T-lymphoblastic lymphoma demonstrating co-expression of CD3 and CD79a cell markers. *Chromosome Res.* **8:** 649–657.

Thomas R., Bridge W., Benke K., and Breen M. 2003a. Isolation and chromosomal assignment of canine genomic BAC clones representing 25 cancer related canine genes. *Cytogenet. Genome Res.* **102:** 249–253.

Thomas R., Smith K.C., Ostrander E.A., Galibert F., and Breen M. 2003b. Chromosome aberrations in canine multicentric lymphomas detected with comparative genomic hybridisation and a panel of single locus probes. *Br. J. Cancer* **89:** 1530–1537.

Thomas R., Fiegler H., Ostrander E.A., Galibert F., Carter N.P., and Breen M. 2003c. A canine cancer-gene microarray for CGH analysis of tumors. *Cytogenet. Genome Res.* **102:** 254–260.

Thomas R., Scott A., Langford C.F., Fosmire S.P., Jubala C.M., Lorentzen T.D., Hitte C., Karlsson E.K., Kirkness E., Ostrander E.A., et al. 2005. Construction of a 2Mb resolution BAC-microarray for CGH analysis of canine tumors. *Genome Res.* (in press).

Tiret L., Blot S., Kessler J.L., Gaillot H., Breen M., and Panthier J.J. 2003. The *cnm* locus, a canine homologue of human autosomal forms of centronuclear myopathy, maps to chromosome 2. *Hum. Genet.* **113:** 297–306.

Tirkkonen M., Tanner M., Karhu R., Kallioniemi A., Isola J., and Kallioniemi O.P. 1998. Molecular cytogenetics of primary breast cancer by CGH. *Genes Chromosomes Cancer* **21:** 177–184.

Trask B., van den Engh G., Pinkel D., Mullikin J., Waldman F., van Dekken H., and Gray J. 1988. Fluorescence in situ hybridization to interphase cell nuclei in suspension allows flow cytometric analysis of chromosome content and microscopic analysis of nuclear organization. *Hum. Genet.* **78:** 251–259.

van de Sluis B., Nanji M.S., Breen M., Pearson P.L., Oost B.A., Cox D.W., and Wijmenga C. 2001. Characterization and chromosomal localization of five canine ATOX1 pseudogenes. *Cytogenet. Cell Genet.* **93:** 105–108.

Veltman J.A., Fridlyand J., Pejavar S., Olshen A.B., Korkola J.E., DeVries S., Carroll P., Kuo W.L., D. Pinkel, D. Albertson, et al. 2003. Array-based comparative genomic hybridization for genome-wide screening of DNA copy number in bladder tumors. *Cancer Res.* **63:** 2872–2880.

Werner M., Wilkens L., Aubele M., Nolte M., Zitzelsberger H., and Komminoth P. 1997. Interphase cytogenetics in pathology: Principles, methods, and applications of fluorescence in situ hybridization (FISH). *Histochem. Cell Biol.* **108:** 381–390.

Wiltshire R.N., Rasheed B.K., Friedman H.S., Friedman A.H., and Bigner S.H. 2000. Comparative genetic patterns of glioblastoma multiforme: Potential diagnostic tool for tumor classification. *Neuro-oncol* **2:** 164–173.

Withrow S.J. and MacEwen E.G. 2001. *Small animal clinical oncology*, 3rd edition. Saunders, Philadelphia.

Wreesmann V.B., Ghossein R.A., Patel S.G., Harris C.P., Schnaser E.A., Shaha A.R., Tuttle R.M., Shah J.P., Rao P.H., and Singh B. 2002. Genome-wide appraisal of thyroid cancer progression. *Am. J. Pathol.* **161:** 1549–1556.

Yang F., O'Brien P.C., Milne B.S., Graphodatsky A.S., Solanky N., Trifonov V., Rens W., Sargan D., and Ferguson-Smith M.A. 1999. A complete comparative chromosome map for the dog, red fox, and human and its integration with canine genetic maps. *Genomics* **62:** 189–202.

Zhao J., Roth J., Bode-Lesniewska B., Pfaltz M., Heitz P.U., and Komminoth P. 2002. Combined comparative genomic hybridization and genomic microarray for detection of gene amplifications in pulmonary artery intimal sarcomas and adrenocortical tumors. *Genes Chromosomes Cancer* **34:** 48–57.

11

The Dog Genome: Sequence, Evolution, and Haplotype Structure

Claire M. Wade
Broad Institute of MIT and Harvard
Cambridge, Massachusetts 02141 and
Center for Human Genetic Research
Massachusettes General Hospital
Boston, Massachusetts 02114

Elinor K. Karlsson, Tarjei S. Mikkelsen, Michael C. Zody,
and Kerstin Lindblad-Toh
Broad Institute of MIT and Harvard
Cambridge, Massachusetts 02141

Mᴀɴ'ѕ ʙᴇѕᴛ ᴀɴɪᴍᴀʟ ꜰʀɪᴇɴᴅ ᴏᴄᴄᴜᴘɪᴇѕ a very special niche in genomics. *Canis familiaris* is just the third fully sequenced mammalian genome and the first representative of a new major mammalian clade, Laurasiatheria. (Fig. 1) As is eloquently shown elsewhere in this book, dogs share many common genetic and acquired diseases with their human owners and also much of their environment. In addition, the unique breeding history of the domestic dog makes this species ideally suited for genetic studies. In 2001, an international consortium wrote a white paper and successfully petitioned the National Human Genome Research Institute (NHGRI) for the creation of a high-quality canine sequence to enhance canine disease-gene mapping and mammalian comparative genomics. Large-scale sequencing began in 2003, and, within a year, a high-quality draft sequence of the dog genome was made publicly available.

The assembled dog sequence covers ~99% of the 2.4-Gb genome of a single female boxer. The remarkably high quality of the genome sequence, and the speed at which it was produced, reflected continuing improvements in sequencing and assembly technology since the initiation of the human genome sequence project, which took approximately 10 years to complete.

The Dog and Its Genome ©2006 Cold Spring Harbor Laboratory Press 0-87969-742-3

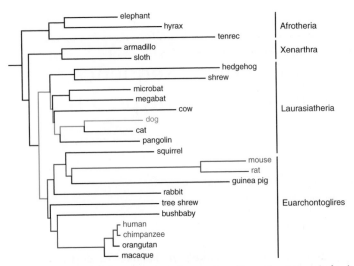

Figure 1. Dog (*red*) is the first sequenced organism from the Laurasiatheria clade. Human, mouse, rat, and chimpanzee (*blue*) were sequenced previously. Note that dog is evolving faster than human but slower than the rodents. Many additional mammals are in the process of being sequenced after dog (*black*), with the purpose of further annotating the human genome.

In conjunction with the sequencing, approximately 2.1 million single-nucleotide polymorphisms (SNPs) were identified in the boxer, 10 additional breeds, and 5 other canids, forming the dense, genome-wide marker set necessary for disease-gene mapping and phylogenetic analyses.

Both the genome and the extensive SNP set complement existing genetic resources in the dog community. The dog's 38 chromosomes, most of which are small and acrocentric, are well delineated (Breen et al. 1999; Chapter 10). Comparative canine:human maps and a modest meiotic linkage map have also been constructed. A radiation hybrid (RH) map of the dog genome includes over 10,000 genes, and 3,000 microsatellite markers and bacterial artificial chromosome (BAC) ends (Breen et al. 2001, 2004; Hitte et al. 2005). In addition, two deep BAC libraries are available (Li et al. 1999), as are a multiplexed set of microsatellite markers that span the genome at 8 cM density (Guyon et al. 2003; Clark et al. 2004), and a low-coverage draft sequence (~1.5✕) from a standard poodle has been published (Kirkness et al. 2003). With the full genome sequence and large SNP set both publicly and freely available, two fundamental new resources in canine research are in place to further facilitate comprehensive studies of gene expression, genome rearrangements, the proteome, RNAi, and the generation of tools for disease-gene mapping and comparative as well as population genetic studies.

Here, we summarize the preliminary analysis of the dog genome and its haplotype structure in the context of the remarkable evolutionary history of domesticated dogs. The first half of this chapter focuses on genome and gene evolution through comparative analysis with the human (Lander et al. 2001) and rodent (Waterston et al. 2002; Gibbs et al. 2004) genomes. The second half focuses on the SNP map and the characterization of linkage disequilibrium and haplotype structure in dog breed populations. Based on these findings, a strategy for canine disease gene mapping is proposed that is expected to yield findings relevant to both human and companion animal health.

A HIGH-QUALITY GENOME SEQUENCE

The high-quality draft assembly covering ~99% of the canine genome was generated from whole-genome shotgun (WGS) reads. The sequence was generated by the Broad Institute and Agencourt Biosciences, yielding 7.5-fold coverage of the genome (every base was sequenced, on average, 7.5 times). In WGS sequencing (Fig. 2), the genome is randomly fragmented and ~700-bp reads are generated from each of the fragments' ends. Using sequence overlaps, the reads are then rejoined into contiguous segments (contigs), much as one would assemble a jigsaw puzzle. Since the process of assembling the genome is more straightforward if the two copies of each chromosome are as alike as possible, a female

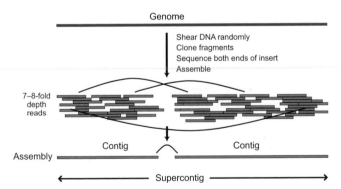

Figure 2. The principle of whole-genome shotgun sequencing. In whole-genome shotgun sequencing, the genome is randomly fragmented. Fragments of different sizes (typically 4 kb, 10 kb, 40 kb, and 200 kb) are cloned into plasmids, Fosmids or BACs, and the inserts are end sequenced. The resulting reads are joined together by sequence overlap (contigs) and linked together into larger structures (supercontigs) using the pairing of the two end reads from single fragments. To achieve coverage of more than 95% of the genome sequence, every position in the genome is sequenced on average 6–8 times.

boxer was chosen based on the low levels of heterozygosity observed in an earlier study (Parker et al. 2004). A female was chosen to ensure equal coverage of the X chromosomes and the autosomes. The resulting genome assembly was of remarkably high quality. Half of the bases in the assembly were found to reside in contigs larger than 180 kb, and for the majority of genes, the full gene sequence was contained within a single contig. In contrast, the contigs in the mouse assembly were on average only 25 kb long. As part of the assembly process, groups of contigs were ordered, oriented, and joined into supercontigs, using the information from paired reads. Fully half the bases in the assembly were found to reside in supercontigs longer than 45 million bases (45 Mb), as compared to the 17-Mb supercontigs seen in the mouse. Thus, just one or two supercontigs are needed to construct most canine chromosomes (Fig. 3). The high quality of the dog genome assembly can be attributed to higher coverage, higher quality data, better assembly algorithms and, critically, a "cooperative genome" with lower levels of repetitive and duplicated sequence than both the mouse and human genomes.

The assembly was anchored to the canine chromosomes using a combination of RH mapping and fluorescence in situ hybridization (FISH) data (Breen et al. 2004; Hitte et al. 2005). Roughly 97% of the assembled sequence was ordered and oriented on the chromosomes, with excellent agreement between the assembly and the RH and FISH data.

Although the bulk of the assembly approaches finished quality, there are a few megabase-sized regions that are clearly unreliable (e.g., over-collapsed). In addition, smaller problematic regions (comprising less than 1% of the genome in total) exist. To fully resolve all these regions, advances in assembly technology or additional clone-based sequencing is needed. To ensure that researchers can identify less reliable sequence, the CanFam2.0 assembly was released with annotations for local data quality (Fig. 3) (http://www.broad.mit.edu/mammals/dog/index.html).

GENOME AND GENE EVOLUTION

Interspecies comparison is one of the most versatile and powerful methods for examining the evolutionary processes that shape genomes, and it has proved effective when applied to organisms as diverse as bacteria, yeast, fruit flies, and mammals (Kellis et al. 2004; Kok et al. 2005; Richards et al. 2005). Comparing the human genome to chimpanzee, a very closely related species, highlights genetic differences between the two species (Lander et al. 2001; Ptak et al. 2005; T.S. Mikkelsen et al., in prep.), whereas using more distantly related mammals (such as rodents) identifies

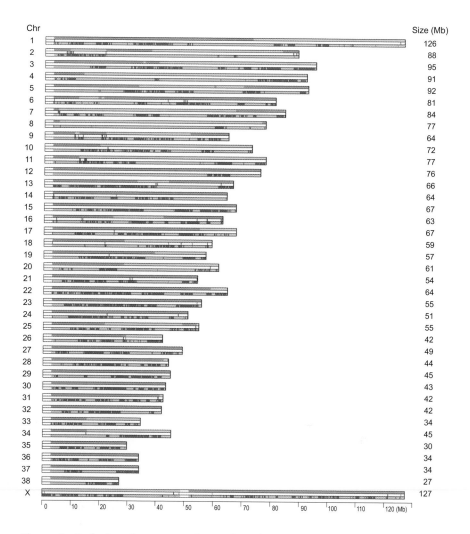

Figure 3. Each dog chromosome has only a few supercontigs and the structural quality is "certified" for over 99% of the genome. Schematic of dog genome chromosomes where the first bar shows consecutive supercontigs in alternating light and dark gray. The second bar shows the position of ∼0.5% of the genome residing in noncertified regions (*red*), where the assembly contains conflicting data. Most regions are smaller than 40 kb in size. The third bar shows regions of low and high heterozygosity in the sequenced boxer.

genomic features that are functionally conserved between the species (Waterston et al. 2002; Gibbs et al. 2004). However, rodent genomes have diverged considerably from the common ancestor of the eutherian mammals, due to a high rate of chromosomal rearrangement, deletions, and accumulation of nucleotide substitutions. The dog genome is the first complete genome sequence from the Laurasiatherian clade (Fig. 1) and thus predates the split of the primate and rodent lineages, offering a fresh perspective on mammalian genome evolution.

CONSERVED SYNTENY AND LARGE-SCALE REARRANGEMENTS

Regions of the dog and human genomes that share a common ancestry are called segments of conserved synteny and are identified by assessing sequence similarity. Their common ancestry reveals similarities and changes in the genomes since their evolutionary split. Large-scale conserved synteny analyses show general relationships across a large region, and are especially useful for pairing orthologous genes (derived from a common

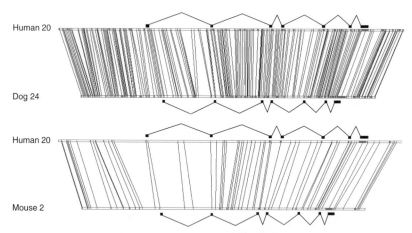

Figure 4. The human and dog genomes are more similar to each other than either one is to mouse. The segments of genomes of dog, human, and mouse that have evolved from the same segment in the common ancestor line up well, as can be seen from a 300-kb region on human 20, dog 24, and mouse 2. Note that more uniquely alignable sequences exist between dog and human (more closely spaced anchors) and that both the dog and mouse genomes are smaller than the human genome. These types of alignments are used to generate maps of conserved synteny between species (covering ~94% of the dog genome) and are very useful for translating information about genes (*black cartoon*) between species.

ancestral gene; Fig. 4) and identifying gene family expansions. Finer-scale synteny maps examine small intrachromosomal changes in detail, allowing examination of evolution at the base-pair level. When pair-wise maps of conserved synteny are made from high-quality genomic alignments of dog, human, mouse, rat, and chicken, the human and rodent maps each cover ~94% of the dog genome, whereas a nonmammalian species, such as the chicken genome, covers only ~76% of the dog genome due to a dearth of regions with clear common ancestry between the two species.

Given that dog is an out-group to the primate and rodent lineages in the evolutionary tree (Kirkness et al. 2003; Thomas et al. 2003; Froenicke 2005), we see clearly that rodents have undergone the highest number of genomic rearrangements relative to humans (with exceptions such as human chromosome 17 and dog chromosome 9). Although dogs have fewer interchromosomal rearrangements than rodents, they still have significantly more than humans (230 dog–human conserved segments at 500-kb resolution, 309 dog–mouse segments). In contrast, the rate of intrachromosomal reshuffling has been similar in the human and dog lineages. The complete dog sequence provides a higher-resolution synteny map than Zoo-FISH (Froenicke 2005) or low-coverage sequencing (Kirkness et al. 2003), which both delineated 80–100 undirected blocks, and comparable resolution to low-coverage sequencing combined with RH mapping (Hitte et al. 2004).

GENOME SIZE IS A RESULT OF GAIN AND LOSS OF DNA

The total size of the sequenced dog genome is ~2.4 gigabases (one gigabase [Gb] is one billion bases). Thus, the euchromatic portion of the dog genome is approximately 500 Mb (18%) smaller than the human genome and 150 Mb (6%) smaller than the mouse genome. This size difference is noticeable within the average segment of conserved synteny (Fig. 4) and can be attributed to two different factors: (1) a lower rate of repeat insertions in the dog genome relative to both human and mouse as well as smaller-sized SINEs in dog and (2) rates of ancestral base deletion that are approximately equal in the dog and human lineages, but higher in mouse (K. Lindblad-Toh et al., in prep.). Consequently, and despite our more recent common ancestry with mouse, the human genome shares approximately 650 Mb more ancestral sequence with dog than with mouse. Due to the low rate of both large-scale deletions and new repeat insertions, the dog genome content is likely closer to the ancestral eutherian mammalian genome, in terms of the proportion of the present-day sequence, than either human or mouse is.

REPEAT CONTENT

In addition to changing the size of a genome, the insertion of repetitive sequence elements plays a role in evolution, changing genomic functions. In fact, up to ~50% of mammalian genomes consist of repeats of three main types, long interspersed nuclear elements (LINEs), short interspersed nuclear elements (SINEs), and long terminal repeats (LTRs). A typical LINE might be 3–4 kb in size, whereas SINEs are usually only a few hundred bases in size. Some LINEs are likely to be involved in the retrotransposition, thereby remodeling the gene content of the organism (Brown 1999). Although the dog genome generally has fewer and older LINE and LTR repeats than human and mouse, Kirkness and coworkers have identified an active carnivore-specific SINE family (defined as SINEC_Cf). Consequently, the presence or absence of a SINE at specific loci is a form of genomic variability occurring much more frequently in dogs than in humans (Kirkness et al. 2003). Not only can these repeats be used as polymorphic markers, but they can also give rise to disease when improperly inserted in the genome (see Chapter 12).

NUCLEOTIDE DIVERGENCE

In a typical region of the genome, the rate of divergence in the dog genome is 20% faster than in the human genome, but 50% slower than in the mouse genome. Thus, human and dog, with an average divergence of ~0.35 substitutions/site, are much more similar to each other than either one is to mouse (Fig. 4). The mean nucleotide divergence rates for the dog, human, and mouse lineages were calculated by aligning all ancestral repeats in syntenic, nonoverlapping, 1-Mb windows, with consensus sequence for the repeats used as an outgroup. The reduced rate of divergence in the human lineage compared to dog, and in the dog lineage compared to mouse, is consistent with the known correlation between lower mutation rates, lower metabolic rates (Martin and Palumbi 1993; Gillooly et al. 2005), and longer generation times (Laird et al. 1969; Li et al. 1987), although the relative contributions of these factors remain unclear (Hwang and Green 2004).

As observed in all other mammalian genomes sequenced to date (Lander et al. 2001; Waterston et al. 2002; Kirkness et al. 2003; Gibbs et al. 2004), there is significant variation in the nucleotide divergence rate across the dog genome (coefficient of variation = 0.11 for 1-Mb windows, compared to 0.024 expected under a Poisson distribution). This regional variation is significantly correlated across conserved synteny in segments of the dog, human, and mouse genomes, but the strength of the correlation appears to decrease with total branch length (thus the correlation in mouse is weaker than that in human and dog). Lineage-specific variation in the regional divergence rates may be correlated with changes in other measures, such as sequence composition (G+C content), gene content, and chromosomal position, but further work needs to be done to understand factors that drive the process (Webber and Pointing 2005; T.S. Mikkelsen et al., K.A. Lindblad-Toh et al.; both in prep.).

G+C content averages 41%, but varies from approximately 25% to 60% (10-kb windows) in the dog and in human genome, whereas mouse has a slightly higher G+C content (42%) and less regional variation. As in other mammals, G+C content correlates with both chromosomal position and divergence rate (Hardison et al. 2003; Rodin and Parkhomchuk 2004; Yang et al. 2004).

FUNCTIONAL CONSERVATION WITHIN MAMMALIAN GENOMES

In the original comparison of the human and mouse genomes (Waterston et al. 2002), slightly more than 5% of the human genome showed greater evolutionary conservation than expected, based on a comparison of the rate of change in the genome as a whole and presumably nonfunctional ancestral repeats. With protein-coding genes comprising just ~1.5% of the genome, this suggested an additional 3.5% contained in unknown functional elements. However, with just two mammals, it was impossible to precisely define the boundaries, extent, or function of most of these elements (Miller et al. 2004). To remedy this, an additional 16 mammals are currently being sequenced to low coverage and aligned to the human genome (Fig. 1) (Margulies et al. 2005).

Adding a full genome from a different order to the comparative genome analysis (dog) has both refined estimates for the proportion of genome under selection and also identified novel functional elements. Using the same approach as was used in the human and rodent comparisons, approximately 5.4% of the orthologous nucleotides between human and dog were found to be under purifying selection. About 96% of these conserved human:dog nucleotides are also conserved in the orthologous mouse sequences, suggesting that most conserved elements are common to all placental mammals.

The positions of the most highly conserved elements between the human and dog genomes tend to be found in clusters that fall near genes involved in developmental regulation and patterning, and correspond to the most highly conserved regions across all vertebrates surveyed to date (Dermitzakis et al. 2004; Ovcharenko et al. 2005; K.A. Lindblad-Toh et al., in prep.).

In a powerful four-genome analysis, a search was conducted for novel functional elements in the conserved regions by surveying for overly conserved k-mers in the promoters and 3′-untranslated regions (UTRs) of ~14,000 orthologous genes in dog, human, mouse, and rat. A total of 180 potential transcription factor-binding sites were identified, half already known (Xie et al. 2005). The 3′UTRs contained ~70 clusters of excessively conserved 8-mers, including ~46% of known microRNA-binding sites. We found an additional ~130 novel microRNAs complementary in sequence to the unknown 8-mers.

Taken together, these measures of conservation are consistent with low dispensability and minimal turnover of ancestral constrained bases in the mammalian genome. However, as comparisons between distantly related mammals may not be sufficiently sensitive to detect weak constraints, 5% may still be an underestimate of the total proportion of the genome under purifying selection.

DOGS HAVE FEWER THAN 20,000 GENES

Defining and examining the protein-coding portion of the dog genome is essential for understanding both the genome itself and its evolutionary relationship to other mammals. For example, the evolution of the species is reflected in both the relative mutation rate of orthologous genes, which can reveal species-specific positive selection, and gene expansions through local genome duplications since the ancestral split.

Analysis of the dog genome identified 19,120 protein-coding genes (date, http//:www.broad.mit.edu/data/dog), considerably fewer than the ~22,000 human genes in the current ENSEMBL gene set (EnsEMBL build 26). By using conserved synteny and the orthologous relationships between genes in the two species, as well as careful scrutiny of genes lacking orthology, we were able to refine the genes in both species, resulting in a count of 20,426 human genes and 18,846 dog genes. The excess of ~1,600 human genes with no dog ortholog reflects reduced gene expansions in dog, suggesting that the dog gene content may best reflect the common ancestor of the currently sequenced placental mammals (K.A. Lindblad-Toh et al., in prep.).

FEW EXPANDED CANINE GENE FAMILIES

Although in general dogs have few gene family expansions, a few notable exceptions have been discovered. The two families with the largest numbers of dog-specific genes are the histone H2Bs and the α-interferons,

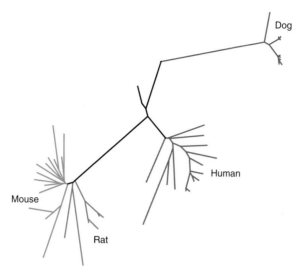

Figure 5. The α-interferon gene family has expanded independently in dog, human, mouse, and rat. This phylogenetic tree of mammalian α-interferon genes was created using a neighbor-joining methodology with Jukes-Cantor distance (A. Heger and C. Ponting, pers. comm.). Here we see independent expansion of the gene family within species, suggesting that each species is evolving independently for these genes. In general, dog has relatively low levels of gene family expansion, which is why they have a lower gene number than human.

which cluster in monophyletic clades when compared to their human homologs (Fig. 5). The gene family expansion is particularly striking for the interferons, since the same pattern is seen in human, mouse, rat, dog, cat, and horse. This homogenization of gene sequences could have arisen from multiple events of in toto gene duplication in each of these six species or by ongoing gene conversion events (Hughes 1995). The histone H2Bs, which are split between two chromosomes in dog, form two distinct monophyletic clades corresponding to their chromosomal location exactly as would be expected through conversion of neighboring paralogous genes. A third well-known case of dog gene expansion, compared to human, is the set of olfactory receptor genes described in Chapter 13.

FEW GENE FAMILIES EVOLVE EXCEPTIONALLY FAST IN DOGS

Genes under particularly strong selection in a species may show a higher relative rate of evolution, as measured by Ka/Ks (mutations that cause amino acid changes vs. those that are silent). We compared the relative evolutionary constraints on the human, dog, and mouse orthologs for ~14,000 genes, the first such comparison across three major mammalian lineages. When genes were grouped according to function, the relative rate of evolution (strength of selection) between the functional groups was highly correlated in the three species. In contrast, the absolute rate of evolution (total number of substitutions) was significantly higher in the dog lineage than in human, but lower than in mouse. This reflects the pattern for rates of neutral evolution discussed above (K.A. Lindblad-Toh et al., in prep.).

Although the signal was quite weak, there was evidence for dog-specific accelerated evolution in some gene families. These were primarily related to metabolism, and they may contain promising candidates for studies of molecular adaptation in carnivores. Some nervous system-related genes appear to have diverged faster in dog and human relative to mouse, but not relative to each other, suggesting similar selection pressures and possibly convergent evolution as suggested previously (Hare et al. 2002). Genes expressed in testis and those functioning in the mitochondrial electron transport chain are uniquely accelerated on the lineage leading to humans.

POPULATION GENETICS AND ITS EFFECT ON THE GENOME

The unique breed structure of the domestic dog population may make disease-gene-mapping studies especially powerful (Ostrander and Kruglyak 2000; Sutter et al. 2004). To fully utilize this potential, researchers need both a large, uniformly spaced marker set and a good understanding of the extent of linkage disequilibrium in their study population. Although many excellent maps and mapping reagents have been produced by the community during the past 15 years (see, e.g., Breen et al. 1999, 2004; Neff et al. 1999; Hitte et al. 2004), the full genome sequence for the first time made it possible to develop an extensive single-nucleotide polymorphism (SNP) map. A set of 2.1 million canine SNPs, designed to be useful for mapping in many breeds, is now publicly available. With this resource, and additional sequence data from many dog breeds and other canids, it has been possible to characterize the haplotype structure and linkage disequilibrium (LD) of the domestic dog, both within single breeds and across the dog population.

A 2.1 MILLION SNP MAP

The genome-wide set of 2.1 million canine SNPs were identified using three distinct methods designed to maximize both the genome-wide

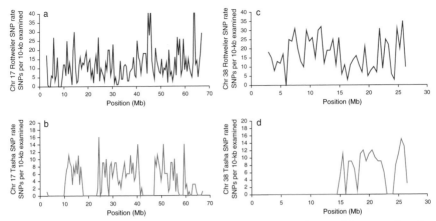

Figure 6. The SNP distribution is strikingly different within and between breeds. The SNP distribution on chromosomes 17 and 38, shown as random reads from a different breed (rottweiler) to that of the boxer assembly (**a**) and within breed by comparing the boxer's two chromosomes (**b**). Within breeds, large homozygous regions where the boxer has two identical haplotypes exist; between breeds different chromosomes are sampled.

marker density and the utility of the SNPs for disease mapping in many breeds. To capture the diversity of the dog population, the boxer sequence was compared to ~100,000 sequence reads from each of 9 diverse dog breeds and ~22,000 sequence reads from each of 4 gray wolves and a single coyote. In addition, the boxer was compared to 1,321 Mb of sequence from the 1.5× poodle assembly (Kirkness et al. 2003). Finally, the boxer's two chromosomes were themselves compared.

The ~440,000 uniquely placed SNPs discovered in 9 canine breeds and 5 canids had a uniform density of about 1 SNP/6 kb across the genome (Fig. 6), or several SNPs per gene. This set is sufficiently dense for regional mapping and should be particularly useful given the diverse set of breeds chosen. The ~1.4 million SNPs discovered by comparing the boxer and the poodle are also uniformly spaced across the genome, whereas the ~770,000 SNPs discovered in the sequenced boxer reflect her inherent pattern of alternating regions of homozygosity and heterozygosity (Fig. 6, see below).

To assess the likely utility of these SNPs as polymorphic markers in a random dog breed, ~1,300 SNPs were genotyped in 20 dogs from 10 breeds distinct from the 2 used in the initial SNP discovery. Twenty dogs provide sufficient power for an 85% probability of discovering 5% alleles within a breed. On average, ~72% of SNPs were polymorphic in any third random breed. Thus, these SNPs will be useful for genome scans and chromosome marker panels, in addition to fine mapping within and across breeds (K.A. Lindblad-Toh et al., in prep.).

HISTORY SUGGESTS LONG HAPLOTYPES WITHIN BREEDS AND SHORT HAPLOTYPES ACROSS BREEDS

The haplotype structure across the genome of modern purebred dogs exhibits a distinctive pattern caused by two strong population bottlenecks. The first, the domestication of dogs from wolves (Vila et al. 1997; Wayne and Ostrander 1999; Savolainen et al. 2002), echoes humanity's own bottleneck during our migration out of Africa. Thus, across the whole dog population, the genome fractures into short haplotype blocks (<100 kb) similar in size to those found in humans (see, e.g., Daly et al. 2001; Gabriel et al. 2002; Wall and Pritchard 2003). The second bottleneck, the creation of breeds from the domestic dog population, mimics the recent breeding of laboratory mouse strains. Thus, within dog breeds, "breed-derived" haplotypes extend for megabases, much like those in modern laboratory mice (Fig. 7) (Grupe et al. 2001; Wade et al. 2002; Wiltshire et al. 2003; Pletcher et al. 2004).

The population bottleneck at domestication is strongly supported by the SNP rates observed in dog breeds (~1 SNP per 900 nucleotides [nt] for a breed vs. the boxer), which is lower than the SNP rate between dog and wolf (1/580 nt) (K.A. Lindblad-Toh et al., in prep.). The only outlier breed in this analysis, the Alaskan malamute (1/790 nt), is known to be of Asian origin (Parker et al. 2004), suggesting that it might have more

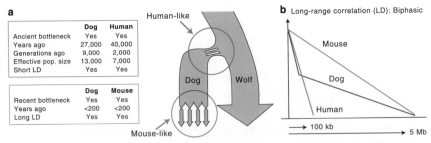

Figure 7. Two bottlenecks, one at dog domestication and one at breed formation, have shaped the linkage disequilibrium (LD) in the dog population. (**a**) The bottleneck at domestication mimics that of the human out-of-Africa bottleneck and is expected to lead to short LD within the population. The recent breed creation bottleneck occurred within the last few hundred years like the creation of laboratory mice and is likely to cause long LD within breeds. (**b**) Taken together, this suggests a bimodal decline of LD within breeds.

diverse wolf ancestry and therefore a higher divergence. The SNP rates for dog compared to wolf (1/580 nt) and coyote (1/420 nt) are considerably smaller than the ~1/70 nt calculated between human and chimpanzee (T.S. Mikkelsen et al., in prep.), reflecting the recent common ancestry of these canid species.

The presence of a breed creation bottleneck, and ensuing extensive LD, are supported by the very low heterozygosity within the boxer (1/1635 nt) and other breeds (Parker et al. 2004) (see Box A) and by preliminary studies of LD (Chapter 9). To examine dog LD in depth, the full genome sequence and SNP map were augmented with carefully designed

---

Box A. Linkage disequilibrium and its importance for mapping

LD describes the association between pairs of loci in the genome. For example, if individuals with allele A1 at locus A are significantly more likely to have allele B1 at locus B, the two loci are in linkage disequilibrium. LD is commonly expressed using two statistics, D' and r^2.

D' calculates the amount of recombination between two loci. D' equals 1 when no recombinations are apparent and the loci appear perfectly linked, and equals zero when extensive recombination has destroyed any association. Although D' can be a useful indicator for LD, it is possible to obtain high values of D' in the absence of real association, especially when loci with rare alleles are used. The r^2 statistic, which also varies from 0 to 1, measures the extent to which the observed allelic associations deviate from the pattern expected through random association. It tends to be a more sensitive measure than D' and is usually lower, particularly when each locus has more than two alleles.

Generally we expect genome-wide LD to be shorter in older populations, because each generation provides an opportunity for recombination which breaks the association. Typically, there is one recombination event on each arm of a chromosome at each meiosis.

However, recent population bottlenecks can abruptly change the extent of LD. If just a few chromosomes are captured from an old population with short LD, as occurred in dog breed creation, the new population will begin anew with no evidence of recombination. The allelic associations inherited on the selected chromosomes will be apparent until sufficient generations pass, resulting in long regions of the genome with high levels of LD. The sets of associated alleles observed across a region of high LD are called haplotypes. A haplotype block is a segment of a chromosome with high LD that is usually flanked by sites of recombination.

The extent of LD is a critical factor in the design of gene-mapping experiments. When LD is very long, few markers are needed to detect the patterns of variation across the genome and find haplotypes associated with the phenotype of interest. However, the genomic regions implicated in disease will also be long, complicating identification of the specific causative mutation. When LD is short, the haplotypes are also much shorter, and many more markers are needed to resolve the patterns of variation. Once a disease-associated region is discovered, however, it will be much shorter and contain just a few genes, significantly simplifying the search for a causative mutation.

sets of resequencing and genotyping data that spanned three magnitudes of scale (kb to Mb) and included 6% (150 Mb) of the genome divided among 10 randomly chosen regions (K.A. Lindblad-Toh et al., in prep.). To assess the diversity of the dog population, 1,283 SNPs found in these regions were genotyped in 24 diverse breeds, with one dog from each breed. To examine within-breed variability, the same SNP set was genotyped in 20 dogs from each of 10 breeds (200 dogs total). Complete resequencing in the first ~10 kb of each region in the 24 diverse breeds, including 509 SNPs across a total of 79 kb, provided sufficient density to capture small haplotype blocks expected across the dog population (see above).

STRIKING HAPLOTYPE PATTERN IN THE BOXER

Using the large data set available from the boxer sequencing, differences between the haplotypes of the two parental chromosomes were measured across the entire genome assembly. The analysis revealed a distinctive pattern that alternated long regions of near total homozygosity with equally long regions of high heterozygosity, suggesting the boxer genome is a composite of haplotypes which are either identical or very different (Figs. 3 and 6). The haplotypes are exceptionally long (most of the sequence resides in haplotypes that are several megabases long) and, in total, more than half of the genome lies in homozygous blocks. The SNP rate in the heterozygous regions matches that observed for other breeds (1/900 bp), suggesting the variation in the sequenced boxer is not unlike that of other dogs.

LONG REGIONS OF HOMOZYGOSITY AND LINKAGE DISEQUILIBRIUM IN MANY BREEDS

The 10 breeds representing different breed groups (Parker et al. 2004) and diverse population histories with breed-creation bottlenecks of varying severity were selected to examine the haplotypes within breeds more closely. For example, the small American population of basenjis, an ancient breed possibly originating in Egypt, is descended primarily from two dogs brought from Africa in the 1940s (Fogel 1995; Wilcox 1995). The popular Labrador retriever, with about 150,000 new dogs registered each year, has a large, diverse breed population, while both the golden retriever and pug, although popular, have limited population diversity due to extensive line breeding and excessive use of popular sires (Ostrander and Kruglyak 2000). The 10 regions described above were genotyped in 200 dogs, 20 dogs from each breed, sufficient to permit detection of alleles with frequencies ≥5%.

Within all 10 breeds, homozygosity extended over long distances, in a manner very similar to that already observed in the boxer. The first 10 kb of a region was completely homozygous in 38% of cases (n = 645), and every single dog examined was homozygous in at least one of the 10 regions. From the total of 244 homozygous 10-kb regions, 46% maintained homozygosity out to 1 Mb, and 17% out to 10 Mb (Fig. 8). Sam-

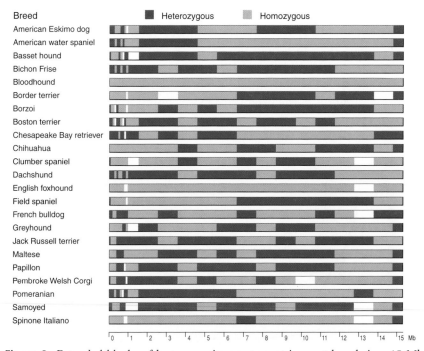

Figure 8. Extended blocks of homozygosity are present in most breeds in a 15-Mb region on chromosome 12. Blocks of continuous homozygosity (*light blue*) and heterozygosity (*dark blue*) within each dog are plotted for 24 individual dogs from different breeds. The first megabase is binned by 100-kb windows and the remaining 14 Mb by 1-Mb windows (which might somewhat overestimate the window sizes). Each dog is a random mosaic of homozygosity and heterozygosity representative of that individual dog and not of the breed. Regions containing genes influencing breed-defining traits might be expected to be under selection and therefore show increased homozygosity within a breed. However, among our 10 regions and 10 breeds we saw no such instances larger than the ancestral haplotype blocks, suggesting either that our 10 regions have not been under particular selective pressure in these breeds, or that the mosaic pattern of ancestral haplotypes makes it harder to detect selection across large regions.

pling the boxer genome at similar intervals showed an almost identical pattern of homozygosity, suggesting that the long haplotypes in the boxer genome are typical of almost all dog breeds (K.A. Lindblad-Toh et al., in prep.).

The extensive regions of homozygosity, and inferred long haplotype blocks, suggest that within breeds, long LD makes whole-genome association mapping feasible. In each dog breed, all 1300 SNPs across the ten 15-Mb regions were genotyped to measure the change in r^2 (a biallelic measure of LD taking allele frequencies into account) with distance. Within each breed, LD initially declined sharply and then plateaued, typically indicating that LD could extend for several more megabases. This extent of LD is roughly 100× longer than seen in human and slightly longer than in inbred mice (Figs. 6 and 9). The only breed with substantially shorter LD was the Labrador retriever, with its mixed etiology and large population size. Among the 10 different regions tested, no major variation in the extent of LD was observed (K.A. Lindblad-Toh et al., in prep.).

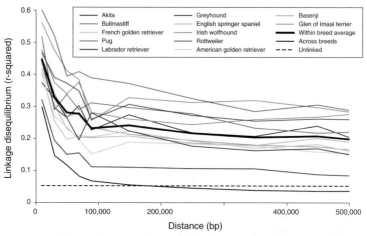

Figure 9. LD is similar for most breeds but also correlates with population history. Note that LD stays above unlinked for all breeds well past the 500 kb shown here. For most breeds LD stays above unlinked as far as 5–15 Mb. Trends in the strength of LD agree with population history, with the most severe bottlenecks among the breeds tested here reported in pugs and Irish wolfhound and less severe bottlenecks in golden retrievers and English springer spaniels. Note that the LD curves in French and American golden retrievers are very similar, suggesting that the original bottleneck at breed creation was more important than later effects of importation to the U.S. or popular sire effects. The clearest difference is seen in Labrador retrievers, which are known to be a breed with mixed ancestry and a large population.

BREED-CREATION BOTTLENECK CORRELATES WITH BREED HISTORY

Through coalescent simulations, the experimental results were recapitulated using a dual-bottleneck population model for dog breeds, assuming commonly accepted mammalian mutation and recombination rates. With an initial domestication bottleneck (inbreeding coefficient F = 0.12) approximately 27,000 years ago (9000 generations) with an effective population size of 13,000, and a subsequent, breed-creation bottleneck, the simulated data successfully mimicked the experimentally determined SNP rates and LD. This model could be further refined with additional data on variation in the wolf population.

Modeling the second breed-creation bottleneck for individual dog breeds, each with its own distinct population history, is considerably more complex. On average, breed creation about 50 generations ago (100–200 years) with a moderate bottleneck (F = 0.12) recapitulates the experimental data and is consistent with historical records. However, Labrador retrievers, and to a lesser extent golden retrievers and English springer spaniels, are more accurately simulated with a slightly less severe breed-creation bottleneck, as might be predicted from their population history.

Although the dual-bottleneck model worked well for most breeds, the Akita was best fitted by a three-bottleneck model that included an intermediate bottleneck ~450 generations ago and a final bottleneck ~30 generations ago. This may reflect its ancient origins as a hunting dog in Japan and subsequent, more recent introduction into the U.S. in conjunction with World War II (http://www.akc.org/breeds/akita/history.cfm). Although other breeds may have analogous histories, apparently only in the Akita was the intermediate bottleneck severe enough or distant enough from the last bottleneck to distinguish it from the normally dominant modern breed-creation event. For example, the American and French golden retrievers (unpublished data) have remarkably similar LD curves, suggesting that initial breed bottleneck dominates more recent influences such as dominant sire effects and importation into the U.S. (Fig. 9). The relatively low diversity present in most breeds after a strong breed-creation bottleneck, equivalent to the effective passage of only about four chromosomes, may explain this phenomenon.

LINKAGE DISEQUILIBRIUM IS SHORT ACROSS BREEDS

Across the whole dog population, as represented by single dogs from each of 24 different breeds, LD is roughly 100-fold shorter than when measured within breeds (see above) (Fig. 9). The rate of decline in LD exceeds that observed in humans (Gabriel et al. 2002). As with the within-breed analysis, little variation in the extent of LD was found among the 10 randomly chosen loci.

ANCESTRAL HAPLOTYPE BLOCKS UNDERLIE
BREED-DERIVED HAPLOTYPES

The intermediate level at which LD plateaus within dog breeds, after the initial rapid decline, suggests that the long breed-derived haplotypes are a mosaic of shorter ancestral haplotype blocks (Fig. 10). Consequently, at any point, different breed-specific haplotypes may share the same ancestral allele, lowering the measured LD. Using the 10 very densely genotyped 10-kb regions, the haplotype pattern within dog breeds was compared to the pattern observed across the domestic dog population, as represented by a single dog from each of 24 breeds. The mosaic pattern is clearly evident when analyzing across the dog population, with each 10-kb region containing a few short ancestral haplotype blocks, each with 4–5 haplotypes. These ancestral blocks are slightly smaller in dogs than in humans, in concordance with observed shorter LD. This small difference is expected, given that dog has a larger population than human, and more generations have passed since dog domestication than since the human migration out of Africa. In each ancestral block the major haplotype dominates with an average frequency

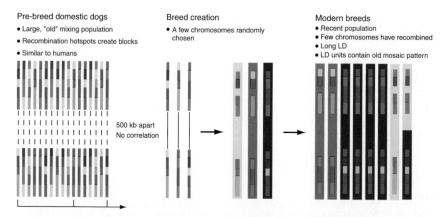

Figure 10. Breed-derived haplotypes consist of shared ancestral haplotype blocks. Within the dog population as a whole, a short-range haplotype block pattern can be expected (ancestral haplotype blocks) based on the long time and large population size since the domestication bottleneck. As modern breeds were created, a random, or semi-random, set of chromosomes was selected from the pool of domestic dogs. The short time since this bottleneck means that haplotypes remain long. These long breed-derived haplotypes, however, retain the mosaic ancestral haplotype pattern from the domestic dog population. (Reproduced from K.A. Lindblad-Toh et al., in prep.)

of 55%, suggesting that many breed-derived haplotypes will share this ancestral haplotype.

HAPLOTYPE DIVERSITY AND SHARING

The presence of several breeds with similar risk factors could enhance gene mapping in dogs, as candidate regions in which the breeds share haplotypes could feasibly contain a disease mutation. The recent formation and geographically related backgrounds of many breeds, and the unsuccessful attempts at resolving a tree structure among the European breeds (Zajc et al. 1997; Koskinen and Bredbacka 2000; Irion et al. 2003), indicates that these breeds are not drawn from distinct lineages. An analysis of haplotype diversity and sharing within breeds confirms this, strongly suggesting that modern breeds can share smaller segments of haplotypes. As mentioned above, the ancestral haplotype blocks identified across 24 diverse breeds include 4–5 haplotypes, of which 1–2 are very common (K.A. Lindblad-Toh et al., in prep.).

In 10-kb windows of genomic sequence, an average of approximately 10 haplotypes occur across all breeds, but only about 4 of these are found within each single breed. Many of the 10-kb haplotypes were common in one breed but are rare in other breeds and less common across the dog population as a whole. Nonetheless, ~80% of haplotypes with a breed frequency of >5% are found in at least one other breed. One of two common haplotypes is seen in the majority of breeds, although sometimes at low frequencies. This pattern persists across longer 100-kb windows, with 30% of haplotypes found in at least 6 of the breeds, often at greatly varying frequencies. This extent of haplotype sharing strongly suggests a founder effect in the dog population.

With extensive haplotype sharing and the relatively recent common ancestry of most modern breeds, disease alleles likely will be frequently shared between breeds. However, detecting the short ancestral block containing the disease allele would require a very dense map of markers, and thus researchers may instead rely on longer co-occurring segments. For example, shared ancestral haplotypes across 9 breeds from different groups have been reported for the multi-drug resistance (MDR1) locus (Neff et al. 1999).

RANDOM MONOMORPHISM VERSUS SELECTION

The expected occurrence of monomorphic loci based on the population structure in dogs is indicated by the rate of heterozygosity within breeds.

With strong, recent selection, large megabase-sized regions may be fixed (Sabeti et al. 2002; Bersaglieri et al. 2004). At the other extreme, monomorphism may be restricted to the ancestral haplotype blocks occasionally shared by breed-derived haplotypes. In a thorough analysis of monomorphism, heterozygosity among 20 individuals of the same breed was measured for haplotypes of greatly varying size, from the small ancestral haplotype blocks (in 3/10 regions) to 15 Mb, the length of the 10 regions. The analysis included 10 diverse breeds, all of which behaved similarly, although the Labrador retrievers were rather more heterozygous than the average and the Irish wolfhound and pugs slightly less so.

For the 100-kb, 1-Mb, and 15-Mb regions, heterozygosity increased with haplotype size as expected, averaging 70%, 80%, and 95%, respectively. In no case did the heterozygosity drop below 20% (K.A. Lindblad-Toh et al., in prep.). Thus, among the 10 regions tested, there was no evidence for sufficiently strong selection to drive the breed-derived haplotype to fixation. Alternatively, an ancestral haplotype, although under selection, might occur on multiple breed-derived haplotypes, thus masking the signature of selection. This suggests that the search for regions under strong selection might be slightly more complex than previously hypothesized. However, the identification of regions associated with traits known to have been under selective pressure, such as coat color, morphology, and behavior, will likely improve the understanding of the signatures of selection in the dog genome.

In contrast, when the ancestral haplotype blocks were examined, within-breed heterozygosity rates as low as 0% (only one haplotype) were observed in 13% of blocks. All breeds, except Labrador retrievers, contained at least one monomorphic region. This reflects the population history, as the fraction of monomorphic loci is proportional to the inbreeding coefficient of each breed (F = 0.12 should result in 12% of loci having coalesced). Thus, the mosaic pattern of the breed-derived haplotypes causes a fraction of the genome to be monomorphic within a breed.

For researchers mapping segregating traits, these monomorphic ancestral blocks are of limited concern. However, should one want to identify all monomorphic/fixed loci in a breed, the small size of the ancestral blocks could make the task quite labor-intensive.

DENSITY OF MARKERS NEEDED FOR A GENOME SCAN

The long LD observed within a breed immediately suggests that roughly 50- to 100-fold fewer markers might be needed for a genome-wide association in dogs (5,000–10,000 SNPs) versus in humans (~500,000 SNPs). However, the moderate level of homozygosity within a breed, as well as the lack of association for some markers based on the underlying shared ancestral haplotypes, might make association mapping within breeds slightly noisier than the long LD would suggest. Using real data from the 10 random regions of the genome (K.A. Lindblad-Toh et al., in prep.), as well as data from coalescent simulations, we were able to show that the absolute majority of SNPs (segregating disease mutations) can be captured by association using this marker density, but that association is detected much more cleanly if haplotypes rather than single SNPs are used as markers for the association. Thus, to be able to form at least 2 SNP haplotypes within 500-kb windows (where little recombination can be expected) a set of 10,000 SNPs or more would be desirable. This marker density would work within the majority of breeds, except Labrador retrievers, where the LD is shorter and the haplotype diversity larger. It is, however, worth noting that this marker density is not sufficient for whole-genome-wide association mapping across the dog population as a whole (K.A. Lindblad-Toh et al., in prep.).

To facilitate whole-genome-wide association and comparison of data between studies, ideally uniform mapping panels of at least 10,000 SNPs should be developed and used uniformly between studies.

DISEASE MAPPING BY GENOME-WIDE ASSOCIATION

Based on the long LD within breeds and the high disease frequency in some breeds as compared to other breeds, suggesting the presence of a major genetic risk factor, it seems feasible that whole-genome-wide association mapping could be performed using the following two-tiered approach (Fig. 11) (K.A. Lindblad-Toh et al., in prep.). First, a breed with a high disease risk and sufficiently long LD should be identified. A genome-wide scan using 10,000–20,000 SNPs could be performed using 100–200 affected and 100–200 unaffected unrelated individuals for many traits conferring a >2-fold increased risk. Such a mapping effort would be expected to yield an associated region of 5–10 Mb in size, where further fine mapping within the breed might be difficult without using large numbers of offspring.

Second, because related breeds that share the disease phenotype may share disease haplotypes, we propose using the shared ancestral haplotypes present on breed-derived haplotypes to further narrow the disease-associated region is the optimal approach. The fact that derived haplotypes of up to 100 kb are frequently shared between breeds suggests that with a limited number of affected and unaffected individuals from two or three additional breeds, one could rapidly narrow the disease-associated region to contain only a few genes and thereby limit the amount of mutation detection needed.

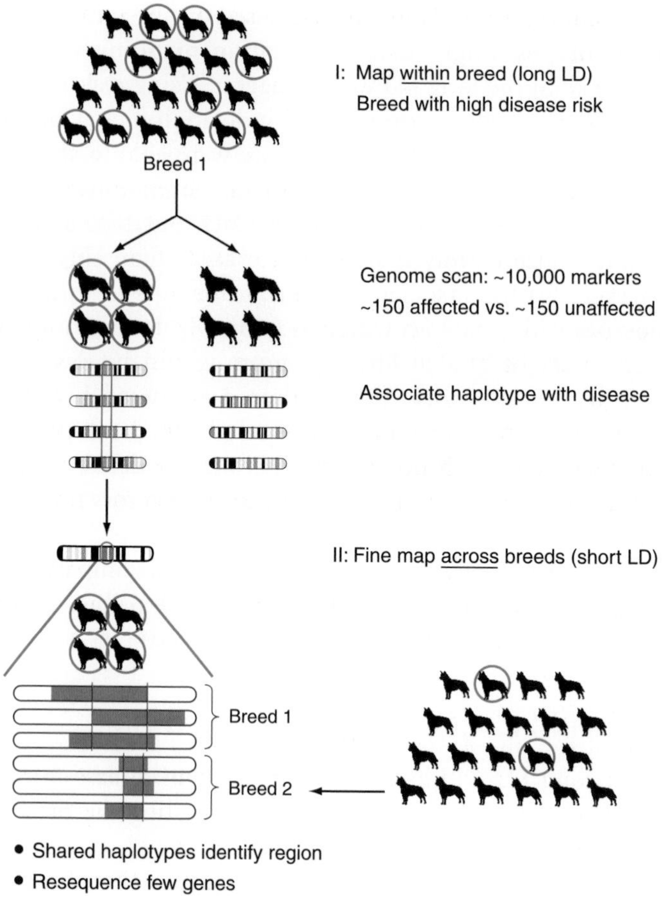

I: Map <u>within</u> breed (long LD)
Breed with high disease risk

Breed 1

Genome scan: ~10,000 markers
~150 affected vs. ~150 unaffected

Associate haplotype with disease

II: Fine map <u>across</u> breeds (short LD)

Breed 1

Breed 2

- Shared haplotypes identify region
- Resequence few genes

Figure 11. Proposed strategy for genome-wide association mapping. We propose a two-tiered mapping strategy. In the first step, whole-genome-wide association mapping using ~10,000–20,000 SNPs is employed in a breed with a high disease risk. Second, fine-mapping of the disease-associated region will be performed in the initial breed together with several related breeds, thus taking advantage of the ancestrally shared disease haplotypes. This should permit rapid narrowing of the region to enable mutation screening of only a few genes.

CONCLUSION

The 2005 dog genome assembly (CanFam2.0) is of very high quality and, as such, will form the basis for studies of mammalian evolution, population genetics, and disease gene mapping for many years to come. From a genomic point of view, anything that can be done in human or mouse, be it arrays for expression studies or large-scale rearrangements, analysis of protein evolution, or tools for disease-gene mapping, can now also be accomplished in dogs.

From an evolutionary standpoint, the dog is an excellent organism. Apart from having a karyotype with a large number of chromosomes, the canine genome is a "typical" mammalian genome with a medium rate of nucleotide evolution and relatively low levels of deletion, insertion, and rearrangement. The high similarity in gene content between human and dog means that it is straightforward to identify the orthologous gene in the other species, if one has a gene with a mutation in one species. However, it is necessary to keep in mind that within gene families (of which human has more than dog) other family members might complement the lost function of a disrupted gene. As the sequencing of many mammals (Margulies et al. 2005) reveals additional noncoding conserved elements, it will also become easier to determine whether mutations reside within such elements and therefore are likely to be functional.

The high observed risks for certain diseases in specific breeds suggest that strong genetic risk factors are present within these breeds. The haplotype structure resulting from the two bottlenecks at domestication from the wolf and at breed formation is very amenable to disease-gene mapping. In particular, linkage is roughly 100-fold longer within dog breeds than in the human population, thus making them a prime candidate for whole-genome association mapping with only ~10,000 SNPs as compared to the ~500,000 SNPs needed in humans. In addition, the small number of ancestral haplotypes and the sharing of ancestral and derived haplotypes between breeds suggest that fine-mapping should utilize multiple breeds that are expected to share the same disease allele.

ACKNOWLEDGMENTS

We thank everyone who contributed to the sequencing and analysis of the dog genome, including the researchers and dog owners who provided DNA samples, the Broad Institute and Agencourt Biosciencies Sequencing Platforms, and the dog community for its work on the White Paper and advice throughout the project. We salute the entire Broad genome assembly team, who produced a sequence of exceptional quality, making this research possible. We are especially grateful to the Breen, Ostrander, and Galibert laboratories, who generated the FISH and RH map data needed to anchor the genome, and to the numerous Broad researchers, including Mike Kamal, Jean Chang, David Jaffe, Michele Clamp, Edward

J Kulbokas III, and Tara Biagi, for the in-depth analysis of the dog genome. Finally, we thank Evan Mauceli, Colleen McCarthy, Leslie Gaffney, and the editors of this book for their assistance with this manuscript, and the NHGRI (HG03067 and HG003069) who funded the dog genome sequencing project in part.

REFERENCES

Bersaglieri T., Sabeti P.C., Patterson N., Vanderploeg T., Schaffner S.F., Drake J.A., Rhodes M., Reich D.E., and Hirschhorn J.N. 2004. Genetic signatures of strong recent positive selection at the lactase gene. *Am. J. Hum. Genet.* **74:** 1111–1120.

Breen M., Hitte C., Lorentzen T.D., Thomas R., Cadieu E., Sabacan L., Scott A., Evanno G., Parker H.G., Kirkness E.F., et al. 2004. An integrated 4249 marker FISH/RH map of the canine genome. *BMC Genomics* **5:** 65.

Breen M., Jouquand S., Renier C., Mellersh C.S., Hitte C., Holmes N.G., Cheron A., Suter N., Vignaux F., Bristow A.E., et al. 2001. Chromosome-specific single-locus FISH probes allow anchorage of an 1800-marker integrated radiation-hybrid/linkage map of the domestic dog genome to all chromosomes. *Genome Res.* **11:** 1784–1795.

Breen M., Langford C.F., Carter N.P., Holmes N.G., Dickens H.F., Thomas R., Suter N., Ryder E.J., Pope M., and Binns M.M. 1999. FISH mapping and identification of canine chromosomes. *J. Hered.* **90:** 27–30.

Brown T.A., ed. 1999. *Genomes.* John Wiley and Sons, New York. pp. 137–140.

Clark L.A., Tsai K.L., Steiner J.M., Williams D.A., Guerra T., Ostrander E.A., Galibert F., and Murphy K.E. 2004. Chromosome-specific microsatellite multiplex sets for linkage studies in the domestic dog. *Genomics* **84:** 550–554.

Daly M.J., Rioux J.D., Schaffner S.F., Hudson T.J., and Lander E.S. 2001. High-resolution haplotype structure in the human genome. *Nat. Genet.* **29:** 229–232.

Dermitzakis E.T., Kirkness E., Schwarz S., Birney E., Reymond A., and Antonarakis S.E. 2004. Comparison of human chromosome 21 conserved nongenic sequences (CNGs) with the mouse and dog genomes shows that their selective constraint is independent of their genic environment. *Genome Res.* **14:** 852–859.

Fogel B., 1995. *The encyclopedia of the dog.* DK Publishing. New York, New York.

Froenicke L. 2005. Origins of primate chromosomes—as delineated by Zoo-FISH and alignments of human and mouse draft genome sequences. *Cytogenet. Genome Res.* **108:** 122–138.

Gabriel S.B., Schaffner S.F., Nguyen H., Moore J.M., Roy J., Blumenstiel B., Higgins J., DeFelice M., Lochner A., Faggart M., et al. 2002. The structure of haplotype blocks in the human genome. *Science* **296:** 2225–2229.

Gibbs R.A., Weinstock G.M., Metzker M.L., Muzny D.M., Sodergren E.J., Scherer S., Scott G., Steffen D., Worley K.C., Burch P.E., et al. 2004. Genome sequence of the Brown Norway rat yields insights into mammalian evolution. *Nature* **428:** 493–521.

Gillooly J.F., Allen A.P., West G.B., and Brown J.H. 2005. The rate of DNA evolution: Effects of body size and temperature on the molecular clock. *Proc. Natl. Acad. Sci.* **102:** 140–145.

Grupe A., Germer, Usuka J., Aud D., Belknap J.K., Klein R.F., Ahluwalia M.K., Higuchi R., and Peltz G. 2001. In silico mapping of complex disease-related traits in mice. *Science* **292:** 1915–1918.

Guyon R., Lorentzen T.D., Hitte C., Kim L., Cadieu E., Parker H.G., Quignon P., Lowe J.K., Renier C., Gelfenbeyn B., et al. 2003. A 1-Mb resolution radiation hybrid map of the canine genome. *Proc. Natl. Acad. Sci.* **100**: 5296–5301.

Hardison R.C., Roskin K.M., Yang S., Diekhans M., Kent W.J., Weber R., Elnitski L., Li J., O'Connor M., Kolbe D., et al. 2003. Covariation in frequencies of substitution, deletion, transposition, and recombination during eutherian evolution. *Genome Res.* **13**: 13–26.

Hare B., Brown M., Williamson C., and Tomasello M. 2002. The domestication of social cognition in dogs. *Science.* **298**: 1634–1636.

Hitte C., Madeoy J., Kirkness E.F., Priat C., Lorentzen T.D., Senger F., Thomas D., Dernen T., Ramirez C., Scott C. et al. 2005. Opinion: Facilitating genome navigation: Survey sequencing and dense radiation-hybrid gene mapping. *Nat. Rev. Genet.* (in press).

Hughes A.L. 1995. The evolution of the type I interferon gene family in mammals. *J. Mol. Evol.* **41**: 539–548.

Hwang D.G. and Green P. 2004. Bayesian Markov chain Monte Carlo sequence analysis reveals varying neutral substitution patterns in mammalian evolution. *Proc. Natl. Acad. Sci.* **101**: 13994–14001.

Irion D.N., Schaffer A.L., Famula T.R., Eggleston M.L., Hughes S.S., and Pedersen N.C. 2003. Analysis of genetic variation in 28 dog breed populations with 100 microsatellite markers. *J. Hered.* **94**: 81–87.

Kellis M., Birren B.W., and Lander E.S. 2004. Proof and evolutionary analysis of ancient genome duplication in the yeast *Saccharomyces cerevisiae*. *Nature* **428**: 617–624.

Kirkness E.F., Bafna V., Halpern A.L., Levy S., Remington K., Rusch D.B., Delcher A.L., Pop M., Wang W., Fraser C.M., and Venter J.C. 2003. The dog genome: Survey sequencing and comparative analysis. *Science* **301**: 1898–1903.

Kok J., Buist G., Zomer A.L., van Hijum S.A., and Kuipers O.P. 2005. Comparative and functional genomics of lactococci. *FEMS Microbiol. Rev.* (in press).

Koskinen M.T. and Bredbacka P. 2000. Assessment of the population structure of five Finnish dog breeds with microsatellites. *Anim. Genet.* **31**: 310–317.

Laird C.D., McConaughy B.L., and McCarthy B.J. 1969. Rate of fixation of nucleotide substitutions in evolution. *Nature* **224**: 149–154.

Lander E.S., Linton L.M., Birren B., Nusbaum C., Zody M.C., Baldwin J., Devon K., Dewar K., Doyle M., FitzHugh W., et al. 2001. Initial sequencing and analysis of the human genome. *Nature* **409**: 860–921.

Li R., Mignot E., Faraco J., Kadotani H., Cantanese J., Zhao B. Lin X., Hinton L., Ostrander E.A., Patterson D.F., and de Jong P.J. 1999. Construction and characterization of an eightfold redundant dog genomic bacterial artificial chromosome library. *Genomics* **58**: 9–17.

Li W.H., Tanimura M., and Sharp P.M. 1987. An evaluation of the molecular clock hypothesis using mammalian DNA sequences. *J. Mol. Evol.* **25**: 330–342.

Margulies E.H., Vinson J.P., Miller W., Jaffe D.B., Lindblad-Toh K., Chang J.L., Green E.D., Lander E.S., Mullikin J.C., Clamp M., and NISC Comparative Sequencing Program. 2005. An initial strategy for the systematic identification of functional elements in the human genome by low-redundancy comparative sequencing. *Proc. Natl. Acad. Sci.* **102**: 4795–4800.

Martin A.P. and Palumbi S.R. 1993. Body size, metabolic rate, generation time, and the molecular clock. *Proc. Natl. Acad. Sci.* **90**: 4087–4091.

Miller W., Makova K.D., Nekrutenko A., and Hardison R.C. 2004. Comparative genomics.

Annu. Rev. Genomics Hum. Genet. **5:** 15–56.

Neff M.W., Broman K.W., Mellersh C.S., Ray K., Acland G.M., Aguirre G.D., Ziegle J.S., Ostrander E.A., and Rine J. 1999. A second-generation genetic linkage map of the domestic dog, *Canis familiaris. Genetics* **151:** 803–820.

Ostrander E.A. and Kruglyak L. 2000. Unleashing the canine genome. *Genome Res.* **10:** 1271–1274.

Ovcharenko I., Loots G.G., Nobrega M.A., Hardison R.C., Miller W., and Stubbs L. 2005. Evolution and functional classification of vertebrate gene deserts. *Genome Res.* **15:** 137–145.

Parker H.G., Kim L.V., Sutter N.B., Carlson S., Lorentzen T.D., Malek T.B., Johnson G.S., DeFrance H.B., Ostrander E.A., and Kruglyak L. 2004. Genetic structure of the purebred domestic dog. *Science* **304:** 1160–1164.

Patterson D.F. 2000. Companion animal medicine in the age of medical genetics. *J. Vet. Intern. Med.* **14:** 1–9.

Pletcher M.T., McClurg P., Batalov S., Su A.I., Barnes S.W., Lagler E., Korstanje R., Wang X., Nusskern D., Bogue M.A., Mural R.J., Paigen B., and Wiltshire T. 2004. Use of a dense single nucleotide polymorphism map for in silico mapping in the mouse. *PLoS Biol.* **2:** e393.

Ptak S.E., Hinds D.A., Koehler K., Nickel B., Patil N., Ballinger D.G., Przeworski M., Frazer K.A., and Paabo S. 2005. Fine-scale recombination patterns differ between chimpanzees and humans. *Nat. Genet.* **37:** 429–434.

Richards S., Liu Y., Bettencourt B.R., Hradecky P., Letovsky S., Nielsen R., Thornton K., Hubisz M.J., Chen R., Meisel R.P., et al. 2005. Comparative genome sequencing of *Drosophila pseudoobscura*: Chromosomal, gene, and *cis*-element evolution. *Genome Res.* **15:** 1–18.

Rodin S.N. and D.V. Parkhomchuk 2004. Position-associated GC asymmetry of gene duplicates. *J. Mol. Evol.* **59:** 372–384.

Sabeti P.C., Reich D.E., Higgins J.M., Levine H.Z., Richter D.J., Schaffner S.F., Gabriel S.B., Platko J.V., Patterson N.J., McDonald G.J., et al. 2002. Detecting recent positive selection in the human genome from haplotype structure. *Nature* **419:** 832–837.

Savolainen P., Zhang Y.P., Luo J., Lundeberg J., and Leitner T. 2002. Genetic evidence for an East Asian origin of domestic dogs. **298:** 1610–1613.

Sutter N.B., Eberle M.A., Parker H.G., Pullar B.J., Kirkness E.F., Kruglyak L., and Ostrander E.A. 2004. Extensive and breed specific linkage disequilibrium in *Canis familiaris. Genome Res.* **12:** 2388–2396.

Thomas J.W., Touchman J.W., Blakesley R.W., Bouffard G.G., Beckstrom-Sternberg S.M., Margulies E.H., Blanchette M., Siepel A.C., Thomas P.J., McDowell J.C., et al. 2003. Comparative analyses of multi-species sequences from targeted genomic regions. *Nature* **424:** 788–793.

van de Sluis B., Rothuizen J., Pearson P.L., van Oost B.A., and Wijmenga C. 2002. Identification of a new copper metabolism gene by positional cloning in a purebred dog population. *Hum. Mol. Genet.* **11:** 165–173.

Vila C., Savolainen P., Maldonado J.E., Amorim I.R., Rice J.E., Honeycutt R.L., Crandall K.A., Lundeberg J., and Wayne R.K. 1997. Multiple and ancient origins of the domestic dog. *Science* **276:** 1687–1689.

Wade C.M., Kulbokas E.J., III, Kirby A.W., Zody M.C., Mullikin J.C., Lander E.S., Lindblad-Toh K., and Daly M.J. 2002. The mosaic structure of variation in the laboratory mouse genome. *Nature* **420:** 574–578.

Wall J.D. and Pritchard J.K. 2003. Haplotype blocks and linkage disequilibrium in the human genome. *Nat. Rev. Genet.* **4:** 587–597.

Waterston R.H., Lindblad-Toh K., Birney E., Rogers J., Abril J.F., Agarwal P., Agarwala R., Ainscough R., Alexandersson M., An P., et al. 2002. Initial sequencing and comparative analysis of the mouse genome. *Nature* **420:** 520–562.

Wayne R.K. and Ostrander E.A. 1999. Origin, genetic diversity, and genome structure of the domestic dog. *Bioessays* **21:** 247–257.

Webber C. and Ponting C.P. 2005. Hot spots of mutation and breakage in dog and human chromosomes. *Genome Res.* (in press).

Wilcox B. and Walkowicz D. 1995. *The atlas of dog breeds*, 5th ed. T.H.F. Publications, Neptune City, New York.

Wiltshire T., Pletcher M.T., Batalov S., Barnes S.W., Tarantino L.M., Cooke M.P., Wu H., Smylie K., Santrosyan A., Copeland N.G., et al. 2003. Genome-wide single-nucleotide polymorphism analysis defines haplotype patterns in mouse. *Proc. Natl. Acad. Sci.* **100:** 3380–3385.

Xie X., Lu J., Kulbokas E.J., Golub T.R., Mootha V., Lindblad-Toh K., Lander E.S., and Kellis M. 2005. Systematic discovery of regulatory motifs in human promoters and 3′ UTRs by comparison of several mammals. *Nature* **434:** 338–345.

Yang S., Smit A.F., Schwartz S., Chiaromonte F., Roskin K.M., Haussler D., Miller W., and Hardison R.C. 2004. Patterns of insertions and their covariation with substitutions in the rat, mouse, and human genomes. *Genome Res.* **14:** 517–527.

Zajc I., Mellersh C.S., and Sampson J. 1997. Variability of canine microsatellites within and between different dog breeds. *Mamm. Genome* **8:** 182–185.

12

SINEs of Canine Genomic Diversity

Ewen F. Kirkness
Department of Mammalian Genomics
The Institute for Genomic Research
Rockville, Maryland 20832

Different breeds of dogs clearly display characteristic phenotypes, such as morphologies, behaviors, and susceptibilities to disease. Because of the extreme variation of these phenotypes between breeds, the dog species provides us with a valuable model system to identify the specific genetic variations that underlie these traits. Until recently, such studies have been hampered by the limited amount of available genomic sequence data from dogs. However, this situation has now changed dramatically, due to the survey-sequencing of a poodle genome, the generation of a draft sequence for a boxer genome, and extensive sampling of sequence data from the genomes of multiple breeds of dogs and wild canids (Kirkness et al. 2003; K. Lindblad-Toh et al., in prep.). In combination, these sequence data provide us with considerable insight into the extent of genomic variation between individual dogs and into the best strategies for identifying those variations that are responsible for traits of interest. In common with humans, the most frequent genomic sequence variations between individual dogs are single-nucleotide polymorphisms (SNPs) and variable numbers of tandem repeats (VNTRs). A distinct type of polymorphism that is much more frequent in dogs than in humans is caused by variable insertions of short interspersed elements (SINEs; Kirkness et al. 2003). Here, we consider the potential phenotypic consequences of these variable SINE insertions. In addition, we consider their utility as abundant, evenly distributed polymorphisms that can be used to better understand relationships between dog breeds and between domesticated dogs and wild canids.

INFLUENCE OF SINEs ON EUKARYOTIC GENOMES

SINEs are retrotransposons that have enjoyed remarkable reproductive success during the course of mammalian evolution. For example, at least 300 Mb (10%) of the human genome is composed of a single family of SINEs, known as Alus (Schmid 1996; Lander et al. 2001; Venter et al. 2001). In dogs, the major family of SINEs is homologous to tRNA-Lys, and these elements comprise ~180 Mb of the poodle genome sequence (Kirkness et al. 2003). SINEs accumulate by a "copy and paste" mechanism. Following transcription by RNA polymerase III, the transcripts can be reverse-transcribed and integrated into the genome at distinct locations (Eickbush 1992; Ohshima et al. 1996). There are no known mechanisms for specific removal of inserted SINEs. Generally, SINEs are dependent on their host genome for transmission through the germ line. In order to retrotranspose, they also require endonuclease and reverse-transcriptase activities that are derived from a distinct family of retro-transposons (e.g., long interspersed elements; LINEs). Because of a dual dependence on the genomic host and other transposable elements, SINEs have been termed a "parasite's parasite" (Schmid 2003). They have also been characterized as examples of "selfish DNA," or simply "junk DNA."

Clearly, SINEs must consume resources of their host for replication, expression, and amplification. In addition, novel transposition events can cause severe disruption of their host's cellular activities (see below). However, it is unclear whether SINEs are primarily intracellular parasites of defenseless host genomes, or whether they are symbionts that are tolerated because of their occasional positive influences on genome evolution. They have certainly been implicated in the dynamics of genome evolution, whereby new functional elements appear and old ones become extinct. First, unequal homologous recombination between Alu elements has clearly contributed to human genomic diversity (Deninger and Batzer 1999). This process appears to underlie the diversification of specific genes (e.g., tropoelastin; Szabo et al. 1999) or of large genomic regions that encompass multiple genes (e.g., segmental duplications; Bailey et al. 2003). Genome modification can also occur during retrotransposition of a donor element, when transcription past its normal terminus can lead to the transduction of 3' sequences that flank the donor element (Good-ier et al. 2000; Pickeral et al. 2000). Transcription of eukaryotic retro-transposons can also interfere with expression of neighboring genes (Han et al. 2004)—a process that may be subject to epigenetic control (Whitelaw and Martin 2001). Furthermore, transcripts of SINEs have been reported to stimulate protein translation in a response to cellular stress (Schmid 1998; Rubin et al. 2002).

Perhaps the most obvious effects of retrotransposition occur when SINEs are inserted within genes and directly influence mRNA splicing and protein expression. Approximately 75% of human genes contain Alus, and there is abundant evidence that retrotransposition of these elements into exons, or close to mRNA splicing signals, can have dramatic effects on the expression of cellular protein activities (Muratani et al. 1991; Wallace et al. 1991; Vidaud et al. 1993; Janicic et al. 1995; Halling et al. 1999; Mustajoki et al. 1999; Sukarova et al. 2001; Claverie-Martin et al. 2003; Ganguly et al. 2003). It is likely that they can also produce more subtle effects. Indeed, it has been estimated that at least 5% of alternatively spliced exons in the human transcriptome are derived from Alus (Sorek et al. 2002), and processes by which intronic Alus can become "exonized" have been described previously (Vervoort et al. 1998; Lev-Maor et al. 2003). Interestingly, Alu-containing transcripts are the predominant substrates for RNA editing in humans (Kim et al. 2004).

Although the human genome contains more than one million Alu elements, the vast majority were inserted prior to divergence of the human and ape lineages and are therefore fixed in the genomes of current primate populations. Our understanding of how Alus have contributed to recent evolution of the human genome has therefore focused on "young" Alus that were born after the divergence of the different primate lineages. These include the ~25% of young Alus that have inserted into the human genome so recently that they are bimorphic with respect to the presence or absence of insertion in different human genomes (Batzer and Deininger 2002).

BIMORPHIC SINE INSERTIONS IN THE GENOME OF A POODLE

Unexpectedly, survey-sequencing of a poodle genome revealed a frequency of bimorphic SINE insertions that is between 10- and 100-fold higher than for humans (Kirkness et al. 2003). This is due largely to a recent and massive expansion of a SINE subfamily, termed SINEC_Cf, in the canine lineage. These elements are ~200 bp in length and are flanked by a characteristic 12–15-bp duplication of the insertion site (Fig. 1). For most examples, PCR amplification across the implied region of SINE insertion in different dogs can verify the polymorphism. To estimate the abundance of bimorphic SINE insertions in the sequenced poodle, a sample of 20,048 SINEC_Cf elements, each flanked by at least 60 bases of nonrepetitive sequence, was searched against the complete collection of survey-sequence data. For 709 (3.5%) of these, there were unique database matches in which the sequence of the SINE flanks are contiguous and the SINE is absent. When the same analysis was performed on an older family of SINEs (SINEC_Cf2), only 0.2% of the sample yielded such matches. The complete survey-sequence data set is predicted to cover ~50% of a 4.8-Gb diploid genome. Consequently, these data indicate that ~7% of SINEC_Cf elements (i.e., ~16,000) are bimorphic in the genome of the sequenced poodle. For comparison, the number of bimorphic SINEs (Alus) in the human population is estimated to be only ~1200 (Batzer and Deininger 2002).

The predicted abundance of bimorphic SINE insertions has been confirmed by comparing the poodle survey-sequence data with the draft

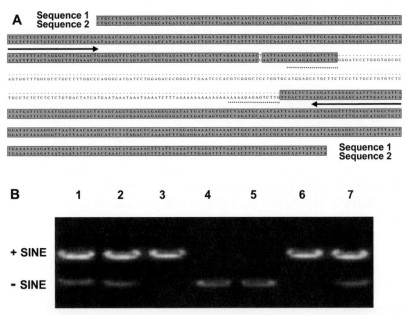

Figure 1. Typical example of a SINEC_Cf insertion. (**A**) Alignment of two genomic sequences, derived from sister chromatids of an individual dog (GenBank Accessions AC113573, AC114890). The SINE insertion within sequence 2 is flanked by a 14-bp duplication (.......). To confirm that the locus is bimorphic for SINE insertions, the region can be amplified using primers indicated by the arrows. (**B**) Products of PCR amplification from the genomic DNA of seven dogs, using the primers described in **A**. The dogs were homozygous for presence of the SINE (*lanes 3, 6*), homozygous for absence of the SINE (*lanes 4, 5*), or heterozygous (*lanes 1, 2, 7*).

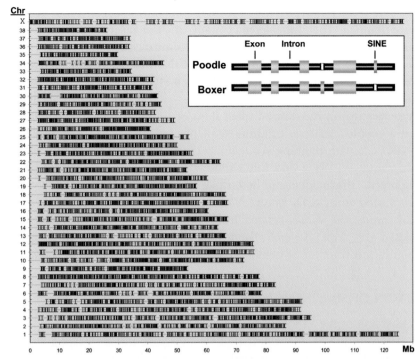

Figure 2. Mapping of 10,263 loci that are bimorphic for SINE insertions between the sequenced poodle and boxer genomes. Each insertion is represented by symbol (I), mapped onto chromosomes of the draft boxer genome sequence (version 1.0). As illustrated in the inset, most insertions are located within introns or intergenic regions.

boxer genome sequence. This has revealed more than 10,000 specific loci that differ between the two genomes by the presence or absence of a SINE_Cf insertion (Fig. 2) (W. Wang and E.F. Kirkness, in prep.). These loci are distributed throughout the genome, with a median separation of 126 kb.

BIMORPHIC SINE INSERTIONS AMONG MULTIPLE DOG BREEDS

Undoubtedly, there are many additional bimorphic loci in the general dog population, and this genetic diversity is likely to be a valuable resource for identifying the ancestral relationships between different dog breeds. This was first examined by amplifying all SINEC_Cf elements that could be identified in a contiguous 425-kb region of a Doberman pinscher genome (Kirkness et al. 2003). Of the 24 SINEs examined, 6 were bimorphic in a sample of 11 dogs from five breeds (i.e., one per 70 kb), and two additional sites displayed variation in related canids. Of the six loci that were bimorphic in dogs, five had minor allele frequencies of more than 0.3. This indicates that many of these insertions are common polymorphisms rather than rare variants. Similarly, an independent study identified two loci that are bimorphic for SINE insertions and have minor allele frequencies in excess of 0.4 (Housley et al. 2004).

The NCBI Trace Archive contains ~1 million short genomic sequences from nine domesticated dogs of different breeds, four wolves, and a coyote. Analysis of these sequences yielded 12,116 examples of SINEC_Cfs with flanking sequence (~1,000 per dog breed and 200 per wild canid; W. Wang and E.F. Kirkness, in prep.). For each of the nine dog breeds, the incidence of detectable bimorphic SINEs (relative to poodle) varied from 6.5% (beagle) to 11.0% (Bedlington terrier). For the wild canids, the values ranged from 9.4% (China gray wolf) to 15.2% (Alaska gray wolf). Overall, the results are consistent with the idea that a significant fraction of canine SINEC_Cf elements are bimorphic.

EFFECTS OF SINE INSERTIONS ON GENE EXPRESSION

At least half of the annotated genes in the boxer genome assembly contain at least one SINEC_Cf element. It is therefore not surprising to find examples that have been inserted at locations where they may influence the transcription and splicing of mRNAs. In common with human Alus, retrotransposition of a SINEC_Cf element to a location that contains mRNA splicing signals can cause alternative splicing of mRNA with severe phenotypic consequences. A notable example is the insertion of a SINEC_Cf element within intron 3 of the gene encoding a hypocretin receptor in narcoleptic Doberman pinschers (Lin et al. 1999). The insertion causes skipping of exon 4 during transcription, and the resulting absence of functional receptors was the first genetic evidence that linked the hypocretin system wth narcolepsy in mammals. Also in common with human Alus, SINEC_Cf elements can be transcribed, either as intact elements within established exons or as new alternative exons. A survey of dog expressed sequence tags (ESTs) in GenBank has identified 121 examples of cDNAs that contain either complete or partial SINEC_Cf elements (W. Wang and E.F. Kirkness, in prep.). Many of these (51%) are located in the "sense" orientation at the 3′ ends of cDNAs, suggesting that the polyadenylation signal within SINEC_Cf elements can cause premature termination of dog mRNAs (Fig. 3). There are also examples of dog cDNAs that have acquired additional exons (relative to their human orthologs) owing to the use of a specific splice acceptor site within SINEC_Cf elements (Fig. 3). Notably, a recent study has revealed that insertion of a SINEC_Cf element is responsible for alternative splicing of

```
>SINEC_Cf
GGGATCCCTGGGTGGCGCAGCGGTTTGGCGCCTGCCTTTGGCCCAGGGCGCGA
TCCTGGAGACCCGGGATCGAATCCCACGTCGGGCTCCCGGTGCATGGAGCCTG
CTTCTCCCTCTGCCTGTGTCTCTGCCTCTCTCTCTCTCTCTGTGTGACTAT
CATAAATAAATAAAAATTAAAAAAAAAAA
```

```
>SINEC_Cf Reverse complement        ↓
TTTTTTTTTTTTAATTTTTATTTATTTATGATAGTCACACAGAGAGAGAGAGA
GAGAGGCAGAGACACAGGCAGAGGGAGAAGCAGGCTCCATGCACCGGGAGCCC
GACGTGGGATTCGATCCCGGGTCTCCAGGATCGCGCCCTGGGCCAAAGGCAGG
CGCCAAACCGCTGCGCCACCCAGGGATCCC
```

Figure 3. Splicing signals within the SINEC_Cf sequence. The upper panel high-lights the location of polyadenylation signals (AATAAA, ATTAAA) within the SINEC_Cf sequence when transcribed in the sense orientation. The lower panel de-picts the reverse complement of the SINEC_Cf sequence and highlights the consen-sus sequence for an intron splice acceptor site, including the branch-site adenosine (A), a pyrimidine-rich tract, and "AG" dinucleotide immediately upstream of the splice site (*arrow*).

the PTPLA mRNA, causing centronuclear myopathy in Labradors that ac-quire two copies of a specific SINEC_Cf insertion (Pele et al. 2005).

The recent expansion of SINEs in the dog genome, reflected by a high frequency of bimorphic SINE insertions, provides a unique opportunity to explore the influence of SINEs on the evolution of a mammalian genome. For many thousands of genes, an individual dog carries two alleles that differ by their content of SINEs. It is therefore possible to assess the impact of SINEs on gene expression patterns within individuals (or even within individual cells) rather than requiring a comparison between multiple individuals or between multiple species. The high frequency of bimorphic SINE insertions in the dog is predicted to provide numerous examples of allele-specific splicing patterns that can be studied further by correlating their potential functional effects with their distribution between dog breeds. The findings have broad scientific impact because so little is known of how periodic amplification of SINEs might influence gene expression patterns, despite the fact that most eukaryotic genomes have experienced periods of massive SINE amplification during their evolution.

EXPLOITING BIMORPHIC SINE INSERTIONS FOR EVOLUTIONARY STUDIES

For evolutionary studies, SINE-insertion polymorphisms offer two ad-vantages over the more common polymorphisms. First, the presence of a SINE element represents identity by descent, since the probability that two different young SINE repeats would integrate independently at the same chromosomal location is very small. Second, the ancestral state of each SINE insertion polymorphism is known to be the absence of the SINE element, and this can be used to root trees of population relationships. In contrast, other types of genetic polymorphisms, such as VNTRs and SNPs, can be identical by state if they have arisen from independent parallel mutations at different times and have not been inherited from a common ancestor. Alu-insertion polymorphisms have therefore been useful for studies of human origins, ancestral relationships, and demography (Batzer et al. 1994). They have also been used to study ancestral relationships between whales and other mammals (Shimamura et al. 1997; Nikaido et al. 1999). Previously, evolutionary studies of canine lineages have focused mainly on variations of mitochondrial DNA or VNTRs (Vila et al. 1997; Savolainen et al. 2002; Koskinen 2003; Parker et al. 2004). These approaches indicate that modern dog breeds were first domesticated from wolves, possibly in East Asia, and that many dog breeds which share morphologies, behaviors, and geographical origins can be segregated by genotype. However, as for all types of markers, mitochondrial DNA variations and VNTRs have their limitations for evolutionary analysis (Ellegren 2000; Sigurgardottir et al. 2000). Bimorphic SINE insertions offer the advantages of identity by descent, and easy typing methodologies, that make these abundant variations a valuable additional resource for identifying the ancestral relationships between different dog breeds, and between domesticated dogs and wild canids.

SUMMARY

Recent efforts to generate sequence data from multiple canine genomes have revealed an unanticipated abundance of genomic loci that are bimorphic for SINE insertions. Identification of these loci provides a rare opportunity for molecular biologists to explore the influence of SINEs on cellular gene expression patterns. For many thousands of genes, an individual dog will carry two alleles that differ by their content of SINEs. In addition, loci that are bimorphic for SINE insertions provide numerous genomic markers that are identical by decent, relatively easy to assay, and readily exploited for high-throughput genotyping of canine genomes. It is likely that these bimorphic SINE insertions will provide us with evidence of how insertion elements can mold a mammalian genome, as well as the means to identify genetic relationships between the diverse collection of current canine populations.

REFERENCES

Bailey J.A., Liu G., and Eichler E.E. 2003. An Alu transposition model for the origin and expansion of human segmental duplications. *Am. J. Hum. Genet.* **73:** 823–834.

Batzer M.A. and Deininger P.L. 2002. Alu repeats and human genomic diversity. *Nat. Rev. Genet.* **3:** 370–379.

Batzer M.A., Stoneking M., Alegria-Hartman M., Bazan H., Kass D.H., Shaikh T.H., Novick G.E., Ioannou P.A., Scheer W.D., Herrera R.J., and Deininger P.L. 1994. African origin of human-specific polymorphic Alu insertions. *Proc. Natl. Acad. Sci.* **91:** 12288–12292.

Claverie-Martin F., Gonzalez-Acosta H., Flores C., Anton-Gamero M., and Garcia-Nieto V. 2003. De novo insertion of an Alu sequence in the coding region of the CLCN5 gene results in Dent's disease. *Hum. Genet.* **113:** 480–485.

Deininger P.L. and Batzer M.A. 1999. Alu repeats and human disease. *Mol. Genet. Metab.* **67:** 183–193.

Eickbush T.H. 1992. Transposing without ends: The non-LTR retrotransposable elements. *New Biol.* **4:** 430–440.

Ellegren H. 2000. Microsatellite mutations in the germline: Implications for evolutionary inference. *Trends Genet.* **16:** 551–558.

Ganguly A., Dunbar T., Chen P., Godmilow L., and Ganguly T. 2003. Exon skipping caused by an intronic insertion of a young Alu Yb9 element leads to severe hemophilia A. *Hum. Genet.* **113:** 348–352.

Goodier J.L., Ostertag E.M., and Kazazian H.H. Jr. 2000. Transduction of 3′-flanking sequences is common in L1 retrotransposition. *Hum. Mol. Genet.* **9:** 653–657.

Halling K.C., Lazzaro C.R., Honchel R., Bufill J.A., Powell S.M., Arndt C.A., and Lindor N.M. 1999. Hereditary desmoid disease in a family with a germline Alu I repeat mutation of the APC gene. *Hum. Hered.* **49:** 97–102.

Han J.S., Szak S.T., and Boeke J.D. 2004. Transcriptional disruption by the L1 retrotransposon and implications for mammalian transcriptomes. *Nature* **429:** 268–274.

Housley D.J., Ritzert E., and Venta P.J. 2004. Comparative radiation hybrid map of canine chromosome 1 incorporating SNP and indel polymorphisms. *Genomics* **84:** 248–264.

Janicic N., Pausova Z., Cole D.E., and Hendy G.N. 1995. Insertion of an Alu sequence in the Ca(2+)-sensing receptor gene in familial hypocalciuric hypercalcemia and neonatal severe hyperparathyroidism. *Am. J. Hum. Genet.* **56:** 880–886.

Kim D.D., Kim T.T., Walsh T., Kobayashi Y., Matise T.C., Buyske S., and Gabriel A. 2004. Widespread RNA editing of embedded alu elements in the human transcriptome. *Genome Res.* **14:** 1719–1725.

Kirkness E.F., Bafna V., Halpern A.L., Levy S., Remington K., Rusch D.B., Delcher A.L., Pop M., Wang W., Fraser C.M., and Venter J.C. 2003. The dog genome: Survey sequencing and comparative analysis. *Science* **301:** 1898–1903.

Koskinen M.T. 2003. Individual assignment using microsatellite DNA reveals unambiguous breed identification in the domestic dog. *Anim. Genet.* **34:** 297–301.

Lander E.S., Linton L.M., Birren B., Nusbaum C., Zody M.C., Baldwin J., Devon K., Dewar K., Doyle M., Fitz Hugh W. et al. 2001. Initial sequencing and analysis of the human genome. *Nature* **409:** 860–921.

Lev-Maor G., Sorek R., Shomron N., and Ast G. 2003. The birth of an alternatively spliced exon: 3′ splice-site selection in Alu exons. *Science* **300:** 1288–1291.

Lin L., Faraco J., Li R., Kadotani H., Rogers W., Lin X., Qiu X., de Jong P.J., Nishino S., and Mignot E. 1999. The sleep disorder canine narcolepsy is caused by a mutation in the hypocretin orexin receptor 2 gene. *Cell* **98:** 365–376.

Muratani K., Hada T., Yamamoto Y., Kaneko T., Shigeto Y., Ohue T., Furuyama J., and Higashino K. 1991. Inactivation of the cholinesterase gene by Alu insertion: Possible mechanism for human gene transposition. *Proc. Natl. Acad. Sci.* **88:** 11315–11319.

Mustajoki S., Ahola H., Mustajoki P., and Kauppinen R. 1999. Insertion of Alu element responsible for acute intermittent porphyria. *Hum. Mutat.* **13:** 431–438.

Nikaido M., Rooney A.P., and Okada N. 1999. Phylogenetic relationships among cetartiodactyls based on insertions of short and long interpersed elements: Hippopotamuses are the closest extant relatives of whales. *Proc. Natl. Acad. Sci.* **96:** 10261–10266.

Ohshima K., Hamada M., Terai Y., and Okada N. 1996. The 3′ ends of tRNA-derived short interspersed repetitive elements are derived from the 3′ ends of long interspersed repetitive elements. *Mol. Cell Biol.* **16:** 3756–3764.

Parker H.G., Kim L.V., Sutter N.B., Carlson S., Lorentzen T.D., Malek T.B., Johnson G.S., DeFrance H.B., Ostrander E.A., and Kruglyak L. 2004. Genetic structure of the purebred domestic dog. *Science* **304:** 1160–1164.

Pele M., Tiret L., Kessler J.L., Blot S., and Panthier J.J. 2005. SINE exonic insertion in the PTPLA gene leads to multiple splicing defects and segregates with the autosomal recessive centronuclear myopathy in dogs. *Hum. Mol. Genet.* **14:** 1417–1427.

Pickeral O.K., Makalowski W., Boguski M.S., and Boeke J.D. 2000. Frequent human genomic DNA transduction driven by LINE-1 retrotransposition. *Genome Res.* **10:** 411–415.

Rieder M.J., Taylor S.L., Clark A.G., and Nickerson D.A. 1999. Sequence variation in the human angiotensin converting enzyme. *Nat. Genet.* **22:** 59–62.

Rubin C.M., Kimura R.H., and Schmid C.W. 2002. Selective stimulation of translational expression by Alu RNA. *Nucleic Acids Res.* **30:** 3253–3261.

Savolainen P., Zhang YP., Luo J., Lundeberg J., and Leitner T. 2002. Genetic evidence for an East Asian origin of domestic dogs. *Science* **298:** 1610–1613.

Schmid C.W. 1996. Alu: Structure, origin, evolution, significance and function of one-tenth of human DNA. *Prog. Nucleic Acid Res. Mol. Biol.* **53:** 283–319.

———. 1998. Does SINE evolution preclude Alu function? *Nucleic Acids Res.* **26:** 4541–4550.

———. 2003. Alu: A parasite's parasite? *Nat. Genet.* **35:** 15–16.

Shimamura M., Yasue H., Ohshima K., Abe H., Kato H., Kishiro T., Goto M., Munechika I., and Okada N. 1997. Molecular evidence from retroposons that whales form a clade within even-toed ungulates. *Nature* **388:** 666–670.

Sigurgardottir S., Helgason A., Gulcher J.R., Stefansson K., and Donnelly P. 2000. The mutation rate in the human mtDNA control region. *Am. J. Hum. Genet.* **66:** 1599–1609.

Sorek R., Ast G., and Graur D. 2002. Alu-containing exons are alternatively spliced. *Genome Res.* **12:** 1060–1067.

Sukarova E., Dimovski A.J., Tchacarova P., Petkov G.H., and Efremov G.D. 2001. An Alu insert as the cause of a severe form of hemophilia A. *Acta Haematol.* **106:** 126–129.

Szabo Z., Levi-Minzi S.A., Christiano A.M., Struminger C., Stoneking M., Batzer M.A., and Boyd C.D. 1999. Sequential loss of two neighboring exons of the tropoelastin gene during primate evolution. *J. Mol. Evol.* **49:** 664–671.

Venter J.C., Adams M.D., Myers E.W., Li P.W., Mural R.J., Sutton GG., Smith H.O., Yandell M., Evans C.A., Holt R.A., et al. 2001. The sequence of the human genome. *Science* **291:** 1304–1351.

Vervoort R., Gitzelmann R., Lissens W., and Liebaers I. 1998. A mutation IVS8+0.6kbdelTC. creating a new donor splice site activates a cryptic exon in an Alu-element in intron 8 of the human beta-glucuronidase gene. *Hum. Genet.* **103:** 686–693.

Vidaud D., Vidaud M., Bahnak B.R., Siguret V., Gispert Sanchez S., Laurian Y., Meyer D., Goossens M., and Lavergne J.M. 1993. Haemophilia B due to a de novo insertion of a human-specific Alu subfamily member within the coding region of the factor IX gene. *Eur. J. Hum. Genet.* **1:** 30–36.

Vila C., Savolainen P., Maldonado J.E., Amorim I.R., Rice J.E., Honeycutt R.L., Crandall K.A., Lundeberg J., and Wayne R.K. 1997. Multiple and ancient origins of the domestic dog. *Science* **276:** 1687–1689.

Wallace M.R., Andersen L.B., Saulino A.M., Gregory P.E., Glover T.W., and Collins F.S. 1991. A de novo Alu insertion results in neurofibromatosis type 1. *Nature* **353:** 864–866.

Whitelaw E. and Martin D.I. 2001. Retrotransposons as epigenetic mediators of phenotypic variation in mammals. *Nat. Genet.* **27:** 361–365.

13

The Dog Olfactory and Vomeronasal Receptor Repertoires

Pascale Quignon, Sandrine Tacher, Maud Rimbault, and
Francis Galibert
UMR CNRS 6061, Génétique et Développement
Université de Rennes 1, France

Dᴜʀɪɴɢ ᴛʜᴇ ᴄᴏᴜʀsᴇ ᴏғ ᴇᴠᴏʟᴜᴛɪᴏɴ, animals have developed several sensory systems that allow them to communicate with the external world. Olfaction allows them to sense odorant molecules and diverse pheromones that enable them to identify prey, dangers, and potential mates. It involves a complicated process in which volatile molecules are first detected through their specific binding to specialized receptors (Buck and Axel 1991; Buck 2000; Firestein 2001). Two organs, the main olfactory and the accessory olfactory organs, have been anatomically described in most mammals, except in primates, which are devoid of the accessory organ (Giorgi and Rouquier 2002; Young et al. 2005). The main olfactory organ is specialized in the identification and recognition of odorants. The accessory organ identifies pheromones (Bargmann 1997) and plays a specific role in behavior among individuals of the same species. In the two systems, specialized neurons express at their surfaces G-protein-coupled receptors that belong to the superfamily of seven-transmembrane-domain proteins. In human and rodent, the receptors of the main olfactory system, termed ORs, are encoded by the largest gene family (1000–1500 copies) (Buck and Axel 1991; Malnic et al. 2004), whereas receptors of the vomeronasal system, termed VRs, do not exceed about 200 (Giorgi and Rouquier 2002; Young et al. 2005).

The availability of sequence data derived from survey sequencing of DNA from a poodle (Kirkness 2003), and of the 7.5× shotgun sequence of DNA from a boxer (K. Lindblad-Toh, in prep.), allowed the recognition of most, if not all, of the ORs and VRs encoded in the dog, whose ability to detect and identify odorant molecules is particularly well developed. We report here the current inventory of the OR and VR repertoires of the dog (Quignon et al. 2003; Young et al. 2005; P. Quignon et al., in prep.) and compare them to those of other mammals (Godfrey et al. 2004; Malnic et al. 2004).

THE DOG OR GENE REPERTOIRE

Identification

OR proteins have a mean length of 315 ± 10 amino acids and share sequence identity that ranges from 34% to 99% within species. In mammals, OR genes comprise two exons, but only one encodes the OR protein. However, in insects and nematodes, OR genes have more than one coding exon. The existence of only one coding exon in mammals greatly facilitated the first OR gene identification by Buck and Axel (1991) and the subsequent in vitro cloning of many OR genes using degenerated primers for PCR amplification or the mining of mammalian genome sequences either by iterative BLAST search or by the recognition of specific amino acid motifs. We searched the 7.5× coverage sequence either assembled (CanFam1.0) or unassembled (ftp://ftp.ncbi.nih.gov/pub/TraceDB) for five amino acid motifs characteristic of the dog OR and retrieved 1014 OR genes from the assembled sequence and 1058 from the nonassembled sequences.

By comparison of these two gene sets, we observed that 1003 OR genes were identified by both approaches, with 55 identified by reads only and 11 identified by the whole-genome assembly only. We also noted that 25 OR genes that have either been identified by in vitro cloning or retrieved from the 1.5× survey sequencing (Kirkness et al. 2003; Quignon et al. 2003) were absent. The lack of these genes could be due either to the fact that 7.5× shotgun sequence covers only about 98% of the genome (K. Lindblad-Toh, in prep.) or to over-collapsing of gene family members sharing up to 99% sequence identity. Presently the lowest estimate of the canine OR repertoire has been calculated at 1094 genes plus 27 additional sequences corresponding, at best, to very highly pseudogenized OR genes.

Genes and Pseudogenes

Analyses of diverse OR repertoires in other organisms have shown that the number of pseudogenes was high and varied between organisms. In human and chimpanzee, up to 60% and 50% of the genes were declared pseudogenes, respectively (Gilad et al. 2005), whereas this percentage is only about 20% for the mouse (Young et al. 2002) and rat (P. Quignon et al., in prep.). It is of interest to note that these pseudogenes are not retrotransposed genes but most probably ancient active genes, where the ORF was interrupted by the accumulation of non-sense or frameshift mutations over time. This is in agreement with the decreasing importance of the olfaction system for those species. Translation of dog OR gene sequences showed that 21.5% were pseudogenes, thus leaving dog with a similar fraction of pseudogenes to rat and mouse. The number of stop mutations in the ORF varied, with 78 of the 222 (35%) dog pseudogenes (for which the sequences encompassed the actual start and stop codons) due to a single mutation, 43 (19%) containing two mutations interrupting the ORF, and 101 (45%) having three or more mutations. Although pseudogenes with more than one extra stop codon are certainly real pseudogenes, those with a single mutation may not always be pseudogenes, as was shown by sequence polymorphism analysis (S. Tacher et al., in prep.).

Genome Localization

Of the 1094 identified dog OR genes (P. Quignon et al., in prep.), 963 have been localized on the dog genome, either by RH mapping (Quignon et al. 2003) or by the assembled sequence CanFam1.0, and only 131 remain unassigned. Figure 1 shows the canine karyotype with the OR positions indicated. In the human genome, all HSAs except 20 and Y contained OR genes (Rouquier et al. 1998; Glusman et al. 2001). In the case of the dog, we observed a similar distribution, given the more fragmented nature of the dog karyotype. All dog genome regions syntenic to human regions in which human OR genes have been found also contain dog OR genes. Genomic clusters contain anywhere between 1 and 200 genes, with the largest cluster on CFA 18. Although pseudogenes are found in nearly all clusters, we noticed the existence of 5 loci consisting of one pseudogene only (Fig. 1). Comparison of the four mammal OR repertoires—human, mouse, rat, and dog—showed that regardless of differences in karyotype and repertoire size, OR genes are distributed in very similar numbers of clusters, as defined by groups of OR separated by at least one megabase (Table 1).

Amino Acid Sequence Comparison

The canine OR amino acid sequences were aligned, and the level of variability at each amino acid position was determined. Apart from the amino-terminal position, no amino acid position is entirely invariant,

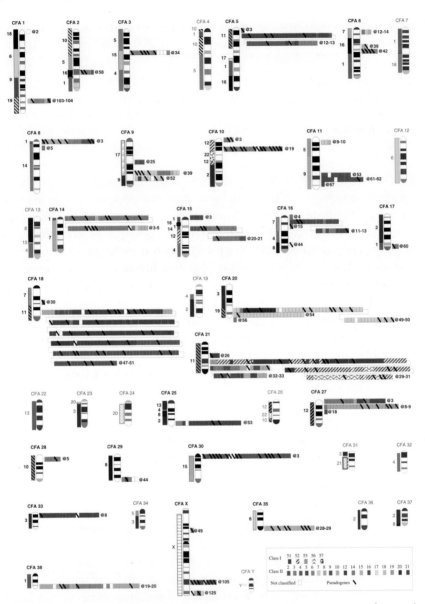

Figure 1. Chromosomal distribution of dog OR genes. Human conserved syntenic regions, according to radiation hybrid data (Hitte et al. 2005) are indicated on the left of the canine chromosome ideograms (CFA1-38, X,Y). On the right of each canine chromosome, canine OR genes are represented by colored squares, each color corresponding to a family. Localization of the loci is indicated with @ followed by the coordinates in megabases.

Table 1. Characteristics of the human, mouse, rat, and dog OR repertoires

	Human[a]	Mouse[b]	Rat[c]	Dog[c]
Number of loci	51	51	56	49
Number of OR per locus	1–116	1–244	1–265	1–211
% Pseudogenes per locus (number of loci with only pseudogenes)	0–100%(13)	7–100%(2)	0–100%(8)	0–100%(5)

[a]Malnic et al. (2004).
[b]Godfrey et al. (2004).
[c]P. Quignon et al. (in prep.).

and only 23 positions are highly conserved (≥90%): They are located throughout the protein (13 in the transmembrane domains [TM], 4 in the extracellular loops [EC], and 6 in the intracellular loops [IC]; see Fig. 2). Additionally, TM IV and V contain a higher proportion of highly variable amino acids, consistent with the role of these domains in ligand recognition and binding (Shepherd 1994; Man et al. 2004).

We used Clustal (Thompson et al. 1994) to compare amino acid sequences and construct phylogenetic trees. The number of families and subfamilies was determined according to previously used thresholds (i.e., 40% and 60% amino acid identity for distinguishing families and

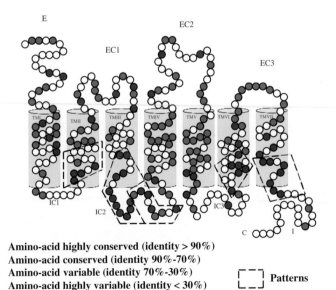

● Amino-acid highly conserved (identity > 90%)
● Amino-acid conserved (identity 90%-70%)
○ Amino-acid variable (identity 70%-30%)
● Amino-acid highly variable (identity < 30%)   ⌐ ¬ Patterns
                                                        L _ ⌐

Figure 2. Positions of conserved and variable amino acids in 1009 dog OR proteins. (E and EC) Extracellular domain; (I and IC) intracellular domain; (TM) transmembrane domain.

Table 2. Family and subfamily composition of the dog OR repertoire

OR family name[a]	Number of subfamilies	Number of OR	% Pseudo
2	12	58	8
3	4	9	11
4	32	156	25
5	50	197	15
6	40	134	34
7	23	87	23
8	4	18	17
9	2	5	40
10	21	46	13
12	2	7	14
14	17	47	19
15	14	24	37
16	5	5	100
17	3	3	33
18	1	1	100
19	1	4	75
20	1	10	20
21	1	3	33
51	31	71	22
52	30	96	15
55	2	2	100
56	3	23	13
57	1	1	0

[a]Families 2–21 belong to class II, families 51–57 belong to class I.

subfamilies; Ben Arie et al. 1994) (Table 2 and P. Quignon et al., in prep.), using the amino acid sequences of all 1009 dog ORs with a complete amino acid sequence. Interestingly, the number of subfamilies does not parallel the size of the repertoires. The high number of subfamilies (294) in human (human olfactory receptor data exploratorium [HORDE] version 40: http://bip.weizmann.ac.il/HORDE) most probably reflects the high percentage of pseudogenes that are without selective pressure and can evolve more rapidly. On the contrary, in dog the larger number of subfamilies (300 compared to 280 and 246 found in the rat and mouse repertoires, respectively) (Godfrey et al. 2004; P. Quignon et al., in prep.) indicated a higher level of diversity, despite the similar percentage of pseudogenes. As a consequence, subfamilies vary considerably in size and are smaller in dog than in rat, with 1–31 genes for the dog and 1–60 genes for the rat (P. Quignon et al., in prep.).

Pseudogenes have been detected in both classes and in all families and subfamilies, but with an uneven distribution. The dog class I (193 OR genes) included fewer pseudogenes (19%) than class II (23%), and even larger variations could be observed in families and subfamilies. For example, in family 6 (class II), which contained 134 genes, pseudogenes accounted for 34%. In family 10 (class II), which contained 46 genes, 13% were identified as pseudogenes. Comparison with phylogenetic trees of the OR repertoires in other organisms made it possible to identify subfamilies that had expanded in one species but not in the other and to identify orthologs that are assumed to encode proteins having similar or identical functions by binding the same ligand. A case in point is subfamily 7A, which contained 31 genes in dog but only 11 genes in rat, or alternatively, subfamily 2K, which includes 11 genes in rat but has not been found in dog. The reverse situation was also observed for dog subfamily 6B, which contained 9 genes; this subfamily was absent from the rat repertoire (P. Quignon et al., in prep.).

As observed with other mammalian repertoires, dog OR genes that belong to the same subfamily tend to be genomically clustered, particularly those with high sequence homology (Glusman et al. 2001). Only 23 dog subfamilies comprising 139 genes were found on more than one chromosome.

Allelic Polymorphism

Sequence analyses of 16 genes from different subfamilies within a panel of 95 dogs of 20 breeds have shown that OR genes are highly polymorphic, with one nucleotide difference per 920 sequenced nucleotides (S. Tacher et al., in prep.). This value is very similar to that observed for the whole genome (K. Lindblad-Toh, in prep.), but markedly higher than that usually found in coding sequences (Masuda et al. 2004), with the exception of the MHC genes. A total of 98 single-nucleotide polymorphisms (SNPs) and 4 insertions/deletions were detected, with 2–11 SNPs per gene. Fifty-five SNPs (56%) induced an amino acid change, resulting in 30 cases where the type of amino acid changed; e.g., a proline into a serine (CfOR0011 position 844) or a glutamic acid into a glycine (CfOR0044 at position 122). Of note, five alleles had an interrupted ORF due to a stop codon or a short indel changing the reading frame. Seventeen dogs were found to be heterozygous and 9 homozygous for one or the other of these five genes, and among the latter, the five analyzed boxers had an interrupted ORF for the two alleles of CfOR0184.

THE DOG VR GENE REPERTOIRES

The receptors expressed by the vomeronasal neurons form two protein families that share some properties with the ORs. They all are G-protein-coupled receptors of the superfamily of seven-transmembrane-domain proteins, and like ORs, they are expressed at the surface of specialized neurons that, like OR neurons, are thought to express only one specific VR (Rodriguez et al. 2002). However, they constitute distinct families from a structural point of view as well as from diverse biological aspects, such as their coupling to a different G α-subunit. Differential screening of cDNA libraries and mining of genome sequencing data have allowed identification of a number of genes of the V1R and V2R families from different mammals (Dulac and Axel 1995; Herrada and Dulac 1997; Matsunami and Buck 1997; Ryba and Tirindelli 1997). In a very recent publication, Young et al. (2005) have compared the V1R repertoire of human, chimpanzee, mouse, rat, and dog, and surprisingly, noted that the dog V1R repertoire was very small with only 8 putative active genes and 54 pseudogenes (Table 3).

We aligned the amino acid sequences of 18 mouse and 8 rat V2R proteins (excluding their amino-terminal extensions) and determined four characteristic motifs with PRATT (Jonassen et al. 1995). We used these motifs to screen the dog database and retrieved only three V2R pseudogenes and no putative intact gene (P. Quignon et al., unpubl.). The fact that most of the vomeronasal genes are pseudogenes may not be so surprising, if one considers that since domestication took place some 15,000 years ago, dogs have ceased to live in ordered pack structures. In that respect, analysis of wolves and other canids living in the wild would be most instructive.

CONCLUSION

The detection and recognition of odorant comprise a very complex suite of biological events that initiate at the surface of specialized neurons by their binding to specific olfactory receptors. There a cascade of enzymatic

Table 3. V1R and V2R repertoires

V1R[a]	Human	Mouse	Rat	Dog
Number of genes	2	165	106	8
Number of pseudogenes	115	165	110	54
V2R (% pseudogenes)	not known	140[b]	154(32%)[c]	3(100%)[c]

[a]Young et al. (2005).
[b]Matsunami and Buck (1997).
[c]P. Quignon et al. (unpubl.).

reactions transduces the signal toward the olfactory bulb before ending in the superior area of the brain where the signals are treated. Dogs have a highly developed capability to detect and identify a large spectrum of odorant molecules, even at minute concentrations; thus, it would be surprising if each of these numerous steps were not optimized, hence the repertoire of receptors. It appears that the dog repertoire is not the largest, as mouse and rat harbored more active genes. However, phylogenetic comparison showed that the dog repertoire is more diverse with more subfamilies. Although few ligands for OR receptor(s) have been identified, there is reason to believe that different subfamilies bind different chemical families, hence the large number of OR subfamilies would support its exquisite odorant-sensing capabilities.

The amount of OR genes with an interrupted ORF (pseudogenes) seemed to be dependent on the origin of the DNA sample analyzed. Our first estimate based on the $1.5\times$ survey sequencing of a poodle DNA was 18%, whereas later a value of 21.8% was obtained from the boxer sample. Of the OR genes common to these two analyses, 17 were pseudogenes in the poodle genome but not in the boxer genome, the reverse being true for 22 genes. Although some of these differences might be attributed to sequencing errors, it is worth noting that 5 of the 16 genes analyzed within the allelic polymorphism study (S. Tacher et al., in prep.) had a pseudogenized allele: 17 animals were heterozygous and 9 homozygous for one of these 5 alleles. In particular, all 5 boxers were homozygous for an allele of CfOR0184 that has an interrupted ORF, and therefore, none of the analyzed boxers had a functional CfOR0184 gene. Thus, it is evident that different animals or breeds may have a different subset of pseudogenes, as has already been noted in human (Menashe et al. 2002, 2003). In addition, it is tempting to link the higher percentage of pseudogenes (21.8%) found in the DNA of the boxer with current views that boxers have a less acute sense of smell than do poodles. Although superior brain functions or individual behavior may be important factors, more extensive studies on allele sequence polymorphism are needed to explore this possibility and others in relation to the observed differences in sensing odorant molecules that might characterize different breeds or animals.

SUMMARY

The availability of the human, mouse, and rat genome sequences had made it possible to retrieve the complete repertoire of 1000–1500 olfactory receptor genes. Here we discuss the recent data obtained from the dog genome. Although very similar to other repertoires, the dog repertoire with 1094 genes (of which some 20% are pseudogenes) appeared by its higher number of subfamilies to be more diversified than the rat repertoire. Canine OR genes appeared also to be frequently polymorphic, as assessed by resequencing 16 OR genes within a set of 95 dogs representing 20 breeds. Comparison of V1R and V2R genes of the accessory system shows that dogs would have far fewer genes than rodents, with many of these being pseudogenes.

ACKNOWLEDGMENTS

We acknowledge the Centre National Recherche Scientifique (CNRS), the Université de Rennes 1, the Conseil Régional de Bretagne, and the Technical Support Working Group (TSWG) for grants to F.G. and for encouragement. We also acknowledge colleagues for helpful discussions.

REFERENCES

Bargmann C.I. 1997. Olfactory receptors, vomeronasal receptors, and the organization of olfactory information. *Cell* **90:** 585–587.

Ben-Arie N., Lancet D., Taylor C., Khen M., Walker N., Ledbetter D.H., Carrozzo R., Patel K., Sheer D., Lehrach H., et al. 1994. Olfactory receptor gene cluster on human chromosome 17: Possible duplication of an ancestral receptor repertoire. *Hum. Mol. Genet.* **3:** 229–235.

Buck L.B. 2000. The molecular architecture of odor and pheromone sensing in mammals. *Cell* **100:** 611–618.

Buck L. and Axel R. 1991. A novel multigene family may encode odorant receptors: A molecular basis for odor recognition. *Cell* **65:** 175–187.

Dulac C. and Axel R. 1995. A novel family of genes encoding putative pheromone receptors in mammals. *Cell* **83:** 195–206.

Firestein S. 2001. How the olfactory system makes sense of scents. *Nature* **413:** 211–218.

Gilad Y., Man O., and Glusman G. 2005. A comparison of the human and chimpanzee olfactory receptor gene repertoires. *Genome Res.* **15:** 224–230.

Giorgi D. and Rouquier S. 2002. Identification of V1R-like putative pheromone receptor sequences in non-human primates. Characterization of V1R pseudogenes in marmoset, a primate species that possesses an intact vomeronasal organ. *Chem. Senses* **27:** 529–537.

Glusman G., Yanai I., Rubin I., and Lancet D. 2001. The complete human olfactory subgenome. *Genome Res.* **11:** 685–702.

Godfrey P.A, Malnic B., and Buck L.B. 2004. The mouse olfactory receptor gene family. *Proc. Natl. Acad. Sci.* **101:** 2156–2161.

Herrada G. and Dulac C. 1997. A novel family of putative pheromone receptors in mammals with a topographically organized and sexually dimorphic distribution. *Cell* **90:** 763–773.

Hitte C., Madeoy J., Kirkness E.F., Priat C., Lorentzen T.D., Senger F., Thomas D., Derrien T., Ramirez C., Scott C. et al. 2005. Facilitating genome navigation: Survey sequencing and dense radiation-hybrid gene mapping. *Nat. Rev. Gen.* **6:** (in press).

Jonassen I., Collins J.F., and Higgins D.G. 1995. Finding flexible patterns in unaligned protein sequences. *Protein Sci.* **4:** 1587–1595.

Kirkness E.F., Bafna V., Halpern A.L., Levy S., Remington K., Rusch D.B., Delcher A.L., Pop M., Wang W., Fraser C.M., and Venter J.C. 2003. The dog genome: Survey sequencing and comparative analysis. *Science* **301:** 1898–1903.

Malnic B., Godfrey P.A., and Buck L.B. 2004. The human olfactory receptor gene family. *Proc. Natl. Acad. Sci.* **101:** 2584–2589.

Man O., Gilad Y., and Lancet D. 2004. Prediction of the odorant binding site of olfactory receptor proteins by human-mouse comparisons. *Protein Sci.* **13:** 240–254.

Masuda K., Hashizume C., Kikusui T., Takeuchi Y., and Mori Y. 2004. Sequencing of canine 5-hydroxytriptamine receptor (5-HTR) 1B, 2A, 2C genes and identification of polymorphisms in the 5-HTR1B gene. *J. Vet. Med. Sci.* **66:** 965–972.

Matsunami H. and Buck L.B. 1997. A multigene family encoding a diverse array of putative pheromone receptors in mammals. *Cell* **90:** 775–784.

Menashe I., Man O., Lancet D., and Gilad Y. 2002. Population differences in haplotype structure within a human olfactory receptor gene cluster. *Hum. Mol. Genet.* **11:** 1381–1390.

———. 2003. Different noses for different people. *Nat. Genet.* **34:** 143–144.

Quignon P., Kirkness E., Cadieu E., Touleimat N., Guyon R., Renier C., Hitte C., Andre C., Fraser C., and Galibert F. 2003. Comparison of the canine and human olfactory receptor gene repertoires. *Genome Biol.* **4:** R80.

Rodriguez I., Del Punta K., Rothman A., Ishii T., and Mombaerts P. 2002. Multiple new and isolated families within the mouse superfamily of V1r vomeronasal receptors. *Nat. Neurosci.* **5:** 134–140.

Rouquier S., Taviaux S., Trask B.J., Brand-Arpon V., van den Engh G., Demaille J., and Giorgi D. 1998. Distribution of olfactory receptor genes in the human genome. *Nat. Genet.* **18:** 243–250.

Ryba N.J. and Tirindelli R. 1997. A new multigene family of putative pheromone receptors. *Neuron* **19:** 371–379.

Shepherd G.M. 1994. Discrimination of molecular signals by the olfactory receptor neuron. *Neuron* **13:** 771–790.

Thompson J.D., Higgins D.G., and Gibson T.J. 1994. CLUSTAL W: Improving the sensitivity of progressive multiple sequence alignment through sequence weighting, position-specific gap penalties and weight matrix choice. *Nucleic Acids Res.* **22:** 4673–4680.

Young J.M., Kambere M., Trask B.J., and Lane R.P. 2005. Divergent V1R repertoires in five species: Amplification in rodents, decimation in primates, and a surprisingly small repertoire in dogs. *Genome Res.* **15:** 231–240.

Young J.M., Friedman C., Williams E.M., Ross J.A., Tonnes-Priddy L., and Trask B.J. 2002. Different evolutionary processes shaped the mouse and human olfactory receptor gene families. *Hum. Mol. Genet.* **11:** 535–546.

14

The Canine Major Histocompatibility Complex

John L. Wagner
Departments of Medicine, Microbiology, and Immunology
Thomas Jefferson University
Philadelphia, Pennsylvania 19107

Rainer F. Storb
Transplantation Biology Program
Fred Hutchinson Cancer Research Center
Department of Medicine
University of Washington
Seattle, Washington 98109

KNOWLEDGE OF THE CANINE IMMUNE SYSTEM is valuable not only for understanding the pathogenesis of a number of human diseases, but also for improving canine health. The dog serves as an important model for drug toxicity trials, solid organ and hematopoietic stem-cell transplants (Thomas and Storb 1999), and a variety of human diseases such as cyclic neutropenia (Weiden et al. 1974), X-linked severe combined immunodeficiency syndrome (SCID) (Felsburg et al. 1999), von Willebrand's disease (Thomas 1996), severe hereditary hemolytic anemia (Weiden et al. 1976), hemophilia (Fogh et al. 1984), gluten-sensitive enteropathy (Hall and Batt 1990), rheumatoid arthritis (Halliwell et al. 1972), systemic lupus erythematosus (Lewis and Schwartz 1971), narcolepsy (Baker et al. 1982), canine atopic dermatitis (Zur et al. 2002), pemphigus vulgaris (Hurvitz and Feldman 1975), and myasthenia gravis (Pflugfelder et al. 1981). Dogs have high rates of spontaneous malignancies and thus have served as models for a variety of cancers including breast cancer (Mol et al. 1999), non-Hodgkin lymphoma (Weiden et al. 1979), and prostate cancer (Navone et al. 1998–1999). Many of these diseases may have an immune basis.

The Dog and Its Genome ©2006 Cold Spring Harbor Laboratory Press 0-87969-742-3

The immune systems of dogs protect them from a variety of infectious agents. In dogs, as in other mammals, the immune system has a complex series of functions including distinguishing self from non-self as well as the ability to "recall" previous foreign antigen exposure (for review, see Klein 1982). An important genetic component of the immune system that facilitates many of these functions is the major histocompatibility complex (MHC), which is highly variable due to polymorphisms at several loci.

OVERVIEW OF THE CANINE MAJOR HISTOCOMPATIBILITY COMPLEX OR DOG LEUKOCYTE ANTIGEN

An understanding of the general structure and function of the MHC is helpful to comprehend its importance in transplantation and disease. The MHC is a tightly linked cluster of genes and gene families and is one of the most extensively studied regions of the genome in several species, including humans and mice. These genes encode proteins that are essential for the regulation of the immune response in vertebrates. In normal physiology, MHC gene products interact with bound peptide ligands and with products of rearranged T-cell receptor (TCR) genes in the thymus that result in positive and negative selection of the peripheral T-cell repertoire. This region of tightly linked genes is responsible for the recognition and regulation of the immune response. MHC molecules perform these roles by binding and presenting peptide antigens to T cells. This antigen presentation can lead to several events, including elimination of infected cells or cellular rejection of transplanted organs.

The genes within the MHC are divided into at least five isochores: extended class I, classic class I, extended class II, classic class II, and class III. The class I and class II molecules are the cell-surface glycoproteins of similar structure involved in antigen presentation to T cells. Allelic variation occurs typically because of polymorphism in or around the peptide-binding site, and in some species there may be more than 100 alleles for a given locus. Class III molecules are structurally unrelated to class I and class II molecules and are not relevant to antigen presentation but can be important in other aspects of the immune system such as complement activation. The complement pathway is an effector mechanism activated by the humoral (antibody) immune response. Class I antigens are expressed on all somatic cells, whereas class II molecules are expressed on antigen-presenting cells such as macrophages. Interestingly, in contrast to mice and humans, canine class II gene products are present on almost all lymphocytes (Doxiadis et al. 1989).

GENOMIC ORGANIZATION OF THE CANINE MHC

Mapping and chromosome assignment in the dog have lagged behind several other mammalian species. Part of the reason is that dog chromosomes are numerous, and many of them are too small and similar to be identified unambiguously by Giemsa banding alone. Using the technique of fluorescence in situ hybridization (FISH), the dog leukocyte antigen (DLA) has been localized to canine chromosome (CFA) 12 (Dutra et al. 1996). Tools that are available to aid in mapping the canine genome include a canine radiation hybrid map (Breen et al. 2004) and a bacterial artificial chromosome (BAC) library (Li et al. 1999). Whereas most DLA loci have been assigned to CFA12, one gene, *DLA-79* (which structurally resembles a class I gene—see later discussion) is located on CFA18 (Mellersh et al. 2000). Two recent publications (Debenham et al. 2005; Wagner et al. 2005) have described portions of the genomic organization of the DLA.

Over 700 kb of the classic and extended class II DLA regions have been sequenced and characterized (Debenham et al. 2005). Forty-five loci have been characterized; 29 are complete (i.e., they do not appear to be pseudogenes and may be transcribed on the basis of analyses of their genomic sequences). Some of the potentially functional DLA class II genes include *DLA-DRA1, -DRB1, -DQA1, -DQB1, -DMA, -DMB, -DOA,* and *-DOB*. The DP region in the dog consists only of pseudogenes, as in the cat, and unlike the human. The genomic organization of the class II region is quite similar to that of the cat and human. This is not surprising, because the orthologous relationship of the class II genes between human and other mammalian species has been well documented (Kumanovics et al. 2003). In general, the class II subregions are smaller in the dog than in the human and, in most cases, smaller than the corresponding subregion in the cat (Debenham et al. 2005).

A 150-kb region at the intersection of the class I and class III regions has also been sequenced (Wagner et al. 2005). Previously characterized class I genes were found in this region as well as some class III genes. The portion of the class III region sequenced in this location shows significant similarities in terms of gene structure and organization to the class III region in humans. The class III genes sequenced in this region include B-associated transcript (BAT1), nuclear factor of kappa light polypeptide gene enhancer in B-cells inhibitor-like 1 (NFKBIL1), vacuolar-ATPase subunit (ATP6G), lymphotoxin A (LTA), and tumor necrosis factor A (TNFα) (initially sequenced in 1994 [Zucker et al. 1994]).

The canine class I genes, which evolved after the radiation of mammals, do not preserve orthologous relationships to class I genes in humans or other mammalian species, and thus their designations are not the same as in the HLA region. In addition, because of the lack of orthologous relationships, the tissue distribution or expression of canine class I genes cannot be inferred from studies of class I loci in other mammalian species such as humans.

DLA POLYMORPHISM

Early understanding of the canine MHC polymorphism involved primarily cellular and serological analyses. On the basis of evidence from early transplantation experiments in various animals, it became apparent that histocompatibility genes could be analyzed by determining their gene products or antigens on the cell-surface membrane. One common method for this determination is the microcytotoxicity test, in which lymphocytes from one animal are incubated with sera and complement from a different animal and, if the animals are histoincompatible, the cells are damaged and can take up dye. The performance of the test depended on a variety of standardized serological reagents, each of which correlated with different allelic forms of various class I or class II molecules. Different class I or class II proteins on the cell surface (based on the allelic polymorphism) react with different antisera. These serological reagents were obtained by immunizing dogs with lymphocytes from histoincompatible dogs and collecting the serum from the immunized animals.

Another method of cellular testing is known as the mixed lymphocyte culture (MLC). In this procedure, lymphocytes are taken from two genetically dissimilar mammals of the same species and cultured together for a number of days. Quantitation of the incorporation of radiolabeled thymidine correlates somewhat with the amount of genetic disparity between the two animals. In dogs, as in humans, specificities are defined by the MLC correlate with class II antigens (Wagner et al. 2002).

Many early experimental histocompatibility studies in the dog have been oriented toward exploration of transplantation immunology. The production of antisera and evidence for the DLA and its importance in the outcomes of hematopoietic stem-cell transplantation was first demonstrated in 1968 (Epstein et al. 1968; Storb et al. 1968).

Evidence for two closely linked DLA loci was described in 1972 (Vriesendorp et al. 1972). Combining serologic typing and selection by MLC became the standard for selecting donor recipient pairs for marrow grafting (Storb et al. 1971). Through the 1970s and the 1980s, a series of international workshops led to a better definition of DLA antigens (Vriesendorp et al. 1973; Deeg et al. 1986; Bull et al. 1987). In the third workshop, the DLA was divided into three serologically defined loci: *DLA-A* (with five specificities), *DLA-B* (with four specificities), and *DLA-C* (with three specificities) (Bull et al. 1987). A fourth locus, *DLA-D* (with ten specificities), was defined by the MLC (Deeg et al. 1986).

BIOCHEMICAL ANALYSES

Biochemical analysis of the DLA in the 1980s followed the serological characterization. The *DLA-A* loci are characterized as class I molecules by their association with β_2-microglobulin (Krumbacher et al. 1986). An immunochemical analysis of glycosylated and nonglycosylated DLA molecules suggests that the products of one predominant DLA class I locus (corresponding to *DLA-A*) are present on the surface of peripheral blood leukocytes, and *DLA-C* gene products are thought to be weakly expressed class I antigens (Van der Feltz and Ploegh 1984; Doxiadis et al. 1986).

Conversely, *DLA-B* gene products, when studied by two-dimensional gel electrophoresis and lysostrip experiments, exhibit typical class II properties with a high level of serological polymorphism in the β-chain and no serological polymorphism in the α-chain (Doxiadis et al. 1989). Based on the current knowledge of the DLA, the *DLA-B* gene product probably represents the protein encoded by the *DLA-DRA* locus (not polymorphic) and the *DLA-DRB1* locus (highly polymorphic). One-dimensional isoelectric focusing and immunoblotting have defined *DLA-A* and *DLA-B* gene products, and there is a high degree of correlation between the biochemically defined antigens and the serological specificities (Kubens et al. 1995). What specific loci the *DLA-A* gene product correlates to is unknown.

The fourth component of complement (*C4*), homologous to the human HLA class III protein of the same name, has been found to be linked to the MHC region in the dog (Grosse-Wilde et al. 1983). In a survey of 291 dogs, five different variants were found, and biochemical analysis suggested that there are two different genes encoding *C4* (Doxiadis et al. 1985).

In summary, biochemical characterization of serologically defined antigens demonstrated class I, II, and III proteins homologous to better-characterized proteins in other mammalian species such as humans, and genomic analyses have confirmed the presence of many of the corresponding genes.

DLA—GENES AND POLYMORPHISM

Class I molecules such as HLA-A, -B, and -C are heterodimeric glycoproteins that were initially defined by antisera. The molecules contain a polymorphic membrane-bound α-chain (45 kD) that is non-covalently associated with β_2-microglobulin (a nonpolymorphic product of a non-MHC-linked gene). Classic class I (Ia) loci, in contradistinction to nonclassic class I (1b) loci, tend to be transcribed at higher levels in more tissues and tend to be more polymorphic than class Ib loci. Class Ia loci are more important in transplantation biology because of their greater contribution to alloreactivity and organ rejection.

The structure of class II loci is well conserved among mammalian species. Class II molecules in humans, such *HLA-DR, -DQ, -DM,* and -*DP,* exist as $\alpha\beta$ heterodimers composed of an α chain (MW 34 kD) and a β chain (MW 29 kD).

Beginning in the late 1980s, a molecular analysis of the canine MHC began. Sarmiento and Storb used various human class II probes to study the number of class IIA and IIB genes on Southern blots (Sarmiento and Storb 1988a,b). The same investigators used a similar approach with an HLA-B7 (a human class I) cDNA probe. By studying the patterns on Southern blots from 40 dogs, Sarmiento and Storb concluded that there are approximately eight canine class I loci (Sarmiento and Storb 1989).

Unfortunately, banding patterns from Southern analysis did not provide definitive information on the number of loci present nor the degree of polymorphism. Therefore, other molecular methods were used. Using reverse transcriptase-polymerase chain reaction (RT-PCR) and primer sequences obtained from corresponding HLA class II loci, Sarmiento et al. found that at least three class II loci, *DLA-DRB, -DQA,* and -*DQB,* are polymorphic (Sarmiento et al. 1990, 1992, 1993). The DRA locus appears to be monomorphic (Wagner et al. 1995).

The next step was to clone and sequence all the genes that were hypothesized to exist based on banding patterns from Southern analyses as well as RT-PCR work. To do this, human probes were used to screen a canine cDNA library. Then the canine cDNA clones were used to probe a canine genomic DNA library. In the case of the class I locus, a canine cDNA library was screened with an *HLA-B7* probe, and only one distinct clone was isolated, designated *I16* (Sarmiento and Storb 1990a). Using *I16* as a probe to screen a genomic library, over 15 clones were found, which inferred that at least seven distinct class I loci exist (Burnett and Geraghty 1995; Burnett et al. 1997).

One locus designated *DLA-79* has shown limited polymorphism, relatively low mRNA expression (in a variety of tissues except skeletal muscle) and, thus, has been designated a class Ib gene (Burnett and Geraghty 1995). The gene was isolated a second time from a genomic library using *HLA-E* as a probe and, like *HLA-E,* it has a long hydrophobic signal sequence (Burnett and Geraghty 1995). As mentioned previously, *DLA-79* is located on a separate chromosome from other DLA loci. Current studies are aimed at elucidating the function of this gene.

Three other class I loci (besides *DLA-79*), termed *DLA-88, DLA-12,* and *DLA-64,* appear to be complete genes by sequence analysis, and all three are transcribed in canine peripheral blood leukocytes (Burnett et al. 1997). DLA-88 appears to be more polymorphic than *DLA-12, DLA-79,* or *DLA-64* (Graumann et al. 1998). Two other genes, termed *DLA-53* and *DLA-12a,* are truncated class I pseudogenes (Burnett et al. 1997). *C1pg-26* is a processed gene located outside the DLA (Burnett et al. 1997). Neither the tissue expression nor the function of any of the class I genes is known at present, although one could infer by analogy from other species that DLA class I gene products could serve as cytotoxic T-lymphocyte targets.

Using methods similar to those described above for class I loci, several class II loci have been characterized. Using a human *DRB* cDNA probe, a canine *DRB* clone called *DRB5* was isolated (Sarmiento and Storb 1990b). This canine cDNA clone was used to screen a genomic library. From these experiments one highly polymorphic *DRB* locus designated *DLA-DRB1* and a pseudogene termed *DLA-DRB2* were cloned and sequenced (Wagner et al. 1996a,c). Similar strategies were used for the *DQ* loci. There is one *DQA* gene (*DLA-DQA1*) with a limited amount of polymorphism (Wagner et al. 1996b). There is one polymorphic *DQB* gene (*DLA-DQB1*) and one *DQB* pseudogene (*DLA-DQB2*) (Wagner et al. 1998b; Debenham et al. 2005). As in the human, some haplotypes do not have the same number of class II loci—the pseudogene *DLA-DRB2* has not been found in all haplotypes (Wagner et al. 1996a). Recently, a complete sequence derived from overlapping BAC clones of the DLA class II region has been published, and newly described genes *DLA-DMA, DLA-DMB, DLA-DOA,* and *DLA-DOB* have been sequenced (Debenham et al. 2005).

Table 1. Selected DLA genes and their polymorphism

Locus name	Class	Number of alleles known
DLA-DRA1	II	1
DLA-DRB1	II	63
DLA-DMA	II	1
DLA-DMB	II	1
DLA-DQA1	II	21
DLA-DQB1	II	55
DLA-DOA	II	1
DLA-DOB	II	1
DLA-88	I?a	55
DLA-79[a]	Ib	3
DLA-12	I?b	5
DLA-64	I?b	5

Not located in the canine MHC, see text.

References: Burnett and Geraghty 1995; Wagner et al. 1995, 1996b,c, 1998a,b, 2000a; Burnett et al. 1997; Graumann et al. 1998; Kennedy et al. 1998, 1999b, 2000, 2001, 2002a; Debenham et al. 2005.

A summary of several class I and class II DLA loci and the number of alleles defined to date are shown in Table 1. With the exception of *DLA-64*, almost all of these polymorphisms resulted in changes in the amino acid sequence, and the location of most of these substitutions was in the putative peptide-binding site (see references at bottom of Table 1). The number of amino acid differences between alleles typically varies from one to five.

Preliminary studies of class II polymorphism have found extensive interbreed, but minimal intrabreed variation of DLA class II alleles and haplotypes in dogs (Kennedy et al. 1999a, 2002b). From large-scale polymorphism studies it has become apparent that the allelic frequencies of genes within the class II region vary from breed to breed (Kennedy et al. 1999a).

HISTOCOMPATIBILITY TYPING

Histocompatibility typing is the process of determining which allele(s) is present for a particular locus. The process involves the application of immunogenetics of the MHC to the field of transplantation. Although much work has been done with the dog outside of transplantation, the initial driving force for a large amount of research into the DLA was/is for transplantation experiments. Each advance in understanding the DLA was accompanied by a refinement in the techniques of histocompatibility typing.

Histocompatibility typing of dogs has several important uses. Not only can it be helpful to identify the degree of histoincompatibility between a donor and recipient, but it also can be used to identify the linkage between a genetic disease and the DLA region. Formerly, histocompatibility typing relied on a combination of cellular and serological methods (Epstein et al. 1968; Storb et al. 1971; Albert et al. 1973; Deeg et al. 1986). Later restriction fragment length polymorphism (RFLP) analysis (Williamson et al. 1989; Lu et al. 1991; Burnett et al. 1994; He et al. 1994a,b; Francino et al. 1997) was used, and more recently, sequence-based DNA typing became standard (Wagner et al. 1998a, 2000a; Kennedy et al. 2002b).

Intrafamilial histocompatibility typing can be done using polymorphic microsatellite satellite markers—located in the class I region (e.g., C.2200 near *DLA-53*) (Burnett et al. 1995) and/or in the class II region (e.g., C.2202 near *DLA-DRB2*) (Wagner et al. 1996d). Both of these aforementioned markers are tetranucleotide repeats of $(GAAA)_n$. Both markers were identified during the sequencing of genomic clones. A polymorphism informational content (PIC) value of 0.804 was obtained for C.2200 (Burnett et al. 1995), and a value of 0.947 for C.2202 (Wagner et al. 1996d). Analysis of both markers in over 100 families (parents and offspring) has shown that each is stable for following Mendelian inheritance through multigeneration families. Because of the high polymorphic index of these markers and their stability, they are useful for determining the inheritance of DLA haplotypes within families. DLA matching of dogs within families determines which littermates are best suited for transplantation. Although a polymorphic marker anywhere on any dog chromosome could potentially be used as a marker, only those within the DLA region will define DLA-matched dogs within families. Recent genomic studies have revealed other microsatellite markers in the DLA region that may be useful for histocompatibility typing (Debenham et al. 2005; Wagner et al. 2005).

Microsatellite markers are not always suitable for finding DLA-matched unrelated dogs for transplantation experiments because they do not identify the genotype of any loci. Genotyping unrelated dogs at each polymorphic locus is an area of current investigation to find matched unrelated dogs for transplants. Currently, typing methods are similar to those used for HLA loci and are based on molecular differences to determine which alleles are present. These methods include PCR–single-stranded conformational polymorphism (SSCP) (Wagner et al. 1998c, 2000a), sequence-specific oligonucleotide probes (SSOP) (Kennedy et al. 1998, 1999a,b, 2000, 2002a,b), and reverse transcriptase (RT)-nested PCR and cycle sequencing (Happ et al. 1999).

Correlation between molecular typing and cellular typing will help elucidate the functional importance of various loci. One such study attempted to correlate *DLA-D* typing and class II polymorphism. This study suggested that *DLA-D* type as assessed by the MLC reaction tended to correlate with *DLA-DRB1* but not necessarily *DLA-DQB1* loci. The contribution of newly potentially expressed polymorphic class II genes such as *DLA-DOA* or *DLA-DOB* was not assessed in this study (Wagner et al. 2002). Potentially unrelated dogs could be DLA matched for class II genes if the situation were similar to that of humans; and *DLA-DRB1*, *DLA-DQA1*, and *DLA-DQB1* were the "only" class II genes that needed to be defined by their alleles in terms of transplantation.

The issues of class I histocompatibility typing are more complicated due to the similarity of the sequences of the known DLA class I genes (Burnett et al. 1997), the lack of knowledge of the full sequence of the DLA class I region, and the lack of knowledge regarding tissue expression and distribution of the various class I loci.

CANINE MHC AND DISEASE—FUTURE DIRECTIONS

As histocompatibility typing methods have evolved, so have efforts been undertaken to determine the relationship between canine diseases and the MHC. This is of interest particularly when the canine disease serves as a model for a human disease that has some association with the HLA. An early study suggested that systemic lupus erythematosus in dogs is associated with the class I serological specificity *DLA-A7* (Teichner et al. 1990), although no attempt to correlate this finding with a specific class I locus or molecular-based typing has been published.

For reasons cited previously, most of the studies have involved the relationship of the class II genes and not the class I genes with disease. For example, canine narcolepsy (Wagner et al. 2000b) and gluten-sensitive enteropathy (Polvi et al. 1998) do not appear to be associated with the MHC as they are in humans. Whether the DLA plays a different role in the development of these diseases than in humans, or whether the expression of the DLA molecules in dog is different from that in humans, awaits further studies. Conversely, canine rheumatoid arthritis has been associated with canine *DLA-DRB1* alleles which contain an epitope that is shared with human DRB1 alleles that have been associated with rheumatoid arthritis (Ollier et al. 2001). Other studies of the DLA have involved diseases that are more prevalent in dogs. For example, a specific *DLA-DRB1* allele has been associated with the development of visceral leishmaniasis (Quinnell et al. 2003).

In summary, the next steps in the study of the DLA should include completing the sequencing of the class I and class III regions. In addition, it is important to understand the expression and tissue distribution of the class I loci. With this information, studies can be carried out to determine the relationship between class I or class III genes and disease.

ACKNOWLEDGMENTS

This work was supported by grants CA78902, CA15704, HL36444, and RR12558 from the National Institutes of Health.

REFERENCES

Albert E.D., Erickson V.M., Graham T.C., Parr M., Templeton J.W., Mickey M.R., Thomas E.D., and Storb R. 1973. Serology and genetics of the DL-A system. I. Establishment of specificities. *Tissue Antigens* **3:** 417–430.

Baker T.L., Foutz A.S., McNerney V., Mitler M.M., and Dement W.C. 1982. Canine model of narcolepsy: Genetic and developmental determinants. *Exp. Neurobiol.* **75:** 729–742.

Breen M., Hitte C., Lorentzen T.D., Thomas R., Cadieu E., Sabacan L., Scott A., Evanno G., Parker H.G., Kirkness E.F., et al. 2004. An integrated 4249 marker FISH/RH map of the canine genome. *BMC Genomics* **5:** 1–11.

Bull R.W., Vriesendorp H.M., Cech R., Grosse-Wilde H., Bijma A.M., Ladiges W.L., Krumbacher K., Doxiadis I., Ejima H., Templeton J., et al. 1987. Joint report of the third international workshop on canine immunogenetics. II. Analysis of the serological typing of cells. *Transplantation* **43:** 154–161.

Burnett R.C. and Geraghty D.E. 1995. Structure and expression of a divergent canine class I gene. *J. Immunol.* **155:** 4278–4285.

Burnett R.C., DeRose S.A., and Storb R. 1994. A simple restriction fragment-length polymorphism assay for MHC class II gene testing of dog families. *Transplantation* **57:** 280–281.

Burnett R.C., DeRose S.A., Wagner J.L., and Storb R. 1997. Molecular analysis of six dog leukocyte antigen DLA class I sequences including three complete genes, two truncated genes, and one full-length processed gene. *Tissue Antigens* **49:** 484–495.

Burnett R.C., Francisco L.V., DeRose S.A., Storb R., and Ostrander E.A. 1995. Identification and characterization of a highly polymorphic microsatellite marker within the canine MHC Class I region. *Mammalian Genome* **6:** 684–685.

Debenham S.L., Hart E. A., Ashurst J.A., Howe K.L., Quail M.A., Ollier W.E.R., and Binns M.M. 2005. Genomic sequence of the class II region of the canine MHC: Comparison with the MHC of other mammalian species. *Genomics* **85:** 48–59.

Deeg H.J., Raff R.F., Grosse-Wilde H., Bijma A.M., Buurman W.A., Doxiadis I., Kolb H.J., Krumbacher K., Ladiges W., Losslein K.L., et al. 1986. Joint report of the third international workshop on canine immunogenetics. I. Analysis of homozygous typing cells. *Transplantation* **41:** 111–117.

Doxiadis G., Rebmann V., Doxiadis I., Krumbacher K., Vriesendorp H.M., and Grosse-Wilde H. 1985. Polymorphism of the fourth complement component in the dog. *Immunobiology* **169:** 563–569.

Doxiadis I., Krumbacher K., Neefjes J.J., Ploegh H.L., and Grosse-Wilde H. 1989. Biochemical evidence that the DLA-B locus codes for a class II determinant expressed on all canine peripheral blood lymphocytes. *Exp. Clin. Immunogenet.* **6:** 219–224.

Doxiadis I., Krumbacher K., Rein R., Neefjes J.J., Doxiadis G., SchoenW., Ploegh H.L., and Grosse-Wilde H. 1986. Canine MHC biochemical definition of class I, class II and class III determinants, similarities and differences to the human and murine systems. *Immunobiology* **173:** 264–265.

Dutra A.S., Mignot E., and Puck J.M. 1996. Gene localization and syntenic mapping by FISH in the dog. *Cytogenet. Cell. Genet.* **74:** 113–117.

Epstein R.B., Storb R., Ragde H., and Thomas E.D. 1968. Cytotoxic typing antisera for marrow grafting in littermate dogs. *Transplantation* **6:** 45–58.

Felsburg P.J., Harnett B.J., Henthorn P.S., Moore P.F., Krakowka S., and Ochs H.D. 1999. Canine X-linked severe combined immunodeficiency. *Vet. Immunol. Immunopathol.* **69:** 127–135.

Fogh J.M., Nygaard L., Andresen E., and Nilsson I.M. 1984. Hemophilia in dogs, with special reference to hemophilia A among German shepherd dogs in Denmark. I: Pathophysiology, laboratory tests and genetics. *Nord. Vet. Med.* **36:** 235–240.

Francino O., Amills M., and Sanchez A. 1997. Canine *Mhc* DRB1 genotyping by PCR-RFLP analysis. *Anim. Genet.* **28:** 41–45.

Graumann M.B., DeRose S.A., Ostrander E., and Storb R. 1998. Polymorphism analysis of four canine class I genes. *Tissue Antigens* **51:** 374–381.

Grosse-Wilde H., Doxiadis G., Krumbacher K., Dekkers-Bijima A., and Kolb H.J. 1983. Polymorphism of the fourth complement component in the dog and linkage to the DLA system. *Immunogenetics* **18:** 537–540.

Hall E.J. and Batt R.M. 1990. Development of wheat-sensitive enteropathy in Irish setters: Morphological changes. *Am. J. Vet. Res.* **51:** 978–982.

Halliwell R.E.W., Lavelle R.B., and Butt K.M. 1972. Canine rheumatoid arthritis: A review and a case report. *J. Small Anim. Pract.* **13:** 239–248.

Happ G.M., Aquilla E., Martick M., Yuncker C., Wojciechowski W., and Fox L. 1999. DLA-DRB1 histocompatibility genotyping using RT-nested PCR and cycle sequencing. *Vet. Immunol. Immunopathol.* **69:** 93–100.

He Y.W., Ferencik S., and Grosse-Wilde H. 1994a. A research on DLA-DRB1 genotyping by PCR-RFLP. I. To select appropriate oligonucleotide primer pair. *J. Tongji Med. Univ.* **14:** 24–28.

———. 1994b. A research on DLA-DRB1 genotyping by PCR-RFLP. II. A study of serology and cellularly defined DLA haplotypes and their segregation. *J. Tongji Med. Univ.* **14:** 29–34.

Hurvitz A.I. and Feldman E. 1975. A disease in dogs resembling human phemphigus vulgaris: Case reports. *J. Am. Vet. Assoc.* **166:** 585–590.

Kennedy L.J., Carter S.D., Barnes A., Bell S., Bennett D., Ollier B., and Thompson W.E. 1999a. Interbreed variation of DLA-DRB1, DQA1 alleles and haplotypes in the dog. *Vet. Immunol. Immunopathol.* **69:** 101–111.

———. 1999b. DLA-DRB1 polymorphisms in dogs defined by sequence-specific oligonucleotide probes (SSOP). *Tissue Antigens* **53:** 184–189.

Kennedy L.J., Carter S.D., Barnes A., Bell S., Bennett D., Ollier W.E., and Thompson W.E. 1998. Nine new dog DLA-DRB1 alleles identified by sequenced based typing. *Immunogenetics* **48:** 296–301.

Kennedy L.J., Barnes A., Happ G.M., Quinnell R.J., Courtenay O., Carter S.D., Ollier W.E., and Thompson W. 2002a. Evidence for extensive DLA polymorphism in different dog populations. *Tissue Antigens* **60**: 43–52.

Kennedy L.J., Hall L.S., Carter S.D., Barnes A., Bell S., Bennett D., Ollier B., and Thompson W.E. 2000. Identification of further DLA-DRB1 and DQA1 alleles in the dog. *Eur. J. Immunogenet.* **27**: 25–28.

Kennedy L.J., Angles J.M., Barnes A., Carter S.D., Francino O., Gerlach J.A., Happ G.M., Ollier W.E.R., Thompson W., and Wagner J.L. 2001. Nomenclature for factors of the dog major histocompatibility system (DLA), 2000: Second report of the ISAG DLA Nomenclature Committee. *Anim. Genet.* **32**: 312–321.

Kennedy L.J., Barnes A., Happ. G.M., Quinnell R.J., Bennett D., Angeles J.M., Day. M.J., Carmichael N., Innes J.F., Isherwood D., et al. 2002b. Extensive interbreed, but minimal intrabreed, variation of DLA class II alleles and haplotypes in dogs. *Tissue Antigens* **59**: 194–204.

Klein J. 1982. *Immunology: The science of self–non-self discrimination.* Wiley-Interscience, New York.

Krumbacher K., van der Feltz M.J.M., Happel M., Gerlach C., Losslein L.K., and Grosse-Wilde H. 1986. Revised classification of the DLA loci by serological studies. *Tissue Antigens* **27**: 262–268.

Kubens B.S., Krumbacher K., and Grosse-Wilde H. 1995. Biochemical definition of DLA-A and DLA-B gene products by one-dimensional isoelectric focusing and immunoblotting. *Eur. J. Immunogenet.* **22**: 199–207.

Kumanovics T., Takada T., and Fischer Lindahl K. 2003. Genomic organisation of the mammalian *MHC. Annu. Rev. Immunol.* **21**: 629–657.

Lewis R.M. and Schwartz R.S. 1971. Canine systemic lupus erythematosus. Genetic analysis of an established breeding colony. *J. Exp. Med.* **134**: 417–438.

Li R., Mignot E., Faraco J., Kadotani H., Cantanese J., Zhao B., Lin X., Hinton L., Ostrander E.A., Patterson D.F., and de Jong P.J. 1999. Construction and characterization of an eight-fold redundant dog genomic bacterial artificial chromosome library. *Genomics* **58**: 9–17.

Lu P., Zucker K.E., Esquenazi V., and Miller J. 1991. Correlation of RFLP typing and MLC reactivity in dogs. *Tissue Antigens* **37**: 224–229.

Mellersh C.S., Hitte C., Richman M., Vignaux F., Priat C., Touquand S., Werner P., Andre C., DeRose S., Patterson D.F., et al. 2000. An integrated linkage-radiation hybrid map of the canine genome. *Mamm. Genome* **11**: 120–130.

Mol J.A., Lantinga-van Leeuwen I.S., van Garderen E., Selman P.J., Oosterlaken-Dijksterhuis M.A., Schalken J.A., and Rijnberk A. 1999. Mammary growth hormone and tumorigenesis—Lessons from the dog. *Vet. Q.* **21**: 111–115.

Navone N.M., Logothetis C.J., von Eschenbach A.C., and Troncoso P. 1998-1999. Model systems of prostate cancer: Uses and limitations. *Cancer Metastasis Rev.* **17**: 361–371.

Ollier W.E., Kennedy L.J., Barnes A.N., Bell S.C., Bennett D., Angels J.M., Innes J.F., and Carter S.D. 2001. Dog MHC alleles containing the human RA shared epitope confer susceptibility to canine rheumatoid arthritis. *Immunogenetics* **53**: 669–673.

Pflugfelder C.M., Cardinet G.H., III, Lutz H., Holliday T.A., and Hansen R.J. 1981. Acquired canine myasthenia gravis: Immunocytochemical localization of immune complexes at the neuromuscular junctions. *Muscle Nerve* **4**: 289–295.

Polvi A., Garden O.A., Houlston R.S., Mäki M., Batt R.M., and Partanen J. 1998. Genetic susceptibility to gluten sensitive enteropathy in Irish setter dogs is not linked to the major histocompatibility complex. *Tissue Antigens* **52**: 543–549.

Priat C., Hitte C., Vignaux F., Renier C., and Jiang Z.H. 1998. A whole-genome radiation hybrid map of the dog genome. *Genomics* **54:** 361–378.

Quinnell R.J., Kennedy L.J., Barnes A., Courtenay O., Dye C., Garcez L.M., Shaw M.A., Carter S.D., Thomson W., and Ollier W.E. 2003. Susceptibility to visceral leishmaniasis in the domestic dog is associated with MHC class II polymorphism. *Immunogenetics* **55:** 23–28.

Sarmiento U.M. and Storb R.F. 1988a. Characterization of class II alpha genes and DLA-D region allelic associations in the dog. *Tissue Antigens* **32:** 224–234.

———. 1988b. Restriction fragment length polymorphism of the major histocompatibility complex of the dog. *Immunogenetics* **28:** 117–124.

———. 1989. RFLP analysis of DLA class I genes in the dog. *Tissue Antigens* **34:** 158–163.

———. 1990a. Nucleotide sequence of a dog class I cDNA clone. *Immunogenetics* **31:** 400–404.

———. 1990b. Nucleotide sequence of a dog DRB cDNA clone. *Immunogenetics* **31:** 396–399.

Sarmiento U.M., Sarmiento J.I., and Storb R. 1990. Allelic variation in the DR subregion of the canine major histocompatibility complex. *Immunogenetics* **32:** 13–19.

Sarmiento U.M., DeRose S., Sarmiento J.I., and Storb R. 1992. Allelic variation in the DQ subregion of the canine major histocompatibility complex: I. DQA. *Immunogenetics* **35:** 416–420.

———. 1993. Allelic variation in the DQ subregion of the canine major histocompatibility complex: II. DQB. *Immunogenetics* **37:** 148–152.

Storb R., Rudolph R.H., and Thomas E.D. 1971. Marrow grafts between canine siblings matched by serotyping and mixed leukocyte culture. *J. Clin. Investig.* **50:** 1272–1275.

Storb R., Epstein R.B., Bryant J., Ragde H., and Thomas E.D. 1968. Marrow grafts by combined marrow and leukocytes infusions in unrelated dogs selected by histocompatibility typing. *Transplantation* **6:** 587–593.

Teichner M., Krumbacher K., Doxiadis I., Doxiadis G., Fournel C., Rigal D., Monier J.C., and Grosse-Wilde H. 1990. Systemic lupus erythematosus in dogs association to the major histocompatibility complex class I antigen DLA-A7. *Clin. Immunol. Immunopathol.* **55:** 255–262.

Thomas E.D. and Storb R. 1999. The development of the scientific foundation of hematopoietic cell transplantation based on animal and human studies. In *Hematopoietic cell transplantation* 2nd edition (ed. E.D. Thomas et al.), pp. 1–11. Blackwell Science, Malden, Massachusetts.

Thomas J.S. 1996. Von Willebrand's disease in the dog and cat. *Vet. Clin. N. Am. Small Anim. Pract.* **26:** 1089–1110.

Van der Feltz M.J.M. and Ploegh H.L. 1984. Immunochemical analysis of glycosylated and nonglycosylated DLA class I antigens. *Immunogenetics* **19:** 95–108.

Vriesendorp H.M., Epstein R.B., D'Amaro J., Westbroek D.L., and van Rood J.J. 1972. Polymorphism of the DLA system. *Transplantation* **14:** 299–307.

Vriesendorp H.M., Westbroek D.L., D'Amaro J., van der Does J.A., van der Steen G.J., van Rood J.J., Albert E., Bernini L., Bull R.W., Cabasson J., et al. 1973. Joint report of 1st international workshop on canine immunogenetics. *Tissue Antigens* **3:** 145–163.

Wagner J.L., Burnett R.C., and Storb R. 1996a. Molecular analysis of the DLA DR subregion. *Tissue Antigens* **48:** 549–553.

Wagner J.L., Creer S.A., and Storb R. 2000a. Dog class I gene DLA-88 histocompatibility typing by PCR-SSCP and sequencing. *Tissue Antigens* **55:** 564–567.

Wagner J.L., Sarmiento U.M., and Storb R. 2002. Cellular, serological, and molecular polymorphism of the canine class I and class II loci of the canine Major Histocompatibility Complex. *Tissue Antigens* **59:** 205–210.

Wagner J.L., Works J.D., and Storb R. 1998a. DLA-DRBB1 and DLA-DQBC1 histocompatibility typing by PCR-SSCP and sequencing. *Tissue Antigens* **52:** 397–401.

Wagner J.L., Burnett R.C., DeRose S.A., and Storb R. 1996b. Molecular analysis and polymorphism of the DLA-DQA gene. *Tissue Antigens* **48:** 199–204.

Wagner J.L., Burnett R.C., Works J.D., and Storb R. 1996c. Molecular analysis of DLA-DRBB1 polymorphism. *Tissue Antigens* **48:** 554–561.

Wagner J.L., DeRose S.A., Burnett R.C., and Storb R. 1995. Nucleotide sequence and polymorphism analysis of canine DRA cDNA clones. *Tissue Antigens* **45:** 284–287.

Wagner J.L., Hayes-Lattin B., Works J.D., and Storb R. 1998b. Molecular analysis and polymorphism of the DLA-DQB genes. *Tissue Antigens* **52:** 397–401.

Wagner J.L., Palti Y., DiDario D., and Faraco J. 2005. Sequence of the canine major histocompatibility complex region containing non-classical class I genes. *Tissue Antigens* (in press).

Wagner J.L., Storb R., Storer B., and Mignot E. 2000b. DLA-DQB1 alleles and bone marrow transplantation experiments in narcoleptic dogs. *Tissue Antigens* **56:** 223–231.

Wagner J.L., Burnett R.C., DeRose S.A., Francisco L.V., Storb R., and Ostrander E.A. 1996d. Histocompatibility testing of dog families with polymorphic microsatellite markers. *Transplantation* **62:** 876–877.

Weiden P.L., Storb R., Graham T.C., and Schroeder M.L. 1976. Severe haemolytic anaemia in dogs treated by marrow transplantation. *Br. J. Haematol.* **33:** 357–362.

Weiden P., Robinett B., Graham T.C., Adamson J.W., and Storb R. 1974. Canine cyclic neutropenia. A stem cell defect. *J. Clin. Investig.* **53:** 950–954.

Weiden P.L., Storb R., Deeg H.J., Graham T.C., and Thomas E.D. 1979. Prolonged disease-free survival in dogs with lymphoma after total-body irradiation and autologous marrow transplantation consolidation of combination-chemotherapy-induced remissions. *Blood* **54:** 1039–1049.

Williamson P., Nicholas F.W., and Stewart G.J. 1989. Restriction fragment length polymorphism analysis of canine class II major histocompatibility complex genes. *Transplant. Proc.* **21:** 3751–3752.

Zucker K., Lu P., Fuller L., Asthana D., Esquenaz V., and Miller J. 1994. Cloning and expression of the cDNA for canine tumor necrosis factor-alpha in *E. coli. Lymphokine Cytokine Res.* **13:** 191–196.

Zur G., Ihrke P.J., White S.D., and Kass P.H. 2002. Canine atopic dermatitis: A retrospective study of 266 cases examined at the University of California, Davis, 1992-1998. Part I. Clinical features and allergy testing results. *Vet. Dermatol.* **13:** 89–102.

15

Breed-specific Hereditary Diseases and Genetic Screening

Urs Giger

Section of Medical Genetics, School of Veterinary Medicine
University of Pennsylvania
Philadelphia, Pennsylvania 19104-6010

David R. Sargan

Department of Clinical Veterinary Medicine
University of Cambridge
Cambridge CB3 0ES, United Kingdom

Elizabeth A. McNiel

Veterinary Clinical Sciences, College of Veterinary Medicine
University of Minnesota
Saint Paul, Minnesota 55108

THERE IS NO MAMMALIAN SPECIES THAT HAS a greater phenotypic diversity than the domestic dog (*Canis familiaris*). This has occurred not because of natural selection, but rather through the domestication of the dog by humans and the associated selection for various uses in different societies. Based on morphological features, such as body size, head and ear form, proportion of body and legs, and coat length and color, the canine species is divided into over 400 distinct breeds (Wilcox and Walkowicz 1995). These breeds also greatly differ with regard to their behavioral traits, such as their suitability for hunting by line of sight or by scent, for retrieving, herding, guarding, and for a number of other specialized roles. Moreover, certain characteristics within a breed may have been further subdivided by geographic separations and selection leading to many additional breeds, which are being recognized by some kennel clubs. Finally, within a breed an individual's ability to be suitable as a pet or a working dog also varies considerably, thereby reflecting the heterogeneic background of dog breeds.

The Dog and Its Genome ©2006 Cold Spring Harbor Laboratory Press 0-87969-742-3

In contrast to many other domesticated species, the evolution and selection of the domestic dog and desired characteristics within a breed were not driven by economic pressures, but rather by humans aiming for suitable companions and working dogs. The selection of a dog as a companion pet by humans depends on many factors and is influenced by dog shows, popularity of a breed, availability, and personal affection.

The domestic dog has become a closely integrated family member whose well-being is important to humans, and thus the dog receives considerable care and a high level of health surveillance from its owners. Proper conditions to raise and maintain dogs have been established and are generally accepted. Balanced nutritional requirements and commercial diets for these carnivores have been developed, and most poisons for humans are also recognized to be toxic for dogs, and hence, their exposure is avoided. In much of the world, vaccines and antimicrobial agents have greatly reduced the occurrence of many canine infectious diseases. Furthermore, other lifesaving advances in veterinary medicine and surgery, which closely followed developments in humans, have increased the chance of survival of companion animals. All these recent developments tend to diminish the frequency of acquired conditions such as nutritional disturbances, intoxications, and infections, and to raise the recognition of genetic defects in dogs. Thereby, many hereditary disorders have been identified in dogs during the past few decades, and every year new ones are being discovered (Patterson 2000; Brooks and Sargan 2001; Giger 2005).

Whereas in humans around 5000 disorders have been described as having a genetic basis (Online Mendelian inheritance in Man [OMIM], http://www.ncbi.nlm.nih.gov/entrez/query.febi?db=OMIM), the second highest number of reported naturally occurring hereditary disorders is seen in the domestic dog ("Inherited Diseases in Dogs" [IDID], http://www.vet.cam.ac.uk/idid). Currently, the literature describes about one-tenth of the number seen in humans, but the disease number in dogs is rapidly rising. These numbers are much higher than in any other species, including food animals, where economic pressures rapidly eliminate these traits and also have prevented the further investigation of diseased animals. Whereas in the past clinical genetics had a large role in only a few canine patients, specific genetic information is becoming critical in small-animal practice for every canine patient and for the future health of many breeds. In addition, many genetic traits in dogs have been recognized as important homologs of human disorders and are helpful in the further study of the pathogenesis and development of new therapeutic strategies. This chapter provides an overview on canine hereditary disorders and genetic disease predispositions, but descriptions of each genetic defect are beyond the scope of this chapter. To that end, separate chapters discuss in more detail specific disease groups as well as therapeutic interventions as disease models for humans.

HEREDITARY DISEASES

Heritable diseases (other than epigenetic diseases) are caused by chromosomal alterations or gene mutations. Mutations occurring in genomic DNA in the germ line alter the expression, structure, or function of the coded protein. Among the disorders entirely or partly caused by genetic factors, three main types are recognized: (1) chromosomal, (2) single gene, and (3) complex or multifactorial disorders. However, any hereditary disease is now recognized to be a continuum involving not only the main disease-causing gene alteration, but also other loci contributing to disease expression, as well as nongenetic factors. In other words, in a single-gene disorder, a specific gene locus has a predominant effect, while other gene loci act as modifiers of the phenotype, whereas in complex disease traits, multiple gene loci may contribute similarly to the phenotype. Nongenetic factors with effects on phenotype include diet and other environmental influences, as well as stochastic factors such as random X-inactivation, immunoglobulin and T-cell receptor gene rearrangements, and also somatic mutations. The latter are extremely important in autoimmune and neoplastic disease processes.

The first canine hereditary disorders were recognized in the early 1900s and included congenital skeletal malformations, such as dysplasias, which had obvious phenotypes. At the beginning of the 20th century, Sir Archibald Garrod, recognizing that major derangements in metabolic pathways led to disorders in people, coined the term "inborn errors of metabolism." Today inborn errors of metabolism include all biochemical disorders due to a genetically determined, specific defect or defects in the structure and/or function of a protein molecule. Aside from the classic enzyme deficiencies, genetic defects in structural proteins, receptors, plasma and membrane proteins, and other proteins covered by this definition result in biochemical disturbances. In fact, the term "inborn errors of metabolism" not only includes the metabolic disorders, but can ultimately be considered to encompass all genetic diseases including congenital malformations as well as increased susceptibility to disease. Nevertheless, hereditary disorders are often grouped into congenital malformations, inborn errors of metabolism, and genetic predisposition to disease, recognizing that there is a considerable degree of overlap (Patterson et al. 1988, 1989; Scriver et al. 2001; Giger 2005).

MODE OF INHERITANCE OF HEREDITARY DISEASES

The mode of inheritance and the degree of heritability of some genetic diseases have been determined for some canine hereditary disorders but not yet for others. The database IDID, which lists inherited disorders by breed as well as by clinical description, contains 1050 genetic disease entities as of early 2005. Approximately 500 disorders of the entries on the IDID are considered to be inherited by a simple monogenic mode, with most of them recessively inherited. As shown in Figure 1 (upper panel), for dogs, autosomal recessive (AR) inheritance of diseases predominates whereas X-linked recessive (XR) and autosomal dominant (AD) diseases are less commonly seen. These data are derived from analysis of patterns of inheritance of disease by disease type, using IDID, and for the lower panel, OMIM. In addition, there are isolated X-chromosomal dominant (XD) traits, including a nephritis in Samoyeds (Zheng et al. 1994), and rare mitochondrial patterns, such as a metabolic myopathy in the Old English sheepdog (Vijavasarathy et al. 1994). Although a proportion of the remaining disorders are likely polygenic (P), many have unknown modes of inheritance (referred to as familial [F] in Fig. 1), and there may be many further simplex or oligogenic disorders discovered. Hence, the proportions shown in Figure 1 have a degree of uncertainty. Any attempt to count modes of inheritance is further confounded because for many canine diseases with important heritable components, the contribution of inheritance has yet to be recognized. Skeletal, circulatory, developmental, autoimmune, and neoplastic diseases with inherited components are among the frequent causes of morbidity and mortality in veterinary practice. Moreover, it is clear that ancient disease-causing mutations occur in multiple breeds (good examples are ivermectin sensitivity in the collies and related breeds [Neff et al. 2004] and progressive rod and cone degeneration, a late-onset progressive retinal atrophy in many western breeds; see Chapter 16), that multiple genetic entities may cause

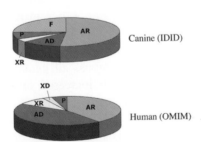

Figure 1. Mode of inheritance of hereditary disorders in dogs (*upper*) and humans (*lower*).

similar diseases in a single breed (e.g., type A and B progressive retinal atrophy [PRA] in miniature schnauzers; see Chapter 16), and also that the same genetic entity has sometimes received more than one clinical description and label, particularly when it occurred against more than one genetic background. All of these factors currently preclude accurate enumeration of genetic disorders.

Figure 1 shows that the proportion of diseases with different modes of inheritance is very different in humans and dogs. OMIM does not distinguish diseases that are familial with mode of inheritance unknown from those that are "sporadic, presumed inherited," so the familial category is not included on the human chart. Nonetheless, it is evident that the proportion of monogenic autosomal recessively inherited diseases in dogs is somewhat higher than in humans, whereas autosomal dominant and X-linked diseases are recorded more rarely in dogs. Dog owners can efficiently select against dominant diseases by not breeding overtly diseased animals, but perhaps of more importance in causing the different proportions of each mode is the role of canine population structure. Kennel clubs in the western world in the late 19th century set out "to maintain a registry of purebred dogs and preserve its integrity" (in the words of the first article of the American Kennel Club Mission Statement) and have placed rigid barriers of genetic isolation around purebred breeds (see Chapters 1–3). These breeds often descend from only a few founder individuals which may themselves have come from relatively isolated subpopulations, bred for distinctive characters over a period of several hundred years. Many of these closed populations are small in number and are subject to popular sire effects and strong selective pressures. Purebred dogs are popular throughout the developed world, and pedigree records suggest considerable interbreeding of geographically distant individuals. All these characteristics tend to increase homozygosity in purebred dog populations and thus the occurrence of recessively inherited disorders.

The proportion of observed X-chromosomal recessive disorders in dogs appears smaller than that which is reported in humans, but no studies have been done to examine the occurrence of these diseases and the age of the dam. Once again, this low rate may be due to selection by owners, as all male carriers of an X-linked disease chromosome, as well as all female homozygotes, can be recognized and removed, making the selection of phenotypes more effective in reducing mutant allele frequencies than for autosomal recessive traits. Moreover, removal of carrier dams of affected animals from breeding is done more effectively than removal of sires, because dams are usually under the control of the affected pup owner, whereas sires are often distant.

FREQUENCY OF HEREDITARY DISORDERS

Although any genetic defect may occur in any dog, many have only been documented in one family or a specific breed of dogs. Practically all of the entries in the IDID are considered to exhibit breed predisposition or to have increased some level of prevalence within certain breeds. In fact, the frequency of a particular disorder and the mutant allele may reach very high proportions in some breeds. This may be due to a founder effect in which one or more of the founders of a small ancestral group was a carrier or even affected, or as observed in several smaller breeds, a popular sire was later determined to be a carrier of a mutant allele.

Several studies have directly shown the effects of breeding practices on allele frequencies: Myotonia congenita causing marked skeletal muscle hypertrophy in the miniature schnauzer is caused by a chloride channel defect and is inherited by an autosomal recessive mode. The mutant allele frequency was determined to be .12 in a screening of >500 miniature schnauzers. The same screening revealed that every identified carrier and deficient animal was related to a popular sire which tested as a carrier himself. His ancestors were deceased by the time the disease was identified in his offspring, and thus it remains unknown whether he was the initial dog with the myotonia-causing mutation in his germ line (Bhalerao et al. 2002).

Severe combined immunodeficiency was found in basset hounds (Henthorn et al. 1994) and Pembroke Welsh corgis (Somberg et al. 1995) and is inherited by an X-chromosomal recessive trait. It is caused by two different mutations in the common gamma chain of the IL-2 receptor and leads to overwhelming infections and death before the first year of life. Both mutations are captured in research colonies, whereas in the breed the defect has been eliminated by identifying carrier females and spaying them. In contrast, hemophilia A in German shepherds causes a relatively mild coagulopathy, and hemophilic males have reached reproductive age. At least one major stud dog has been used extensively for breeding, making this a widespread X-chromosomal recessive disorder (Fogh 1988).

In people, a disease occurrence of 1 in <500 individuals is considered a high-frequency disorder, whereas many hereditary diseases appear to occur in 1–10% of dogs in certain breed populations. Inbreeding practices that were initially commenced by the "canine fancy" in the 19th century to preserve and make more uniform the desirable traits in certain breeds do favor the persistence and occurrence of recessively inherited diseases, but contrary to the general public's belief, they do not cause genetic defects. Unfortunately, genetic disease frequencies in the domestic dog breeds are generally not available or are severely biased because of data collection methods. Data may be gathered by registries, breed clubs, testing laboratories, and in clinical studies, or simply by individuals (see Chapter 3). All these sources depend on the owner's willingness to test and report the disease in a particular dog. Hence, registries are likely to overrepresent disease-free and mutant-allele-free dogs, whereas clinical studies and surveys likely favor participation of owners with affected and related dogs. Large-scale randomized screening programs and open registries with data on genetic diseases in certain breeds have been established only for a small number of diseases. Because there are fees involved, affected and carrier animals are often not registered, although some registries will register affected animals without charge. Similarly, some genetic screening laboratories do not charge when a dog is found to be affected. Moreover, surveys do not generally differentiate between pet and breeding animals, despite the fact that they have vastly different impacts on future generations. Unfortunately, most such studies depend on the accuracy of clinical records, and most concentrate on a single disease or a single breed.

There is only one randomized but relatively small survey of an entire breeding stock for a monogenic trait by a DNA test. Phosphofructokinase (PFK) deficiency causes intermittent hemolytic anemia and metabolic myopathy in English springer spaniels (Fig. 2). PFK deficiency was first described in 1985, and the disease-causing mutation was identified in 1991 (Giger et al. 1991; Smith et al. 1996), but available screening has not been embraced by the breed club. Based on DNA testing results, the frequency of carrier and affected dogs ranged between 8% and 16% and 3% and 5%, respectively. In 1998, a randomized survey of AKC championship English springer spaniels of both Field and Show (conformational) lines in the U.S. that had produced offspring was performed to screen for the mutant allele and, surprisingly, still found 3% carriers (Giger et al. 2000; Kimmel 2001).

In recent years there have been a number of studies on complex inherited disease prevalence in dogs with varied results. For instance, the prevalence of hip dysplasia may differ greatly depending on methods used to reach a diagnosis, and whether a registry requires or only encourages recording of every examined animal. However, these studies, together with data from open registries, show that such disorders can be common: hip dysplasia occurs in nearly 75% of boxers in OFA records and more than 25% of dogs in each of the 20 most commonly affected breeds. Using the PennHIP standard protocol for laxity testing and the

Figure 2. Phosphofructokinase-deficient English springer spaniel with severe hemolytic crisis resulting in icterus, anemia, and metabolic myopathy.

required registration of any dog tested for hip dysplasia, more meaningful results and trends over time can be obtained (Bliss et al 2002; Paster et al. 2005; see Chapter 21).

For several common breeds, more than 40 inherited diseases have been reported, although most defects are probably rare in any one breed (http://www.vet.cam.ac.uk/idid, http://www.angis.org.au/databases/BR IX/omia; Gough and Thomas 2004). For less common breeds, peer-reviewed disease reports are rare, but this does not necessarily mean that there are fewer diseases. The number of clinically relevant genetic defects (those that cause morbidity or mortality in a significant proportion of animals) may be around 6–8 in each breed, although accurate estimates, especially of complex disease traits, are difficult to develop. The overall genetic burden can also show up in studies of mortality and morbidity of different breeds. Breeds differ greatly in average life expectancy, with extremes ranging from below 7 years in several giant breeds (in the majority of studies) to above 12 years in a range of small to mid-sized dogs (Patronek et al. 1997; Michell 1999; Proschowsky et al. 2003). Although differences in morbidity and mortality cannot always be ascribed to particular genetic diseases, they are profoundly influenced by genetic factors. The influence of genetic factors on litter size is also profound, and may be affected by genetic diseases.

The examples above show that real shifts in the frequency of deleterious alleles can only be reached with the full cooperation of breeders. Systematic analyses of hip dysplasia and of ocular diseases in Finland confirm this (Leppanen and Saloniemi 1998, 1999). Nonetheless, such targeted reduction or removal of diseases has occurred many times, when breed clubs and owners have actively promoted removal of disease carriers (e.g., progressive retinal atrophy in U.K. setter breeds). That DNA-based testing makes disease elimination easier is clear for recessive monogenic disease, and is also borne out in many ongoing DNA testing programs that have received kennel club or breed society support in many parts of the world. To improve the health of canine breeds, many breed clubs or even, in some cases, national kennel clubs, refuse registration of dogs if they have not been screened and found to be free of those diseases for which DNA testing is available.

CHARACTERIZATION OF HEREDITARY DISORDERS

Hereditary defects can involve any gene and hence any organ. The clinical manifestations of hereditary diseases are therefore extremely variable, ranging from benign to debilitating and lethal, but some characteristic features may raise our suspicion of a genetic disorder. Genetic defects often cause clinical signs early in life and are usually chronic and progressive. Although the term congenital implies that the disease is present at birth, it does not necessarily mean it is inherited. Fetal losses and stillborns may be caused by genetic traits and are reflected by lack of pregnancy or small litter sizes. A number of studies have shown strong correlations between litter size and the dam's body weight (Robinson 1973; Willis 1989). The actual litter size for certain breeds (bulldog, Pomeranian, cavalier King Charles spaniel, Newfoundland) is smaller than predicated based on body weight, indicating a genetic basis for reduced litter size.

Most fading puppies and puppy losses are seen during the first week of life, soon after the maternal homeostatic system can no longer compensate for any serious endogenous defect. Little is known about the metabolic and molecular bases of the fading puppy and mortality complex, as these puppies are seldom shown to veterinarians, are rarely further examined, and are considered to be usual losses by breeders, yet various genetic defects are likely responsible for these deaths (Jezyk 1983; Giger and Jezyk 1992).

Congenital skeletal malformations were among the first diseases recognized as having a genetic basis as a result of their noticeable phenotypes such as proportionate dwarfs (German shepherd pituitary dwarfism) or disproportionate growth retardation (various chondrodysplasias) with leg deformities and gait abnormalities, but also varied facial and other dysmorphias. They include many orthopedic problems such as elbow dysplasia and hip dysplasia, and it is clear that joint laxity also contributes to these diseases (see Chapter 21). Many congenital malformations are also seen as a part of a multisystemic syndrome, as in lysosomal storage disorders (Fig. 3) (Haskins and Giger 1997).

Aside from skeletal deformities, there are numerous other congenital malformations involving internal organs, such as hereditary eye anomalies, congenital heart defects, brain developmental defects, juvenile renal dysplasia, hepatic shunts, and pancreatic atrophies. The prevalence of a number of congenital (although not exclusively hereditary) abnormalities has been estimated through case studies. For instance, the prevalence of 7 cardiac malformations per 1,000 cases was found in a comprehensive review of more than 35,000 case records from the Veterinary Hospital of the University of Pennsylvania over the period 1953–1965 (Patterson 1971). Anatomic subclassification of the various defects revealed strong breed associations, subsequently confirmed by other survey studies. For example, a study of subaortic stenosis (SAS) found that breeds at significantly increased relative risk included the Newfoundland (odds

Figure 3. German shepherd puppy with mucopolysaccharidosis type VII with severe multisystemic skeletal, eye, and other abnormalities.

ratio, 88.1, breed incidence of SAS of greater than 10%), rottweiler (odds ratio, 19.3), boxer (odds ratio, 8.6), and golden retriever (odds ratio, 5.5) (Kienle et al. 1994).

Many hereditary disorders may mimic acquired disorders and thus require specific tests to determine the genetic basis. Radiology and other imaging methods, such as fundic exam of the eye, ultrasound of the abdomen, echocardiography, brain magnetic resonance imaging, and computer tomography are helpful in revealing skeletal, eye, and cardiac, as well as many other internal organ, anomalies. Routine tests such as complete blood cell counts, chemistry screen, and urinalysis, as well as advanced clinical gastrointestinal and other function studies may suggest some specific hematological, metabolic, or specific internal organ disorders and can rule out many acquired disorders. However, pathological examination of tissues or a necropsy may only give the first clue as to presence of a genetic disorder.

Inborn errors typically produce a metabolic block in a biochemical pathway leading to product deficiency, substrate accumulation, and/or production of unusual substances via alternative pathways. These derangements in metabolites may be first discovered in urine and other fluid and tissue specimens and identify a failing biological system. Once the failing system has been identified, the defective step in a pathway and specific protein involved can often be determined (Giger and Jezyk 1992). These protein assays include the functional, such as classic enzyme activity tests, as well as quantitative assays by immunological studies. Because most enzymes are present in abundant amounts, no major functional abnormalities are observed unless the protein activity is severely reduced, usually to less than 20% of normal values for the recessive disorders. Hence the homozygously affected dogs have very low protein function, often ranging between 0 and 10%. Typically the low activity correlates with quantitative protein assays, but there are some severe deficiencies with normal amounts of cross-reacting material (CRM) positive but dysfunctional protein. Hereditary hemostatic disorders are a group of disorders in which all three types of assays (functional, quantitative for protein, and DNA-based mutational) are used. Many coagulopathies occur in dogs and are typically assessed by functional plasma clotting assays. von Willebrand disease is generally diagnosed by an ELISA test measuring the reduced plasma protein concentration of von Willebrand factor (Fig. 4) (Brooks 2000). In contrast, in coagulation factor VII deficiency in beagles, caused by single missense mutation of a glycine to a glutamine, a full-length protein is produced, but the clotting activity in deficient dogs is less than 4% of normal (Callan et al. 2005).

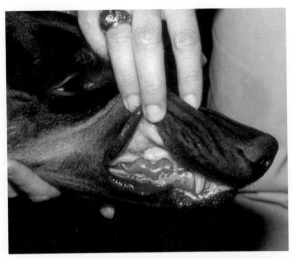

Figure 4. Doberman pinscher puppy with excessive gum bleeding while changing his deciduous teeth due to von Willebrand disease type I.

Once a specific protein deficiency is suspected, a search for the disease-causing mutation is warranted and likely to be successful. The cloning and sequencing of mutant and normal canine genes has relied on comparisons with characterized genes in other species. The recent availability of the complete canine genome sequence permits a more direct approach with sequencing of the candidate gene in an affected animal based on known canine sequence. A large variety of mutations have been identified, including missense, insertions, deletions, splicing defects, and inversions, just as in humans, but only rarely is the identical mutation seen in human patients with the same disease. These recent advances have led to the discovery of many disease-causing mutations and should allow the characterization of the molecular defect of most known single-gene defects within this decade where a candidate gene has been identified. The Section of Medical Genetics at the University of Pennsylvania has established a National Institutes of Health-supported national referral center for the identification of new hereditary diseases in dogs as models of human diseases (http://www.vet.upenn.edu/penngen). In addition, there are several disease-specific laboratories that provide assistance in the diagnosis of hereditary disorders, such as the Comparative Hemostasis Laboratory at Cornell University and the Comparative Muscle Laboratory at the Salk Institute. Although human laboratories have also been helpful in the further pursuit of novel disease-causing mutations, these laboratories may not accept any canine samples due to the stringent requirements for full accreditation of laboratories for humans.

The molecular defect of single-gene defects in dogs is currently known for nearly 50 hereditary diseases in dogs. They include an ever-expanding spectrum of inherited disorders and are listed in Table 1. Virtually all mutations recognized thus far are breed-specific. That is, one mutation causes specific disease in a certain breed or a restricted group of (usually related) breeds. For example, canine leukocyte adhesion deficiency, first identified in Irish setters, is also observed in the red and white setters. Furthermore, the same specific mutation may also cause disease in mixed breed dogs, particularly if there was a mother–son (or father–daughter) mating as described in an Irish setter mixed breed dog (Foureman et al. 2002). However, different mutations in the same gene have been found to cause the same protein deficiency in different breeds. For instance, erythrocytic pyruvate kinase deficiency is caused by different mutations in basenjis, West Highland white terriers, beagles, and several other breeds (Whitney et al. 1994; Skelly et al. 1999; Giger 2000). Thus, the various disease-causing mutations are breed-specific, and there is clearly heterogeneity of the molecular basis of diseases among breeds. The prcd gene defect seems to be an exception and may have been present at or close to the time of canine domestication (Chapter 16). It has no effect on fitness until after breeding age and is located at a flat spot for recombination. It will be interesting to see whether it is also linked to a gene selected for in domestication. In contrast, in "outbred" human populations different mutations inherited from both parents are typically responsible for the occurrence of recessive disorders.

GENETIC SCREENING FOR KNOWN MONOGENIC DISEASES

Screening for monogenic disorders involves the recognition of affected as well as asymptomatic carrier animals (Giger and Jezyk 1992). Diseased animals may be readily identified by breeders and companion animal owners, although the definitive diagnosis may only be reached by a veterinarian or a specific diagnostic test. Numerous diagnostic tests including imaging and laboratory methods have been developed or adopted for screening of dogs and made available to confirm specific monogenic disorders. Veterinary associations and also kennel clubs and breed societies have expended considerable efforts in instituting unified methods for the clinical assessment of a number of skeletal, ophthalmic, and other diseases at national and sometimes international levels.

Table 1. Hereditary disorders characterized at the molecular genetic level

Disease	Gene	Mutation	Breed	Reference
Hereditary blood disorders				
A. Hemostatic disorders				
Factor VII deficiency	*F7*	missense G443A, Gly140Glu	beagle	Callan et al. (2005)
Hemophilia A (Factor VIII deficiency)	*F8*	inversion in exon 22	mixed breed research colony (Guelph)	Hough et al. (2002)
		inversion and replacement of last 4 exons	mixed breed research colony (Chapel Hill)	Lozier et al. (2002)
Hemophilia B (Factor IX deficiency)	*F9*	point mutation	mixed breed research colony (Chapel Hill)	Evans et al. (1989) Deletion Lhasa apso Mause et al. (1996)
		deletion	Labrador retriever and mixed breed	Brooks et al. (1997)
		missense mutation	mixed breed research colony (Chapel Hill)	Chao and Walsh (1999)
		1.5-kb insertion in intron 5	German wirehaired pointer	Brooks et al. (2003); Gu et al. (1999)
von Willebrand disease (vWD) Type I	*VWF*	frameshift	Bernese mountain dog, Doberman pinscher, Kerry blue terrier, Manchester terrier, papillon, Pembroke Welsh corgi; poodles, West Highland white terrier	http://www.vetgen.com/vwdrpt.html
von Willebrand disease (vWD) Type II	*VWF*	missense, exon 28 Asn>Ser	German shorthaired and wirehaired pointer	Kramer et al. (2004)

Disease	Gene	Mutation	Breed	Reference
von Willebrand disease Type III	VWF	deletion	Scottish terrier	Venta et al. (2000)
		donor splice site defect intron 16 (TGgtaagt—>TGataagt)	Dutch Kooiker	Rieger et al. (1998); van Oost et al. (2004)
Glanzmann's type thrombasthenia	GpaIIb	insertion (14 bp), exon 13 and splice site defect (intron 13)	Great Pyrenees	Lipscomb et al. (2000)
	GpaIIb	missense D398H	otterhound	Boudreaux and Lipscomb (2001)
B. Primary immunodeficiencies				
Complement C3 deficiency	C3	1-bp deletion at position 2136	Brittany spaniel	Ameratunga et al. (1998)
Canine leukocyte adhesion deficiency (CLAD)	ITGB-2	missense Cys36Ser	Irish setter, Irish red and white setters	Kijas et al. (1999) Foureman et al. (2002)
Severe combined immunodeficiency (autosomal recessive)	PRKDC	nonsense	Jack Russell terriers	Ding et al. (2002)
Severe combined immunodeficiency (X-linked SCID)	IL2RG	4-bp deletion	basset hound	Henthorn et al. (1994)
Leishmania sensitivity	NRAMP1	1-bp insertion G-rich block in promoter	Pembroke Welsh corgi beagle	Somberg et al. (1994) Altet et al. (2002)
Cyclic hematopoesis (cyclic neutropenia)	AP3B1	1-bp insertion in A run and exon 11 splicing variant	collie	Benson et al. (2003)

(Continued)

Table 1. (*Continued*)

Disease	Gene	Mutation	Breed	Reference
C. Erythrocyte enzymopathies				
Pyruvate kinase (PK) deficiency	*PKLR*	1-bp deletion unpublished	basenji Eskimo toy, beagle, dachshund	Whitney et al. (1994) U. Giger (unpubl.)
		6-bp insertion causing splicing defect	West Highland white terrier	Skelly et al. (1999)
Phosphofructokinase (PFK) deficiency	*PFKM*	nonsense (stop) codon (G2228A)	English springer spaniel, American cocker spaniel, mixed breed	Giger et al. (1991); Smith et al. (1996)
Elliptocytosis	*EPB41*	63-bp del of spectrin-actin binding domain	mixed breed dog	Conboy et al. (1991)
Central nervous system disorders and lysosomal storage diseases				
GM1 gangliosidosis	*GLB1*	missense G200A, Arg60His	Portuguese water dog	Wang et al. (2000)
GM1 gangliosidosis	*GLB1*	deletion C1668	Shiba Inu	Yamato et al. (2002)
GM1 gangliosidosis	*GLB1 (GALB)*	19-bp duplication exon 15	Alaskan husky	Kreutzer et al. (2005)
Epilepsy (Lafora type)	*NHLRC1 (EPM2)*	tandem 12-bp repeat expansion	miniature wirehaired dachshund	Bradbury (2005); Lohi et al. (2005)
Shaking puppy (generalized tremor)	*PLP1*	missense His36Pro results in overexpression of immature (DM-20) isoform	English springer spaniel	Nadon et al. (1990); Tosic et al. (1997)
Narcolepsy	*HCRTR2*	SINE insertion deletion gives exon skipping with frameshift	Doberman pinscher Labrador retriever	Riehl et al. (1998)
		missense Glu54Lys	dachschund	Lin et al. (1999)

Disease	Gene	Mutation	Breed	Reference
Mucopolysaccharidosis I (Hurler-Scheie syndrome)	IDUA	splicing (G > A in intron 1 donor site)	Plott hound	Stoltzfus et al. (1992); Menon et al. (1992)
Mucopolysaccharidosis type IIIA	SGSH	3-bp deletion, Thr 246 lost	wirehaired dachshund	Aronovich et al. (2000)
	SGSH	insertion C at 708–709	New Zealand huntaway dog	Yogalingam et al. (2002)
Mucopolysaccharidosis type IIIB	NAGLU	poly AAA insertion	schipperke	Ellinwood et al. (2003)
Mucopolysaccharidosis type VI	ARSB	missense at the 3' end of the exon 5	miniature pinscher	Foureman et al. (2004)
		26 bp of the 5' UTR and the first 10 codons of the open reading frame	miniature schnauzer	U. Giger et al. (unpubl.)
Mucopolysaccharidosis (MPS) VII	GUSB	missense Arg166His	mixed breed, German shepherd	Ray et al. (1998); Silverstein Dombrowski et al. (2004)
Fucosidosis	FUCA	Del last 14 bp exon 1	English springer spaniel	Skelly et al. (1996); Occhiodoro and Anson (1996)
Ceroid lipofuscinosis	CNL8	missense T>C gives Leu164Pro	English setters	Katz et al. (2005)
Myotonia congenita	ClC-1	missense Thr268Met	miniature schnauzer	Rhodes et al. (1999)
Globoid cell leukodystrophy (Krabbe disease)	GALC	A>C transversion Tyr158Ser	West Highland white terriers; cairn terriers	Victoria et al. (1996)
Pharmacogenetic problems				
Drug sensitivity (Ivermectin)	MDR1 (ATPBCB1)	4-bp deletion	multiple collie-related breeds	Mealey et al. (2001); Roulet et al. (2003); Neff et al. (2004)
Malignant Hyperthermia[4]	RYR1	missense T1640C, Val 547Ala	greyhound	Roberts et al. (2001)

(Continued)

Table 1. (*Continued*)

Disease	Gene	Mutation	Breed	Reference
Gastrointestinal, liver, and endocrine disorders				
Imerslund-Grasbeck syndrome (cobalamin malabsorption)	AMN	deletion 33-bp exon 10	giant schnauzer	Fyfe et al. (2004)
		missense G3A abolishes initiation codon	Australian shepherd	He et al. (2005)
Congenital hypothyroidism with goiter	TPO	nonsense C331T stop codon	toy fox terrier	Fyfe et al. (2003)
Copper toxicosis	MURR1	large deletion exon 2	Bedlington terrier	van de Sluis et al. (2002)
Catalase deficiency	CAT	missense G>A gives Ala327Thr	beagle	Nakamura et al. (2000)
Glycogen storage disease 1a (von Gierke disease)	G6PC	missense G450C gives Met121Ile	maltese terrier	Kishnani et al. (1997)
Muscular and skeletal disorders, including cardiomyopathy				
Centronuclear myopathy	PTPLA	SINE element insertion	Labrador retriever	Pele et al. (2005)
X-linked dystrophin muscular dystrophy	DMD	single-base mutation, splice site mutation intron 6 acceptor leads to exon 7 skipping	golden retriever	Sharp et al. (1992)
		Complete gene deletion	German shorthaired pointer	Schatzberg et al. (1999)
Osteogenesis imperfecta (dominant)	COL1A2	likely de novo 5-bp insertion	beagle	Campbell et al. (2001)
		likely de novo missense G1276C Gly208Ala	golden retriever	Campbell et al. (2000)

Disease	Gene	Mutation	Breed	Reference
Tail short (brachyury)	*T* (unnamed T-box gene)	heterozygous missense Ile63Met (homozygous embryonic lethal)	Pembroke Welsh corgi	Haworth et al. (2001)
Dilated cardiomyopathy	*DES*	linkage	Doberman pinscher	Stabej et al. (2004)
Genodermatoses				
X-linked ectodermal dysplasia	*EDA*	splice acceptor defect	German shepherd dog	Casal et al. (2005)
Epidermolysis bullosa (dystrophic form)	*COL7A1*	missense G1906S	golden retriever	Baldeschi et al. (2003)
Epidermolysis bullosa (junctional form)	*LAMA3*	Insertion (4818 + 207ins 6.5 kb) of repetitive satellite DNA within intron 35	German shorthaired pointer	Capt et al. (2005)
Renal Disorders				
Cystinuria	*SLC3A1*	nonsense C663T,Arg>* unpublished	Newfoundland	Henthorn et al. (2000) P.S. Henthorn (unpubl.)
Alport syndrome (hereditary nephritis)	*COL4a5*	deletion 10 bp in exon 9	Labrador retriever mixed breed dog	Cox et al. (2003)
		nonsense G>T gives Gly1027*	Samoyed	Zheng et al. (1994)
Renal cystadenocarcinoma and nodular dermatofibrosis (RCND)	*FLCN*	Missense, dominant His255Arg; probably recessive lethal	German shepherd dog	Lingaas et al. (2003)
Renal dysplasia		linkage	Lhasa apso; soft-coated wheaten terrier; shih tzu	G. Brewer (unpubl.)

For hereditary retinal eye diseases, see Chapter 16.

Many, but not all, laboratory tests are suitable to identify affecteds even before clinical signs occur, but they may not be able to identify carriers. However, in order to assure the health of future generations, it is important not only to identify affecteds, but also to screen for carriers. Several protein assays are capable to detect carriers, as heterozygotes typically have intermediate activity or quantity (30–75%) at the protein level. Unfortunately, protein assays are often labor-intense and require submission of appropriate fresh tissue or blood samples under special conditions to a specialized laboratory. Furthermore, the lability of proteins makes differentiation between the various genotypes sometimes difficult, even when a control sample collected under the same condition is shipped along for direct comparison. Any damage to a sample may reduce the protein activity/level and hence misclassify the dog as carrier or affected when it would be normal. Hence the need for sturdy and simple standardized test methods is clearly recognized.

Although only a small proportion of inherited disorders are, as yet, understood at the molecular level, these disorders offer the best opportunities to control their frequency in a breed (Table 1; updated http://www.vet.upenn.edu/penngen). Once the disease-causing mutation is identified, specific screening tests for the presence of the mutant allele can be designed. Any mutation can be identified by a single routine method (sequencing through the mutation), although other methods based on the type of mutation have also been applied to screen dogs by several laboratories. Currently there are some research laboratories as well as for-profit commercial and nonprofit laboratories that offer genetic testing for the more common disorders with known mutations. These tests are most accurate in the diagnosis of all three genotypes for a particular mutation. However, misidentification of dogs and samples, mispaternity, and human errors in the laboratory may rarely be responsible for inaccuracies.

A number of genetic tests, in particular those for copper toxicosis in the Bedlington terrier (Chapter 17) and for progressive retinal rod-cone degeneration in many breeds (Chapter 16), have also been based on the tight linkage of one or more polymorphic markers (microsatellites or SNP) to the disease gene. Linkage-based tests in general have the disadvantage that mutations inevitably occur on a preexisting genetic background, and descendants of the pre-mutation chromosome may still exist in the population. Hence, the test will only tell that an individual is of a haplotype that includes affected individuals, but not with certainty that they themselves are affected. In addition, recombinations will occur at some finite rate between any marker and a proximal mutation unless an allele of the marker itself is in fact causative of the disease. Thus, there is always a possibility of misassignment of disease status. Linkage-based tests are certainly useful, particularly when there are affected animals in a family, but they must be used with great care and only with the most tightly linked markers.

COMPLEX GENETIC DISEASES OR POLYGENIC TRAITS

Many of the inherited disorders with highest incidences in the canine population do not segregate in a manner consistent with a single-gene pattern. Although in some cases the environment may be a modifier causing segregation distortion, the majority involve additional or modifier genes. Various abnormalities of the circulatory, musculoskeletal, nervous, immunoregulatory, urogenital, and reproductive systems may fall in this category. Catalogs of these disorders may be found at the IDID and at On Line Mendelian Inheritance in Animals (OMIA, http://www.angis.org.au/Databases/BIRX/omia/) Web sites, whereas the Canine Inherited Disorders Database (http://www.upei.ca/~cidd/intro.htm) offers fuller clinical descriptions of many of these diseases, but less genetic information. In studying these complex disease traits, several principles can be considered.

Documenting the Disease Problem

A difficulty in researching any complex genetic disease is to ensure that one is dealing with one discrete disease entity. This applies particularly in polygenic disease, where disease often results from being at the extreme of a range for a given QTL trait. Methods for consideration and study of polygenic disease have traditionally used the concept of a disease threshold within a given variable range. This method works well for many quantitative disease variables (e.g., skeletal measurements) and is ideal for partitioning of variances, and data examination by principal component analysis. It is sometimes less useful when one considers traits that are qualitative, such as differentiation of tumor type in analysis of a breed predisposition to tumors. In these cases, it is important that any development of a quantitative scale is appropriate. It is essential that all clinical assessments in any study (whether it is of a quantitative or qualitative trait) are conducted in a similar manner.

"Simulation of Mendelism"

The categorization of an inherited disorder as simplex or polygenic often depends on segregation analysis. Segregation of the disease in litters where full information is available may well fit the simple model that two alleles of a single gene are present, one of which is associated with disease. It should be noted, however, that data which are compatible with a single-gene model do not prove such a model. Even in large data sets, it is always possible to find more complex models that also fit the data. Occam's razor makes clear that the simpler model should be pursued until it is refuted, but in mapping studies it is worth continuing to accumulate material to allow work on more complex models.

An example where mapping has suggested a genetic structure under-lying disease that is more complicated than segregation data suggest is the conotruncal defect (CTD) of keeshonds. In an inbred keeshond line affected with CTD crossed onto beagles (and in F_2 and backcrosses), this defect appears to segregate as a single autosomal recessive gene with modifiers affecting only the severity, but not the presence, of disease (Patterson et al. 1993). This analysis involved more than 300 offspring. Yet in mapping this defect by parametric linkage analysis in 101 back-cross offspring from the same group of dogs, three different loci have been implicated on CFA9, 2, and 15. In this family, genetic heterogeneity in the origin of the defect can be ruled out. LOD scores at the three loci range from 2.71 to 3.7, with two loci just under the significance level of 3 and one slightly above, but the authors show that a model in which disease alleles are needed at any two of these loci to cause CTD fits best with their data (Werner et al. 2005).

Mapping Considerations

As described elsewhere in this volume, the breeding structures of purebred breeds lead to relative homogeneity of genetic background, and also to the segregation of chromosomes that show long regions of linkage disequilibrium across groups of neighboring polymorphic markers. This population structure has two advantages for mapping studies to try to define the genetic basis of polygenic disease: Within a single breed or line, the lack of background genetic variability favors the identification of single genes with major effect on phenotype, where a more diverse genetic background would mask their effects. Actual mapping of QTLs in the dog is just beginning, but the promise of mapping within breeds is already apparent. Examples in this book include the studies of skeletal size and of hip dysplasia described in Chapters 4, 5, and 21. The marker density used can be reduced without much information loss because of the length of regions in linkage disequilibrium. In the dog, such regions are very much longer than in humans: See the chapters on hip dysplasia

(Chapter 21), on copper toxicosis (Chapter 17), and on linkage disequilibrium within breeds (Chapter 9).

Set against these advantages, there is a reduction in marker polymorphism within canine breeds when compared with levels seen between breeds, reducing the power of mapping. The level of this reduction depends on the effective population size, but even when there are few founder chromosomes this reduction is unlikely to be sufficient to prevent mapping. A study by Brouillette and Venta (2002), designed to measure within-breed reductions in SNP heterozygosity, found only those reductions that are expected in reduced sample sizes and did not detect inbreeding. A number of other studies have shown moderate reduction in the number of microsatellite alleles or in haplotype diversity within breeds, and also the presence of "private" SNP loci variable only in single breeds or groups of breeds (Irion et al. 2003; Koskinen 2003; Parker et al. 2004; Sutter et al. 2004). Typically, the F statistic measure of loss of heterozygosity is about 0.15–0.25 in these studies. Compensating increases in numbers of individuals and of markers may well be feasible, but where all affected animals are drawn from a set of closely related lines within a breed, even greater reductions in marker polymorphism may present substantial losses of analytical power.

Clearly, long regions of LD mean that fine mapping of the region containing a disease locus is not possible, where recombination events are not available. However, it has often proved possible to use admixtures obtained by crossing breeds, or out-group comparison techniques, both to increase levels of marker polymorphism and to minimize the mapping interval for a given mutation. A now-classic example employing both of these strategies came in mapping narcolepsy in dogs. Narcolepsy phenotypes in both Doberman pinscher and Labrador retriever are due to noncomplementing mutations at a locus designated *canarc-1* (Foutz et al. 1979). Initial mapping concentrated on an admixture route to increase genetic variation, using backcrosses from a colony of narcoleptic Doberman pinscher × Labrador retriever dogs. The location was further refined by comparison with narcoleptic dachshunds from the pet population, as well as additional pet Labradors, leading to the identification of the *canarc-1* locus as encoding the hypocretin (orexin) receptor 2 gene (Lin et al. 1999).

Simplifying the Disease Model

Where it is thought that many factors contribute to the disorder seen (in additive models with many contributors of small effect), mapping of all contributors at once may be difficult. It is sometimes possible to find phenotypic traits that can be individually mapped and that each contribute to (or correlate with) the overall disorder. An example of quantitative developmental phenotypes that have strong correlation with the later disease onset and partial penetrance is hip dysplasia (Chapter 21).

Although these polygenic and QTL disease traits are likely only the beginning of a wave of gene discoveries associated with complex diseases, other canine studies have been less successful, such as investigations on dilated cardiomyopathies (Dukes-McEwan and Jackson 2002) (Chapter 19); the limited availability of samples from large enough families, and/or the difficulty in correctly assigning the disease genotype, are likely reasons. In particular where pedigree-based linkage analyses are used, the availability of complete sib groups and of both parents and grandparents may be difficult to arrange, especially in adult-onset diseases and when the analyzed samples are collected from the pet population. The pet population is also a source of particular problems in stratification of assignment data unless systematic veterinary surveillance is available. Complicating the issue is the potential inaccuracy of a clinicopathological diagnosis reached by clinicians. Chapter 19 gives an up-to-date review of the genetics of cardiomyopathy.

GENETIC PREDISPOSITION TO DISEASE

As in humans, the genetic basis of susceptibility to diseases is also being increasingly recognized in dogs. Although dogs affected with a typical inborn error of metabolism will inevitably develop the disease within a narrow, predetermined age range, genetic predisposition has more variable expressivity and is greatly influenced by environmental factors. Genetic predispositions can be involved in a variety of diseases, including the development of infections, inflammatory/allergic conditions, cancer, bleeding tendencies, behavior problems, and adverse drug reactions. For some disease predispositions, simple mutations have been identified, whereas for others, polygenic traits form the basis. Various coagulation deficiencies have been identified in different breeds which lead to remarkably different bleeding tendencies: Hemophiliac males from many breeds commonly experience severe hemorrhagic episodes associated with deficiencies in factors VIII or IX, whereas deficiencies in factors I, VII, and XI result in a more mild to moderate bleeding tendency. Various primary immunodeficiencies can predispose to infectious diseases: Puppies with severe X-linked combined immunodeficiency succumb within months to any form of viral, bacterial, or parasitic diseases; some bassets and miniature schnauzers have a specific predisposition to develop avian tuberculosis later in life (mode of inheritance unknown); foxhounds and beagles seem to be predisposed to leishmaniasis (Giger and Greene 2000). Pharmacogenetics, the genetic variation in response to treatment, has only recently been recognized as relevant to dogs. The best example is the mutation in the MDR gene in a large number of Border collies, Australian cattle dogs, and other related breeds of dogs (Fig. 5) (Mealey et al. 2001; Neff et al. 2004). These dogs have a defect in the blood-brain barrier and may experience life-threatening reactions to certain drugs such as ivermectin, vincristine, and loperamide. Similarly, Doberman pinschers and other red and tan dogs seem to be predisposed to develop a polyarthropathy and other immunological reactions when exposed to sulfonamides (Giger et al. 1985). Dogs may also exhibit predisposition to the inflammatory and immune-mediated disorders affecting the skin, joints, intestine, kidney, and blood. The genetic predisposition to develop hip dysplasia is covered in Chapter 21. Finally, dogs frequently develop cancer, and for many of them increased risk to develop certain forms of cancer has been documented. This genetic predisposition to cancer is discussed subsequently and in Chapters 22 and 23.

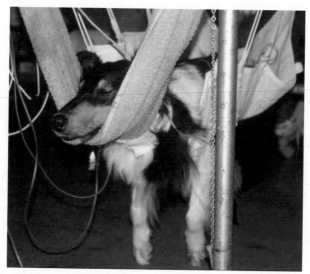

Figure 5. Ivermectin-induced toxicity in a collie due to a glycoprotein P deficiency causing life-threatening neurologic and cardiopulmonary compromise (courtesy of Kate Hopper, BVSc).

GENETIC PREDISPOSITION TO CANCER

Cancers result from germ-line and somatic gene alterations acting in concert with microenvironmental factors to promote loss of differentiation and unregulated cell growth. Thus, the recent advancements in characterizing the canine genome provide the opportunity to increase our knowledge of the complex basis of cancer occurring in the dog. Cancer is not only a major health problem in dogs, but their spontaneously occurring cancers are also emerging as an important model system for cancer in humans. Malignancies are clinically well-characterized in dogs and are not only disease models, but true homologs of cancers occurring in humans in terms of morphology, behavior, and molecular characteristics. In particular, dogs share large body size, longevity, a heterogenic genetic and immunological background, and aspects of cell growth such as expression of telomerase activity with humans. Therefore, the dog is an excellent model in which to study tumor etiology and development and for the assessment of the safety and efficacy of novel therapies for humans that also benefit dogs.

In contrast to the sophisticated surveillance mechanisms established to track cancer incidence and mortality in humans, epidemiological information on canine cancer is primarily descriptive and is often dated, and canine cancer databases are poorly developed. Despite limitations in the data, it is apparent that cancer is very common in dogs and a frequent cause of death or reason for euthanasia in pets. Based on study of necropsy submissions, canine mortality due to cancer is 23% overall and 45% in dogs older than 10 years (Dorn 1976). Although necropsies are only performed in a small proportion of dogs, several owner surveys, as well as insurance records, concur that cancer is the single most important disease cause of deaths (Eichelberg and Seine 1996; Michell 1999; Dobson et al. 2002; Proschowsky et al. 2003), and hence cancer in dogs is at least as common as in humans. One aspect of cancer in dogs that has been apparent for many years is the strong association between breed and the occurrence of cancer (Priester and Mantel 1971; Dorn and Schneider 1976). Breed-associated cancers have been observed in many generations of dogs and exhibit widespread geographic distribution. Certain breeds of dogs, such as boxers, appear to be particularly susceptible to many types of cancers (Priester 1967; Peters 1969; Priester et al. 1977). In addition, specific tumor types occur more frequently in certain breeds, including histiocytic sarcoma in Bernese mountain dogs (Moore 1984; Padgett et al. 1995) and flat-coated retrievers (Morris et al. 2000, 2002; King 2003), gastric carcinoma in chow chows (King 2003), and renal cancer in German shepherds

(Suter et al. 1983; Lium and Moe 1985; Moe and Lium 1997; Jonasdottir et al. 2000; Lingaas et al. 2003). Table 2 gives a more complete list of breed-associated malignancies. For some of these cancers, breed-associated relative risks are very high, giving the potential opportunity to map predisposing genes. For example, about 50% of all malignant tumors in flat-coated retrievers and more than 25% of *all* tumors in Bernese mountain dogs are histiocytic sarcomas (Padgett et al. 1995; Morris et al. 2002). This tumor is familial in both breeds and closely resembles the most common soft-tissue sarcoma of adulthood in humans, malignant fibrous histiocytoma. Aggressive gastric carcinoma in chow chows and closely related breeds shows an odds ratio of risk ranging from 13 to 23 compared to dogs of other breeds (Fig. 6). This cancer occurs within certain families and appears to be inherited by an autosomal dominant trait. Morphologically, gastric cancer in chow chows is indistinguishable from the diffuse form of gastric carcinoma in humans, which also exhibits a familial occurrence (McNiel et al. 2004a), and may thus provide insight into the carcinogenesis and tissue specificity of tumor growth (McNiel et al. 2004a and unpubl.). Transitional cell carcinoma (TCC) of the urinary bladder and urethra is the most common neoplasm of the canine urinary tract and accounts for 1–2% of all cancers in dogs (Mutsaers et al. 2003). Canine TCC shares histopathological characteristics, biological behavior, molecular features, and response to medical therapy with invasive human bladder cancer (Knapp et al. 2000). TCCs most commonly occur in older dogs (median age of 11 years) and are typically located in the trigone region of the bladder, but also often involve the urethra and/or prostate and have a relatively high rate of metastasis (Knapp et al. 2000; Mutsaers et al. 2003). Certain breeds are clearly predisposed to TCC, including Scottish terriers, West Highland white terriers, and Shetland sheepdogs. The odds risk ratio for Scottish terriers is 18, suggesting that genetic determinants may be particularly important in causing this cancer (Knapp et al. 2000). Despite the potential for a strong genetic predisposition, epidemiologic studies have also demonstrated that environmental influences such as obesity, insecticides, and herbicides may contribute to the increased risk of TCC in dogs (Glickman et al. 1989, 2004; Knapp et al. 2000; Raghavan et al. 2004). Because it is exceedingly difficult to elucidate gene–environment interactions in humans, this model may be particularly useful and may possibly define preventive strategies.

Although there are a great number of breed-associated cancers in the dog, only hereditary renal carcinoma in German shepherd dogs has been well characterized genetically (Jonasdottir et al. 2000; Lingaas et al. 2003).

Table 2. Examples of breed-associated or familial tumors in dogs

Tumor type	Breed(s)
Epithelial	
gastric	chow chow (McNiel et al. 2004a)
mammary	beagle (Schafer et al. 1998), dachshund, spaniel (springer, cocker), German shepherd
nasal	Airedale, basset hound, collie, Scottish terrier, Shetland sheepdog, German shorthaired pointer (Wilson and Dungworth 2002)
renal	German shepherd (Lium and Moe 1985, Jonasdottir et al. 2000; Lingaas et al. 2003)
urinary bladder	beagle, Scottish terrier, Shetland sheepdog, West Highland white terrier, wirehaired fox terrier (Knapp et al. 2000)
skin	
squamous cell carcinoma	basset hound, standard poodle (Goldschmidt and Shofer 1992)
squamous cell carcinoma (subungual)	giant schnauzer, Gordon setter, Kerry blue terrier, standard poodle, standard schnauzer, dachshund, Scottish terrier, rottweiler, Labrador retriever (Goldschmidt and Shofer 1992)
basal cell tumors	Kerry blue terrier, bichon frise, cockapoo, Shetland sheepdog, Siberian husky, cocker spaniel, miniature poodle, West Highland white terrier, English springer spaniel (Goldschmidt and Shofer 1992)
perianal gland tumors (hepatoid tumors)	Siberian husky, Samoyed, Pekinese, cocker spaniel, vizsla, Brittany spaniel, Lhasa apso, cockapoo, shih tzu, beagle (Goldschmidt and Shofer 1992)
ceruminous gland carcinoma	cocker spaniel, German shepherd (Goldschmidt and Shofer 1992)
anal sac adenocarcinoma	English cocker spaniel, dachshund, Alaskan malamute, English springer spaniel, German shepherd (Goldschmidt and Shofer 1992)
intracutaneous cornifying epithelioma	Norwegian elkhound, Lhasa apso, Yorkshire terrier, German shepherd, standard poodle (Goldschmidt and Shofer 1992)
trichoepithelioma	basset hounds, bullmastiff, Irish setter, standard poodle, English springer spaniel, golden retriever, English setter, Airedale terrier, miniature schnauzer (Goldschmidt and Shofer 1992)

(*Continued on facing page.*)

Table 2. (*Continued*)

Tumor type	Breed(s)
Endocrine	
thyroid	beagle, boxer, golden retriever (Hayes and Fraumeni 1975)
pheochromocytoma	wirehair fox terrier; Airedale (McNiel and Husbands 2005)
pancreatic islet cell	standard poodle (Priester 1974)
Mesenchymal	
brain tumors	boxer, Boston terrier (Glioblastoma multiforme)
histiocytic sarcoma	Bernese mountain dog (Moore 1984; Moore and Rosin 1986; Padgett et al. 1995; Paterson et al. 1995), Flat-coat retriever (Morris et al. 2000, 2002; King 2003), golden retriever (Affolter and Moore 2000, 2002)
hemangiosarcoma	German shepherd (Prymak et al. 1988), golden retriever (Goldschmidt and Hendrick 2002)
melanoma (oral)	chow chow, cocker spaniel, golden retriever, miniature poodle, Pekinese/poodle cross (Ramos-Vara et al. 2000)
osteosarcoma	large breed dogs: rottweiler, Great Dane, Irish setter, Irish wolfhound, greyhound, Great Pyrenees, Saint Bernard, bullmastiff, mastiff, golden retriever, borzoi, Newfoundland, Laborador retriever (Misdorp and Hart 1979; Ru et al. 1998)
histiocytoma (cutaneous, benign)	Scottish terrier, boxer, Boston terrier, English cocker spaniel, Jack Russell terrier, American pit bull terrier, bull terrier, rottwieler, pug, shar-pei, West Highland white terrier, bulldog, Doberman pinscher, dalmatian, Labrador retriever, miniature schnauzer, cocker spaniel, English springer spaniel (Goldschmidt and Shofer 1992)
Hematopoietic	
mast cell tumors	breeds with bulldog ancestry (Boston terrier, boxer, bull terrier) (Peters 1969), pug (McNiel et al. 2004b), Labrador and golden retriever, shar-pei (Miller 1995)
lymphoma/leukemias	many breeds; see Chapter 17
plasmacytoma (cutaneous)	cocker spaniel, Airedale terrier, Scottish terrier, standard poodle (Goldschmidt and Shofer 1992)

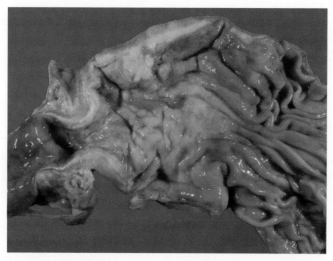

Figure 6. Stomach removed from a chow chow affected by gastric carcinoma at necropsy. The wall of the pyloric antrum demonstrates marked thickening and the mucosal folds are flattened, characteristic of diffuse tumor infiltration.

This syndrome was first described in Switzerland with subsequent reports from other countries (Suter et al. 1983). Affected dogs initially develop nodular dermatofibrosis and then bilateral renal cystadenocarcinoma. Examination of pedigrees from Norway identified a male founder (Lium and Moe 1985; Moe and Lium 1997), and in a breeding colony an autosomal dominant mode of inheritance with complete penetrance was confirmed. The syndrome was mapped to CFA5 (Jonasdottir et al. 2000) and subsequently tightly linked (no recombinants seen) to a change in an evolutionarily conserved amino acid in folliculin (Lingaas et al. 2003). Mutations in this gene have been associated with Birt Hogg Dube syndrome, which is a rather similar disease syndrome in humans.

In conclusion, the availability of genetic tools should permit progress in the genetic study of canine cancers and other predispositions. Technological molecular advances alone will not suffice, however. Clinical and epidemiological tools with sample banks and enhanced systems to track disease occurrence are essential. In addition, phenotyping of diseases must be improved and standardized. Many companion animals do not undergo adequate diagnostic testing and confirmed phenotyping. The reasons for this include expense to pet owners and limited availability of noninvasive diagnostic techniques. Thus, progress in the study of hereditary traits in dogs will ultimately require a coordinated effort between classic and molecular geneticists, clinicians, pathologists, and epidemiologists.

SUMMARY

This chapter addresses the common occurrence of hereditary disorders and genetic disease predispositions in dogs. Many of these diseases are breed-specific, and many diseases that show a high incidence in particular breed do so because of inbreeding practices. These include congenital malformations and dysplasias, inborn errors of metabolism, as well as increased susceptibility to, for instance, infection and cancer. The recent advances in clinical veterinary medicine and molecular genetics have allowed the characterization of the metabolic and molecular bases of many monogenic disorders and the development of DNA-based screening tests to control the disease spread in a breed population. Diseases for which such tests are currently available are monogenic recessive traits, whereas many other traits have a polygenic/complex mode of inheritance. With the completion of the canine genome sequence and the availability of appropriate markers, these latter genetic traits can now also be elucidated. Many of these diseases serve as models for human disorders and are particularly helpful in the development and the assessment of efficacy and safety of novel therapeutic strategies.

ACKNOWLEDGMENTS

The scientific contributions by our colleagues in our laboratories are greatly appreciated. The author's studies performed at the university were supported in part by the National Institutes of Health (02512), the Canine Health Foundation, the Muscular Dystrophy Association, and the National MPS Society.

REFERENCES

Affolter V.K. and Moore P.F. 2000. Canine cutaneous and systemic histiocytosis: Reactive histiocytosis of dermal dendritic cells. *Am. J. Dermatopathol.* **22:** 40–48.

———. 2002. Localized and disseminated histiocytic sarcoma of dendritic cell origin in dogs. *Vet. Pathol.* **39:** 74–83.

Altet L., Francino O., Solano-Gallego L., Renier C., and Sanchez A. 2002. Mapping and sequencing of the canine NRAMP1 gene and identification of mutations in leishmaniasis-susceptible dogs. *Infect. Immun.* **70:** 2763–2771.

Ameratunga R., Winkelstein J.A, Brody L., Binns M., Cork L.C., Colombani P., and Valle D. 1998. Molecular analysis of the third component of canine complement (C3) and identification of the mutation responsible for hereditary canine C3 deficiency. *J. Immunol.* **160:** 2824–2830.

Aronovich E.L., Carmiclael K.P., Morizona H., Koutlas I.G., Deanching M., Hganson G., Fisher A., and Whitley C.B. 2000. Canine heparan sulfate sulfamidase and the molecular pathology underlying Sanfilippo syndrome type A in Dachshunds. *Genomics* **68:** 80–84.

Baldeschi C., Gache Y., Rattenholl A., Bouille P., Danos O., Ortonne J.P., Bruckner-Tuderman L., and Meneguzzi G. 2003. Genetic correction of canine dystrophic epidermolysis bullosa mediated by retroviral vectors. *Hum. Mol. Genet.* **12:** 1897–1905.

Benson K.F., Li F.Q., Person R.E., Albani D., Duan Z., Wechsler J., Meade-White K., Williams K., Acland G.M., Niemeyr G., Lothrop C.D., and Horwitz M. 2003. Mutations associated with neutropenia in dogs and humans disrupt intracellular transport of neutrophil elastase. *Nat. Genet.* **35:** 90–96.

Bhalerao D.P., Rajpurohit Y., Vite C., and Giger U. 2002. Mutation screening for myotonia congenita among 372 Miniature Schnauzers and identification of a common carrier ancestor. *Am. J. Vet. Res.* **63:** 1443–1447.

Bliss S., Todhunter R.J., Quaas R., Casella G., Wu R., Lust G., Williams A.J., Hamilton S., Dykes N.L., Yeager A., Gilbert R.O., Burton-Wurster N.I., and Acland G.M. 2002. Quantitative genetics of traits associated with hip dysplasia in a canine pedigree constructed by mating dysplastic Labrador Retrievers with unaffected Greyhounds. *Am. J. Vet. Res.* **63:** 1029–1035.

Boudreaux M.K. and Lipscomb D.L. 2001. Clinical, biochemical, and molecular aspects of Glanzmann's thrombasthenia in humans and dogs. *Vet. Pathol.* **38:** 249–260.

Bradbury J. 2005. Canine epilepsy gene mutation identified. *Lancet Neurol.* **4:** 143.

Brooks M. 2000. von Willebrand Disease. In *Schalm's veterinary hematology*, 5th edition (ed. B.F. Feldman et al.), pp. 509–515. Lippincott Williams & Wilkins Philadelphia, Pennsylvania.

Brooks M. and Sargan D.R. 2001 Genetic aspects of disease in dogs. In *The genetics of the dog* (ed. A. Ruvinsky and J. Sampson), pp. 191–266. CABI Publishing, New York.

Brooks M.B., Gu W., and Ray K. 1997. Complete deletion of factor IX gene and inhibition of factor IX activity in a labrador retriever with hemophilia B. *J. Am. Vet. Med. Assoc.* **211:** 1418–1421.

Brooks M.B., Gu W., Barbas J.L., Ray J., and Ray K. 2003. A Line 1 insertion in the Factor IX gene segregates with mild hemophilia B in dogs. *Mamm. Genome* **14:** 788–795.

Brouillette J.A. and Venta P.J. 2002. Within-breed heterozygosity of canine single nucleotide polymorphisms identified by across-breed comparison. *Anim. Genet.* **33:** 464–467.

Callan M.B., Aljamali M.N., Griot-Wenk M.E., Pollak E.S., Werner, P., Giger U., and High K.A. 2005. Molecular characterization of hereditary factor VII deficiency in the Beagle *J. Vet. Intern. Med.* (in press).

Campbell B.G., Wootton J.A., MacLeod J.N., and Minor R.R. 2000. Sequence of normal canine COL1A1 cDNA and identification of a heterozygous α1(I) collagen Gly208Ala mutation in a severe case of canine osteogenesis imperfecta. *Arch. Biochem. Biophys.* **384:** 37–46.

———. 2001. Canine COL1A2 mutation resulting in C-terminal truncation of pro-α2(I) and severe osteogenesis imperfecta. *J. Bone Miner. Res.* **16:** 1147–1153.

Capt A., Spirito F., Guaguere E., Spadafora A., Ortonne J.P., and Meneguzzi G. 2005. Inherited junctional epidermolysis bullosa in the German pointer: Establishment of a large animal model. *J. Invest. Dermatol.* **124:** 530–535.

Casal M.L., Scheidt J.L., Rhodes J.L., Henthorn P.S., and Werner P. 2005. Mutation identification in a canine model of X-linked ectodermal dysplasia. *Mamm. Genome* (in press).

Chao H. and Walsh C.E. 1999. Endogenous canine FIX antigen in Chapel Hill strain hemophilia B canine. *Thromb Haemostasis.* **82:** 1378.

Conboy J.G., Shitamoto R., Parra M., Winardi R., Kabra A., Smith J., and Mohandas N. 1991. Hereditary elliptocytosis due to both qualitative and quantitative defects in membrane skeletal protein 4.1. *Blood* **78:** 2438–2443.

Cox M.L., Lees G.E., Kashtan C.E., and Murphy K.E. 2003. Genetic cause of X-linked Alport syndrome in a family of domestic dogs. *Mamm. Genome* **14:** 396–403.

Ding Q., Bramble L., Yuzbasiyan-Gurkan V., Bell T., and Meek K. 2002. DNA-PKcs mutations in dogs and horses: Allele frequency and association with neoplasia. *Gene* **283:** 263–269.

Dobson J.M., Samuel S., Milstein H., Rogers K., and Wood J.L. 2002. Canine neoplasia in the UK: Estimates of incidence rates from a population of insured dogs. *J. Small Anim. Pract.* **43:** 240–246.

Dorn C.R. 1976. Epidemiology of canine and feline tumors. *Compend. Contin. Educ. Pract. Vet.* **12:** 307–312.

Dorn C.R. and Schneider R. 1976. Inbreeding and canine mammary cancer: A retrospective study. *J. Natl. Cancer Inst.* **57:** 545–548.

Dukes-McEwan J. and Jackson I.J. 2002. The promises and problems of linkage analysis by using the current canine genome map. *Mamm. Genome* **13:** 667–672.

Eichelberg H. and Seine R. 1996. Life expectancy and cause of death in dogs. I. The situation in mixed breeds and various dog breeds. *Berl. Muench. Tieraerztl. Wochenschr.* **109:** 292–303.

Ellinwood N.M., Henthorn P.S., Giger U., and Haskins M.E. 2003. Mucopolysaccharidosis type IIIB: Identification of the causative mutation in the canine model. *Am. J. Hum. Genet.* **73:** 449.

Evans J.P., Brinkhous K.M., Brayer G.D., Reisenr H.M., and High K.A. 1989. Canine hemophilia B resulting from a point mutation with unusual consequences. *Proc. Natl. Acad. Sci.* **86:** 10095–10099.

Fogh J.M. 1988. A study hemophilia A in German Shepherd dogs in Denmark. *Vet. Clinics N. Am. Sm. Anim. Pract.* **18:** 245–254.

Foureman P., Whiteley M., and Giger U. 2002. Canine leukocyte adhesion deficiency: Presence of the Cys36Ser beta-2 integrin mutation in an affected Irish Setter crossbreed dog and in US Irish Red and White Setters. *J. Vet. Intern. Med.* **16:** 518–523.

Foureman P., Berman L., Stieger K., Van Hoeven M., Ellinwood N.M., Haskins M.E., Kirkness E., and Giger U. 2004. Mucopolysaccharidosis Type VI in Miniature Pinschers: Screening for the mutation. *J. Vet. Intern. Med.* **18:** 408–409.

Foutz A.S., Mitler M.M., Cavalli-Sforza L.L., and Dement W.C. 1979. Genetic factors in canine narcolepsy. *Sleep.* **1:** 413–421.

Fyfe J.C., Kampschmidt K., Dang V., Poteet B.A., He Q., Lowrie C., Graham P.A., and Fetro V.M. 2003. Congenital hypothyroidism with goiter in toy fox terriers. *J. Vet. Intern. Med.* **17:** 50–57.

Fyfe J.C., Madsen M., Hojrup P., Christensen E.I., Tanner S.M., de la Chapelle A., He Q., and Moestrup S.K. 2004. The functional cobalamin (vitamin B$_{12}$) intrinsic factor receptor is a novel complex of cubilin and amnionless. *Blood* **103:** 1573–1579.

Giger U. 2000. Erythrocyte phosphofructokinase and pyruvate kinase deficiencies. In *Schalm's veterinary hematology,* 5th edition (ed. B.F. Feldman et al.), pp. 1020–1025. Lippincott Williams & Wilkins Philadelphia, Pennsylvania.

———. 2005. Clinical genetics. In *Textbook of veterinary internal medicine,* 6th edition (ed. S.J. Ettinger and E.C. Feldman), pp. 264–268. Elsevier Saunders, St. Louis, Missouri.

Giger U. and Jezyk P.F. 1992. *Diagnosis of inborn errors of metabolism in small animals.* In *Current veterinary therapy XI* (ed. R.W. Kirk), pp. 18–22. Saunders, Philadelphia, Pennsylvania.

Giger U. and Greene C. 2000. Immunodeficiency and infection. In *Infectious diseases of the dog and cat,* 2nd edition (ed. C.E. Greene), pp. 683–693. Saunders, Philadelphia, Pennyslvania.

Giger U., Werner L.L., Millichamp N.R., and Gorman N.T. 1985. Sulfadiazine-induced allergy in six Doberman Pinschers. *J. Am. Vet. Med. Assoc.* **186:** 479–485.

Giger U., Kimmel A., Overley D.E., Schwartz L.T., Smith B.F., and Rajpurohit Y. 2000. Frequency of phosphofructokinase (PFK) deficiency in English Springer Spaniels: A longitudinal and randomized study. *J. Vet. Intern. Med.* **14:** 360.

Giger U., Smith B.F., Griot-Wenk M., Rajpurohit Y., McCully K., Haskins M.E., and Stedman H. 1991. Phosphofructokinase deficiency: Molecular defect and bone marrow transplantation. *Blood* **78:** 365a.

Glickman L.T., Raghavan M., Knapp D.W., Bonney P.L., and Dawson M.H. 2004. Herbicide exposure and the risk of transitional cell carcinoma of the urinary bladder in Scottish Terriers. *J. Am. Vet. Med. Assoc.* **224:** 1290–1297.

Glickman L.T., Schofer F.S., McKee L.J., Reif, J.S., and Goldschmidt M.H. 1989. Epidemiologic study of insecticide exposures, obesity, and risk of bladder cancer in household dogs. *J. Toxicol. Environ. Health* **28:** 407–414.

Gough A. and Thomas A. 2004. *Breed predisposition to disease in dogs and cats.* Blackwell, Oxford, United Kingdom.

Goldschmidt M.H. and Hendrick M.J. 2002. Tumors of the skin and soft tissue. In *Tumors in domestic animals,* 4th edition (ed. D.J. Meuten), pp. 45–118. Iowa State Press, Ames.

Goldschmidt M.H. and Shofer F.S. 2002. Skin tumors of the dog and cat. Butterworth-Heinemann, Oxford, England.

Gu W., Brooks M., Catalfamo J., Ray J., and Ray K. 1999. Two distinct mutations cause severe hemophilia B in two unrelated canine pedigrees. *Thromb. Haemostasis* **82:** 1270–1275.

Haskins M.E. and Giger U. 1997. Lysosomal storage diseases. In *Clinical biochemistry of domestic animals,* 5th edition (ed. J.J. Kaneko et al.), pp. 741–760. Academic Press, San Diego, California.

Haworth K., Putt W., Cattanach B., Breen M., Binns M., Lingaas F., and Edwards Y.H. 2001. Canine homolog of the T-box transcription factor T; failure of the protein to bind to its DNA target leads to a short-tail phenotype. *Mamm. Genome* **12:** 212–218.

Hayes H.M., Jr. and Fraumeni J.F., Jr. 1975. Canine thyroid neoplasms: Epidemiologic features. *J. Natl. Cancer Inst.* **55:** 931–934.

He Q., Madsen M., Kilkenney A., Gregory B., Christensen E.I., Vorum H., Hojrup P., Schaffer A.A., Kirkness E.F., Tanner S.M., de la Chapelle A., Giger U., Moestrup S.K., and Fyfe J.C. 2005. Amnionless function is required for cubilin brush-border expression and intrinsic factor-cobalamin (vitamin B_{12}) absorption in vivo. *Blood* (in press).

Henthorn P.S., Somberg R.L., Fimiani V.M., Puck J.M., Patterson D.F., and Felsburg P.J. 1994. IL 2R gamma gene microdeletion demonstrates that canine X-linked severe combined immunodeficiency is a homologue of the human disease. *Genomics* **23:** 69–74.

Henthorn P.S., Liu J., Gidalevich T., Fang J., Casal M.L., Patterson D.F., and Giger U. 2000. Canine cystinuria: Polymorphism in the canine SLC3A1 gene and identification of a nonsense mutation in cystinuric dogs. *Hum. Genet.* **107:** 295–303.

Hough C., Kamisue S., Cameron C., Notley C., Tinlin S., Giles A., and Lillicap D. 2002. Aberrant splicing and premature termination of transcription of the FVIII gene as a cause of severe canine hemophilia A: Similarities with the intron 22 inversion mutation in human hemophilia. *Thromb. Hemat.* **87:** 659–665.

Irion D.N., Schaffer A.L., Famula T.R., Eggleston M.L., Hughes S.S., and Pedersen N.C. 2003. Analysis of genetic variation in 28 dog breed populations with 100 microsatellite markers. *J. Hered.* **94:** 81–87.

Jezyk P.F. 1983. Metabolic diseases—An emerging area of veterinary pediatrics. *Compend. Contin. Educ. Pract.* **6:** 1026–1034.

Jonasdottir T.J., Mellersh C.S., Moe L., Heggebo, R., Gamlem H., Ostrander E.A., and Lingaas F. 2000. Genetic mapping of a naturally occurring hereditary renal cancer syndrome in dogs. *Proc. Natl. Acad. Sci.* **97:** 4132–4137.

Katz M.L., Khan S., Awano T., Shahid S.A., Siakotos A.N., and Johnson G.S. 2005. A mutation in the CLN8 gene in English Setter dogs with neuronal ceroid-lipofuscinosis. *Biochem. Biophys. Res. Commun.* **327:** 541–547.

Kienle R.D., Thomas W.P., and Pion P.D. 1994. The natural clinical history of canine congenital subaortic stenosis. *J. Vet. Intern. Med.* **8:** 423–431.

Kijas J.M., Bauer T.R., Jr., Gafvert S., Marklund S., Trowald-Wigh G., Johannisson A., Hedhammar A., Binns M., Juneja R.K., Hickstein D.D., and Andersson L. 1999. A missense mutation in the β-2 integrin gene (*ITGB2*) causes canine leukocyte adhesion deficiency. *Genomics* **61:** 101–107.

Kimmel A. 2001. "Frequency of the mutant allele for M-type phosphofructokinase deficiency in the English Springer Spaniel population in the United States." Ph.D. thesis, University of Zürich, Switzerland.

King T.M. 2003. Heritability and mode of inheritance of cancer in the flat-coated Retriever. In *Proceedings of the 3rd annual canine cancer conference: Genes, dogs, and cancer,* Seattle, Washington.

Kishnani P.S., Bao Y., Wu J.Y., Brix A.E., Lin J.L., and Chen Y.T. 1997. Isolation and nucleotide sequence of canine glucose-6-phosphatase mRNA: Identification of mutation in puppies with glycogen storage disease type Ia. *Biochem. Mol. Med.* **61:** 168–177.

Knapp D.W., Glickman N.W., DeNicola D.B., Bonney P.L., Lin T.L., and Glickman L.T. 2000. Naturally-occurring canine transitional cell carcinoma of the urinary bladder: A relevant model of human invasive bladder cancer. *Urol. Oncol.* **5:** 59.

Koskinen M.T. 2003. Individual assignment using microsatellite DNA reveals unambiguous breed identification in the domestic dog. *Anim. Genet.* **34:** 297–301.

Kramer J.W., Venta P.J., Klein S.R., Cao Y., Schall W.D., and Yuzbasiyan-Gurkan V. 2004. A von Willebrand's factor genomic nucleotide variant and polymerase chain reaction diagnostic test associated with inheritable type-2 von Willebrand's disease in a line of German shorthaired pointer dogs. *Vet. Pathol.* **41:** 221–228.

Kreutzer R., Leeb T., Muller G., Moritz A., and Baumgartner W. 2005. A duplication in the canine β-galactosidase gene GLB1 causes exon skipping and GM1-gangliosidosis in Alaskan Huskies. *Genetics* (in press).

Leppanen M. and Saloniemi H.1998. Screening and controlling canine inherited ocular diseases in Finland: Epidemiological, economical and health promotional aspect. *Vet. Ophthalmol.* **1:** 203–210.

———. 1999. Controlling canine hip dysplasia in Finland. *Prev. Vet. Med.* **42:** 121–131.

Lin L., Faraco J., Li R., Kadotani H., Rogers W., Lin X, Qiu X., de Jong P.J., Nishino S., and Mignot E. 1999. The sleep disorder canine narcolepsy is caused by a mutation in the hypocretin (orexin) receptor 2 gene. *Cell* **98:** 365–376.

Lingaas F., Comstock K.E., Kirkness E.F., Sorensen A., Aarskaug T., Hitte C., Nickerson M.L., Moe L., Schmidt L.S., Thomas R., et al. 2003. A mutation in the canine BHD gene is associated with hereditary multifocal renal cystadenocarcinoma and nodular dermatofibrosis in the German Shepherd dog. *Hum. Mol. Genet.* **12:** 3043–3053.

Lipscomb D.L., Bourne C., and Boudreaux M.K. 2000. Two genetic defects in alphaIIb are associated with type I Glanzmann's thrombasthenia in a Great Pyrenees dog: A 14-base insertion in exon 13 and a splicing defect of intron 13. *Vet. Pathol.* **37:** 581–588.

Lium B. and Moe L. 1985. Hereditary multifocal renal cystadenocarcinomas and nodular dermatofibrosis in the German Shepherd dog: Macroscopic and histopathologic changes. *Vet. Pathol.* **22:** 447–455.

Lohi H., Young E.J., Fitzmaurice S.N., Rusbridge C., Chan E.M., Vervoort M., Turnbull J., Zhao X.C., Ianznao L., Paterson A.D., et al. 2005. Expanded repeat in canine epilepsy. *Science* **307:** 81.

Lozier J.N., Durta A., Pak E., Zhou N., Zheng Z., Nichols T.C., Bellinger D.A., Read M., and Morgan R.A. 2002. The Chapel Hill hemophilia A dog colony exhibits a factor VIII gene inversion. *Proc. Natl. Acad. Sci.* **99:** 12991–12996.

Mause A.E., Whitlark J., Whitney K.M., and Lothrop C.D., Jr. 1996. A deletion mutation causes hemophilia B in Lhasa Apso dogs. *Blood* **88:** 3451–3455.

McNiel E.A. and Husbands B. 2005. Pheochromocytoma. In *Textbook of veterinary internal medicine*, 6th edition (ed. S.N. Ettinger and B.F. Feldman). Saunders, Philadelphia, Pennsylvania.

McNiel E.A., Mickelson J.R., and Huntsman D.G. 2004a. Genetic contributions to gastric carcinoma in Chow Chows. In *Proceedings of the Third Annual Frontiers in Cancer Prevention research conference*, Seattle, Washington.

McNiel E.A., Prink A., and O'Brien T D. 2004b. Biologic behavior of mast cell tumors in pug dogs. In *Proceedings of the Veterinary Cancer Society annual conference*, Kansas City, Missouri.

Mealey K.L., Bentjen S.A., Gay J.M., and Cantor G.H. 2001. Ivermectin sensitivity in collies is associated with a deletion mutation of the mdr1 gene. *Pharmacogenetics* **11:** 727–733.

Menon K.P., Tieu P.T., and Neufeld E.F. 1992. Architecture of the canine IDUA gene and mutation underlying canine mucopolysaccharidosis I. *Genomics* **14:** 763–768.

Michell A.R. 1999. Longevity of British breeds of dog and its relationships with sex, size, cardiovascular variables and disease. *Vet. Rec.* **145:** 625–629

Miller D.M. 1995. The occurrence of mast cell tumors in young Shar-Peis. *J. Vet. Diagn. Invest.* **7:** 360–363.

Misdorp W. and Hart A.A. 1979. Some prognostic and epidemiologic factors in canine osteosarcoma. *J. Natl. Cancer Inst.* **62:** 537–545.

Moe L. and Lium B. 1997. Hereditary multifocal renal cystadenocarcinomas and nodular dermatofibrosis in 51 German shepherd dogs. *J. Sm. Anim. Pract.* **38:** 498–505.

Moore P.F. 1984. Systemic histiocytosis of Bernese mountain dogs. *Vet. Pathol.* **21:** 554–563.

Moore P.F. and Rosin A. 1986. Malignant histiocytosis of Bernese mountain dogs. *Vet. Pathol.* **23:** 1–10.

Morris J.S., Bostock D.E., McInnes E.F., Hoather T.M., and Dobson J.M. 2000. Histopathological survey of neoplasms in flat-coated retrievers, 1990 to 1998. *Vet. Rec.* **147:** 291–295.

Morris J.S., McInnes E.F., Bostock D.E., Hoather T.M., and Dobson J.M. 2002. Immunohistochemical and histopathologic features of 14 malignant fibrous histiocytomas from Flat-Coated Retrievers. *Vet. Pathol.* **39:** 473–479.

Mutsaers A.J., Widmer W.R., and Knapp D.W. 2003. Canine transitional cell carcinoma. *J. Vet. Intern. Med.* **17:** 136–144.

Nadon N.L., Duncan I.D., and Hudson L.D. 1990. A point mutation in the proteolipid protein gene of the 'shaking pup' interrupts oligodendrocyte development. *Development* **110:** 529–537.

Nakamura K., Watanabe M., Takanaka K., Sasaki Y., and Ikeda T. 2000. cDNA cloning of mutant catalase in acatalasemic beagle dog: Single nucleotide substitution leading to thermal-instability and enhanced proteolysis of mutant enzyme. *Int. J. Biochem. Cell Biol.* **32:** 1183–1193.

Neff M.W., Robertson K.R. Wong A.K. Safra N., Broman K.W., Slatkin M., Mealey K.L., and Pederson N.C. 2004. Breed distribution and history of canine *mdr1-1Δ*, a pharmacogenetic mutation that marks the emergence of breeds from the collie lineage. *Proc. Natl. Acad. Sci.* **101:** 11725–11730.

Occhiodoro T. and Anson D.S. 1996. Isolation of the canine alpha-L-fucosidase cDNA and definition of the fucosidosis mutation in English Springer Spaniels. *Mamm. Genome* **7:** 271–274.

Padgett G.A., Madewell B.R., Keller E.T., Jodar L., and Packard M. 1995. Inheritance of histiocytosis in Bernese mountain dogs. *J. Sm. Anim. Pract.* **36:** 93–98.

Parker H.G., Kim L.V., Sutter N.B., Carlson S., Lorentzen T.D., Malek T.B., Johnson G.S., DeFrance H.B., Ostrander E.A., and Kruglyak L. 2004. Genetic structure of the purebred domestic dog. *Science* **304:** 1160–1164.

Paster E.R., LaFond E., Biery D.N., Iriye A., Gregor T.P., Shofer F.S., and Smith G.K. 2005. Estimates of prevalence of hip dysplasia in Golden Retrievers and Rottweilers and the influence of bias on published prevalence figures. *J. Am. Vet. Med. Assoc.* **226:** 387–392.

Paterson S., Boydell P., and Pike R. 1995. Systemic histiocytosis in the Bernese mountain dog. *J. Sm. Anim. Pract.* **36:** 233–236.

Patronek G.J., Waters D.J., and Glickman L.T. 1997. Comparative longevity of pet dogs and humans: Implications for gerontology research. *J. Gerontol. Biol. Sci. A* **52:** B171–B178 (Abstr.).

Patterson D.F. 1971. Canine congenital heart disease: Epidemiology and etiological hypotheses. *J. Sm. Anim. Pract.* **12:** 263–287.

———. 2000. Companion animal medicine in the age of medical genetics. *J. Vet. Intern. Med.* **14:** 1–9.

Patterson D.F., Pexieder T., Schnarr W.R., Navratil T., and Alaili R. 1993. A single major-gene defect underlying cardiac conotruncal malformations interferes with myocardial growth during embryonic development: Studies in the CTD line of keeshond dogs. *Am. J. Hum. Genet.* **52:** 388–397.

Patterson D.F., Aguirre G.D., Fyfe J.C., Giger U., Green P.L., Haskins M.E., Jezyk P.F., and Meyers-Wallen V.N. 1989. Is this a genetic disease? *J. Sm. Anim. Pract.* **30:** 127–139.

Patterson D.F., Haskins M.E., Jezyk P.F., Giger U., Meyers-Wallen V.N., Aguirre G.D, Fyfe J.C., and Wolfe J.H. 1988. Research on genetic diseases: Reciprocal benefits to animals and man. *J. Am. Vet. Med. Assoc.* **193:** 1131–1144.

Pele M., Tiret L., Kessler J.L., Blot S., and Panthier J.J. 2005. A SINE exonic insertion in the PTPLA gene leads to multiple splicing defects and segregates with the autosomal recessive centronuclear myopathy in dog. *Hum. Mol. Genet.* **14:** 1417–1427.

Peters J.A. 1969. Canine mastocytoma: Excess risk as related to ancestry. *J. Natl. Cancer Inst.* **42:** 435–443.

Priester W.A. 1967. Canine lymphoma: Relative risk in the boxer breed. *J. Natl. Cancer Inst.* **39:** 833–845.

———. 1974. Pancreatic islet cell tumors in domestic animals. Data from 11 colleges of veterinary medicine in the United States and Canada. *J. Natl. Cancer Inst.* **53:** 227–229.

Priester W.A. and Mantel N. 1971. Occurrence of tumors in domestic animals. Data from 12 United States and Canadian colleges of veterinary medicine. *J. Natl. Cancer Inst.* **47:** 1333–1344.

Priester W.A., Goodman D.G., and Theilen G.H. 1977. Nine simultaneous primary tumors in a Boxer dog. *J. Am. Vet. Med. Assoc.* **170:** 823–826.

Proschowsky H.F., Rugbjerg H., and Ersboll A.K. 2003. Mortality of purebred and mixed-breed dogs in Denmark. *Prev. Vet. Med.* **58:** 63–74.

Prymak C., McKee L.J., Goldschmidt M.H., and Glickman L.T. 1988. Epidemiologic, clinical, pathologic, and prognostic characteristics of splenic hemangiosarcoma and splenic hematoma in dogs: 217 cases (1985). *J. Am Vet. Med. Assoc.* **193:** 706–712.

Raghavan M., Knapp D.W., Dawson M.H., Bonney P.L., and Glickman LT. 2004. Topical flea and tick pesticides and the risk of transitional cell carcinoma of the urinary bladder in Scottish Terriers. *J. Am Vet. Med. Assoc.* **225:** 389–394.

Ramos-Vara J.A., Beissenherz M.E., Miller M.A., Johnson G.C., Pace L.W., Fard A., and Kottler S.J. 2000. Retrospective study of 338 canine oral melanomas with clinical, histologic, and immunohistochemical review of 129 cases. *Vet. Pathol.* **37:** 597–608.

Ray J., Bouvet A., DeSanto C., Fyfe J.C., Xu D. Wolfe J.H., Agurirre G.D., Patterson D.F., Haskins M.E., and Henthorn P.S. 1998. Cloning of the canine beta-glucuronidase cDNA, mutation identification in canine MPS VII, and retroviral vector-mediated correction of MPS VII cells. *Genomics* **48:** 248–253.

Rhodes T.H., Vite C.H., Giger U., Patterson D.F., Fahlke C., and George A.L., Jr. 1999. A missense mutation in canine ClC-1 causes recessive myotonia congenita in the dog. *FEBS Lett.* **456:** 54–58.

Rieger M., Schwarz H.P., Turecek P.L., Dorner F., van Mourik J.A., and Mannhalter C. 1998. Identification of mutations in the canine von Willebrand factor gene associated with type III von Willebrand disease. *Thromb. Haemost.* **80:** 332–337.

Riehl J., Nishino S., Cederberg R., Dement W.C., and Mignot E. 1998. Development of cataplexy in genetically narcoleptic Dobermans. *Exp. Neurol.* **152:** 292–302.

Roberts M.C., Mickelson J.R., Patterson E.E., Nelson T.E., Armstrong P.J., Brunson D.B., and Hogan K. 2001. Autosomal dominant canine malignant hyperthermia is caused by a mutation in the gene encoding the skeletal muscle calcium release channel (RYR1). *Anesthesiology* **95:** 716–725.

Robinson R. 1973. Relationship between litter size and weight of dam in the dog. *Vet. Rec.* **92:** 221–223.

Roulet A., Puel O., Gesta S., Lepage J.F., Drag M., Soll M., Alvinerie M., and Pineau T. 2003. MDR1-deficient genotype in Collie dogs hypersensitive to the P-glycoprotein substrate ivermectin. *Eur. J. Pharmacol.* **460:** 85–91.

Ru G., Terracin B., and Glickman L.T. 1998. Host related risk factors for canine osteosarcoma. *Vet. J.* **156:** 31–39.

Schafer K.A., Kelly G., Schrader R., Griffith W.C., Muggenburg B.A., Tierney L.A., Lechner J.F., Janovitz E.B., and Hahn F.F. 1998. A canine model of familial mammary gland neoplasia. *Vet. Pathol.* **35:** 168–177.

Schatzberg S.J., Olby N.J., Breen M., Anderson L.V., Langford C.F., Dickens H.F., Wilton S.D., Zeiss C.J., Binns M.M., Kornegay J.N., Morris G.E., and Sharp N.J. 1999. Molecular analysis of a spontaneous dystrophin "knockout" dog. *Neuromuscul. Disord.* **5:** 289–295.

Scriver C.R., Beaudet A.L., Sly W.S., and Valle D. 2001. *Molecular and metabolic basis of hereditary diseases.* McGraw Hill, New York.

Sharp N.J., Kornegay J.N., Van Camp S.D., Herbstreith M.H., Secore S.L., Kettle S., Hung W.Y., Constantinou C.D., Dykstra M.J., and Roses A.D., et al. 1992. An error in dystrophin mRNA processing in golden retriever muscular dystrophy, an animal homologue of Duchenne muscular dystrophy. *Genomics* **13:** 115–121.

Silverstein Dombrowski D.C., Carmichael K.P., Wang P., O'Malley T.M., Haskins M.E., and Giger U. 2004. Mucopolysaccharidosis type VII in a German Shepherd dog. *J. Am. Vet. Med. Assoc.* **224:** 553–557.

Skelly B.J., Sargan D.R., Herrtage M.E., and Winchester B.G. 1996. The molecular defect underlying canine fucosidosis. *J. Med. Genet.* **33:** 284–288.

Skelly B.J., Wallace M., Rajpurohit Y.R., Wang P., and Giger U. 1999. Identification of a 6 base pair insertion in West Highland White Terriers with erythrocyte pyruvate kinase deficiency. *Am. J. Vet. Res.* **60:** 1169–1172.

Smith B.F., Stedman H., Rajpurohit Y., Henthorn P.S., Wolfe J.H., Patterson D.F., and Giger U. 1996. Molecular basis of canine muscle type phosphofructokinase deficiency. *J. Biol. Chem.* **16:** 20070–20074.

Somberg R.L., Robinson J.P., and Felsburg P.J. 1994. T lymphocyte development and function in dogs with X-linked severe combined immunodeficiency. *J. Immunol.* **153:** 4006–4015.

Somberg R.L., Pullen R.P., Casal M.L., Patterson D.F., Felsburg P.J., and Henthorn P.S. 1995. A single nucleotide insertion in the canine interleukin-2 receptor gamma chain results in X-linked severe combined immunodeficiency disease. *Vet. Immunol. Immunopathol.* **47:** 203–213.

Stabej P., Leegwater P.A., Imholz S., Loohuis M., Mandigers P., Dimanjko-Petric A., Stokhof A.A., and van Oost B.A. 2004. The giant myofilament protein titin associated with dilated cardiomyopathy in the Dobermann. In *The 2nd International Conference: Advances in canine and feline genomics: Comparative genome anatomy and genetic disease*, Utrecht, The Netherlands.

Stoltzfus L.J., Sosa-Pineda B., Moskowitz S.M., Menon K.P., Dlott B., Hooper L., Teplow D.B., Shull R.M., and Neufeld E.F. 1992. Cloning and characterization of cDNA encoding canine α-L-iduronidase. mRNA deficiency in mucopolysaccharidosis I dog. *J. Biol. Chem.* **267:** 6570–6575.

Suter M., Lott-Stolz G., and Wild P. 1983. Generalized nodular dermatofibrosis in six Alsatians. *Vet. Pathol.* **20:** 632–634.

Sutter N.B., Eberle M.A., Parker H.G., Pullar B.J., Kirkness E.F., Kruglyak L., and Ostrander E.A. 2004. Extensive and breed specific linkage disequilibrium in *Canis familiaris*. *Genome Res.* **14:** 2388–2396.

Tosic M., Matthey B., Gow A., Lazzarini R.A., and Matthieu J.M. 1997. Intracellular transport of the DM-20 bearing shaking pup (shp) mutation and its possible phenotypic consequences. *J. Neurosci. Res.* **50:** 844–852.

van de Sluis B., Rothuizen J., Pearson P.L., van Oost B.A., and Wijmenga C. 2002. Identification of a new copper metabolism gene by positional cloning in a purebred dog population. *Hum. Mol. Genet.* **11:** 165–173.

van Oost B.A., Versteeg S.A., and Slappendel R.J. 2004. DNA testing for type III von Willebrand disease in Dutch Kooiker dogs. *J. Vet. Intern. Med.* **18:** 282–288.

Venta P.J., Li J., Yuzbasiyan-Gurkan V., Brewer G.J., and Schall W.D. 2000. Mutation causing von Willebrand's disease in Scottish Terriers. *J. Vet. Intern. Med.* **14:** 10–19.

Victoria T., Rafi M.A., and Wenger D.A. 1996. Cloning of the canine GALC cDNA and identification of the mutation causing globoid cell leukodystrophy in West Highland White and Cairn terriers. *Genomics* **33:** 457–462.

Vijavasarathy C., Giger U., Prociuk U., Patterson D.F., Breitschwerdt E.B., and Avadhani N.G. 1994. Canine mitochondrial myopathy associated with reduced mitochondrial mRNA and altered cytochrome C oxidase activities in fibroblasts and skeletal muscle. *Comp. Biochem. Physiol.* **109A:** 887–894.

Wang Z.H., Zeng B., Shibuya H., Johnson G.S., Alrov J., Pastores G.M., Raghavan S., and Kolodny E.H. 2000. Isolation and characterization of the normal canine β-galactosidase gene and its mutation in a dog model of GM1-gangliosidosis. *J. Inherit. Metab. Dis.* **23:** 593–606.

Werner P., Raducha M.G., Prociuk U., Ostrander E.A., Spielman R.S., Kirkness E.F., Patterson D.F., and Henthorn P.S. 2005. The keeshond defect in cardiac conotruncal development is oligogenic(1). *Hum Genet.* **116:** 368–377.

Whitney K.M., Goodman S.A., Bailey E.M., and Lorthrop C.D., Jr. 1994. The molecular basis of canine pyruvate kinase deficiency. *Exp. Hematol.* **22:** 866–874.

Wilcox B. and Walkowicz C. 1995. *The atlas of dog breeds of the world*, 5th edition. T.F.H. Publications, Neptune City, New Jersey.

Willis M.B. 1989. *Genetics of the dog.* Howell Book House, New York, pp. 37–50.

Wilson D.W. and Dungworth D.L. 2002. Tumors of the respiratory tract. In *Tumors in domestic animals*, 4th edition (ed. D.J. Meuten), pp. 365–400. Iowa State Press, Ames, Iowa.

Yamato O., Endoh D., Kobayashi A., Masuoka Y., Yonemura M., Hatakeyama A., Satoh H., Tajima M., Yamasaki M., and Maede Y. 2002. A novel mutation in the gene for canine acid β-galactosidase that causes GM1-gangliosidosis in Shiba dogs. *J. Inherit. Metab. Dis.* **25:** 525–526.

Yogalingam G., Pollard T., Gliddon B., Jolly R.D., and Hopwood J.J. 2002. Identification of a mutation causing mucopolysaccharidosis type IIIA in New Zealand Huntaway dogs. *Genomics* **79:** 150–153.

Zheng K., Thorner P.S., Marrano P., Baumal R., and McInnes R.R. 1994. Canine X chromosome-linked hereditary nephritis: A genetic model for human X-linked hereditary nephritis resulting from a single base mutation in the gene encoding the α5 chain of collagen type IV. *Proc. Natl. Acad. Sci.* **91:** 3989–3993.

WWW RESOURCES

http://www.angis.org.au/databases/BRIX/omia Online Mendelian Inheritance in Animals database, Faculty of Veterinary Science, University of Sydney, Australia.

http://www.ncbi.nlm.nih.gov/OMIM Online Mendelian Inheritance in Man [OMIM].

http://www.upei.ca/~cidd/intro.htm Canine Inherited Disorders Database. A. Crook, B. Hill, S. Dawson, authors. Joint effort of Sir James Dunn Animal Welfare Centre at the Atlantic Veterinary College, University of Prince Edward Island, and the Canadian Veterinary Medical Association.

http://www.vet.cam.ac.uk/idid Inherited Diseases in Dogs [IDID], Cambridge Veterinary School, University of Cambridge.

http://www.vetgen.com/vwdrpt.html DNA Studies in Doberman von Willebrand's Disease. G.J. Brewer, Department of Human Genetics and Internal Medicine, University of Michigan Medical School.

http://www.vet.upenn.edu/penngen Section of Medical Genetics, University of Pennsylvania, School of Veterinary Medicine.

16

Models, Mutants, and Man: Searching for Unique Phenotypes and Genes in the Dog Model of Inherited Retinal Degeneration

Gustavo D. Aguirre

Section of Medical Genetics
Department of Clinical Studies
School of Veterinary Medicine
University of Pennsylvania
Philadelphia, Pennsylvania 19104-6010

Gregory M. Acland

James A. Baker Institute for Animal Health
College of Veterinary Medicine
Cornell University
Ithaca, New York 14853

THE PAST 15 YEARS HAS SEEN A VIRTUAL EXPLOSION of gene and phenotype discovery in medical ophthalmology. Since the mapping of an autosomal dominant retinitis pigmentosa (ADRP) locus to the long arm of human chromosome 3 in the region of the rhodopsin (*RHO*) gene (McWilliam et al. 1989; Farrar et al. 1990), and the subsequent identification of the first disease-associated mutation (Dryja et al. 1990), there has been very rapid progress in identification of multiple loci and genes that are responsible for inherited visual dysfunction or blindness occurring alone or as part of more complex syndromes. Approximately 150 retinal disease loci have been mapped to date; of these, disease-causing mutations have been found in approximately 100 genes (Fig. 1; http://www.sph.uth.tmc.edu/RetNet/). Even for the *RHO* gene, over 100 disease-causing mutations have been identified, and, although most are inherited as autosomal dominant (AD) and cause RP, autosomal recessive

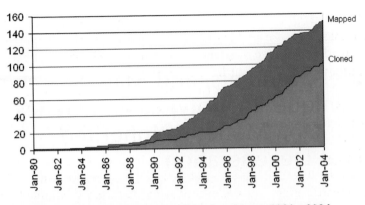

Mapped and Cloned Retinal Disease Genes 1980 - 2004

Figure 1. Graph illustrating the cumulative number of human retinal disease genes mapped and cloned. Yearly updates are available through the RetNet Retinal Information Network Web page (http://www.sph.uth.tmc.edu/RetNet/). RetNet provides tables of genes causing inherited retinal diseases, such as retinitis pigmentosa, macular degeneration, and Usher syndrome, and related information, and is a service of the Laboratory for the Molecular Diagnosis of Inherited Eye Diseases, a joint program of The Hermann Eye Center, Department of Ophthalmology and Visual Science, and The Human Genetics Center, School of Public Health at The University of Texas–Houston Health Science Center. (Reprinted, with permission, from Daiger 2005 [http://www.sph.uth.tmc.edu/RetNet/sum-dis.htm#D-graph].)

RP (ARRP) and AD congenital stationary nightblindness (CSNB) forms also are recognized (Rivolta et al. 2002). Thus, allelic and genetic heterogeneity are the rule when considering the inherited diseases of the retina. Despite this complexity, rapid progress has been possible because of the very thorough earlier studies that critically defined the phenotypic characteristics of many of these disorders in man (for review, see Berson 1993; see also Berson et al. 1968, 1969; Kemp et al. 1988; Cideciyan and Jacobson 1993; Jacobson et al. 1994).

Dogs also suffer from various forms of inherited retinal blindness, the principal one termed progressive retinal atrophy (PRA). This is a heterogeneous inherited disorder affecting many breeds, and all affected dogs show the same general clinically recognizable ocular abnormalities. Associated with these findings are behavioral signs indicative of visual deficits; the disease is progressive, and end-stage retinal atrophy and blindness are inevitable. PRA was first identified as a clinical entity in 1909 by Hilding Magnusson (1886–1957), who worked at the Veterinary Bacteriologic Laboratory in Malmo, Sweden, and it was reported in the veterinary and medical literature soon after (Magnusson 1909, 1910, 1911, 1917). Magnusson was the first to recognize that PRA was similar to RP in man, and the diseases in both species shared genetic, clinical, and pathological features. Magnusson brought to veterinary ophthalmology a broad perspective in science, using genetic, clinical, and pathological approaches to carry out his studies.

The major advances in the early studies of PRA were carried out by Herbert B. Parry during his tenure at the Canine Research Station of the Animal Health Trust, Newmarket, U.K. Parry served this organization as a Senior Scientific Officer between 1947 and 1954, and did his research on canine retinopathies during that time. In 1949, PRA was recognized in the red Irish setter (Hodgman et al. 1949), and, within a short time, a major research effort was under way in the U.K. to study and eliminate this major inherited disease of dogs. Parry published a series of landmark papers between 1951 and 1955. These described the development and use of electroretinography (ERG) to study normal retinal function in the dog (Parry et al. 1951, 1953) and to assess the retinal dysfunction that occurs in PRA affected animals (Parry et al. 1955). These ERG studies served as the platform for the future use of electroretinography for the early diagnosis of inherited retinal diseases in dogs (for review, see Aguirre 1995). Parry's studies of generalized PRA in Irish setters stand out because of their scientific rigor, detail, and thoroughness (Parry 1953a). He established this form of PRA as a model for studies of a primary photoreceptor disease, comparable to what was then being described in the *rd* mouse (Tansley 1951). Parry also contributed to our understanding of the retinal pathology associated with glaucoma (Parry 1953b) and canine distemper (Parry 1954). These established the foundation for future research studies on the inherited and acquired retinopathies of dogs.

In this chapter, we present a summary of research leading to our current understanding of the genes and mutations responsible for inherited diseases of the retina in dogs. The first gene defect, responsible for PRA in Irish setters, was found in 1993 (Clements et al. 1993; Suber et al. 1993). At the time, the tools available for molecular genetic and genomic studies in dogs were nonexistent, and identification of disease-associated genes was based on the candidate gene approach that had been quite successful in human patients (Dryja 1997). Soon after, publication of the first meiotic linkage map of the dog (Mellersh et al. 1997) occurred together with the development of canine–rodent somatic hybrid cell lines (Langston et al. 1997); these provided the initial tools necessary for mapping and identifying some of the unique phenotypic traits and diseases of dogs (Acland et al. 1998). Through several iterations, the linkage map has been increased in marker coverage and depth, and integrated with the radiation hybrid map (Neff et al. 1999; Werner et al. 1999b; Mellersh et al. 2000; Breen et al. 2001; Guyon et al. 2003), a BAC library has become available (Li et al. 1999), and, in 2003, a 1.5× genome sequence of the dog was published (Kirkness et al. 2003). Now that the 6.5× sequence is in the public domain (http://www.ncbi.nlm.nih.gov/genome/guide/dog), these resources will foster continued cutting-edge genomic studies in the dog. From the humble beginnings of veterinary molecular ophthalmology in 1993, it is hard to believe that such progress could be possible in such a short period of time.

INHERITED RETINAL DISEASES IN DOGS: OVERVIEW

Table 1 presents a limited selection of inherited retinal diseases of dogs for which there is extensive clinical, pathological, biochemical, and/or molecular information. The diseases and models were selected based on the authors' research interest and their clinical experience in evaluating dogs for inherited retinal diseases, the value of the models for comparative cell and molecular studies of photoreceptor disease, frequency of the disorder within the population at risk, and potential for the use of the models in therapy studies that may eventually be applicable to human patients. A more comprehensive listing of diseases is available (ACVO 1999; Gelatt 1999) among other sources.

The diseases are divided into broad clinical groupings. Both PRA and cone-rod dystrophies (CRD) are progressive disorders that affect the retinal photoreceptor cells primarily, or possibly secondary to defects in the retinal pigment epithelial (RPE) cell layer. In general, dogs affected with any one of the PRA group of diseases show damage first to the rod photoreceptors and subsequently to the cones; hence the reason that night blindness is the predominant clinical finding prior to severe visual dysfunction under both dim and bright light conditions. Associated with these defects, the dogs show very characteristic changes in the fundus that are visible with an ophthalmoscope—the blood vessels become thin, the tapetal layer is increased in reflectivity secondary to retinal thinning, and the optic nerve becomes pale. In late stages of the disease, most dogs develop secondary cataracts.

As the name implies, CRDs are disorders predominantly of cones, with rods being affected later and to a lesser extent, at least initially. In many cases, affected dogs show extensive impairment of visual function under both bright and dim light conditions, and the identification of a more severe cone abnormality is based on quite detailed ERG testing (Kijas et al. 2004). The CRD group of diseases have ophthalmoscopic retinal changes that are similar to PRA, and for that reason, many of these

Table 1. Selected inherited diseases of the retina for which there is clinical, pathological, biochemical, genetic, and/or molecular information

Disease class	Disease name	Breed	Gene locus	Defective gene	References
Progressive retinal atrophy (PRA)					
	rod-cone dysplasia 1	Irish setter	rcd1	PDE6B	Parry (1953); Aguirre et al. (1978); Farber et al. (1992); Clements et al. (1993); Suber et al. (1993); Ray et al. (1994)
	rod-cone dysplasia 1[a]	Sloughi	rcd1[a]	PDE6B	Dekomien et al. (2000)
	rod-cone dysplasia 2	collie	rcd2	N.D.	Woodford et al. (1982); Acland et al. (1989); Wang et al. (1999a)
	rod-cone dysplasia 3	Cardigan Welsh corgi	rcd3	PDE6A	Petersen-Jones et al. (1999)
	photoreceptor dysplasia	miniature schmauzer	Type A PRA	pd[a]	Parshall et al. (1991)
	rod dysplasia[b]	Norwegian elkhound	rd	N.D.	Aguirre (1978)
	early retinal degeneration[c]	Norwegian elkhound	erd	N.D.	Acland and Aguirre (1987); Acland et al. (1989, 1999); Kukekova et al. (2003)
	progressive rod-cone degeneration	many breeds[d]	prcd	N.D.[e]	Aguirre et al. (1982a); Aguirre and Acland (1988); Acland et al. (1998); Sidjanin et al. (2003)

(Continued)

Table 1. (*Continued*)

Disease class	Disease name	Breed	Gene locus	Defective gene	References
	X-linked PRA	Siberian husky Samoyed	XLPRA	RPGR	Acland et al. (1994); Zeiss et al. (1999); Zeiss et al. (2000); Zhang et al. (2001, 2002) Kijas et al. (2002, 2003)
	autosomal dominant PRA	English mastiff bullmastiff	RHO	Rho	
Cone-rod dystrophy	cone-rod dystrophy 1, 2, and 3	pit bull terrier pit bull terrier Glen of Imaal	crd1 crd2 crd3	N.D.[f]	Kijas et al. (2004)
	cone-rod dystrophy 4[g]	miniature longhaired dachshund	crd4		Curtis and Barnett (1993)
Stationary disorders	cone degeneration	Alaskan malamute German shorthaired pointer	cd	CNGB3[h]	Rubin et al. (1967); Long and Aguirre (1991); Sidjanin et al. (2002b)
	congenital stationary nightblindness[i]	Briard	csnb	RPE65	Narfström et al. (1989); Wrigstad (1994); Aguirre et al. (1998); Veske et al. (1999)

Abnormal development				
oculo-skeletal dysplasia	Labrador retriever Samoyed	osd1 osd2	N.D.[j]	Meyers et al. (1983); Carrig et al. (1988); Acland and Aguirre (1995)
collie eye anomaly	collie, Shetland sheepdog, Australian shepherd, Border collie	cea	N.D.[k]	Lowe et al. (2003)

N.D. = Not defined.

[a]The original dogs used to characterize the pd gene defect were not available for the molecular studies. As there are two non-allelic forms of PRA in the breed, the first identified mutation was termed Type A PRA.

[b]The original colony of rd dogs is no longer extant.

[c]erd has been mapped to the SHARP1 region on CFA27.

[d]The following breeds have been found affected with prcd: American cocker spaniel, American Eskimo, Australian cattle dog, Chesapeake Bay retriever, English cocker spaniel, Entlebucher, Finnish lapphund, Labrador retriever, miniature poodle, Nova Scotia duck tolling retriever, Portuguese water dog, toy poodle. The prcd disease haplotype is also found in Chinese crested, silky terrier, giant schnauzer, German shorthaired pointer.

[e]prcd mapped to an 85-kb interval on CFA9. The gene and causal mutation have been found but not published (B. Zangerl et al., unpubl.).

[f]crd1 and crd2 are non-allelic disorders in pit bull terrier-derived breeds.

[g]A mutation in RPGRIP1 has been found to cause the disease (C.S. Mellersh et al., pers. comm.).

[h]Different mutations in the CNGB3 gene account for the clinically identical disorders.

[i]The disease in Briard has had a number of different terms, e.g., congenital stationary nightblindness, retinal dystrophy, and canine Lebers congenital amaurosis (cLCA). Because the mutation is in the RPE65 gene of dogs, and human has the same clinical phenotype, the appropriate name for the disease is cLCA.

[j]These are non-allelic disorders in Labrador retrievers and Samoyeds (Acland and Aguirre 1995; G.M. Acland and G.D. Aguirre, unpubl.).

[k]The disease has been mapped to a 3.9-cM region on CFA37. The gene and causal mutation have been found but not published (E.A. Ostrander et al., unpubl.).

disorders were initially misclassified. Of the four recognized CRDs, three show extensive retinal disease before 1 year of age (*crd1, crd2, crd4*) (Curtis and Barnett 1993; Kijas et al. 2004) whereas *crd3* is a late-onset, slowly progressive disease (Kijas et al. 2004).

The "progressive" designation of both PRA and CRD is based on both the increased severity of the vision problems with time and the progressive retinal changes observed, either clinically with an ophthalmoscope, or at the cellular level by histopathology. These give a very definitive indication of the inexorable progression of the disorders until complete blindness ensues. In contrast, the "stationary" *cd* and *csnb* disorders are characterized by pathology that affects the cones (*cd*) or rod/cone (*csnb*) cells without showing further deterioration in vision or retinal abnormalities from the time of initial diagnosis in young animals. In the case of *csnb*, there has been much debate regarding both the naming of the disease and the progressive or nonprogressive nature of the disorder (see, e.g., Narfström 1999; Aguirre 2000). There is no dispute, however, in the change in clinical phenotype that has occurred with this disease since first recognized; initially, affected dogs were primarily night-blind, hence the congenital stationary night blindness designation. However, all affected dogs now are completely blind or have very severe visual impairment (Narfström et al. 1989, 1994; Aguirre et al. 1998).

The retina is also affected by developmental disorders. Retinal dysplasia in dogs (not to be confused with the photoreceptor dysplasias that affect dogs with the early-onset form of PRA [see below]) is a defect of retinal differentiation. The generalized form is characterized by separation from the retinal pigment epithelium of the abnormally developed retina at birth, or during the first 6 weeks of life, the time period when the retina undergoes the major stages of postnatal retinal maturation. The disease has been reported as an inherited trait in the Labrador retriever (Kock 1974) and in Sealyham (Ashton et al. 1968), Yorkshire (Stades 1978), and Bedlington (Rubin 1968) terriers. Syndromic retinal dysplasia is present in the Labrador retriever and Samoyed breeds, where it represents nonallelic oculoskeletal disorders (oculoskeletal dysplasia; *osd1, osd2* [Meyers et al. 1983; Nelson and MacMillan 1983; Acland and Aguirre 1995]). Homozygous affected dogs exhibit short-limbed dwarfism and a constellation of ocular changes, which, in the most severe form, are characterized by complete retinal dysplasia with detachment and cataracts. Heterozygous dogs have lesions limited to the retina that consist of unilateral or bilateral focal/multifocal dysplastic changes (Carrig et al. 1977, 1988; Acland and Aguirre 1995).

Another inherited developmental defect of the retina is collie eye anomaly (*cea*); the disorder affects the retina/choroid/scleral complex and is characterized by regional hypoplasia of the highly vascular choroid, and defects of the sclera consisting of colobomatous lesions of variable size and severity. The retina is not a primary target of the gene defect, as it has normal structure and function. Retinal involvement is secondary to the other ocular defects, particularly those that affect the sclera and optic disc and result in retinal detachment, which occurs in approximately 10% of affected dogs at a young age. The disease affects collies worldwide, and Shetland sheepdogs mainly in England and other parts of Europe (Roberts 1969; Barnett and Stades 1979). In both breeds the incidence of the disorder ranges from 50% to 90% in Europe. In contrast, the incidence of disease for Shetland sheepdogs in the U.S. is less than 1% (data for 1991–1999, Canine Eye Registry Foundation). Other breeds affected are the Border collie and Australian shepherd, and the frequency of the disease in these is lower (Bedford 1982; Rubin et al. 1991).

PROGRESSIVE RETINAL ATROPHY: A COMPLEX AND DIVERSE GROUP OF RETINAL DISEASES

Early clinical studies of PRA in different breeds indicated a broad age range for the clinical manifestations of the disease. Even though all affected dogs showed the same ophthalmoscopic and clinical abnormalities that were used to establish early diagnosis, the age at which these abnormalities were evident was variable, and, for the most part, breed-specific (Table 2). This suggested early on that PRA could be the result of the same gene defect in all breeds, and that the observed variability resulted from differing "genetic background" that was breed-specific. Alternatively, the differences could result from having mutations at multiple gene loci causing retinal diseases that were phenotypically similar or even indistinguishable. Research in our laboratory over the past 30 years has shown that both alternatives occur in PRA.

PRA: Early-Onset Diseases

The early-onset group of photoreceptor diseases is expressed in dogs between 2 and 6 weeks of age, the period of postnatal differentiation of the retina. In these diseases, the rod and cone photoreceptors fail to develop normally and subsequently show extensive disease, degeneration,

Table 2. Progressive retinal atrophy; comparison of age of diagnosis using ophthalmoscopy and electroretinography in selected breeds

Breed	Disease locus	Age of diagnosis	
		ophthalmoscopy	ERG
Irish setter	rcd 1	16 wks	6 wks
Collie	rcd 2	16 wks	6 wks
Norwegian elkhound	erd	6–8 ms	35 days
Norwegian elkhound	rd	1–1.5 yrs	6 wks
Miniature schnauzer	pd	>1.5 yrs	6 wks
Siberian husky	XLPRA1	1.5–2 yrs	1 yr
English mastiff	Rho	2–5 yrs	10–11 ms
Poodle: toy, miniature	prcd	3–5 yrs	9 ms
American cocker spaniel	prcd	3–5 yrs	9 ms
Portuguese water dog	prcd	3–5 yrs	1.5 yrs
Labrador retriever	prcd	4–6 yrs	1.5 yrs
English cocker spaniel	prcd	8–12 yrs	>2.5 yrs

and cell death. This is readily detectable by electrophysiological and morphological methods, the latter using aldehyde/osmium fixation and plastic embedding to preserve the structural integrity of the very delicate photoreceptor cells (Aguirre 1978; Aguirre et al. 1982a; Acland and Aguirre 1987; Parshall et al. 1991). Other than night blindness and ERG abnormalities that are present early in affected breeds with early-onset PRA, most dogs do not show ophthalmoscopic abnormalities until there is extensive loss of retinal neurons. Because the photoreceptor cells develop abnormally, these diseases are termed photoreceptor dysplasias, and the target cell(s) is used in naming the disease and establishing the gene locus designation (e.g., rod dysplasia, rod-cone dysplasia 1).

PRA: Late-Onset Diseases

The late-onset diseases are degenerations of photoreceptor cells that developed normally; as such, they represent defects in pathways critical for the long-term maintenance of normal photoreceptor function and viability. Three genetically distinct disease classes are recognized now in this category: the autosomal recessive group that represent mutations at the progressive rod-cone degeneration (prcd) locus (Aguirre and Acland 1988), X-linked recessive retinal degeneration found in the Siberian husky and Samoyed breeds (XLPRA1; Acland et al. 1994; Zeiss et al. 1999, 2000; Zhang et al. 2002), and autosomal dominant PRA caused by a Rho mutation (Kijas et al. 2002).

ALLELIC AND GENETIC HETEROGENEITY IN PRA AND THE EFFECT OF GENETIC AND ENVIRONMENTAL MODIFIERS

Non-allelism of Early-Onset Diseases

The structural and functional methods described in the previous section were very specific in phenotypically distinguishing the different forms of early-onset PRA. However, this specificity had limitations, and, prior to the time when the specific genes and disease-causing mutations were identified, we could not distinguish whether the different phenotypes resulted from different mutations of the same gene, i.e., allelic heterogeneity. Since the diseases were autosomal recessive, and the obligate carriers showed no abnormalities, it was possible to address the issue of genetic heterogeneity by carrying out interbreed crosses between dogs of different breeds affected with PRA. Figure 2 shows an example of one such set of matings where dogs informative for the *rcd1*, *rcd2*, and *erd* forms of PRA were intercrossed. The production of normal progeny confirms that these diseases are nonallelic (Acland et al. 1989; Wang et al. 1999a).

Allelic Forms of *prcd*

We also have addressed the issue of genetic heterogeneity in *prcd*, the late-onset form of PRA, by performing interbreed crosses of affected dogs. Initially, we crossed PRA-affected English and American cocker spaniels with miniature poodles whose form of PRA had been characterized as *prcd* (Aguirre et al. 1982a). In all cases, the progeny that survived to the age of diagnosis were affected with the same retinal degenerative process, an indication that the diseases were allelic (Aguirre and Acland 1988). We have extended this study by using other breeds of dogs (Labrador retriever, Australian cattle dog, basenji, Border collie, Italian greyhound, Nova Scotia duck tolling retriever, Portuguese water dog) affected with PRA. The results for Labrador retriever, Australian cattle dog, Nova Scotia duck tolling retriever, and Portuguese water dog were the same as for the English and American cocker spaniels, and established the allelic nature of *prcd* in these breeds (Pearce-Kelling et al. 2002). In contrast, normal progeny was produced when PRA-affected dogs of the Border collie, basenji and Italian greyhound breeds were mated to *prcd*-affected dogs. This indicates that mutations in one or more different genes are responsible for these other late-onset inherited retinal degenerations. Figure 3 illustrates the strategy used with the Labrador retriever. A dog affected with PRA was outcrossed to normal beagles, and the F_1 generation was backcrossed to the affected Labrador

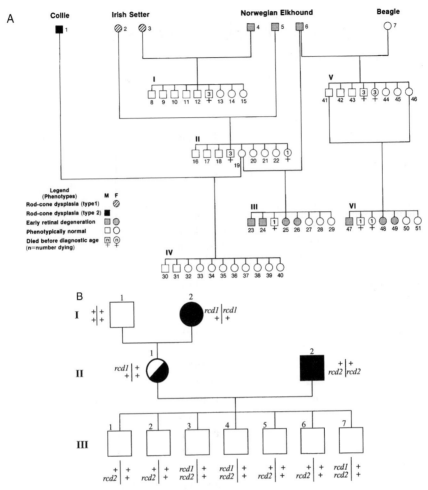

Figure 2. (**A**) Pedigrees used to illustrate nonallelism between *rcd1*, *rcd2*, and *erd*, respectively, in Irish setters, collie, and Norwegian elkhounds. Dogs were homozygous affected for the breed-specific form of PRA. (**B**) To confirm that *rcd2* and *rcd1* are nonallelic, a homozygous affected *rcd2* male was bred to a heterozygous *rcd1* female, and the progeny were tested for the *rcd1* mutation in *PDE6B*. Dogs 3, 4, and 7 were phenotypically normal and heterozygous for both *rcd1* and *rcd2*, thus confirming nonallelism of two disorders. (Reprinted, with permission, from Acland et al. 1989; Wang et al. 1999a [©Elsevier].)

to confirm that the disease in this breed is autosomal recessive (Fig. 3). When mated to a *prcd*-affected English cocker spaniel–miniature poodle crossbred, the resultant progeny were all affected with *prcd* form of PRA, an indication that the diseases were allelic and represented a mutation at the same gene locus.

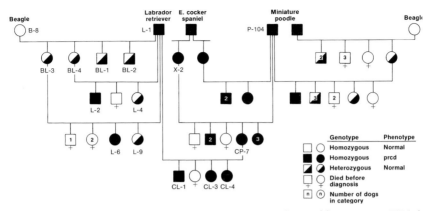

Figure 3. Pedigree used to illustrate that mutation at the *prcd* locus causes PRA in Labrador retriever, English cocker spaniel, and miniature poodle. Dogs were homozygous affected for the breed-specific form of PRA.

Nonallelism of Early- and Late-Onset Diseases

Three crossbreeding studies have been carried out to determine whether early- and late-onset diseases represent mutations at different gene loci. The first studies in dogs with XLPRA are discussed in the following section (allelic heterogeneity). In the second study, we bred a *prcd*-affected miniature poodle to a Norwegian elkhound affected with rod dysplasia (*rd*), a unique disorder of this breed that no longer appears to be present in the population (Aguirre 1976, 1978; Acland and Aguirre 1987; Aguirre and Acland 1988). In the third instance, *prcd*-affected dogs were bred to *erd*-affected Norwegian elkhound crossbreeds (Ray et al. 1996). In both situations, all the progeny resulting from the crosses to *prcd*-affected dogs were normal by ERG and morphology when examined at an age beyond which it is possible to make the earliest diagnosis for each of the disorders. We can conclude, therefore, that these early- and late-onset diseases are nonallelic.

Allelic Heterogeneity in X-linked PRA

The X-linked form of PRA shows marked allelic heterogeneity that is mutation-specific (Zhang et al. 2002). XLPRA2 is found in mongrel-derived dogs and is an early-onset developmental disease resulting from a 2-nucleotide deletion of *RPGR ORF15*. The deletion causes a frameshift that replaces many acidic glutamic acid residues with basic arginine residues. This change causes a toxic gain of function that severely compromises the photoreceptors in the early stages of development. In contrast, the

5-nucleotide deletion in *RPGR ORF15* present in XLPRA1-affected Siberian huskies and Samoyeds results in a premature stop, and truncation of the carboxyl terminus of the protein. The mutation affects only the photoreceptors, and these degenerate in young adults after developing and functioning normally (Zeiss et al. 1999). The severity of the retinal disease phenotype by mutation is the same in hemizygous males as homozygous females. On the other hand, females that are compound heterozygote for the XLPRA1 and XLPRA2 alleles show a more severe disease than is present in either of the two diseases, suggesting that when the XLPRA2 allele is expressed in the already compromised retina of a compound heterozygote the disease is very severe (Zhang et al. 2002).

Genetic and Environmental Modifiers in PRA

In the XLPRA1-affected Siberian husky-derived dogs we have found a variation in the severity of the retinal disease phenotype that presumably results from one or more modifying gene loci. To map the disease, we initially created outbred pedigrees in which the mutant X chromosome was derived from one purebred affected male dog (Zeiss et al. 1999, 2000; Zhang et al. 2002). Despite having the same single mutant X chromosome present in all XLPRA1-affected males that make this highly outbred colony, there is variability in onset and severity of the retinal phenotype (Zeiss et al. 1999). As the goal of the mapping study was to produce a sufficient number of informative dogs, we selected for breeding dogs whose disease phenotype could be readily identified at 6 months of age, or if carrier females were used, would produce more severely affected males. Despite this selection for a more severe disease phenotype, we produced sibships with distinct phenotypes (*mild, moderate, severe*) and found that carrier females could produce progeny that have *mild, moderate,* or *severe* phenotypes in the same or different litters. Small numbers of matings between severely affected males and severely affected females produced progeny with the severe phenotype; breeding between moderately affected males and severely affected females produced both moderate and severe phenotypes in the progeny. Because every animal in the colony has the one mutant X chromosome contributed by the founder male, a dog that had a very severe disease phenotype, and all progeny have the same stable microdeletion, these findings indicate that the variation in disease phenotype does not result from allelic heterogeneity at the XLPRA1 locus. The phenotype modifier suggests a single-locus, autosomal semi-dominant trait, but this needs to be validated experimentally.

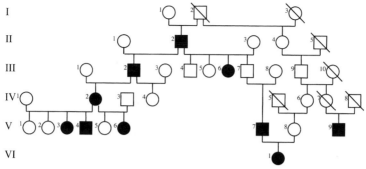

Figure 4. A six-generation subset pedigree of animals from the English mastiff pet population. Affected animals were diagnosed by clinical examination. Animals unavailable for analysis are indicated using a diagonal line. On the left side of the pedigree, the disease behaves as autosomal dominant; on the right side of the pedigree, the disease behaves as autosomal recessive. Molecular analysis confirmed the clinical status of all dogs except for male III-7 that had the *Rho* mutation but was clinically normal at the time of examination. (Reprinted, with permission, from Kijas et al. 2003 [©American Genetic Association].)

A similar variation in disease phenotype is present in the English mastiff breed with a *Rho* mutation (Kijas et al. 2002, 2003). Initial examination of the pedigree available from the pet population could not resolve whether the disease was autosomal dominant or recessive (Fig. 4), and there was a wide variability in disease onset and severity within the population. To address the question of mode of inheritance, we outcrossed an affected female dog to a genetically normal male beagle (Kijas et al. 2002). Once affected male and female progeny were produced in that outcross, autosomal dominant inheritance was established. Recent studies have indicated that the modifier is environmental rather than genetic, and is determined by the intensity and duration of light exposure (A. Cideciyan et al., in prep.).

A dramatic variation in disease phenotype is also present in *prcd* (Aguirre and Acland 1988). In some breeds, e.g., miniature and toy poodles and Portuguese water dogs, affected dogs are diagnosed by clinical examination between 3 and 5 years of age. In Labrador retrievers, diagnosis is made between 4 and 6 years, and in English cocker spaniels diagnosis is usually between 8 and 12 years of age (Table 2). In contrast, breeds like American Eskimo, Nova Scotia duck tolling retriever, and Australian cattle dogs have a less uniform age of onset, and the age of diagnosis varies greatly within each breed.

Until now, we could not establish whether the phenotypic differences were allelic in origin, i.e., different mutations at the same gene locus, or influenced by genetic modifiers. Unpublished information from our laboratory has now shown that all *prcd*-affected dogs regardless of breed have a common disease haplotype in the linkage disequilibrium region (Goldstein et al. 2004). This led to the positional cloning and identification of the gene and disease-causing mutation (B. Zangerl et al., unpubl.). We have found that all the affected dogs tested have the same mutation, even though the severity of the retinal disease phenotype is variable. Because most of the dogs under study are maintained in the same laboratory animal facility, exposed to same light level, and fed the same diet, these results suggest that genetic rather than environmental modifiers determine the phenotypic severity. These are under investigation in our laboratory.

SEARCHING FOR GENES AND MUTATIONS THAT CAUSE RETINAL DEGENERATION IN DOGS

Since the causative mutation responsible for *rcd1* in Irish setters was identified in 1993, the genes and disease-causing mutations have been identified for 11 diseases; publications describe 7 of these mutations, and the remaining are not yet published (Table 1). This work has been done using candidate gene and/or linkage mapping and positional cloning strategies. Although this is a significant accomplishment given the paucity of resources needed for gene discovery in dogs, at least until recently, it is but a small subset of the number of inherited retinal diseases that exist in dogs. If we extrapolate from the medical ophthalmology field, there are at least 150 mapped loci where genes important for retinal function are located (Fig. 1). Although mutations in some of these genes may not be associated with a recognizable phenotype in dogs, e.g., mutations in the cone opsin genes that cause color vision deficits in humans without affecting visual acuity (Nathans et al. 1986a,b), mutations in the majority of these genes would be expected to be associated with a recognizable phenotype. Coupled with the large number of breeds that are known to have retinal disorders which are presumed to be inherited (ACVO 1999), this is indeed a fertile field for continued molecular genetic studies. In the following sections, we review the progress made using the gene-based and linkage mapping and positional cloning strategies.

CANDIDATE OR GENE-BASED STUDIES

Early studies in dogs utilized this approach because of the lack of resources to do linkage mapping and positional cloning. Candidate genes were selected for analysis based on the known role of the gene product

in photoreceptor structure or function, a priori knowledge of the biochemical pathway or defect involved, or information from other species,
particularly humans and mice, that such gene defects were associated
with inherited retinopathies. In addition, novel candidates were identified by subtraction techniques between normal and mutant retinas that
identified genes which were under- or overexpressed, depending on the
"direction" of the subtraction. Such subtraction techniques were successfully used in mice to identify the *rds* (Travis and Sutcliffe 1988; Travis
et al. 1989) and *rd* (Bowes et al. 1989, 1990) disorders, and were subsequently applied in dogs (Zhang et al. 1998; Zangerl et al. 2002b). Success
in mice, however, was primarily dependent on prior information of the
chromosomal location of the retinal disease locus, and the availability of
somatic cell hybrid lines or RH panels that could identify the chromosomal
location of the clones isolated by subtraction methods. Until recently,
comparable resources were not available for use in dogs. Figure 5 provides

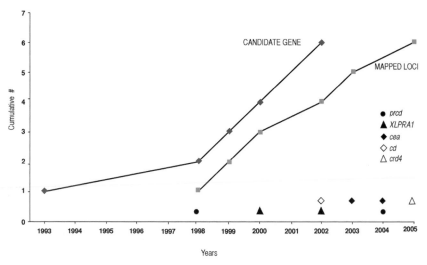

Figure 5. Summary of the cumulative number of retinal disease genes found by the
candidate gene and linkage/positional cloning approaches. For candidate genes, the
following diseases are included by the year when the gene and disease-causing
mutation were identified, or a DNA-based test became available. *rcd1*, 1993; *csnb*, 1998;
rcd3, 1999; *rcd1^a*, 2000; Type A PRA, 2002; *Rho*, 2002. The following diseases are
included by the year when they were mapped. *prcd*,1998; *erd*, 1999; *XLPRA1*, 2000; *cd*,
2002; *cea*, 2003; *crd4*, 2005. The lower part of the figure illustrates those diseases that
were first mapped, and the gene found subsequently (*prcd, XLPRA1, cea*); pairs of
solid symbols indicate the two time points. The open symbols refer to the diseases in
which the mapping and mutation identification were reported concurrently (*cd, crd4*).

a summary of the cumulative number of retinal disease genes found by the candidate gene approach.

Success in Finding the *rcd1* Gene Defect

The success in identifying the *PED6B* defect in *rcd1* is the best example of a directed candidate gene approach in dogs. The disease had been well characterized in terms of inheritance, function, and pathology (Parry 1953a; Aguirre and Rubin 1975). A biochemical defect had been identified in 1978 which pointed to defective activity of retinal cyclic guanosine monophosphate phosphodiesterase (cGMP-PDE) (Aguirre et al. 1978, 1982b). Defective enzyme activity resulted in elevation of retinal cGMP to levels that were 10-fold higher than normal and were causally associated with the abnormal development and subsequent rapid degeneration of the visual cells (Fig. 6) (Lolley et al. 1977; Chader 1991). The disease in dogs was phenotypically and biochemically very similar to the corresponding disease in the *rd* mouse strain (Farber and Lolley 1974). Because the PDE complex is a heterotetramer consisting of catalytically active α and β subunits and two inhibitory γ subunits, the search was directed to the genes coding for these proteins (*PDE6A, PDE6B, PDE6G*).

Once the *PDE6B* gene was found to be mutated in the *rd* mouse (Bowes et al. 1990), a candidate gene strategy was used to establish

PATTERN OF CYCLIC GMP DEVELOPMENT IN CONTROL AND AFFECTED RETINAS

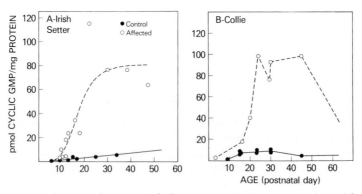

Figure 6. Developmental pattern of changes in cGMP concentration with age in retinas from *rcd1* (Irish setter) and *rcd2* (collie) affected dogs. Normal values (*solid circles*) from controls are included in each figure. In both *rcd1* and *rcd2*, levels of cGMP increase early in the postnatal period to values eight- to tenfold higher than normal. (Modified, with permission, from Chader et al. 1988.)

whether a similar defect was present in *rcd1*. This approach was strengthened by finding that, as in the *rd* mouse, the mRNA levels for *PDE6B* were selectively decreased in the early stages of photoreceptor development and prior to the onset of degeneration (Farber et al. 1992). Soon after, a stop mutation resulting from a G to A transition in codon 807 of exon 21 of *PDE6B* was found in both the pet and research dog populations (Clements et al. 1993; Suber et al. 1993; Ray et al. 1994). This is the only mutation so far recognized in early-onset PRA in Irish setters, and it has a carrier frequency of approximately 7% in the U.S., but much less than 1% in the U.K. (Petersen-Jones et al. 1995; Aguirre et al. 1999).

Failure in Finding the *rcd2* Gene Defect

Like Irish setters with *rcd1*, collies have an early-onset inherited retinal degeneration that is phenotypically identical and has been designated as *rcd2*. Using several criteria for comparison, the diseases are identical (see Wolf et al. 1978; Santos-Anderson et al. 1980; Woodford et al. 1982; Chader 1987, 1991). Both diseases showed a dramatic and early increase in retinal cGMP levels (Fig. 6), and the only identified difference between the two disorders was the dependency on calmodulin for PDE activation; PDE was calmodulin-independent in *rcd2*, but calmodulin-dependent in *rcd1* (Liu et al. 1979; Woodford et al. 1982).

PDE6B was presumptively ruled out by the interbreed crosses discussed above (Nonallelism of Early-Onset Diseases), and the strategy illustrated in Figure 2a. Because the crosses involved matings between PRA-affected collies, and phenotypically normal dogs that were heterozygous for both *erd* and *rcd1*, there was a low probability that none of the progeny in this mating inherited the mutant *rcd1* allele (Acland et al. 1989). The study subsequently was repeated, and the *rcd1* genotype status of all dogs was confirmed by mutation testing. We found that dogs that were compound heterozygotes for both *rcd1* and *rcd2* were normal, thus ruling out *PDE6B* from causal association with the disease (Wang et al. 1999a).

Having ruled out one of the PDE subunits, we then directed our attention to the remaining subunits, as these would be likely candidates for the disease. In addition to the better-known *PDE6A* and *PDE6G*, we examined the recently identified *PDE6D* (Wang et al. 1999b). Using a combination of approaches, all three of these subunits were excluded from causal association (Wang et al. 1999a). Additionally, several other photoreceptor-specific genes were ruled out as candidates for *rcd2*. The methods used for analysis varied, but the results were the same in that there was no association between the gene and the disorder. The genes examined and methods were: peripherin/*rds*, rod β-transducin, and arrestin by heteroduplex analysis; rod α-transducin and ROM-1 by linkage analysis; opsin and rod γ-transducin by identity by descent (Wang 1999).

Some Success Stories, but Too Many Failures

A second mutation in exon 21 of *PDE6B* was found in the Sloughi breed (Dekomien et al. 2000), and one in *PDE6A* in the Cardigan Welsh corgi (Petersen-Jones et al. 1999). In the latter study, screening of an intron 8 *PDE6A* polymorphism was done in a large informative pedigree, and linkage of this polymorphism with the disease directed sequence analysis and discovery of 1-bp deletion in codon 616. The T4R mutation in *Rho* responsible for autosomal dominant PRA was found using a candidate gene approach (Kijas et al. 2002). Once test breedings confirmed autosomal dominant inheritance, a list of potential candidates was made and testing was started while awaiting development of informative pedigrees useful for linkage analysis or genome scans. *Rho* was the second gene on the list, and a mutation was found in the first exon.

The difficulties with candidate gene analysis are clearly illustrated with the *csnb* disorder in briards which is caused by a mutation in *RPE65* (Aguirre et al. 1998; Veske et al. 1999). Andreas Gal's group carried out exhaustive analysis of several candidate genes that were photoreceptor specific; e.g., arrestin, E isoform of retinal guanylate cyclase (*cGC-E*), rod photoreceptor cGMP-gated cation channel α-subunit gene, *PDE6A*, *Rho*, peripherin/*rds*, *ROM1*, *PDE6B*, but all gave negative results (Veske et al. 1997a,b,c, 1998). Only when the *RPE65* gene was examined was a mutation identified (Veske et al. 1999). Once this gene and mutation were known, it was perfectly clear that the salient abnormalities of the disease, i.e., profound blindness, absence of retinal function as measured by ERG testing, structurally intact retina, and accumulation of lipoidal inclusions in the RPE, could only be explained by a defect in retinoid processing in the RPE. In retrospect, the search should have focused on the smaller number of potential candidate genes involved in the visual cycle (Aguirre et al. 1998; Redmond et al. 1998).

One of the problems with the candidate gene approach is that a priori selection of the likely causative gene is often based on insufficient or incomplete knowledge of the disease and/or of the putative function of the gene in the retina. A classic example of this is in the identification of the pre-mRNA splicing gene *PRPF31* that, when mutated, causes a form of RP in man termed RP11, a disease with a dominant mode of inheritance with incomplete penetrance (Vithana et al. 2001). The disease had been mapped to HSA19q13.4, and a BAC PAC contig made of the critical region. Because it would be *highly unlikely* that a mutation in a pre-mRNA splicing factor gene could cause a selective photoreceptor degeneration, *PRPF31* was the last gene in the interval examined. Subsequent elegant studies have explained the mechanism of disease and the process of incomplete penetrance in RP11-affected families (Vithana et al. 2003); since then, two other pre-mRNA splicing factor genes (*PRPC8*, *HPRP3*) have been shown to be causally associated with ADRP (McKie et al. 2001; Chakarova et al. 2002). These two genes, as well as *PRPF31*, are ubiquitously expressed, but disease is limited to the retinal photoreceptors.

An additional complication for candidate gene studies in dogs is that, with few exceptions, most of the eye diseases tend to be breed-specific and lack a common founder that is represented in more than one breed, especially if the breeds are unrelated. Both the *Rho* mutation causing ADPRA, and the *RPGR ORF15* stop mutation causing XLPRA, are present in related breeds: English mastiff and bullmastiff for *Rho*, and Siberian husky and Samoyed for *RPGR*. The one exception is *prcd*, where a common mutation is present in at least 15 different breeds of dogs and is likely to represent an ancient mutation in a common founder before these breeds became isolated breeding populations. Thus, searching for the gene responsible for a specific retinal disease in a breed using a candidate gene approach is likely to come up with negative results, as the number of potential genes, although finite, is very large.

The approach for candidate gene studies in canine retinal diseases is to select a candidate gene and test it in a diverse population of affected and carrier dogs of different breeds in the hopes of identifying mutations in the tested genes in one or more breeds/diseases. Using this strategy, many breeds have been tested for mutations in different retinal disease genes. Some examples are *RPE65* (Dekomien and Epplen 2003b), PDE6A (Dekomien and Epplen 2000), PDE6D and *PDE6G* (Dekomien and Epplen 2003a), *PDE6B* (Aguirre et al. 1999; Dekomien et al. 2000), arrestin (Dekomien and Epplen 2002c), peripehrin/*rds* (Ray et al. 1996; Runte et al. 2000), *ROM1* (Gould et al. 1997; Klein et al. 1998), phosducin (Lin et al. 1998; Zhang et al. 1998; Dekomien and Epplen 2002a), opsin (Gould et al. 1995; Ray et al. 1999), and recoverin (Dekomien and Epplen 2002b). Despite this considerable effort, and the testing of multiple breeds, only one disease-associated mutation has been found, and this was in the Sloughi breed (Dekomien et al. 2000). Such low return for such considerable effort should raise concerns about the applicability and effectiveness of the candidate gene approach in the context of these diseases.

LINKAGE MAPPING AND POSITIONAL CLONING STRATEGIES

Linkage mapping and positional cloning bypass the limitations of the candidate gene studies in that there is no a priori selection of the candidate gene(s) to be tested, and the approach relies on prioritizing the selection of genes located in the mapped interval. Obviously, the density of genes in this interval, their tissue expression, and their putative retinal function guide the prioritization. Increasing the number of informative individuals and identification of recombinants can markedly reduce the searched interval and facilitate progress. When the disease of interest is in a region of recombination suppression, alternative strategies, e.g., LD mapping, are needed to reduce the critical disease interval. The success of this approach is evident by the large number of disease loci mapped to date, and the identification of disease genes and mutations for most of them (Fig. 5).

Now that adequate resources for genomic studies in dogs are available, these approaches in dogs are readily doable provided sufficiently large informative pedigrees are available. For most of the eye diseases of interest to our laboratory, sufficiently large, informative pedigrees are not available in the general population. The reasons for this vary, and include low heterozygosity for tested markers secondary to high inbreeding coefficient, low frequency of the disease in the population, difficulties in phenotype ascertainment because diagnostic criteria are not understood by the examining ophthalmologists, phenocopy issues, and lack of cooperation among different dog owners or breeders to bring critical dogs for examination, resulting in gaps in the test pedigrees. Because most of the retinal disorders under study are recessive, and most of the matings are between non-affected carrier animals, the pedigrees generally lack sufficient power as the genotypes of most of the non-affected dogs produced cannot be established. Although some of these issues are manageable, it is often easier to create disease informative pedigrees which are then examined by the same investigators using very rigorous and well-defined criteria; this increases the pedigree power and eliminates ascertainment bias. Moreover, as these pedigrees are outcrossed, they have a high degree of heterozygosity, and most of the markers tested are informative.

To map and positionally clone retinal disease genes of interest, our lab has produced several outcrossed pedigrees which have been used for genome-wide scans to determine the map locations of these loci. The diseases include *prcd* (Acland et al. 1998), *XLPRA* (Zeiss et al. 2000; Zhang et al. 2001, 2002), *erd* (Acland et al. 1999; Kukekova et al. 2003), *cd* (Sidjanin et al. 2002b), and *cea* (Lowe et al. 2003). Mapping pedigrees have also been developed for *rcd2* and *osd1* and *osd2*, and mapping of the *rcd2* locus is completed (A.V. Kukekova et al., unpublished). However, it is not essential to develop an outcross pedigree for each mapping study. In the case of *prcd* in the American Eskimo breed, investigators used samples from 66 dogs, 53 normal and 13 affected, and used homozygosity mapping to detect linkage between two markers on CFA9, and the *prcd* locus (see below). Even though some dogs used in the analysis were related, linkage was still observed with one marker using only those dogs known to be related by at least three generations (K. Murphy, pers. comm.). The following sections summarize this work.

Mapping the *prcd* Locus

In parallel with developing the first meiotic linkage map of the dog (Mellersh et al. 1997), we undertook the mapping of *prcd*, the first autosomal disease locus in dogs assigned to a canine chromosome (Acland et al. 1998). The disease was mapped to the centromeric end of CFA9, homologous to HSA17q, at that time the best-characterized canine autosome because it contained *BRCA1* and candidate genes for congenital heart disease (Fig. 7) (Werner et al. 1997, 1998, 1999a). For this work, we used nine related three-generation families that had 70 *prcd*-informative progeny and found tight linkage (LOD score of 10.8) with zero recombinations with *TK1* and *GALK1*. The nearest recombinant to the disease locus was *MYL4*, which represented a large genomic interval. Progress during the subsequent years was directed to reducing the disease interval by finding recombinant animals on the telomeric and centromeric ends of the disease locus, and identifying and testing putative positional candidate genes. These candidate genes and markers included a linked RAPD marker (Gu et al. 1998), *APOH* (Gu et al. 1999), and two of the tubulin family genes that mapped to the homologous region in HSA17q, *TUBG1* and *TUBD* (Sidjanin et al. 2001, 2002a). These and other candidate genes from the region did not show any disease-associated sequence changes.

Because of the conservation in gene content and, presumably, gene order between CFA9 and HSA17q, we identified polymorphisms in genes and markers that flanked the disease locus, and tested these in the outcross mapping pedigree, and in a purebred Nova Scotia duck tolling retriever pedigree. Centromeric and telomeric recombinants, respectively, were found, thus reducing the critical disease interval. The whole-genome radiation hybrid panels that became available did not significantly affect progress as both the Research Genetics (Huntsville, AL) $RH08_{3000}$

canine–hamster panel and the Rennes 5000 rad panel (Priat et al. 1998) were *TK1* selected; as *TK1* is tightly linked to the *prcd* locus, it was difficult to order positional candidates in the critical disease interval. Despite this, we could order several candidate genes to the *prcd* region by RH mapping (Fig. 7) (Sidjanin et al. 2003).

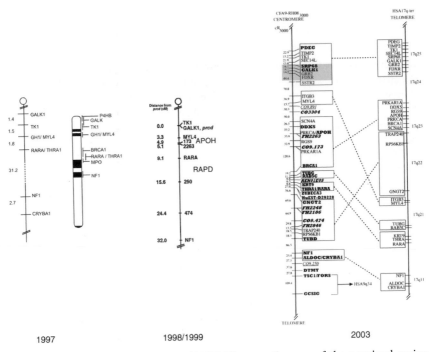

1997   1998/1999   2003

Figure 7. Evolution of CFA9 map. (**1997**) The genetic map of the proximal region of CFA9 gives the distances between loci in cM. The Giemsa-banded idiogram of CFA9 shows the FISH localization of 10 loci. (**1998/1999**) Linkage map of the *prcd* interval. The location is shown relative to that of nine loci corresponding to type I and II markers. Order and distances shown are based on 2-point linkage analyses. The relative positions of a RAPD marker (Gu et al. 1998) and *APOH* (Gu et al. 1999) are added based on subsequent studies in 1998 and 1999, respectively. (**2003**) RH map of CFA9-RH08$_{3000}$ (*left*) and HSA17 (*right*). The gene order for all genes from HSA17, except *PDE6G*, was generated from http://genome.ucsc.edu/. The location for human *PDE6G* was assigned from GB4 RH map (http://www.ncbi.nlm. nih.gov/genemap/map.cgi?MAP=GB4&BIN=536&MARK=sts-X62025). The boxed genes represent clusters of conserved synteny identified with HSA17q-ter, and one cluster on the telomeric end of CFA9 syntenic with HSA9q34. The shaded area represents genes (*SRP68-GALK1-GRB2-FDXR*) that were identified in the same order on the BAC physical map. (*1997* [*left*], Modified, with permission, from Werner et al. 1997 [©Elsevier]; *1998/1999* [*middle*] from Acland et al. 1998 [©National Academy of Sciences]; *2003* [*right*] from Sidjanin et al. 2003 [©Elsevier].)

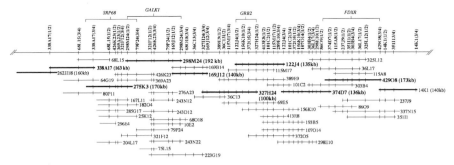

Figure 8. Schematic representation of a physical map of 50 BAC clones and 49 new STSs generated and assembled into a contig with an average depth of 6.4 BACs per STS marker. Clones that form a minimal tiling path across ~1.5 Mb are shown with a thicker line (not drawn to scale). Low-pass 3.2× sequence of ~1.2 Mb from eight BAC clones (*in bold*) from the minimal tiling path has been generated and analyzed. Brackets above the contig identify BACs where *SRP68*, *GALK1*, *GRB2*, *FDXR* were identified, confirming the gene order as initially established on the CFA9-RH08$_{3000}$ map (see Fig. 7). (Reprinted from Sidjanin et al. 2003 [©Elsevier].)

To overcome these limitations, we created a physical map of the area by selecting BACs containing *GRB2*, a gene in the zero recombination region, and BAC end-STSs were used to extend the contig (Fig. 8). Ten successive bidirectional walks were done, resulting in 50 BAC clones assembled in a contig. Of these, 10 BACs were selected to create a minimal tiling path of the *prcd minimal disease interval* that was sequenced to provide ~3.2× coverage (Sidjanin et al. 2003). This sequence information was used to identify SNPs in genes located both within and flanking the disease interval, and was used to generate an extended haplotype that was unique to the disease chromosome. Because the disease has now been shown to be present in many different breeds (Table 1), we took advantage of ancestral recombinations within genetically isolated populations to identify a disease-specific LD region. Beginning with a 1.9-Mb target region that contained ~50 genes, we carried out five successive steps to reduce the LD region (Fig. 9). The first step reduced the LD interval to 834 kb; by the fifth step, the interval was 87 kb and contained 3 genes (Goldstein et al. 2004). This led to the testing of the positional candidates and identification of the gene and disease-causing mutation (B. Zangerl et al., unpubl.).

Mapping Other Retinal Degeneration Loci, and Identification of Gene Defects

Five other retinal disease loci subsequently have been mapped: *erd* (Acland et al. 1999), *XLPRA1* (Zeiss et al. 2000; Zhang et al. 2001), *cd*

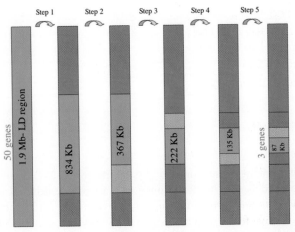

Figure 9. Changes in *prcd* LD. 18 non-affected chromosomes from 12 different breeds, and 34 affected chromosomes from 12 different breeds, were screened. 167 regions in the LD interval were analyzed, generating a total of 155 polymorphisms that generate a 96 SNP fine-scale affected haplotype common to all 12 affected breeds. This common haplotype reduces the LD region to an 87-Kb interval. (O. Goldstein et al., unpubl.)

(Sidjanin et al. 2002b), *cea* (Lowe et al. 2003), and *crd4* (C.S. Mellersh et al., pers. comm.). *XLPRA1*, like *prcd*, proved challenging because of the few canine-specific markers and genes on the X chromosome that had been characterized or mapped. Because of the conservation in genes and gene order between the human and canine X chromosomes (Zhang et al. 2001), and the development of additional microsatellite markers for the region where the disease locus was assumed to map, it was possible to fine-map the disease (Zhang et al. 2001; Zangerl et al. 2002a). This led to testing the most likely positional candidate, *RPGR*, and identification of two different disease-causing mutations (Zhang et al. 2002). For both *cd* and *cea*, genome-wide screens localized the disease locus to a specific chromosomal region, and the disease gene was identified (Sidjanin et al. 2002b; Lowe et al. 2003; E.A. Ostrander et al., unpubl.).

SUMMARY

Progress in veterinary molecular ophthalmology over the past 6–8 years has been dramatic, and a number of new retinal disease loci and genes have been identified. These diseases are truly disease homologs for the comparable human disorders. With the increased availability of genomic resources, the rate of progress over the coming years will be greatly increased. The dog retinal diseases are also developing as a model system to develop and test novel therapies applicable to man. Because of the similarities in disease phenotype and eye structure and function between dog and man, this species is ideally suited for testing potential gene therapy interventions (Acland et al. 2001). Thus, the transition from gene discovery to gene therapy will be readily possible.

ACKNOWLEDGMENTS

This work was made possible by the contribution of many organizations and individuals. In particular, the authors acknowledge NEI/NIH grants EY-01244, EY06855, EY13132, EY13729, The Foundation Fighting Blindness, The Morris Animal Foundation/The Seeing Eye Inc., The ONCE International Prize for Research & Development in Biomedicine and New Technologies for the Blind, and The Van Sloun Fund for Canine Genetic Research for their generous support. We are indebted to Sue Pearce-Kelling for her many contributions in all aspects of the research, and to our scientific colleagues—research scientists, postdoctoral fellows, graduate students, and research technicians—for their invaluable contributions. Drs. Aguirre and Acland are founding members and owners of *OptiGen LLC*, a company that carries out DNA testing for a large number of inherited eye diseases in dogs.

REFERENCES

Acland G.M. and Aguirre G.D. 1987. Retinal degenerations in the dog: IV. Early retinal degeneration (*erd*) in Norwegian elkhounds. *Exp. Eye Res.* **44:** 491–521.

———. 1995. Oculoskeletal dysplasias in Samoyed and Labrador retriever dogs: Nonallelic disorders akin to Stickler-like syndromes affecting humans. In *2nd International DOGMAP Meeting*, Cambridge, United Kingdom.

Acland G.M., Blanton S.H., Hershfield B., and Aguirre G.D. 1994. XLPRA: A canine retinal degeneration inherited as an X-linked trait. *Am. J. Med.Genet.* **52:** 27–33.

Acland G., Fletcher R., Gentleman S., Chader G., and Aguirre G. 1989. Non-allelism of three genes (*rcd1, rcd2, erd*) for early-onset hereditary retinal degeneration. *Exp. Eye Res.* **49:** 983–998.

Acland G.M., Ray K., Mellersh C.S., Langston A.A., Rine J., Ostrander E.A., and Aguirre G.D. 1999. A novel retinal degeneration locus identified by linkage and comparative mapping of canine early retinal degeneration. *Genomics* **59:** 134–142.

Acland G.M., Ray K., Mellersh C.S., Gu W., Langston A.A., Rine J., Ostrander E.A., and Aguirre G.D. 1998. Linkage analysis and comparative mapping of canine progressive rod-cone degeneration (*prcd*) establishes potential locus homology with retinitis pigmentosa (RP17) in humans. *Proc. Natl. Acad. Sci.* **95:** 3048–3053.

Acland G.M., Aguirre G.D., Ray J., Zhang Q., Aleman T.S., Cideciyan A.V., Pearce-Kelling S.E., Anand V., Zeng Y., Maguire A.M., et al. 2001. Gene therapy restores vision in a canine model of childhood blindness. *Nat. Genet.* **28:** 92–95.

ACVO (American College of Veterinary Ophthalmologists). 1999. *Ocular disorders presumed to be inherited in purebred dogs*, 3rd. edition. Canine Eye Registration Foundation, Purdue University, West Lafayette, Indiana.

Aguirre G.D. 1978. Retinal degenerations in the dog. I. Rod dysplasia. *Exp. Eye Res.* **26**: 233–253.

———. 1976. Inherited retinal degenerations in the dog. *Trans. Am. Acad. Ophthalmol. Otolaryngol.* **81**: 667–676.

———. 1995. Electroretinography—Are we misusing an excellent diagnostic tool? *Vet. Comp. Ophthalmol.* **5**: 2–5.

———. 2000. Correspondence. *Vet. Ophthalmol.* **3**: 49.

Aguirre G.D. and Acland G.M. 1988. Variation in retinal degeneration phenotype inherited at the *prcd* locus. *Exp. Eye Res.* **46**: 663.

Aguirre G.D. and Rubin L.F. 1975. Rod-cone dysplasia (Progressive Retinal Atrophy) in Irish setters. *J. Am. Vet. Med. Assoc.* **166**: 157–164.

Aguirre G., Alligood J., O'Brien P., and Buyumihci N. 1982a. Pathogenesis of progressive rod cone degeneration in miniature poodles. *Investig. Ophthalmol. Vis. Sci.* **23**: 610–630.

Aguirre G.D., Baldwin V., Weeks K.M., Acland G.M., and Ray K. 1999. Frequency of the codon 807 mutation in the cGMP phosphodiesterase beta-subunit gene in Irish setters and other dog breeds with hereditary retinal degeneration. *J. Hered.* **90**: 143–147.

Aguirre G.D., Lolley R., Farber D., Fletcher T., and Chader G.J. 1978. Rod-cone dysplasia in Irish setters: A defect in cyclic GMP metabolism in visual cells. *Science* **201**: 1133–1135.

Aguirre G.D., Baldwin V., Pearce-Kelling S., Narfstrom K., Ray K., and Acland G.M. 1998. Congenital stationary night blindness in the dog: Common mutation in the RPE65 gene indicates founder effect. *Mol. Vis.* **4**: 23–29.

Aguirre G., Farber D., Lolley R., O'Brien P., Alligood J., Fletcher R.T., and Chader G. 1982b. Retinal degeneration in the dog. III. Abnormal cyclic nucleotide metabolism in rod-cone dysplasia. *Exp. Eye Res.* **35**: 625–642.

Ashton N., Barnett K., and Sachs D. 1968. Retinal dysplasia in the Sealyham terrier. *J. Pathol. Bacteriol.* **96**: 269–272.

Barnett K. and Stades F. 1979. Collie eye anomaly in the Shetland sheepdog in the Netherlands. *J. Small Anim. Pract.* **20**: 321–329.

Bedford P.G.C. 1982. Collie eye anomaly in the border collie. *Vet. Rec.* **111**: 34–35.

Berson E.L. 1993. Retinitis pigmentosa: The Friedenwald lecture. *Investig. Ophthalmol. Vis. Sci.* **34**: 1659–1676.

Berson E., Gouras P., and Gunkel R. 1968. Rod responses in retinitis pigmentosa, dominantly inherited. *Arch. Ophthalmol.* **80**: 58–67.

Berson E., Gouras P., and Hoff M. 1969. Temporal aspects of the electroretinogram. *Arch. Ophthalmol.* **81**: 207–214.

Bowes C., Danciger M., Kozak C., and Farber D. 1989. Isolation of a candidate cDNA for the gene causing retinal degeneration in the *rd* mouse. *Proc. Natl. Acad. Sci.* **86**: 9722–9726.

Bowes C., Li T., Danciger M., Baxter L.C., Applebury M.L., and Farber D.B. 1990. Retinal degeneration in the *rd* mouse is caused by a defect in the β subunit of rod cGMP-phosphodiesterase. *Nature* **347**: 677–680.

Breen M., Jouquand S., Renier C., Mellersh C.S., Hitte C., Holmes N.G., Chéron A., Suter N., Vignaux F., Bristow A.E., et al. 2001. Chromosome-specific single-locus FISH probes allow anchorage of an 1800-marker integrated radiation-hybrid/linkage map of the domestic dog genome to all chromosomes. *Genome Res.* **11:** 1784–1795.

Carrig C., Sponenberg D., Schmidt G., and Tvedten H. 1988. Inheritance of associated ocular and skeletal dysplasia in Labrador retrievers. *J. Am. Vet. Med. Assoc.* **193:** 1269–1272.

Carrig C., MacMillan A., Brundage S., Pool R., and Morgan J. 1977. Retinal dysplasia associated with skeletal abnormalities in Labrador retrievers. *J. Am. Vet. Med. Assoc.* **170:** 49–57.

Chader G. 1987. Biochemical studies of retinal degeneration in animal models and in the human. *Adv. Biosci.* **62:** 303–322.

———. 1991. Animal mutants of hereditary retinal degeneration: General considerations and studies on defects in cyclic nucleotide metabolism. *Prog. Vet. Comp. Ophthalmol.* **1:** 109–126.

Chader G.J., Aguirre G.D., and Sanyal S. 1988. Studies on animal models of retinal degeneration. In *Retinal diseases—biomedical foundations and clinical management* (ed. M.O.M. Tso), pp. 80–99. Lippincott, Philadelphia.

Chakarova C.F., Hims M.M., Bolz H., Abu-Safieh L., Patel R.J., Papaioannou M.G., Inglehearn C.F., Keen T.J., Willis C., Moore A.T., et al. 2002. Mutations in *HPRP3*, a third member of pre-mRNA splicing factor genes, implicated in autosomal dominant retinitis pigmentosa. *Hum. Mol. Genet.* **11:** 87–92.

Cideciyan A.V. and Jacobson S.G. 1993. Negative electroretinograms in retinitis pigmentosa. *Investig. Ophthalmol. Vis. Sci.* **34:** 3253–3263.

Clements P.J.M., Gregory C.Y., Peterson-Jones S.M., Sargan D.R., and Bhattacharya S.S. 1993. Confirmation of the rod cGMP phosphodiesterase β subunit (PDEβ) nonsense mutation in affected rcd-1 Irish setters in the UK and development of a diagnostic test. *Curr. Eye Res.* **12:** 861–866.

Curtis R. and Barnett K.C. 1993. Progressive retinal atrophy in miniature longhaired dachshund dogs. *Br. Vet. J.* **149:** 71–85.

Daiger S.P. 2005. RetNet: Summaries of Genes Causing Retinal Diseases. The University of Texas Health Science Center, Houston [http://www.sph.uth.tmc.edu/RetNet/sum-dis.]

Dekomien G. and Epplen J.T. 2000. Exclusion of the PDE6A gene for generalised progressive retinal atrophy in 11 breeds of dog. *Anim. Genet.* **31:** 135–139.

———. 2002a. The canine *Phosducin* gene: Characterization of the exon-intron structure and exclusion as a candidate gene for generalized progressive retinal atrophy in 11 dog breeds. *Mol. Vis.* **8:** 138–142.

———. 2002b. The canine recoverin (*RCV1*) gene: A candidate gene for generalized progressive retinal atrophy. *Mol. Vis.* **8:** 436–441.

———. 2002c. Screening of the arrestin gene in dogs afflicted with generalized progressive retinal atrophy. *BMC Genet.* **3:** 12.

———. 2003a. Analysis of PDE6D and PDE6G genes for generalised progressive retinal atrophy (gPRA) mutations in dogs. *Genet. Sel. Evol.* **35:** 445–456.

———. 2003b. Evaluation of the canine RPE65 gene in affected dogs with generalized progressive retinal atrophy. *Mol. Vis.* **9:** 601–605.

Dekomien G., Runte M., Godde R., and Epplen J.T. 2000. Generalized progressive retinal atrophy of Sloughi dogs is due to an 8-bp insertion in exon 21 of the PDE6B gene. *Cytogenet. Cell Genet.* **90:** 261–267.

Dryja T.P. 1997. Gene-based approach to human gene-phenotype correlations. *Proc. Natl. Acad. Sci.* **94:** 12117–12121.

Dryja T., McGee T., Hahn L., Cowley G., Olsson J., Reichel E., Sandberg M., and Berson E. 1990. Mutations within the rhodopsin gene in patients with autosomal dominant retinitis pigmentosa. *N. Engl. J. Med.* **323:** 1302–1307.

Farber D. and Lolley R. 1974. Cyclic guanosine monophosphate: Elevation in degenerating photoreceptor cells of the C3H mouse retina. *Science* **186:** 449–451.

Farber D.B., Danciger J.S., and Aguirre G. 1992. The beta subunit of cyclic GMP phosphodiesterase mRNA is deficient in canine rod-cone dysplasia 1. *Neuron* **9:** 349–356.

Farrar G.J., McWilliam P., Bradley D.G., Kenna P., Lawler M., Sharp E.M., Humphries M.M., Eiberg H., Conneally P.M., Trofatter J.A., et al. 1990. Autosomal dominant retinitis pigmentosa: Linkage to rhodopsin and evidence for genetic heterogeneity. *Genomics* **8:** 35–40.

Gelatt K.N., ed. 1999. *Veterinary ophthalmology.* Lippincott Williams & Wilkins, Baltimore, Maryland.

Goldstein O., Nelson J.L., Kijas J.W., Sidjanin D.J., Acland G.M., and Aguirre G.D. 2004. A linkage disequilibrium map of the progressive rod cone degeneration interval. *Invest. Ophthalmol. Vis. Sci.* **45:** E4756 (Abstr.).

Gould D.J., Petersen-Jones S.M., Lin C.T., and Sargan D.R. 1997. Cloning of canine rom-1 and its investigation as a candidate gene for generalized progressive retinal atrophies in dogs. *Anim. Genet.* **28:** 391–396.

Gould D.J., Petersen-Jones S.M., Sohal A., Barnett K.C., and Sargan D.R. 1995. Investigation of the role of opsin gene polymorphism in generalized progressive retinal atrophies in dogs. *Anim. Genet.* **26:** 261–267.

Gu W., Acland G.M., Langston A.A., Ostrander E.A., Aguirre G.D., and Ray K. 1998. Identification of a RAPD marker linked to progressive rod-cone degeneration in dogs. *Mamm. Genome* **9:** 740–744.

Gu W., Ray K., Pearce-Kelling S., Baldwin V.J., Langston A.A., Ray J., Ostrander E.A., Acland G.M., and Aguirre G.D. 1999. Evaluation of the APOH gene as a positional candidate for prcd in dogs. *Investig. Ophthalmol. Vis. Sci.* **40:** 1229–1237.

Guyon R., Lorentzen T.D., Hitte C., Kim L., Cadieu E., Parker H.G., Quignon P., Lowe J.K., Renier C., Gelfenbeyn B., et al. 2003. A 1-Mb resolution radiation hybrid map of the canine genome. *Proc. Natl. Acad. Sci.* **100:** 5296–5301.

Hodgman S., Parry H., Rosbridge W., and Steel J. 1949. Progressive retinal atrophy in dogs 1. The disease in Irish setters (red). *Vet. Rec.* **61:** 185–189.

Jacobson S.G., Kemp C.M., Cideciyan A.V., Macke J.P., Sung C.-H., and Nathans J. 1994. Phenotypes of stop codon and splice site rhodopsin mutations causing retinitis pigmentosa. *Investig. Ophthalmol. Vis. Sci.* **35:** 2521–2534.

Kemp C., Jacobson S., and Faulkner D. 1988. Two types of visual dysfunction in autosomal dominant retinitis pigmentosa. *Investig. Ophthalmol. Vis. Sci.* **29:** 1235–1241.

Kijas J.W., Miller B.J., Pearce-Kelling S.E., Aguirre G.D., and Acland G.M. 2003. Canine models of ocular disease: Outcross breedings define a dominant disorder present in the English mastiff and bull mastiff dog breeds. *J. Hered.* **94:** 27–30.

Kijas J.W., Zangerl B., Miller B., Nelson J., Kirkness E.F., Aguirre G.D., and Acland G.M. 2004. Cloning of the canine ABCA4 gene and evaluation in canine cone-rod dystrophies and progressive retinal atrophies. *Mol. Vis.* **10:** 223–232.

Kijas J.W., Cideciyan A.V., Aleman T.S., Pianta M.J., Pearce-Kelling S.E., Miller B.J., Jacobson S.G., Aguirre G.D., and Acland G.M. 2002. Naturally occurring rhodopsin mutation in the dog causes retinal dysfunction and degeneration mimicking human dominant retinitis pigmentosa. *Proc. Natl. Acad. Sci.* **99:** 6328–6333.

Kirkness E.F., Bafna V., Halpern A.L., Levy S., Remington K., Rusch D.B., Delcher A.L., Pop M., Wang W., Fraser C.M., and Venter J.C. 2003. The dog genome: Survey sequencing and comparative analysis. *Science* **301:** 1898–1903.

Klein W., Dekomien G., Holmes N., and Epplen J.T. 1998. Evaluation of ROM1 as a candidate gene in generalised progressive retinal atrophy in dogs. *Anim. Genet.* **29:** 316–318.

Kock E. 1974. Retinal dysplasia: A comparative study in human beings and dogs. In *Ophthalmology.* Karolinska Institutet, Stockholm, Sweden.

Kukekova A.V., Aguirre G.D., and Acland G.M. 2003. Cloning and characterization of canine *SHARP1* and its evaluation as a positional candidate for canine early retinal degeneration (*erd*). *Gene* **312:** 335–343.

Langston A.A., Mellersh C.S., Neal C.L., Ray K., Acland G.M., Gibbs M., Aguirre G.D., Fournier R.E., and Ostrander E.A. 1997. Construction of a panel of canine-rodent hybrid cell lines for use in partitioning of the canine genome. *Genomics* **46:** 317–325.

Li R., Mignot E., Faraco J., Kadotani H., Cantanese J., Zhao B., Lin X., Hinton L., Ostrander E.A., Patterson D.F., and de Jong P.J. 1999. Construction and characterization of an eightfold redundant dog genomic bacterial artificial chromosome library. *Genomics* **58:** 9–17.

Lin C.T., Petersen-Jones S.M., and Sargan D.R. 1998. Isolation and investigation of canine phosducin as a candidate for canine generalized progressive retinal atrophies. *Exp. Eye Res.* **67:** 473–480.

Liu Y.P., Krishna G., Aguirre G., and Chader G.J. 1979. Involvement of cyclic GMP phosphodiesterase activator in an hereditary retinal degeneration. *Nature* **280:** 62–64.

Lolley R., Farber D., Rayborn M., and Hollyfield J. 1977. Cyclic GMP accumulation causes degeneration of photoreceptor cells: Simulation of an inherited disease. *Science* **196:** 664–666.

Long K.O. and Aguirre G.D. 1991. The cone matrix sheath in the normal and diseased retina: Cytochemical and biochemical studies of peanut agglutinin-binding proteins in cone and rod-cone degeneration. *Exp. Eye Res.* **52:** 699–713.

Lowe J.K., Kukekova A.V., Kirkness E.F., Langlois M.C., Aguirre G.D., Acland G.M., and Ostrander E.A. 2003. Linkage mapping of the primary disease locus for collie eye anomaly. *Genomics* **82:** 86–95.

Magnusson H. 1909. Om nattblindhet hos hund sasom foljd af släktskapfsafvel. (On night blindness in the dog following inbreeding). *Sven. Vet. Tidskr.* **14:** 462–466.

———. 1910. Retinitis pigmentosa och konsangvinitet hos hund. (Retinitis pigmentosa and consanguinity in the dog.). *Sven. Vet. Tidskr.* **15:** 378–380.

———. 1911. Über retinitis pigmentosa und Konsanguinität beim hunde. *Arch. Vergleichende Ophth* **2:** 147–163.

———. 1917. Noch ein fall von nachtblindheit beim hunde. *Graefe's Arch. Ophthalmol.* **93:** 404–411.

McKie A.B., McHale J.C., Keen T.J., Tarttelin E.E., Goliath R., van Lith-Verhoeven J.J.C., Greenberg J., Ramesar R.S., Hoyng C.B., Cremers F.P.M., et al. 2001. Mutations in the pre-mRNA splicing factor gene *PRPC8* in autosomal dominant retinitis pigmentosa (RP13). *Hum. Mol. Genet.* **10:** 1555–1562.

McWilliam P., Farrar G., Kenna P., Bradley D., Humphries M., Sharp E., McConnell D., Lawler M., Shells D., Ryan C., et al. 1989. Autosomal dominant retinitis pigmentosa (ADRP): Localisation of an ADRP gene to the long arm of chromsome 3. *Genomics* **5:** 619–622.

Mellersh C.S., Langston A.A., Acland G.M., Fleming M.A., Ray K., Wiegand N.A., Francisco L.V., Gibbs M., Aguirre G.D., and Ostrander E.A. 1997. A linkage map of the canine genome. *Genomics* **46:** 326–336.

Mellersh C.S., Hitte C., Richman M., Vignaux F., Priat C., Jouquand S., Werner P., André C., DeRose S., Patterson D.F., et al. 2000. An integrated linkage-radiation hybrid map of the canine genome. *Mamm. Genome* **11:** 120–130.

Meyers V.N., Jezyk P.F., Aguirre G.D., and Patterson D.F. 1983. Short-limbed dwarfism and ocular defects in the samoyed dog. *J. Am. Vet. Med. Assoc.* **183:** 975–979.

Narfström K. 1999. Retinal dystrophy or 'congenital stationary night blindness' in the Briard dog. *Vet. Ophthalmol.* **2:** 75–76.

Narfström K., Wrigstad A., and Nilsson S.E.G. 1989. The Briard dog: A new animal model of congenital stationary night blindness. *Br. J. Ophthalmol.* **73:** 750–756.

Narfström K., Wrigstad A., Ekesten B., and Nilsson S.E.G. 1994. Hereditary retinal dystrophy in the briard dog: Clinical and hereditary characteristics. *Vet. Comp. Ophthalmol.* **4:** 85–92.

Nathans J., Thomas D., and Hogness D.S. 1986a. Molecular genetics of human color vision: The genes encoding blue, green, and red pigments. *Science* **232:** 193–202.

Nathans J., Piantanida T.P., Eddy R.L., Shows T.B., and Hogness D.S. 1986b. Molecular genetics of inherited variation in human color vision. *Science* **232:** 203–210.

Neff M.W., Broman K.W., Mellersh C.S., Ray K., Acland G.M., Aguirre G.D., Ziegle J.S., Ostrander E.A., and Rine J. 1999. A second-generation genetic linkage map of the domestic dog, *Canis familiaris*. *Genetics* **151:** 803–820.

Nelson D. and MacMillan A. 1983. Multifocal retinal dysplasia in field trial Labrador retrievers. *J. Am. Anim. Hosp. Assoc.* **19.**

Parry H.B. 1953a. Degenerations of the dog retina. II. Generalized progressive atrophy of hereditary origin. *Br. J. Ophthalmol.* **37:** 487–502.

———. 1953b. Degenerations of the dog retina. III. Retinopathy secondary to glaucoma. *Br. J. Ophthalmol.* **37:** 670–679.

———. 1954. Degenerations of the dog retina. IV. Retinopathies associated with dog distemper-complex virus infections. *Br. J. Ophthalmol.* **38:** 295–309.

Parry H.B., Tansley K., and Thomson L.C. 1951. The electroretinogram in the normal dog. *Proc. Physiol. Soc.* **6.**

———. 1953. The electroretinogram of the dog. *J. Physiol.* **120:** 28–40.

———. 1955. Electroretinogram during development of hereditary retinal degeneration in the dog. *Br. J. Ophthalmol.* **39:** 349–352.

Parshall C., Wyman M., Nitroy S., Acland G., and Aguirre G. 1991. Photoreceptor dysplasia: An inherited progressive retinal atrophy of miniature schnauzer dogs. *Prog. Vet. Comp. Ophthalmol.* **1:** 187–203.

Pearce-Kelling S.E., Nickle A., Kijas J.W., Sidjanin D.J., Miller B.J., Aguirre G.D., and Acland G.M. 2002. Test matings confirm allelism of *prcd* across many dog breeds. *Invest. Ophthalmol. Vis. Sci.* **43:** E3673 (Abstr.).

Petersen-Jones S.M., Entz D.D., and Sargan D.R. 1999. cGMP phosphodiesterase-alpha mutation causes progressive retinal atrophy in the cardigan Welsh corgi dog. *Investig. Ophthalmol. Vis. Sci.* **40:** 1637–1644.

Petersen-Jones S.M., Clements P.J.M., Barnett K.C., and Sargan D.R. 1995. Incidence of the gene mutation causal for rod-cone dysplasia type-1 in Irish setters in the UK. *J. Small Anim. Pract.* **36:** 310–314.

Priat C., Hitte C., Vignaux F., Renier C., Jiang Z., Jouquand S., Chéron A., André C., and Galibert F. 1998. A whole-genome radiation hybrid map of the dog genome. *Genomics* **54:** 361–378.

Ray K., Acland G.M., and Aguirre G.D. 1996. Nonallelism of *erd* and *prcd* and exclusion of the canine RDS/peripherin gene as a candidate for both retinal degeneration loci. *Investig. Ophthalmol. Vis. Sci.* **37:** 783–794.

Ray K., Baldwin V.J., Acland G.M., Blanton S.H., and Aguirre G.D. 1994. Cosegregation of codon 807 mutation of the canine rod cGMP phosphodiesterase β gene and *rcd1*. *Investig. Ophthalmol. Vis. Sci.* **35:** 4291–4299.

Ray K., Wang W., Czarnecki J., Zhang Q., Acland G.M., and Aguirre G.D. 1999. Strategies for identification of mutations causing hereditary retinal diseases in dogs: Evaluation of opsin as a candidate gene. *J. Hered.* **90:** 133–137.

Redmond T.M., Yu S., Lee E., Bok D., Hamasaki D., Chen N., Goletz P., Ma J.-X., Crouch R.K., and Pfeifer K. 1998. RPE65 is necessary for production of 11-cis-vitamin A in the retinal visual cycle. *Nat. Genet.* **20:** 344–351.

Rivolta C., Sharon D., DeAngelis M.M., and Dryja T.P. 2002. Retinitis pigmentosa and allied diseases: Numerous diseases, genes, and inheritance patterns. *Hum. Mol. Genet.* **11:** 1219–1227.

Roberts S. 1969. The collie eye anomaly. *J. Am. Vet. Med. Assoc.* **155:** 859–878.

Rubin L. 1968. Heredity of retinal dysplasia in bedlington terriers. *J. Am. Vet. Med. Assoc.* **152:** 260–262.

Rubin L., Bourns T., and Lord L. 1967. Hemeralopia in dogs: Heredity of hemeralopia in Alaskan malamutes. *Am. J. Vet. Res.* **28:** 355–357.

Rubin L.F., Nelson E.J., and Sharp C.A. 1991. Collie eye anomaly in Australian shepherd dogs. *Prog. Vet. Comp. Ophthalmol.* **1:** 105–108.

Runte M., Dekomien G., and Epplen J.T. 2000. Evaluation of *RDS/Peripherin* and *ROM1* as candidate genes in generalised progressive retinal atrophy and exclusion of digenic inheritance. *Anim. Genet.* **31:** 223–227.

Santos-Anderson R., Tso M., and Wolf E. 1980. An inherited retinopathy in collies. A light and electron microscopic study. *Investig. Ophthalmol. Vis. Sci.* **19:** 1281–1294.

Sidjanin D.J., Zangerl B., Johnson J.L., Xue F., Mellersh C., Ostrander E.A., Acland G., and Aguirre G.D. 2002a. Cloning of the canine delta tubulin cDNA (*TUBD*) and mapping to CFA9. *Anim. Genet.* **33:** 161–162.

Sidjanin D.J., Lowe J.K., McElwee J.L., Milne B.S., Phippen T.M., Sargan D.R., Aguirre G.D., Acland G.M., and Ostrander E.A. 2002b. Canine *CNGB3* mutations establish cone degeneration as orthologous to the human achromatopsia locus *ACHM3*. *Hum. Mol. Genet.* **11:** 1823–1833.

Sidjanin D.J., Xue F., McElwee J., Johnson J.L., Holmgren C., Mellersh C., Ostrander E., Acland G., and Aguirre G.D. 2001. Cloning of canine γ-tubulin (*TUBG1*) cDNA and mapping to CFA9. *Anim. Genet.* **32:** 328–329.

Sidjanin D.J., Miller B., Kijas J., McElwee J., Pillardy J., Malek J., Pai G., Feldblyum T., Fraser C., Acland G., and Aguirre G. 2003. Radiation hybrid map, physical map, and low-pass genomic sequence of the canine prcd region on CFA9 and comparative mapping with the syntenic region on human chromosome 17. *Genomics* **81:** 138–148.

Stades F. 1978. Hereditary retinal dysplasia (RD) in a family of Yorkshire terriers. *Tijdschr. Diergeneeskd.* **103:** 1087–1090.

Suber M.L., Pittler S.J., Qin N., Wright G.C., Holcombe V., Lee R.H., Craft C.M., Lolley R.N., Baehr W., and Hurwitz R.L. 1993. Irish setter dogs affected with rod/cone dysplasia contain a nonsense mutation in the rod cGMP phosphodiesterase β-subunit gene. *Proc. Natl. Acad. Sci.* **90:** 3968–3972.

Tansley K. 1951. Hereditary degeneration of the mouse retina. *Br. J. Ophthalmol.* **35:** 573–582.

Travis G.H. and Sutcliffe J.G. 1988. Phenol emulsion-enhanced DNA-driven subtractive cDNA cloning: Isolation of low-abundance monkey cortex-specific mRNAs. *Proc. Natl. Acad. Sci.* **85:** 1696–1700.

Travis G.H., Brennan M.B., Danielson P.E., Kozak C.A., and Sutcliffe J.G. 1989. Identification of a photoreceptor-specific mRNA encoded by the gene responsible for retinal degeneration slow (*rds*). *Nature* **338:** 70–73.

Veske A., Nilsson S.E.G., and Gal A. 1997a. Characterization of canine rod photoreceptor cGMP-gated cation channel alpha-subunit gene and exclusion of its involvement in the hereditary retinal dystrophy of Swedish briards. *Gene* **202:** 115–119.

Veske A., Nilsson S.E., and Gal A. 1998. Organization of the canine gene encoding the E isoform of retinal guanylate cyclase (cGC-E) and exclusion of its involvement in the inherited retinal dystrophy of the Swedish Briard and Briard-beagle dogs. *Biochim. Biophys. Acta* **1372:** 69–77.

Veske A., Nilsson S.E.G., Narfström K., and Gal A. 1999. Retinal dystrophy of Swedish Briard/Briard-beagle dogs is due to a 4-bp deletion in RPE65. *Genomics* **57:** 57–61.

Veske A., Narfström K., Finckh U., Sargan D.R., Nilsson S.E.G., and Gal A. 1997b. Isolation of canine retinal arrestin cDNA and exclusion of three candidate genes for Swedish Briard retinal dystrophy. *Curr. Eye Res.* **16:** 270–274.

Veske A., Nilsson S.E.G., Finckh U., Narfström K., Petersen-Jones S., Gould D., Sargan D., and Gal A. 1997c. Hereditary retinal dystrophy of Swedish briard dogs. Exclusion of six candidate genes by molecular genetic analysis. In *Degenerative retinal diseases* (ed. M.M. LaVail et al.), pp. 81–87. Plenum Press, New York.

Vithana E.N., Abu-Safieh L., Pelosini L., Winchester E., Hornan D., Bird A.C., Hunt D.M., Bustin S.A., and Bhattacharya S.S. 2003. Expression of *PRPF31* mRNA in patients with autosomal dominant retinitis pigmentosa: A molecular clue for incomplete penetrance? *Investig. Ophthalmol. Vis. Sci.* **44:** 4204–4209.

Vithana E.N., Abu-Safieh L., Allen M.J., Carey A.H., Papaioannou M., Chakarova C., Al-Maghtheh M., Ebenezer N.D., Willis C., Moore A.T., et al. 2001. A human homolog of yeast pre-mRNA splicing gene, *PRP31*, underlies autosomal dominant retinitis pigmentosa on chromosome 19q13.4 (*RP11*). *Mol. Cell* **8:** 375–381.

Wang W. 1999. Molecular analysis of photoreceptor genes for causal association with rod-cone dysplasia 2 (*rcd2*), a canine model of abnormal retinal cGMP metabolism. In *Graduate School, Cornell University*, pp. 1–216. Cornell University, Ithaca, New York.

Wang W., Acland G.M., Ray K., and Aguirre G.D. 1999a. Evaluation of cGMP-phosphodiesterase (PDE) subunits for causal association with rod-cone dysplasia 2 (*rcd2*), a canine model of abnormal retinal cGMP metabolism. *Exp. Eye Res.* **69:** 445–453.

Wang W., Zhang Q., Acland G.M., Mellersh C., Ostrander E.A., Ray K., and Aguirre G.D. 1999b. Molecular characterization and mapping of canine cGMP-phosphodiesterase delta subunit (PDE6D). *Gene* **236:** 325–332.

Werner P., Raducha M.G., Prociuk U., Henthorn P.S., and Patterson D.F. 1997. Physical and linkage mapping of human chromosome 17 loci to dog chromosomes 9 and 5. *Genomics* **42:** 74–82.

Werner P., Raducha M.G., Prociuk U., Budarf M., Henthorn P.S., and Patterson D.F. 1999a. Comparative mapping of the DiGeorge region in the dog and exclusion of linkage to inherited canine conotruncal heart defects. *J. Hered.* **90:** 494–498.

Werner P., Mellersh C.S., Raducha M.G., DeRose S., Acland G.M., Prociuk U., Wiegand N., Aguirre G.D., Henthorn P.S., Patterson D.F., and Ostrander E.A. 1999b. Anchoring of canine linkage groups with chromosome-specific markers. *Mamm. Genome* **10:** 814–823.

Werner P., Raducha M.G., Prociuk U., Lyons L.A., Kehler J.S., Henthorn P.S., and Patterson D.F. 1998. RXRA and HSPA5 map to the telomeric end of dog chromosome 9. *Anim. Genet.* **29:** 220–223.

Wolf E.D., Vainisi S.J., and Santos-Anderson R. 1978. Rod-cone dysplasia in the collie. *J. Am. Vet. Med. Assoc.* **173:** 1331–1333.

Woodford B., Liu Y., Fletcher R., Chader G., Farber D., Santos-Anderson R., and Tso M. 1982. Cyclic nucleotide metabolism in inherited retinopathy in collies: A biochemical and histochemical study. *Exp. Eye Res.* **34:** 703–714.

Wrigstad A. 1994. Hereditary dystrophy of the retina and the retinal pigment epithelium in a strain of briard dogs: A clinical, morphological and electrophysiological study. In *Linköping University Medical Dissertations*, No. 423. Linkoping, Sweden.

Zangerl B., Zhang Q., Acland G.M., and Aguirre G.D. 2002a. Characterization of three microsatellite loci linked to the canine RP3 interval. *J. Hered.* **93:** 70–73.

Zangerl B., Zhang Q., Pearce-Kelling S.E., and Aguirre G.D. 2002b. Molecular cloning, characterization and mapping of the canine glucocorticoid receptor DNA binding factor 1 (*GRLF1*). *Gene* **294:** 167–176.

Zeiss C.J., Acland G.M., and Aguirre G.D. 1999. Retinal pathology of canine X-linked progressive retinal atrophy, the locus homologue of RP3. *Investig. Ophthalmol. Vis. Sci.* **40:** 3292–3304.

Zeiss C.J., Ray K., Acland G.M., and Aguirre G.D. 2000. Mapping of X-linked progressive retinal atrophy (XLPRA), the canine homolog of retinitis pigmentosa 3 (RP3). *Hum. Mol. Genet.* **9:** 531–537.

Zhang Q., Acland G.M., Parshall C.J., Haskell J., Ray K., and Aguirre G.D. 1998. Characterization of canine photoreceptor phosducin cDNA and identification of a sequence variant in dogs with photoreceptor dysplasia. *Gene* **215:** 231–239.

Zhang Q., Acland G.M., Zangerl B., Johnson J.L., Mao Z., Zeiss C.J., Ostrander E.A., and Aguirre G.D. 2001. Fine mapping of canine XLPRA establishes homology of the human and canine RP3 intervals. *Investig. Ophthalmol. Vis. Sci.* **42:** 2466–2471.

Zhang Q., Acland G.M., Wu W.X., Johnson J.L., Pearce-Kelling S., Tulloch B., Vervoort R., Wright A.F., and Aguirre G.D. 2002. Different RPGR exon ORF15 mutations in Canids provide insights into photoreceptor cell degeneration. *Hum. Mol. Genet.* **11:** 993–1003.

17

Copper Toxicosis in Bedlington Terriers

Bart van de Sluis and Cisca Wijmenga
Complex Genetics Section, DBG-Department of Medical Genetics
UMC Utrecht, 3508 AB, The Netherlands

Bernard A. van Oost
Department of Animals, Science and Society
Faculty of Veterinary Medicine, Utrecht University
3508 TD Utrecht, The Netherlands

IT IS BELIEVED THAT THE BEDLINGTON TERRIER was derived from the rough-coated Scotch terrier. In the late 18th century, a number of these dogs were brought from Scotland to England, where a strain of terriers was bred in Northumberland (northeast England), known as "the land of the Bedlington terrier." A dog called Flint, born in either 1782 or 1792, is believed to be one of the early founders of the breed, and most pedigrees of today's Bedlington terriers can be traced back to Flint (Bounden 1990). Originally the Bedlington terrier was a breed for companionship and work: The dogs were used for hunting rabbits, hares, and even rats, but lately the Bedlington terrier has become a popular show breed.

Many dog breeds are afflicted with hereditary defects, and the Bedlington terrier is no exception; the Bedlington terrier has a particularly high prevalence of copper toxicosis. The exact reason for this is not known, but often the enrichment of a certain disease in a specific breed can be explained by a founder or popular sire effect and/or population bottlenecks (Ostrander and Kruglyak 2000). The high frequency of copper toxicosis in Bedlington terriers has been observed worldwide, including in Australia (Robertson et al. 1983), Belgium (Rothuizen et al. 1999), Finland (Eriksson 1983), Germany (Rothuizen et al. 1999), the Netherlands (Meulenaar et al. 1983), the U.K. (Kelly et al. 1984), and the U.S. (Hardy and Stevens 1978). Estimates of the incidence of copper toxicosis vary from 34% in the U.K. (Herrtage et al. 1987) and 37% in the Netherlands and Belgium (Rothuizen et al. 1999) to 66% in the U.S. (Hardy and Stevens 1978). The current prevalence of copper toxicosis is unknown but is presumed to be significantly lower since screening with a linked DNA marker began (Yuzbasiyan-Gurkan et al. 1997).

Copper storage disorders have also been observed in other dog breeds, such as West Highland white terriers, Skye terriers, dalmatians, keeshonden, and Doberman pinschers (Johnson et al. 1982; Thornburg and Crawford 1986; Thornburg et al. 1986; Haywood et al. 1988; Thornburg 2000; Webb et al. 2002). Copper toxicosis in Bedlington terriers is, however, the best-characterized disorder, since there are no detailed studies available for the other breeds. We do not yet know whether the gene underlying copper toxicosis in Bedlington terriers also causes the copper storage disorders in the other breeds. Inherited copper accumulation disorders have also been observed in humans. The best-studied copper toxicity disorder is Wilson disease (OMIM 309400), which is an autosomal recessive disorder characterized by reduced copper excretion into the bile, resulting in hepatic copper accumulation and leading to liver cirrhosis. Bull et al. (1993) identified mutations in the ATP7B gene, coding for a P-type ATPase-dependent cation transporter, causing Wilson disease. Other hepatic copper storage disorders have been described (OMIM 215600) in Indian childhood cirrhosis (ICC) (Tanner 1998), in endemic Tyrolean infantile cirrhosis (ETIC) (Muller et al. 1996), and in sporadic cases occurring worldwide and referred to collectively as idiopathic copper toxicosis (ICT) (Scheinberg and Sternlieb 1996; Muller et al. 1998).

Because of the phenotypic similarities between copper toxicosis in Bedlington terriers and idiopathic copper toxicosis, we hypothesized that copper toxicosis in Bedlington terriers would be a good animal model for studying copper storage disorders in man. We felt that isolating and characterizing the gene involved in copper toxicosis in Bedlington terriers would enrich our knowledge about copper metabolism in humans and animals and help identify additional genes that are implicated in human copper storage disorders for which the causal gene is still unknown.

COPPER TOXICOSIS IN BEDLINGTON TERRIERS

Copper toxicosis in Bedlington terriers is an autosomal recessive disorder (Johnson et al. 1980; Owen and Ludwig 1982) and was first reported in the U.S. (Hardy et al. 1975). The disorder is characterized by immense copper accumulation (>1000 μg/g dry weight of liver) in the liver caused by a defect in the copper excretion into the bile. Studies of the biliary copper excretion showed that the copper level in affected Bedlington terriers was at least half that in normal dogs (Su et al. 1982a,b). Furthermore, affected Bedlington terriers had increased renal copper, but no more copper was absorbed from the intestinal tract than in normal dogs (Su et al. 1982a,b). This hepatic copper accumulation causes chronic hepatitis and eventually liver cirrhosis and rarely results in acute hemolytic crises (Rothuizen et al. 1999). The copper toxicity does not always lead to overt disease because some affected Bedlington terriers with a high copper level in the liver do not become ill. The accumulated hepatic copper is seen as electron-dense granules in the lysosomes and is mainly present centrilobularly (Owen and Ludwig 1982; Nederbragt et al. 1984; Hultgren et al. 1986).

The age of onset of copper toxicosis is between 2 and 6 years (Twedt et al. 1979; Herrtage et al. 1987), and treatment options are currently limited (Rolfe and Twedt 1995). Reduction of intestinal copper intake can be achieved by giving the dogs a low copper diet or a zinc therapy (Brewer et al. 1992). Increased copper excretion can be accomplished by the lifelong administration of copper chelators, such as D-penicillamine (Magne and Chiapella 1986). Since the identification of the copper-toxicosis-linked marker, the dogs can be diagnosed at a young age, long before the disease is manifest. Early diagnosis can be useful in preventing hepatic copper accumulation by giving the terriers a low-copper diet.

CLINICAL DIAGNOSES OF COPPER TOXICOSIS IN BEDLINGTON TERRIERS

There are currently two ways to diagnose canine copper toxicosis, namely (1) by measuring the hepatic copper concentration in a liver biopsy or (2) by molecular means, either by assaying for a particular variant in a DNA marker known to be in linkage disequilibrium with the copper toxicosis mutation, or by assaying for the exon 2 deletion of the copper toxicosis gene *MURR1* directly.

Hepatic copper concentration measurements can be performed either quantitatively by irradiating the hepatic biopsies and measuring the induced ^{64}Cu radioactivity, or semi-quantitatively by histochemical or cytochemical analysis (Johnson et al. 1984; Thornburg et al. 1985; Teske et al. 1992). Liver biopsies can cause complications and are only diagnostic in adult dogs. A hepatic copper concentration higher than 850 μg/g per dry weight of liver (150–500 μg/g) is indicative of copper toxicosis in Bedlington terriers (Twedt et al. 1979; Owen and Ludwig

1982; Johnson et al. 1984, Thornburg 2000), but levels of more than 7000 μg/g per dry weight of liver can be found (Haywood et al. 1996). When the results are intermediate, i.e., a copper concentration between 500 μg/g and 850 μg/g, a second biopsy must be performed at least 6 months after the first (Ubbink et al. 2000).

To semi-quantify the hepatic copper concentration, a histochemical grading system was developed based on the number and distribution of copper-containing granules and their rhodamine staining intensity, with grades 3, 4, or 5 corresponding to states of copper toxicity (Johnson et al. 1984), to which morphological criteria were added (Hultgren et al. 1986). Haywood et al. (2001a) eventually divided the Bedlington terriers into three groups: healthy, intermediate, and affected. The copper toxicosis diagnosis made by directly measuring hepatic copper concentrations or semi-quantitatively by histochemistry can only recognize homozygously affected Bedlington terriers, and only when they have accumulated copper.

In 1997, Yuzbasiyan-Gurkan and colleagues identified an anonymous microsatellite marker closely linked to the copper toxicosis locus in Bedlington terriers. Using 75 Bedlington terriers registered with the American Kennel Club, of which 25 were affected, linkage was found with DNA marker C04107, with a LOD score of 5.96 and at a recombination fraction of zero. Allele 2 (167 bp) of the C04107 locus was shown to be in complete linkage with the copper toxicosis phenotype, whereas allele 1 (163 bp) was never found in association with the disease, suggesting that allele 2 was in strong linkage disequilibrium with the disease allele, and thereby was a good prognostic marker for the copper toxicosis status. Subsequently, the C04107 marker was evaluated in different Bedlington terrier populations in different countries: Belgium, Denmark, Germany, the Netherlands, and the U.K. (Holmes et al. 1998; Rothuizen et al. 1999; Proschowsky et al. 2000). These studies confirmed the use of the C04107 marker as an accurate diagnostic test for copper toxicosis in these Bedlington terriers. However, Holmes et al. (1998) and Haywood et al. (2000) described the rare observation of association of copper toxicosis with allele 1, implying the possible presence of recombination events that had separated the true disease gene from C04107 allele 2. In large surveys, almost all affected Bedlington terriers were found to have the genotype 2,2 for marker C04107 (Yuzbasiyan-Gurkan et al. 1997; Rothuizen et al. 1999); thus far, only a few affected terriers with the C04107 genotypes 1,1 and 1,2 have been reported (Holmes et al. 1998; Haywood et al. 2001a; Coronado et al. 2003; Hyun and Filippich 2004).

IDENTIFICATION OF A COPPER TOXICOSIS
GENE IN BEDLINGTON TERRIERS

The linkage disequilibrium of copper toxicosis to C04107 provided an excellent opportunity to clone and identify the disease-causing gene. Because of the strong linkage disequilibrium, we and other workers hypothesized that all affected Bedlington terriers must carry an identical copper toxicosis mutation in homozygous state inherited from this founder dog (Fig. 1; see Box 1). All the affected Bedlington terriers would then share a copper toxicosis chromosomal region that is identical by

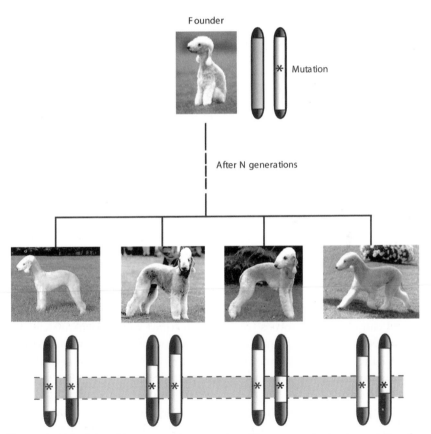

Figure 1. Principle of homozygosity mapping. The copper toxicosis mutation has occurred in one of the founder Bedlington terriers. The haplotype on the founder chromosome is indicated in white. In time, the haplotype of the founder chromosome will become fragmented due to recombination events. Thus, all the affected Bedlington terriers share two distinct fragments of the founder chromosome that are homozygous and identical by descent. The shared region is indicated by a light gray bar.

Box 1. Linkage disequilibrium mapping in the dog

Advances in human genome analysis should also be applicable to other genomes, including the canine genome. Population resources for genetic mapping in dogs are potentially enormous. Today more than 400 breeds are officially recognized, and every year more than 1 million purebred dogs are registered in the U.S. (http://www.akc.org/index.cfm) and about 260,000 in the U.K. (http://www.the-kennel-club.org.uk/). These breeds are partially inbred genetic isolates (Jonasdottir et al. 2000; Ostrander and Kruglyak 2000; Ostrander et al. 2000; Sutter and Ostrander 2004). More than 400 different genetic diseases have been documented (Ostrander and Kruglyak 2000; Patterson 2000), of which 50% are breed-specific.

Due to the highly selective inbreeding of specific dog populations, they are comparable to isolated human populations, but they have an even higher degree of genetic isolation and, in some cases, exhibit narrower genetic bottlenecks. In addition, there are much better genealogical dog records available for performing genetic studies, which can eventually lead to the identification of disease-causing genes in specific dog traits. In principle, purebred dogs having a breed-specific trait all share a disease locus from one founder. It should therefore be possible to determine the approximate chromosomal location of such a trait by linkage analysis in extended pedigrees, and further fine mapping of the disease candidate region by linkage disequilibrium (LD) mapping should be profitable.

descent (IBD) (Fig. 1). It should therefore be feasible to delineate more precisely the copper toxicosis region by homozygosity mapping and to clone the copper toxicosis gene by positional cloning (see Box 2).

Positional Cloning of a Copper Toxicosis Gene: *MURR1*

In 1997 there were only limited resources from the dog genome for performing genetic linkage studies. A standardized karyogram of the 21 largest canine chromosomes (of the 39 in total) was available (Switonski et al. 1996) and two different linkage maps had been constructed, but the C04107 had not been mapped (Lingaas et al. 1997; Mellersh et al. 1997). Furthermore, a total canine genomic bacterial artificial chromosome (BAC) library (Li et al. 1999) and a radiation hybrid panel, the canine T72 RH panel (Research Genetics, Huntsville, AL) were available.

For linkage disequilibrium mapping, we were able to collect DNA samples from 23 related Bedlington terriers of Belgian origin (9 affected, 12 carriers, 2 unaffected) and from 10 unrelated affected Bedlington terriers of Belgian origin. In addition, DNA from 10 English Bedlington terriers (4 affected, 3 carriers, 1 unaffected, 2 unknown) was collected, and their disease status was described by Holmes et al. (1998).

Box 2. Positional cloning

Positional cloning is the identification of disease genes solely based on the approximate position of a disease gene on a chromosome and not on any preconceived functional information (Collins 1995). The approximate chromosomal location of a disease gene in a family or population is usually made by linkage analysis in extended pedigrees, whereas the fine mapping is often achieved by linkage disequilibrium (LD) mapping (Jorde 2000). In LD mapping, one scans for those polymorphic markers, which are non-randomly associated with a disease locus either in families or populations (Jorde 2000). Accordingly, LD mapping using additional and more informative markers across a given region can substantially reduce the candidate region of a disease gene in families or populations sharing the same ancestral mutation. Houwen et al. (1994) delineated a candidate region for the BRIC gene by identifying a chromosomal region that is identical by descent and shared by all affected individuals from a Dutch population isolate. This method, also called homozygosity mapping, has been successfully applied in consanguineous families (Neufeld et al. 1997; Wang et al. 1997; Wijmenga et al. 1998) and in isolated human populations, e.g., in the Finnish population (Peltonen et al. 1999).

Based on the available data and resources at that time, our positional cloning strategy to identify the copper toxicosis gene (van de Sluis et al. 1999, 2000, 2002) consisted of four steps: (1) chromosomal assignment of the copper toxicosis gene, (2) construction of physical maps by radiation hybrid mapping and by the construction of a BAC contig covering the copper toxicosis region, (3) genetic fine mapping by homozygosity mapping, and (4) construction of a potential gene map using the human comparative genome data and BAC clone sequences.

Chromosomal Assignment of the Copper Toxicosis Gene

The chromosomal localization of the copper toxicosis gene was determined by fluorescence in situ hybridization (FISH) using the BAC clone N21-27, containing the anonymous marker C04107, as a probe on dog metaphase chromosomes. This BAC clone was mapped to canine chromosome 10, region q26 (CFA10q21-26; CFA is the symbol for *Canis familiaris*) (van de Sluis et al. 1999). Random sequencing of the BAC clone N21-27 revealed a sequence homologous to the human *MURR1* gene, which was located on human chromosome 2, region p13-21 (HSA2p13-21; HSA is the symbol for *Homo sapiens*) (van de Sluis et al. 1999), indicating that this region was syntenic to CFA10q26. Based on the synteny between CFA10q21-26 and HSA2p13-21, we could exclude the known copper transporter genes *ATP7B*, *CTR1*, *CTR2*, and *ATOX1* as candidates for canine copper toxicosis, because none of these genes was mapped to HSA2p13-21. This exclusion was confirmed by dual color FISH of BAC clone N21-27 with the BAC clones containing *ATP7B, CTR1, CTR2,* or *ATOX1* (Dagenais et al. 1999; Nanji and Cox 1999; van de Sluis et al. 1999).

Construction of the Two Physical Maps and Fine Mapping

Construction of a radiation hybrid map (Fig. 2A) and the BAC contig (Fig. 2B) covering the copper toxicosis region was needed because of the limited mapping data of the canine chromosome 10, region q21-26. Linkage maps of the canine genome mapped the C04107 marker to linkage group 8 (Neff et al. 1999) or to canine chromosome 10 (Werner et al. 1999). Using the whole-genome radiation hybrid panel, the canine T72 RH panel (Research Genetics, Huntsville, AL), we constructed a contiguous BAC map covering the copper toxicosis region. Accordingly, these BAC clones were used for isolating new polymorphic markers and identifying putative coding sequences by random shotgun sequencing. The isolated putative canine coding sequences were then used to construct a well-defined comparative gene map between dog and man by mapping these sequences onto the human genome map and, conversely, by mapping human genes and expressed sequence tags (ESTs) to the canine map. The new polymorphic DNA markers were used to delineate the candidate region as closely as possible using homozygosity mapping.

Yuzbasiyan-Gurkan et al. (1997) were unable to determine the distance between the C04107 marker and the copper toxicosis gene because no recombinants were observed between the copper toxicosis gene and the C04107 DNA marker in their pedigrees. We constructed a radiation hybrid map that covered a region of at least 9 cM on both sides of the C04017 marker. Combining mapping information of the first whole-genome radiation hybrid map (Priat et al. 1998), linkage maps (Mellersh et al. 1997; Neff et al. 1999), and marker information (Holmes et al. 1993; Yuzbasiyan-Gurkan et al. 1997; Schelling et al. 1998; Werner et al. 1999), a high-resolution hybrid map of the dog chromosome region 10q26-21 was generated. This map consisted of 10 polymorphic markers and 6 genes (*CALM2, FSHR, SPTBN1, MURR1, SLC1A4,* and *PLEK*) from the comparative region on HSA2p13-21 (van de Sluis et al. 2000). The final radiation map contained a total of 23 markers (11 polymorphic markers, 8 genes and 4 ESTs) (Fig. 2A). The construction of the radiation hybrid map of dog chromosome region 10q21-26 showed that the genes between *CALM2* and *PLEK* were completely conserved between CFA10q21-26 and HSA2p13-21.

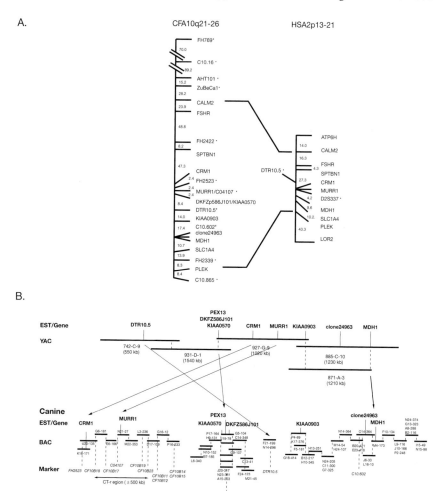

Figure 2. (A) Status of the radiation hybrid map of the canine chromosome region 10q21-26 and human chromosome region 2p13-21 from the year 2001. These maps were utilized to clone the *MURR1* gene. The distance between subsequent loci are indicated in centiray (cR$_{3000}$). The polymorphic markers used in this study are indicated with an asterisk. **(B)** Physical map of the canine copper toxicosis region and the human homologous region from the year 2001, which were used to clone the *MURR1* gene. Mapped genes and ESTs in the copper toxicosis region are shown in bold and in the upper part, the BAC and YAC clones in the middle part, and the microsatellite markers in italic and in the lower part. The refined candidate region of ±500 kb is indicated with a horizontal arrow. The arrows show the chromosomal inversion between man and dog.

A contiguous BAC map (Fig. 2B) was constructed starting from different points. All the genes, ESTs, and polymorphic markers, which had been mapped to the RH map, were isolated in BAC clones. This BAC contig was also used as a resource for the isolation of new polymorphic markers and for the construction of the high-resolution physical map. Combining the different maps, including the RH map and BAC contig information, a chromosomal inversion in this syntenic region was observed. This inversion probably occurred during the separate evolution of man and dog (Fig. 2B). Based on the genetic analysis to delineate the copper toxicosis region by homozygosity mapping, 10 new polymorphic markers were isolated from the BAC clones (Fig. 2B) and confined the copper toxicosis region to approximately 500 kb (Figs. 2B and 3) (van de Sluis et al. 2002).

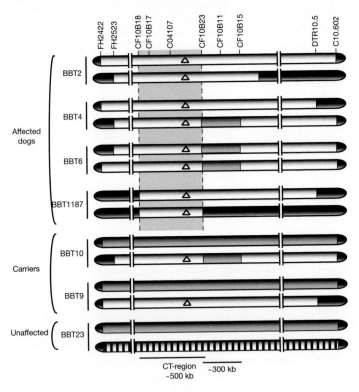

Figure 3. Pairs of haplotypes of a selected number of affected (n = 4), obligate carriers (n = 2) and unaffected (n = 1) Bedlington terriers. The possible copper toxicosis founder haplotypes (identical by descent) are indicated in white. The shared copper toxicosis haplotypes between the affected dogs are indicated with a light gray bar. Wild-type haplotypes are indicated in gray or with stripes. *MURR1* mutation is indicated with Δ. The small haplotype block of 300 kb is indicated in gray.

Construction of a Preliminary Gene Map

This 500-kb copper toxicosis region was shown to contain six ESTs/genes from the corresponding human genome sequence of the human orthologous "copper toxicosis region" (http://www.ensembl.org and http://www.ncbi.nlm.nih.gov). Because the human orthologous "copper toxicosis region" was incompletely sequenced at that time, BACs covering the copper toxicosis region were randomly sequenced. This revealed additional putative coding sequences based on genomic sequence homology between dog and man. The final map of the copper toxicosis region contained 16 putative coding sequences (van de Sluis et al. 2002). Screening for mutations in known genes located in the copper toxicosis region identified a truncated transcript of *MURR1* cDNA in mRNA from an affected Bedlington terrier. Sequence comparison to wild-type sequence revealed a homozygous deletion of exon 2 of the *MURR1* gene (van de Sluis et al. 2002). Obligate carriers were heterozygous for this mutation (Fig. 3). The exon 2 deletion of the *MURR1* transcript in all affected Bedlington terriers is caused by a genomic deletion of 39.7 kb (O.P. Forman et al., in prep.). The deletion of exon 2 in Bedlington terriers results in an in-frame deletion and the production of a predicted truncated protein of 94 amino acid residues instead of the predicted 188 amino acids of the wild-type allele, but immunoblotting did not identify the truncated MURR1 protein in liver from an affected Bedlington terrier, indicating that the mutation causes a loss of function of MURR1 (Fig. 4) (Klomp et al. 2003).

In affected Bedlington terriers, hepatic copper excess has been seen in lysosomes, most likely caused by inefficient copper excretion via the bile. In hepatocytes, ATP7B is localized to the trans-Golgi network, where copper is incorporated into apoceruloplasmin to form ceruloplasmin. If there is excess copper in the hepatocytes, ATP7B is redistributed to vesicles for copper efflux into the bile canaliculi. Normal levels of ceruloplasmin and reduced copper excretion have been observed in affected Bedlington terriers (Su et al 1982b). It is therefore very tempting to speculate that MURR1 has a role downstream of ATP7B, which is further supported by the observed interaction between MURR1 and ATP7B (Tao et al. 2003). The increase of cellular copper levels in RNAi-mediated *MURR1* knock-down cells demonstrates again the importance of MURR1 in copper excretion (Burstein et al. 2004).

THE *MURR1* GENE

The mouse *Murr1* was first identified by Nabetani et al. (1997) and given its name because of its genomic location toward the mouse *U2af1-rs1* gene. *Murr1* stands for "gene that locates in the <u>m</u>ouse <u>U2af1-rs1</u> region"

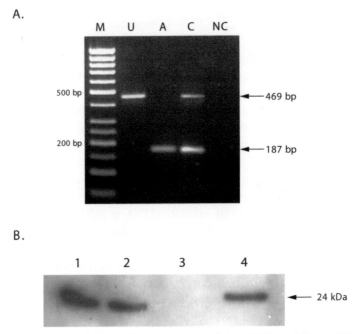

Figure 4. Mutation analysis of *MURR1* gene. **(A)** RT-PCR of liver cDNA from unaffected (U), affected (A), carrier (C), and water as a negative control (NC) for the RT-PCR. The size marker indicated by M is a 50-bp ladder. **(B)** Immunoblot analysis of MURR1. Cell lysates from hepatocytes from mouse (1 and 2), Bedlington terrier with copper toxicosis (3), and healthy beagle (4) were separated by SDS/PAGE, transferred to nitrocellulose, and visualized with anti-MURR1 antibody, followed by goat anti-rabbit peroxidase-conjugated antibody.

(Nabetani et al. 1997). The canine *MURR1* gene consists of three exons: exon 1 is at least 226 bp in length, exon 2 is 284 bp, and exon 3 measures 1011 bp. The human and mouse *MURR1* genes also consist of 3 exons, although an isoform of *Murr1* in mice has been seen (Wang et al. 2002). The canine *MURR1* gene spans a genomic region of approximately 170 kb (www.genome.ucsc.edu). The linked C04107 marker is located within intron 1 of canine *MURR1* approximately 3.5 kb from the *MURR1* deletion (O.P. Forman et al., in prep.). High homology is observed between the proximal site of the deleted fragment and the adjacent sequence of the distal breakpoint. In both fragments the sequence showed a $(CA)_2(GA)_{13}$ repeat, which is probably the causative factor for the genomic deletion observed in the affected Bedlington terriers (O.P. Forman et al., in prep.). The exact length of exon 1 is still unknown, since the transcription start site has yet to be determined and the promoter has not been identified.

The open reading frame of 564 bp encodes a predicted protein consisting of 188 amino acids. *MURR1* is ubiquitously expressed, indicating more functions for MURR1 (see below) (van de Sluis et al. 2002; Klomp et al. 2003). The canine MURR1 protein shows high homology with the human (accession no. NP_689729) and mouse proteins (accession no. CAC24865), with 87% and 86% amino acid identity, respectively. Analyzing the MURR1 protein sequence revealed no copper-binding domains such as have been shown in the copper transporters ATP7B, CTR1, and ATOX1. In addition, no other specific domains were identified that could help with the prediction of the putative function of MURR1.

MOLECULAR AND FUNCTIONAL CHARACTERIZATION OF MURR1

To investigate the function of the MURR1 protein, polyclonal antibodies were generated that recognized MURR1 as a single protein with a molecular mass of 23 kD (Klomp et al. 2003). MURR1 was mainly detected as a cytosolic protein, but some protein was also associated with a membranous cell fraction. Subcellular localization showed MURR1 being associated with an unidentified intracellular vesicular compartment (Klomp et al. 2003). Interestingly, Tao and colleagues (2003) identified an interaction between MURR1 and the copper transporter ATP7B. These results would fit in with our hypothesis that MURR1 cooperates with ATP7B in the biliary copper excretion (Wijmenga and Klomp 2004). Recently, interactions between MURR1 and various proteins (X-linked inhibitor of apoptosis, ameloride-induced sodium current, and NK-κB) have been observed, suggesting a pleiotropic function of MURR1 (Ganesh et al. 2003; Biasio et al. 2004; Burstein et al. 2004).

DIAGNOSTIC DISCREPANCIES USING THE C04107 MARKER

Interestingly, the closely linked C04107 marker, which is used as a diagnostic tool, is located within intron 1 of *MURR1* at a distance of approximately 13.5 kb from the proximal breakpoint of the *MURR1* deletion (O.P. Forman et al., in prep.). Given the close proximity of the C04107 marker to the MURR1 deletion, the marker should be very useful in predicting copper toxicosis. However, several groups have recently published possible recombinants between the C04107 marker and the *MURR1* mutation. A test for the *MURR1* mutation will therefore be the ultimate predictive diagnostic test.

There are two types of recombinants: (1) Healthy dogs that are homozygous for allele 2 of marker C04107 (Yuzbasiyan-Gurkan et al. 1997; Coronado et al. 2003; van de Sluis et al. 2003; Hyun et al. 2004); these healthy dogs have been studied by three different groups (Coronado et al. 2003; van de Sluis et al. 2003; Hyun et al. 2004) and have been shown to be homozygous for the wild-type MURR1 allele. (2) Affected dogs that are heterozygous or homozygous for allele 1 of marker C04107 (Holmes et al. 1998; Haywood et al. 2001a; Coronado et al. 2003; Hyun et al. 2003); the affected dogs studied by Coronado and coworkers (2003) and Hyun et al. (2004) appeared to be homozygous for the wild-type allele or heterozygous for the *MURR1* deletion. This means that, in the first case, allele 2 can also be associated with wild-type *MURR1* and, in the second case, allele 1 is linked to the deletion of MURR1 exon 2. Different groups (Coronado et al. 2003; van de Sluis et al. 2003; Hyun et al. 2004) have given an explanation for these possible recombinants. Three new haplotypes in Bedlington terriers from American, Finnish, and Australian populations were not associated with copper toxicosis in Bedlington terriers. Using microsatellite markers (including C04107), *MURR1* exon 2 deletion, and SNPs covering a genomic region of 600 kb, we identified a new haplotype in unaffected Bedlington terriers from the U.S. containing the allele 2 of the marker C04107 and wild-type *MURR1* allele (van de Sluis et al. 2003). Coronado et al. (2003) and Hyun et al. (2004) also identified a haplotype containing C04107 allele 2 and wild-type *MURR1* allele in unaffected dogs. Whether the haplotypes (allele 2 C04107-wild-type *MURR1*) identified by these groups are identical is not known, because the groups used different polymorphisms. However, Coronado et al. (2003) and Hyun et al. (2004) used a SNP located within the 3′ acceptor consensus sequences of exon 2 and identified similar alleles, which points to identical haplotypes. Uncoupling of allele 2 for the C04107 marker and the copper toxicosis mutation has also been seen in the dog population used by Yuzbasiyan-Gurkan et al. (1997).

The identified new haplotypes could be explained by the existence of a different genetic background in the American, Finnish, and Australian Bedlington terrier populations, as the present Bedlington terriers were either derived from one or more different breed lineages or evolved by outcrossing with other breeds into the Bedlington terrier. It seems that, since the 1960s, three distinct Bedlington terrier lines have been developed by a combination of line breeding, in-breeding, and outcross breeding (Bounden 1990), thereby explaining the observed genetic heterogeneity.

How do we explain the 1-2 and 1-1 allele Bedlington terriers diagnosed as affected and having hepatitis (Coronado et al. 2003; Hyun et al. 2004)? They all had a high copper concentration (>1000 µg/g), but no homozygous exon 2 deletion or other mutation in the coding region of *MURR1* gene was seen. It is tempting to speculate that these dogs have an as yet unidentified mutation within the *MURR1* gene (e.g., promoter region) or a mutation in an entirely different gene. The latter is more likely, knowing that other copper transporters give an almost similar phenotype. In addition, copper storage disorders have also been seen in other dog breeds and in human populations (Muller et al. 1996, 1998; Scheinberg and Sternlieb 1996; Tanner 1998), in which the underlying cause is still unknown (Muller et al. 2003), indicating the presence of other, unidentified genes causing copper toxicosis in dogs.

Another explanation for the misconception of recombinants between C04107 allele 1 and the *MURR1* mutation could also be misdiagnosis of the disease status. In different studies, the copper toxicosis status was determined by a hepatic copper concentration of ≥ 1000 µg/g dry weight of liver (van de Sluis et al. 2002; Coronado et al. 2003; Hyun et al. 2004). If dogs have a copper concentration between 300 and 1000 µg/g dry weight of liver, the measurements have to be repeated as shown by Ubbink et al. (2000), since normal hepatic copper concentrations in dogs vary widely among dogs and may exceed 400 µg/g dry weight of liver (Thornburg 2000). Haywood and colleagues (2001a) observed dogs with hepatic copper up to 1000 µg/g that had marginal or no liver damage, in contrast to dogs with active hepatitis having normal hepatic copper levels. Similarly, healthy dogs homozygous for allele 2 could also be misinterpreted if they were diagnosed at a young age (under 1 year old) or were fed a low-copper diet. Young dogs (<1 year of age) or dogs with a low copper intake can have low hepatic copper levels (Hyun and Filippich 2004).

Since the C04107 marker is not always predictive for copper toxicosis in American, Finnish, and Australian Bedlington terriers, the only reliable diagnostic test for MURR1-associated copper toxicosis is the analysis of the exon 2 deletion. A direct PCR-based diagnostic test to determine the loss of exon 2 at the level of genomic DNA could be invaluable in screening the Bedlington terrier breeding stock.

The discovery of the MURR1 deletion in Bedlington terriers with copper toxicosis makes *MURR1* an interesting candidate gene for copper toxicosis in other dog breeds, in human patients, and in certain breeds of sheep, where copper toxicosis is common, although *MURR1* has been excluded as the causative gene for non-Wilsonian hepatic copper toxicosis in man (Muller et al. 2003). Nevertheless, the unraveling of the function of MURR1 will provide us with better insight into the regulation of copper homeostasis in mammals, and new proteins will be discovered when the MURR1 pathway is unraveled. These proteins will be good candidates for non-Wilsonian hepatic copper toxicosis, copper toxicosis in North Ronaldsay sheep (Haywood et al. 2001b) and in other dog breeds, e.g., Doberman pinscher, West Highland terrier, and dalmatian.

Since the copper toxicosis gene has been identified in Bedlington terriers, new and interesting data have been published about the possible function of MURR1 in several cellular pathways (Ganesh et al. 2003; Tao et al. 2003; van de Sluis et al. 2003; Burstein et al. 2004; Greene 2004). Future experiments will have to elucidate the exact role of *MURR1* in all these different pathways.

SUMMARY

This chapter describes the isolation and characterization of the copper toxicosis gene *MURR1* in Bedlington terriers. Copper toxicosis in Bedlington terriers is an autosomal recessive disorder characterized by an inefficient excretion of copper via the bile, resulting in accumulation of copper in the liver, and leading to chronic hepatitis and, ultimately, cirrhosis. The worldwide prevalence of copper toxicosis is unique to this breed and very high. Due to the intense selection seen in the breeding of Bedlington terriers, we hypothesized that all the affected Bedlington terriers would carry an identical copper toxicosis mutation in homozygous state inherited from a common ancestor. We thought it should therefore be possible to delineate the copper toxicosis region by linkage disequilibrium mapping. Positional cloning was used to clone the copper toxicosis gene in Bedlington terriers. The positional cloning approach consisted of four steps: (1) chromosomal localization of the copper toxicosis gene, (2) construction of physical maps by radiation hybrid mapping and by construction of a BAC contig covering the copper toxicosis region, (3) genetic fine mapping by homozygosity mapping, and (4) construction of a gene map.

With linkage disequilibrium mapping, we were able to confine the copper toxicosis region to approximately 500 kb. Screening for mutation in ESTs and genes located in this candidate region, we found that exon 2 of the MURR1 gene was deleted in both alleles of all affected Bedlington terriers and in single alleles in obligate carriers. This genomic deletion results in a loss of function of the MURR1 protein because no MURR1 was detected in liver cell lysates of affected Bedlington terriers. MURR1 is expressed ubiquitously and MURR1 expression has been observed in different cell lines. Down-regulation of MURR1 via RNAi does increase the copper level in mammalian cell lines. Furthermore, interaction between the MURR1 protein and the Wilson disease protein, ATP7B, was identified, underlining the role of *MURR1* in copper homeostasis.

With the identification of the MURR1 mutation, it will now be possible to set up a reliable diagnostic test for MURR1-associated copper toxicosis. This diagnostic test can be useful in setting up a breeding program to obtain a copper-toxicosis-free Bedlington terrier stock. Although MURR1 has already been excluded for non-Wilsonian hepatic copper toxicosis, MURR1 will be a good candidate gene for other copper toxicosis disorders in which the causal gene is still unknown. Copper toxicosis has been observed in man, sheep, and other dog breeds. In addition, the unraveling of the function of *MURR1* and its pathway in copper homeostasis will subsequently lead to the identification of new genes and thus will be good candidates for copper toxicosis with unknown etiology.

ACKNOWLEDGMENTS

This study was financially supported by grants from the International Copper Association (TPT0551-98) and the Netherlands Organization for Scientific Research (NOW 902-23-254). We thank the British Bedlington terrier club for their financial support and Jackie Senior for improving the text.

REFERENCES

Biasio W., Chang T., McIntosh C.J., and McDonald F.J. 2004. Identification of Murr1 as a regulator of the human delta epithelial sodium channel. *J. Biol. Chem.* **279:** 5429–5434.

Bounden K. 1990. *The Bedlington terrier*, 1st edition. Dickson Price Publisher, Brookland, Romney Marsh, Kent, United Kingdom.

Brewer G.J., Dick R.D., Schall W., Yuzbasiyan-Gurkan V., Mullaney T.P., Pace C., Lindgren J., Thomas M., and Padgett G. 1992. Use of zinc acetate to treat copper toxicosis in dogs. *J. Am. Vet. Med. Assoc.* **201:** 564–568.

Bull P.C., Thomas G.R., Rommens J.M., Forbes J.R., and Cox D.W. 1993. The Wilson disease gene is a putative copper transporting P-type ATPase similar to the Menkes gene. *Nat. Genet.* **5:** 327–337.

Burstein E., Ganesh L., Dick R.D., van De Sluis B., Wilkinson J.C., Klomp L.W., Wijmenga C., Brewer G.J., Nabel G.J., and Duckett C.S. 2004. A novel role for XIAP in copper homeostasis through regulation of MURR1. *EMBO J.* **23:** 244–254.

Collins F.S. 1995. Positional cloning moves from perditional to traditional. *Nat. Genet.* **9:** 347–350.

Coronado V.A., Damaraju D., Kohijoki R., and Cox D.W. 2003. New haplotypes in the Bedlington terrier indicate complexity in copper toxicosis. *Mamm. Genome* **14:** 483–491.

Dagenais S.L., Guevara-Fujita M., Loechel R., Burgess A.C., Miller D.E., Yuzbasiyan-Gurkan V., Brewer G.J., and Glover T.W. 1999. The canine copper toxicosis locus is not syntenic with ATP7B or ATX1 and maps to a region showing homology to human 2p21. *Mamm. Genome* **10:** 753–756.

Eriksson J. 1983. Copper toxicosis in Bedlington terriers. *Acta Vet. Scand.* **24:** 148–152.

Ganesh L., Burstein E., Guha-Niyogi A., Louder M.K., Mascola J.R., Klomp L.W., Wijmenga C., Duckett C.S., and Nabel G.J. 2003. The gene product Murr1 restricts HIV-1 replication in resting CD4+ lymphocytes. *Nature* **426:** 853–857.

Greene W.C. 2004. How resting T cells deMURR HIV infection. *Nat. Immunol.* **5:** 18–19.

Hardy R.M. and Stevens J.B. 1978. Chronic progressive hepatitis in Bedlington terriers. *Proc. Am. Anim. Hosp. Assoc.* **45:** 187–190.

Hardy R.M., Stevens J.B., and Stowe C.M. 1975. Chronic progressive hepatitis in Bedlington terriers associated with elevated liver copper concentrations. *Minn. Vet.* **15:** 13–24.

Haywood S., Fuentealba I.C., and Kemp S.J. 2000. Copper toxicosis in Bedlington terriers. *Vet. Rec.* **146:** 383–384.

Haywood S., Rutgers H.C., and Christian M.K. 1988. Hepatitis and copper accumulation in Skye terriers. *Vet. Pathol.* **25:** 408–414.

Haywood S., Fuentealba I.C., Foster J., and Ross G. 1996. Pathobiology of copper-induced injury in Bedlington terriers: Ultrastructural and microanalytical studies. *Anal. Cell. Pathol.* **10:** 229–241.

Haywood S., Fuentealba I.C., Kemp S.J., and Trafford J. 2001a. Copper toxicosis in the Bedlington terrier: A diagnostic dilemma. *J. Small Anim. Pract.* **42:** 181–185.

Herrtage M.E., Seymour C.A., White R.A.S., Small G.M., and Wight D.G.D. 1987. Inherited copper toxicosis in the Bedlington terrier: The prevalence in asymptomatic dogs. *J. Small Anim. Pract.* **28:** 1141–1151.

Holmes N.G., Herrtage M.E., Ryder E.J., and Binns M.M. 1998. DNA marker C04107 for copper toxicosis in a population of Bedlington terriers in the United Kingdom. *Vet. Rec.* **142:** 351–352.

Holmes N.G., Humphreys S.J., Binns M.M., Curtis R., Holliman A., and Scott A.M. 1993. Characterization of canine microsatellites. *EXS* **67:** 415–420.

Houwen R.H., Baharloo S., Blankenship K., Raeymaekers P., Juyn J., Sandkuijl L.A., and Freimer N.B. 1994. Genome screening by searching for shared segments: Mapping a gene for benign recurrent intrahepatic cholestasis. *Nat. Genet.* **8:** 380–386.

Hultgren B.D., Stevens J.B., and Hardy R.M. 1986. Inherited, chronic, progressive hepatic degeneration in Bedlington terriers with increased liver copper concentrations: Clinical and pathologic observations and comparison with other copper-associated liver diseases. *Am. J. Vet. Res.* **47:** 365–377.

Hyun C. and Filippich L.J. 2004. Inherited canine copper toxicosis in Australian Bedlington terriers. *J. Vet. Sci.* **5:** 19–28.

Hyun C., Lavulo L.T., and Filippich L.J. 2004. Evaluation of haplotypes associated with copper toxicosis in Bedlington terriers in Australia. *Am. J. Vet. Res.* **65:** 1573–1579.

Hyun C., Filippich L.J., Lea R.A., Shepherd G., Hughes I.P., and Griffiths L.R. 2003. Prospects for whole genome linkage disequilibrium mapping in domestic dog breeds. *Mamm. Genome* **14**: 640–649.

Johnson G.F., Zawie D.A., Gilbertson S.R., and Sternlieb I. 1982. Chronic active hepatitis in Doberman pinschers. *J. Am. Vet. Med. Assoc.* **180**: 1438–1442.

Johnson G.F., Gilbertson S.R., Goldfischer S., Grushoff P.S., and Sternlieb I. 1984. Cytochemical detection of inherited copper toxicosis of Bedlington terriers. *Vet. Pathol.* **21**: 57–60.

Johnson G.F., Sternlieb I., Twedt D.C., Grushoff P.S., and Scheinberg I. 1980. Inheritance of copper toxicosis in Bedlington terriers. *Am. J. Vet. Res.* **41**: 1865–1866.

Jonasdottir T.J., Mellersh C.S., Moe L., Heggebo R., Gamlem H., Ostrander E.A., and Lingaas F. 2000. Genetic mapping of a naturally occurring hereditary renal cancer syndrome in dogs. *Proc. Natl. Acad. Sci.* **97**: 4132–4137.

Jorde L.B. 2000. Linkage disequilibrium and the search for complex disease genes. *Genome Res.* **10**: 1435–1444.

Kelly D.F., Haywood S., and Bennet A.M. 1984. Copper toxicosis in Bedlington terriers in the United Kingdom. *J. Small Anim. Pract.* **25**: 293–298.

Klomp A.E., van de Sluis B., Klomp L.W., and Wijmenga C. 2003. The ubiquitously expressed MURR1 protein is absent in canine copper toxicosis. *J. Hepatol.* **39**: 703–709.

Li R., Mignot E., Faraco J., Kadotani H., Cantanese J., Zhao B., Lin X., Hinton L., Ostrander E.A., Patterson D.F., and de Jong P.J. 1999. Construction and characterization of an eightfold redundant dog genomic bacterial artificial chromosome library. *Genomics* **58**: 9–17.

Lingaas F., Sorensen A., Juneja R.K., Johansson S., Fredholm M., Wintero A.K., Sampson J., Mellersh C., Curzon A., Holmes N.G., Binns M.M., Dickens H.F., Ryder E.J., Gerlach J., Baumle E., and Dolf G. 1997. Towards construction of a canine linkage map: Establishment of 16 linkage groups. *Mamm. Genome* **8**: 218–221.

Magne M. and Chiapella A.M. 1986. Medical management of canine chronic hepatitis. *Comp. Cont. Edu. Vet. Pract.* **8**: 915.

Mellersh C.S., Langston A.A., Acland G.M., Fleming M.A., Ray K., Wiegand N.A., Francisco L.V., Gibbs M., Aguirre G.D., and Ostrander E.A. 1997. A linkage map of the canine genome. *Genomics* **46**: 326–336.

Meulenaar H., van den Ingh T.S., and Rothuizen J. 1983. Copper storage in the liver, a hereditary problem in Bedlington terriers (in Dutch). *Tijdschr. Diergeneeskd.* **108**: 916–919.

Muller T., Muller W., and Feichtinger H. 1998. Idiopathic copper toxicosis. *Am. J. Clin. Nutr.* **67**: 1082S–1086S.

Muller T., Feichtinger H., Berger H., and Muller W. 1996. Endemic Tyrolean infantile cirrhosis: An ecogenetic disorder. *Lancet* **347**: 877–880.

Muller T., van de Sluis B., Zhernakova A., van Binsbergen E., Janecke A.R., Bavdekar A., Pandit A., Weirich-Schwaiger H., Witt H., Ellemunter H., et al. 2003. The canine copper toxicosis gene MURR1 does not cause non-Wilsonian hepatic copper toxicosis. *J. Hepatol.* **38**: 164–168.

Nabetani A., Hatada I., Morisaki H., Oshimura M., and Mukai T. 1997. Mouse U2af1-rs1 is a neomorphic imprinted gene. *Mol. Cell. Biol.* **17**: 789–798.

Nanji M.S. and Cox D.W. 1999. The copper chaperone Atox1 in canine copper toxicosis in Bedlington terriers. *Genomics* **62**: 108–112.

Nederbragt H., van den Ingh T.S., and Wensvoort P. 1984. Pathobiology of copper toxicity. *Vet. Q.* **6:** 179–185, 235.

Neff M.W., Broman K.W., Mellersh C.S., Ray K., Acland G.M., Aguirre G.D., Ziegle J.S., Ostrander E.A., and Rine J. 1999. A second-generation genetic linkage map of the domestic dog, *Canis familiaris. Genetics* **151:** 803–820.

Neufeld E.J., Mandel H., Raz T., Szargel R., Yandava C.N., Stagg A., Faure S., Barrett T., Buist N., and Cohen N. 1997. Localization of the gene for thiamine-responsive megaloblastic anemia syndrome, on the long arm of chromosome 1, by homozygosity mapping. *Am. J. Hum. Genet.* **61:** 1335–1341.

Ostrander E.A. and Kruglyak L. 2000. Unleashing the canine genome. *Genome Res.* **10:** 1271–1274.

Ostrander E.A., Galibert F., and Patterson D.F. 2000. Canine genetics comes of age. *Trends Genet.* **16:** 117–124.

Owen C.A., Jr. and Ludwig J. 1982. Inherited copper toxicosis in Bedlington terriers: Wilson's disease (hepatolenticular degeneration). *Am J. Pathol.* **106:** 432–434.

Patterson D.F. 2000. Companion animal medicine in the age of medical genetics. *J. Vet. Intern. Med.* **14:** 1–9.

Peltonen L., Jalanko A., and Varilo T. 1999. Molecular genetics of the Finnish disease heritage. *Hum. Mol. Genet.* **8:** 1913–1923.

Priat C., Hitte C., Vignaux F., Renier C., Jiang Z., Jouquand S., Cheron A., Andre C., and Galibert F. 1998. A whole-genome radiation hybrid map of the dog genome. *Genomics* **54:** 361–378.

Proschowsky H.F., Jepsen B., Jensen H.E., Jensen A.L., and Fredholm M. 2000. Microsatellite marker C04107 as a diagnostic marker for copper toxicosis in the Danish population of Bedlington terriers. *Acta Vet. Scand.* **41:** 345–350.

Robertson H.M., Studdert V.P., and Reuter R.E. 1983. Inherited copper toxicosis in Bedlington terriers. *Aust. Vet. J.* **60:** 235–238.

Rolfe D.S. and Twedt D.C. 1995. Copper-associated hepatopathies in dogs. *Vet. Clin. N. Am. Small Anim. Pract.* **25:** 399–417.

Rothuizen J., Ubbink G.J., van Zon P., Teske E., van den Ingh T.S., and Yuzbasiyan-Gurkan V. 1999. Diagnostic value of a microsatellite DNA marker for copper toxicosis in West-European Bedlington terriers and incidence of the disease. *Anim. Genet.* **30:** 190–194.

Scheinberg I.H. and Sternlieb I. 1996. Wilson disease and idiopathic copper toxicosis. *Am. J. Clin. Nutr.* **63:** 842S–845S.

Schelling C., Stranzinger G., Dolf G., Schlapfer J., and Switonski M. 1998. Assignment of the canine microsatellite ZuBeCa1 to canine chromosome 10q22-q24. *Anim. Genet.* **29:** 398.

Su L.C., Owen C.A., Jr., Zollman P.E., and Hardy R.M. 1982a. A defect of biliary excretion of copper in copper-laden Bedlington terriers. *Am. J. Physiol.* **243:** G231–G236.

Su L.C., Ravanshad S., Owen C.A., Jr., McCall J.T., Zollman P.E., and Hardy R.M. 1982b. A comparison of copper-loading disease in Bedlington terriers and Wilson's disease in humans. *Am. J. Physiol.* **243:** G226–G230.

Sutter N.B. and Ostrander E.A. 2004. Dog star rising: The canine genetic system. *Nat. Rev. Genet.* **5:** 900–910.

Switonski M., Reimann N., Bosma A.A., Long S., Bartnitzke S., Pienkowska A., Moreno-Milan M.M., and Fischer P. 1996. Report on the progress of standardization of the G-banded canine (*Canis familiaris*) karyotype. Committee for the Standardized Karyotype of the Dog (*Canis familiaris*). *Chromosome Res.* **4:** 306–309.

Tanner M.S. 1998. Role of copper in Indian childhood cirrhosis. *Am. J. Clin. Nutr.* **67:** 1074S–1081S.

Tao T.Y., Liu F., Klomp L., Wijmenga C., and Gitlin J.D. 2003. The copper toxicosis gene product Murr1 directly interacts with the Wilson disease protein. *J. Biol. Chem.* **278:** 41593–41596.

Teske E., Brinkhuis B.G., Bode P., van den Ingh T.S., and Rothuizen J. 1992. Cytological detection of copper for the diagnosis of inherited copper toxicosis in Bedlington terriers. *Vet. Rec.* **131:** 30–32.

Thornburg L.P. 2000. A perspective on copper and liver disease in the dog. *J. Vet. Diagn. Invest.* **12:** 101–110.

Thornburg L.P. and Crawford S.J. 1986. Liver disease in West Highland white terriers. *Vet. Rec.* **118:** 110.

Thornburg L.P., Beissenherz M., Dolan M., and Raisbeck M.F. 1985. Histochemical demonstration of copper and copper-associated protein in the canine liver. *Vet. Pathol.* **22:** 327–332.

Thornburg L.P., Shaw D., Dolan M., Raisbeck M., Crawford S., Dennis G.L., and Olwin D.B. 1986. Hereditary copper toxicosis in West Highland white terriers. *Vet. Pathol.* **23:** 148–154.

Twedt D.C., Sternlieb I., and Gilbertson S.R. 1979. Clinical, morphologic, and chemical studies on copper toxicosis of Bedlington terriers. *J. Am. Vet. Med. Assoc.* **175:** 269–275.

Ubbink G.J., Van den Ingh T.S., Yuzbasiyan-Gurkan V., Teske E., Van de Broek J., and Rothuizen J. 2000. Population dynamics of inherited copper toxicosis in Dutch Bedlington terriers (1977–1997). *J. Vet. Intern. Med.* **14:** 172–176.

van de Sluis B., Peter A.T., and Wijmenga C. 2003. Indirect molecular diagnosis of copper toxicosis in Bedlington terriers is complicated by haplotype diversity. *J. Hered.* **94:** 256–259.

van de Sluis B., Rothuizen J., Pearson P.L., van Oost B.A., and Wijmenga C. 2002. Identification of a new copper metabolism gene by positional cloning in a purebred dog population. *Hum. Mol. Genet.* **11:** 165–173.

van de Sluis B., Kole S., van Wolferen M., Holmes N.G., Pearson P.L., Rothuizen J., van Oost B.A., and Wijmenga C. 2000. Refined genetic and comparative physical mapping of the canine copper toxicosis locus. *Mamm. Genome* **11:** 455–460.

van de Sluis B.J., Breen M., Nanji M., van Wolferen M., de Jong P., Binns M.M., Pearson P.L., Kuipers J., Rothuizen J., Cox D.W., Wijmenga C., and van Oost B.A. 1999. Genetic mapping of the copper toxicosis locus in Bedlington terriers to dog chromosome 10, in a region syntenic to human chromosome region 2p13-p16. *Hum. Mol. Genet.* **8:** 501–507.

Wang C.Y., Hawkins-Lee B., Ochoa B., Walker R.D., and She J.X. 1997. Homozygosity and linkage-disequilibrium mapping of the urofacial (Ochoa) syndrome gene to a 1-cM interval on chromosome 10q23-q24. *Am. J. Hum. Genet.* **60:** 1461–1467.

Wang Y., Joh K., and Mukai T. 2002. Identification of a novel isoform of Murr1 transcript, U2mu, which is transcribed from the portions of two closely located but oppositely oriented genes. *Genes Genet. Syst.* **77:** 377–381.

Webb C.B., Twedt D.C., and Meyer D.J. 2002. Copper-associated liver disease in Dalmatians: A review of 10 dogs (1998–2001). *J. Vet. Intern. Med.* **16:** 665–668.

Werner P., Mellersh C.S., Raducha M.G., DeRose S., Acland G.M., Prociuk U., Wiegand N., Aguirre G.D., Henthorn P.S., Patterson D.F., and Ostrander E.A. 1999. Anchoring of canine linkage groups with chromosome-specific markers. *Mamm. Genome* **10:** 814–823.

Wijmenga C. and Klomp L.W. 2004. Molecular regulation of copper excretion in the liver. *Proc. Nutr. Soc.* **63:** 31–39.

Wijmenga C., van den Heuvel L.P., Strengman E., Luyten J.A., van der Burgt I.J., de Groot R., Smeets D.F., Draaisma J.M., van Dongen J.J., De Abreu R.A., Pearson P.L., et al. 1998. Localization of the ICF syndrome to chromosome 20 by homozygosity mapping. *Am. J. Hum. Genet.* **63:** 803–809.

Yuzbasiyan-Gurkan V., Blanton S.H., Cao Y., Ferguson P., Li J., Venta P.J., and Brewer G.J. 1997. Linkage of a microsatellite marker to the canine copper toxicosis locus in Bedlington terriers. *Am. J. Vet. Res.* **58:** 23–27.

18

Cystinuria

Paula S. Henthorn and Urs Giger
Section of Medical Genetics
Department of Clinical Studies
University of Pennsylvania School of Veterinary Medicine
Philadelphia, Pennsylvania 19104-6010

CYSTINURIA IS A DISORDER OF THE AMINO ACID TRANSPORT of cystine and the dibasic amino acids lysine, ornithine, and arginine, which normally occurs in the gut and in the renal tubules after glomerular filtration. Although this defect is evident in epithelial cells of both the kidney and the gut, the clinical manifestations result from a failure of the kidney to reabsorb cystine, leading to excess cystine in the urine. Cystine, but not the dibasic amino acids, has low solubility in acidic urine. Only small rises above the normal concentration exceed the saturation point in acidic urine (Treacher 1966), predisposing the individual to the formation of cystine crystals, uroliths, and urinary obstruction. Cystine calculi can be found from the kidney to the bladder and urethra. The impaired intestinal absorption and renal resorption of cystine and dibasic amino acids do not appear to lead to any clinical deficiency states in dogs and humans, as these are not essential amino acids (Palacin et al. 2001).

Cystinuria was first described in humans nearly 200 years ago when Wollaston reported a new type of renal stone, which he referred to as cystic oxide (Wollaston 1810; Segal and Thier 1995; Palacin et al. 2001). Cystinuria received special recognition as one of the four diseases that Sir Archibald Garrod proposed, in 1908, as "inborn errors of metabolism," caused by lack of specific metabolic enzymes (Garrod 1908). Although it is now known that an enzyme deficiency is not the defect in cystinuria, the disease has served as the prototype for hereditary renal transport disorders.

The postmortem discovery of a cystine stone in a dog was first reported in 1823, and the first documented case of cystinuria in a dog with evidence of a metabolic defect (elevated cystine levels in the urine accompanied by stone formation) appeared in 1935 (Lassaigne 1823; Morris et al. 1935). Brand and colleagues' investigation of this Irish terrier family and their breeding studies demonstrated a genetic basis for cystinuria in dogs (Green et al. 1936). Since then, numerous breeds with increased risk for formation of cystine calculi have been documented.

Major advances in the understanding of the underlying metabolic and molecular basis of the disease have been made in the past decade since the first discovery of mutations associated with cystinuria in human patients (Calonge et al. 1994). Indeed, recent studies have identified mutations in two different genes involved in cystine transport that account for cystinuria in more than 95% of human patients (Font-Llitjós et al. 2005). Despite these achievements, the genetics of cystinuria are complex, with aspects concerning variations in cystinuria phenotypes remaining to be explained (Palacin et al. 2005). Although mutations in one of the cystine transporter genes have been found to cause cystinuria in some dogs, the genetic basis of canine cystinuria appears to be even more complex (Casal et al. 1995; Henthorn et al. 2000).

This chapter focuses on the genetic basis of cystinuria in the dog, with emphasis on genetic heterogeneity in the disease and comparison to the current understanding of cystinuria in humans. With apologies to those whose work has gone uncited here, several important aspects of the disease, such as cystinuria in other species, metabolic studies, and treatment options, are not discussed.

SUMMARY OF GENETIC AND MOLECULAR ASPECTS OF CYSTINURIA IN HUMANS

Human cystinuria was first classified phenotypically into two subtypes by Harris in 1955, based on the presence or absence of aminoaciduria among close relatives of affected individuals (Harris et al. 1955). Type I referred to the group with recessive inheritance, in which carriers had normal urinary amino acid concentrations. Type II cystinuria displayed incompletely dominant inheritance; heterozygotes excreted excess cystine and lysine but did not form stones. Although the classification scheme has been modified during the intervening years, in particular by dividing type II cystinuria into two types referred to as types II and III (Rosenberg et al. 1966; Segal and Thier 1995), the scheme is still a reasonable approximation of the current classification system in which patients are categorized on the basis of urine amino acid analyses of their parents. Today, the cystinuria type I phenotype still refers to cases in which parents

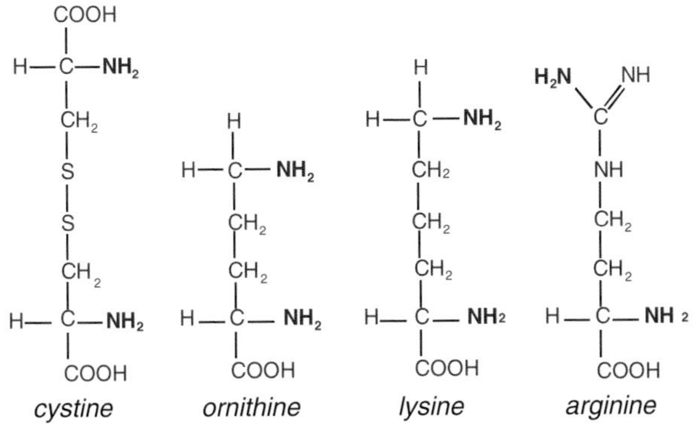

COOH
|
H—C—NH₂
|
CH₂
|
S
|
S
|
CH₂
|
H—C—NH₂
|
COOH

cystine

H
|
H—C—NH₂
|
CH₂
|
CH₂
|
H—C—NH₂
|
COOH

ornithine

H
|
H—C—NH₂
|
CH₂
|
CH₂
|
CH₂
|
H—C—NH₂
|
COOH

lysine

H₂N   NH
 \\   /
   C
   ‖
   NH
   |
   CH₂
   |
   CH₂
   |
H—C—NH₂
   |
   COOH

arginine

Figure 1. Structures of cystine and the dibasic amino acids, commonly referred to as COLA.

and other carriers have a normal urinary amino acid pattern. The second pattern, called type non-I cystinuria, refers to patients whose parents show moderately elevated levels of cystine and the dibasic amino acids, lysine, arginine, and ornithine (Fig. 1). Today, this phenotype is considered to exhibit autosomal dominant inheritance with incomplete penetrance based on the rare incidence of cystine stone formation in obligate carrier individuals who have only moderately elevated urinary cystine (Palacin et al. 2001).

In the mid to late 1990s, the two genes SLC3A1 and SLC7A9 were found to encode the polypeptide subunits of the $b^{0,+}$ amino acid transporter (for reviews, see Palacin et al. 1998, 2001, 2005). SLC3A1 encodes a protein called rBAT, whereas the protein encoded by SLC7A9 is referred to as $b^{0,+}AT$. The system $b^{0,+}$ amino acid transport is one of several known heteromeric amino acid transporters, which are composed of a heavy chain of the SLC3 (solute carrier family 3) family, of which there are two known members, and a light chain of the SLC7 family, of which there are nine identified members. The SLC7 subunits are composed of 12 transmembrane domains and are not glycosylated, and the amino as well as carboxyl ends of the polypeptides appear to reside in the cytoplasm. The SLC7 subunit provides the specific amino acid transport activity and is linked to the heavy subunit by a single disulfide bond. The SLC3 subunits have a single transmembrane domain with a cytoplasmic amino terminus, and the majority of this heavy chain is extracellular and heavily glycosylated. The SLC3 heavy chain appears to be essential for localization of the heteromeric transporter to the plasma membrane (Palacin et al. 2005). Biochemical analyses of human and mouse kidney brush border membranes indicate that $b^{0,+}AT$

heterodimerizes exclusively with rBAT, and that rBAT is produced in excess to $b^{0,+}AT$ (Fernandez et al. 2002).

Over 100 different mutations in SLC3A1 and nearly 70 different mutations in SLC7A9 have been identified in cystinuric human patients (Font-Llitjós et al. 2005). With the exception of a partial gene duplication (Schmidt et al. 2003), all of the SLC3A1 mutations are associated with type I cystinuria. Whereas nearly 85% of the SLC7A9 mutations are associated with the type non-I urinary amino acid pattern, the remaining 15% of SLC7A9 mutations are associated with type-I cystinuria (Dello Strologo et al. 2002; Palacin et al. 2005). In approximately 15% of the cystinuric patients, the disease-causing mutations remain unidentified (Font-Llitjós et al. 2005). These unexplained cystinuria cases may be the result of mutations in the promoter or intronic regions, the combined results of SLC3A1 or SLC7A9 polymorphisms in combination with specific cystinuria-causing mutations, or they may be caused by mutations in as-yet-unidentified genes (Font-Llitjós et al. 2005).

With the large number of completely genotyped cystinuric patients, clinical correlations between phenotype and molecular defects are feasible. Studies utilizing the multinational database assembled by the International Cystinuria Consortium revealed that neither the clinical outcome nor the urinary amino acid pattern was affected by which allele of either SLC3A1 or SLC7A9 was mutated (Dello Strologo et al. 2002). They also showed that cystinuria is more severe in males than in females (first calculi occurred earlier in life and stone formation recurred more frequently) and that frequency of stone formation does not directly correlate with the degree of urinary amino acid excretion. As mentioned above, urolithiasis has been observed in individuals who are obligate carriers for type non-I cystinuria (generally due to a mutant SLC7A9 allele). In a recent study, 1.7% of all type non-I carriers reported urolithiasis (Font-Llitjós et al. 2005).

A mouse model for type I cystinuria exists as a result of a mouse mutagenesis experiment (Peters et al. 2003). The mice have a D140G amino acid substitution in the extracellular domain of the SLC3A1 gene, and a phenotype similar to human and canine type I cystinuria. In addition, a mouse knock-out of the SLC7A9 has been produced, providing a model for type non-I cystinuria that exhibits lower hyperexcretion in heterozygotes and a lower incidence of stone formation in homozygotes compared to humans. Nevertheless, these, as well as the canine cystinurias described below, should prove useful in the study of novel therapies.

CANINE CYSTINURIA

Stone Analysis

In dogs, cystinuria has primarily been diagnosed by analysis of the chemical composition of uroliths that have been removed from dogs suffering from some degree of urinary tract blockage. Due to the availability of free stone analysis at the Universities of Minnesota and California, a large amount of data has been accumulated. Cystine stones (Fig. 2a), like other types of uroliths except phosphate, were originally, and continue to be, found predominantly in male dogs (Weaver 1970; Brown et al. 1977; Case et al. 1992; Ling et al. 1998b; Osborne et al. 1999a). The small size of cystine calculi and anatomical differences between males and females, with males more susceptible to urethral obstruction, particularly at the os penis, make cystine stones less likely to cause obstruction in females, where the urethra is relatively wider and more dilatable than in the male dog (Weaver 1970). Furthermore, because the formation of calculi primarily occurs in dogs after they reach sexual maturity, it was initially concluded that the formation of cystine calculi involved the sex hormones (Wallerstrom and Wagberg 1992). Cystine stones currently account for approximately 1% of all uroliths in dogs in the U.S. (Ling et al. 1998b; Osborne et al. 1999a), with incidences as high as 20% reported from stone analysis laboratories in Europe (see, e.g., Wallerstrom and Wagberg 1992; Hoppe and Denneberg 2001). The mean age of dogs when cystine calculi are removed is between 4.8 years and 5.6 years, which is younger than the average age for other stone types (Brown et al. 1977; Ling et al. 1998b; Hoppe and Denneberg 2001).

Figure 2. (**a**) Cystine uroliths. (**b**) Urinary cystine crystal, with characteristic hexagonal shape. (**c**) Urine cyanide nitroprusside screening test.

Biochemical Characterization

Cystinuric dogs can also be identified by urine analysis for cystine. The presence of urinary cystine crystals, which have a characteristic hexagonal shape (Fig. 2b), is a useful diagnostic sign (Osborne et al. 1986). The cyanide nitroprusside reaction is used as a simple qualitative screening test for cystinuria and gives a positive result when at least 75–125 mg cystine/g creatinine are present in a urine sample (Fig. 2c). Quantification of urinary amino acids by high-pressure liquid chromatography and reabsorption studies can document the degree of the cystine and dibasic amino acid transport defect. In dogs, quantitation is often performed on single urine specimens collected without regard to time since most recent feeding, in contrast to human medicine where a 24-hour urine sample is analyzed. However, when amino acid concentrations are adjusted to the urinary creatinine concentration (Treacher 1966), there is a strong correlation between these cystine values (Hoppe et al. 1993a).

Evidence of Genetic Basis

Cystine stones have been recovered from dogs of approximately 70 breeds (Ling et al. 1998a; Osborne et al. 1999a). There is no evidence of an acquired form of cystinuria, and the following findings have been taken as evidence of a genetic basis, with a summary of breeds with documented evidence in Table 1:

1. Documentation of increased incidence or risk of cystine stone formation in the breed determined by stone analysis laboratories
2. Detection of excess urinary cystine and dibasic amino acids in stone-forming dogs and/or relatives
3. Multiple related cystinuric dogs or test matings that produce affected dogs
4. Stone formation documented in a female dog

For a number of breeds, cystinuria has been documented among close relatives, including the early studies of Irish terriers, where a cystinuric nephew of the propositus was identified (Brand and Cahill 1936). Additional matings among Irish terriers produced 300 dogs, of which 12 were cystinuric (Brand et al. 1940). Interestingly, in a mating of a cystinuric male to two of his sisters, cystine urinary excretion was reported as slightly higher than normal in several dogs of one mating, whereas the other mating produced two cystinuric males and one cystinuric female. It was concluded that the inheritance of cystinuria was recessive, and although most affected dogs were males, an autosomal trait could not be excluded.

The analysis of a Scottish terrier pedigree is suggestive of X-linked recessive inheritance based on the urinary amino acid excretion patterns found in cystine stone-forming dogs (Tsan et al. 1972). A single female of this breed has been reported to have produced a cystine stone (Osborne et al. 1999a). In a similar study, only male basset hounds were cystinuric (Albrecht 1974). However, all unaffected dams were siblings or daughters of cystinuric dogs; hence, autosomal recessive inheritance could not be excluded.

Studies beginning with a cystinuric Newfoundland dog included informative breedings that excluded the X-linked recessive mode of inheritance in this breed. The production of an affected female demonstrated that cystinuria in Newfoundland dogs is inherited in an autosomal recessive fashion. Aminoaciduria is present in males and females at equal frequencies, with clinical signs predominating in males due to anatomical differences (Casal et al. 1995).

Cystinuria exists in another canid species, *Chrysocyon brachyurus*, the maned wolf of South America, where it is a major and well-recognized health problem in captive populations and appears to also have a high incidence in the native population (Bush and Bovee 1978; Bovee et al. 1981). Cystinuria in the maned wolf affects males and females equally and involves cystine and the dibasic amino acids, thus resembling human type I and Newfoundland cystinuria. More recently, cystinuria has also been documented in cats (DiBartola et al. 1991), servals (*Felis serval*), and ferrets (*Mustela putorius furo*) (U. Giger et al., unpubl.).

Canine Type I Cystinuria

Urine amino acid analysis and metabolic studies demonstrated that cystinuria in Newfoundland dogs closely resembles the type I disease in humans (Casal et al. 1995). Affected dogs excreted excessive cystine and the dibasic amino acids in urine at levels similar to cystinuric humans, whereas obligate heterozygotes had no clinical signs and normal urinary cystine and dibasic amino acid excretion patterns, as in human type I cystinuria (Table 2). Affected male dogs formed cystine stones as early as 4–6 months of age, whereas females formed calculi less commonly and later in life. At the time of our studies, only the SLC3A1 gene was known to be associated with cystinuria in humans and thus was an obvious candidate gene. Cloning and sequencing the canine SLC3A1 cDNA and gene from normal and affected animals revealed that affected dogs were

Table 1. Evidence for genetic component by breed

Breed	Urolithiasis in female	Increased risk	Pedigree/Breeding studies	Urine amino acid analysis or renal clearance study
Australian cattle dog		Case et al. (1992)		
Australian shepherd		Case et al. (1992)		
Basenji		Case et al. (1992)		
Basset hound		Ling et al. (1998a); Osborne et al. (1999b); Wallerstrom et al. (1992); Weaver (1970)	Albrecht (1974)	Bovee et al. (1974); Hoppe et al. (1993a)
Brussels griffon	Ling et al. (1998a)			
Boxer				Treacher (1963) cys, lys
Bullmastiff		Case et al. (1992)		
Chihuahua		Case et al. (1992); Ling et al. (1998a)		Hoppe et al. (1993a)
Dachshund		Case et al. (1992); Ling et al. (1998a); Osborne et al. (1999a); Wallerstrom et al. (1992)		Bovee et al. (1974); Hoppe et al. (1993a); Tsan et al. (1972)
English bulldog	Ling et al. (1998a)	Case et al. (1992); Bartges et al. (1994); Ling et al. (1998a); Osborne et al. (1999b)		Bovee et al. (1974); Cornelius et al. (1967); Treacher (1963); Tsan et al. (1972)

Breed				
Labrador retriever				Crane and Turner (1956)
Mastiff		Case et al. (1992); Osborne et al. (1999b)		Tsan et al. (1972)
Newfoundland		Osborne et al. (1999b)	Casal et al. (1995)	Casal et al. (1995)
Pug		Case et al. (1992)		Tsan et al. (1972)
Scottish deerhound	Casal et al. (1995)			
Shetland sheepdog				
Shih Tzu				Hoppe et al. (1993a)
Spaniel, Tibetan				Hoppe et al. (1993a)
Terrier, Irish		Weaver (1970)	Morris et al. 1935; Brand et al. 1940; Tsan et al. (1972)	Hoppe et al. (1993b)
Terrier, Scottish	Osborne et al. (1999a)		Tsan et al. (1972)	
Terrier, silky				
Terrier, Staffordshire		Case et al. (1992); Osborne et al. (1999b); Wallerstrom et al. (1992)		Hoppe et al. (1993a)
Welsh corgi				Bovee et al. (1974)

Table 2. Urine aminoaciduria in canine and human cystinuria

	Cystine[a]	Ornithine[a]	Lysine[a]	Arginine[a]	COLA[a]
Dogs of 9 breeds (not Newfoundland) (Hoppe et al. 1993b) mean (S.D.; range)					
Cystinuric (9 breeds)	368 (291; 17–1115)	152 (215; 18–1062)	1283 (1558; 169–6859)	177 (157; 12–655)	
normal	39 (26; 8–92)	39 (16; 9–73)	190 (58; 56–286)	84 (96; 14–335)	
Newfoundland dogs (Casal et al. 1995) mean ± S.D.					
Cystinuric	1081 ± 446	1930 ± 2414	3494 ± 3667	4552 ± 5173	
normal relatives	54 ± 38	71 ± 36	143 ± 102	83 ± 86	
unrelated normal dogs	<179	<202	<464	<452	
Human (Dello Strologo et al. 2002) mean ± S.D.[b]					
Affected SLC3A1 −/−	2013 ± 1418	2644 ± 2518	7753 ± 4403	2863 ± 2185	15489 ± 8618
SLC3A1 −/+ carrier	70 ± 46	26 ± 26	171 ± 123	24 ± 27	300 ± 153
Affected SLC7A9 −/−	1704 ± 1003	2068 ± 1434	6269 ± 3427	2976 ± 1825	12781 ± 5633
SLC7A9 +/− carrier	475 ± 413	172 ± 216	1692 ± 1309	85 ± 142	2435 ± 1798
normal	52 ± 26	28 ± 23	181 ± 151	17 ± 24	277 ± 172

Although these values were not obtained from the same laboratory, examination of the normal values for each group indicates that they are roughly comparable.
[a]Expressed as nmole/g creatinine.
[b]See also Palacin et al. (2001) for a similar summary.

homozygous for a nonsense mutation in exon 2 of the gene, demonstrating that cystinuria in the Newfoundland is indeed a true homolog of human type I cystinuria. This mutation results in a severely truncated protein of 197 amino acids compared to the normal polypeptide length of 700 amino acids (Henthorn et al. 2000). We recently identified, in cystinuric Labrador retrievers, homozygosity for a different SLC3A1 mutation which also produces a truncated protein and a disease phenotype that closely resembles the Newfoundland type I cystinuria (P. Henthorn et al., in prep.).

Variability and Evidence for Genetic Heterogeneity in Canine Cystinuria: Results from Urine Analysis of Stone-forming Dogs

Whereas Newfoundland and Labrador retrievers clearly have a severe type I form, the clinical and biochemical expression of cystinuria in other breeds seems to be quite variable. Although some differences may be related to degree of clinical and laboratory investigation as well as methodological differences, multiple reports of variant aspects of phenotypes have emerged from independent sources over the years.

Amino Acid Excretion Other Than Cystine

Early studies reported cystinuria both with and without elevated excretion of any of the dibasic amino acids, but increased urinary lysine values were most consistently detected (Crane and Turner 1956; Cornelius et al. 1967; Treacher 1963, 1964a,b; Holtzapple et al. 1971; Bovee et al. 1974). The most comprehensive and most contemporary urine amino acid analysis, on a series of 24 cystinuric dogs of 12 different breeds, reports that cystine excretion is most highly correlated ($p < .001$) with excretion of lysine, ornithine, arginine, and 1-methyl-histidine, and significantly correlated with the excretion of five additional amino acids. Isolated cystinuria was not observed (Hoppe et al. 1993b).

Variable Cystine Excretion in Stone-forming Dogs

Tsan recognized the difficulty in diagnosing aminoaciduria from a single urine sample, when he stated "A low cystine value from a single urinary sample does not prove that the dog is not cystinuric" (Tsan et al. 1972). There may be multiple factors contributing to this phenomenon, such as diet and diurnal variation in cystine secretion (Lindell et al. 1995). This issue is well-demonstrated in the study of cystine stone-forming dogs, of which some had normal cystine excretion over the initial 24-hour time period, whereas excess cystine excretion was apparent in urine analyzed on subsequent occasions (Hoppe et al. 1993b). In addition, there is great variation in the amount of cystine excretion between cystine stone-forming dogs, ranging from 18 to 2610 nmole cystine/mg creatinine (Hoppe and Denneberg 2001). Variation in the underlying renal transport defect is evident from studies measuring the cystine reabsorption in renal clearance studies, which can vary from 50% of the normal reabsorption to active secretion within a breed and between breeds (Bovee et al. 1974).

There have been multiple reports of dogs that have formed cystine stones, but in which excess urine cystine levels were not detected (Holtzapple et al. 1971; Hoppe et al. 1993a). Whether or not these cases represent a nongenetic form of cystinuria, where cystine stone formation occurs in the absence of an amino acid transport defect, they do suggest that high urinary cystine is not the only factor to consider as a cause of cystine urolith formation (Hoppe and Denneberg 2001).

Age-related Variability

Although the mean age at which cystine urolithiasis occurs is about 5 years, 11% of dogs form stones at less than 2 years of age, and there is a tendency for decreased calculus formation in older dogs (Wallerstrom et al. 1992), with huge variation between canine breeds and within individual cystinuric dogs of a breed (Hoppe and Denneberg 2001). There are also instances of stone-forming dogs with documented high cystine values that decreased with age to within the normal range (Hoppe et al. 1993a). Apparent spontaneous resolution of cystinuria has also been observed in a dachshund (Osborne et al. 1999a).

Genetic Heterogeneity

Cystinuria in Newfoundlands (and also Labrador retrievers) clearly differs from the "average" phenotype of cystinuria in other dog breeds (Casal et al. 1995), including higher incidence of stone formation in female dogs, and the juvenile age of stone formation in male dogs (Osborne et al. 1999a). In non-Newfoundland dogs, cystine and the dibasic amino acids appear to be excreted in the urine at lower levels than in Newfoundlands. Table 1 shows the urine cystine and dibasic amino acid levels in Newfoundland dogs (Casal et al. 1995) and those seen in 24 cystinuric dogs of other breeds (Hoppe et al. 1993b). Although the ranges of these two sets of values overlap, mean values for Newfoundland dogs are at least three times as high as for the group of the other 24 cystinuric dogs. How these values relate to human cystinuria is unclear. However, it is interesting to note that the values reported by Hoppe et al. (1993a) are reminiscent of values in obligate carriers of type non-I cystinuria in humans (Table 1) (Palacin et al. 2001).

There is also molecular evidence for genetic heterogeneity in canine cystinuria (Henthorn et al. 2000). Stone-forming dogs of six breeds (Welsh corgi, German short haired pointer, Irish setter, Jack Russell terrier, dachshund, and Swedish laphund) did not have the Newfoundland SLC3A1 mutation. These dogs were also examined for polymorphism in the SLC3A1 gene, at a highly polymorphic DNA sequence within the SLC3A1 gene. Four of the purebred dogs were heterozygous, indicating that their cystinuria was not due to homozygosity for a breed-specific SLC3A1 mutant allele. The two dogs that were homozygous at this polymorphic site, the dachshund and Swedish laphund, did not have any apparent disease-causing mutations in the exons of the SLC3A1 gene. Although sequencing the exons of the SLC3A1 gene would not detect some types of mutations, taken together with clinical and biochemical data, these data argue for more than one molecular mechanism in canine cystinuria (Henthorn et al. 2000).

SUMMARY

Based on the breed predilections and varied clinical features and urinary amino acid patterns, there are at least two types of cystinuria in the dog. Type I cystinuria, documented in Newfoundlands and Labrador retrievers, is homologous to type I cystinuria in humans with mutations in the SLC3A1 gene. With its early age of cystine calculus formation in males, stone occurrence in females, and marked urinary excretion of cystine and the dibasic amino acids, it is clinically distinct from cystinuria in many other breeds. Cystinuria in many other breeds is characterized by cystine urolith formation later in life (not before adulthood) and moderately elevated excretion of cystine and dibasic amino acids, but the association to the type non-I cystinuria in humans is less clear, as no mutations in the SLC7A9 gene have yet been found. Cystinuria in these other dog breeds shows some phenotypic similarities to human type non-I *carriers*, individuals who carry a single mutant allele of the SLC7A9 gene, or rarely, of the SLC3A1 gene. Type non-I carriers have mildly to moderately elevated cystine and dibasic amino acid excretion, and a small percentage of these individuals form cystine stones. Testing of the hypothesis that late-onset cystine stone formation in dogs is due to heterozygosity for particular mutant alleles of either the SLC3A1 or SLC7A9 gene is certainly possible from the molecular level but will require additional genetic and biochemical evidence from within breeds. As much as cystinuria was one of the first examples of an inborn error of metabolism, canine cystinuria also exemplifies the genetic complexity of a metabolic disorder.

ACKNOWLEDGMENTS

The authors thank Drs. Margret L. Casal, Peter F. Jezyk, Kenneth Bovee, and Stanton Segal, as well as Junlong Liu, Adam Seng, Tanya Gidalevitz, and Ping Wang for their contributions to the various canine cystinuria studies performed at the University of Pennsylvania. The generous support for these studies from the National Institutes of Health (RR02512) and the Canine Health Foundation is also acknowledged.

REFERENCES

Albrecht F. 1974. Symptoms, diagnosis, and treatment of cystinuria in the dog [German]. *Kleintier-Praxis* **19:** 202, 204–206, 209–211.

Bartges J.W., Osborne C.A., Lulich J.P., Unger L.K., Koehler L.A., Bird K.A., Clinton C.W., and Davenport M.P. 1994. Prevalence of cystine and urate uroliths in bulldogs and urate uroliths in dalmatians. *J. Am. Vet. Med. Assoc.* **204:** 1914–1918.

Bovee K.C., Thier S.O., Rea C., and Segal S. 1974. Renal clearance of amino acids in canine cystinuria. *Metab. Clin. Exp.* **23:** 51–58.

Bovee K.C., Bush M., Dietz J., Jezyk P., and Segal S. 1981. Cystinuria in the maned wolf of South America. *Science* **212:** 919–920.

Brand E. and Cahill G.F. 1936. Canine cystinuria iii. *J. Biol. Chem.* **114:** xv–xvi.

Brand E., Cahill G.F., and Kassell B. 1940. Canine cystinuria v. Family history of two cystinuric Irish terriers and cystine determinations in dog urine. *J. Biol. Chem.* **133:** 431–436.

Brown N.O., Parks J.L., and Greene R.W. 1977. Canine urolithiasis: Retrospective analysis of 438 cases. *J. Am. Vet. Med. Assoc.* **170:** 414–418.

Bush M. and Bovee K.C. 1978. Cystinuria in a maned wolf. *J. Am. Vet. Med. Assoc.* **173:** 1159–1162.

Calonge M.J., Gasparini P., Chillaron J., Chillon M., Gallucci M., Rousaud F., Zelante L., Testar X., Dallapiccola B., Di Silverio F., et al. 1994. Cystinuria caused by mutations in *rBAT*, a gene involved in the transport of cystine [see comments]. *Nat. Genet.* **6:** 420–425.

Casal M.L., Giger U., Bovee K.C., and Patterson D.F. 1995. Inheritance of cystinuria and renal defect in Newfoundlands. *J. Am. Vet. Med. Assoc.* **207:** 1585–1589.

Case L.C., Ling G.V., Franti C.E., Ruby A.L., Stevens F., and Johnson D.L. 1992. Cystine-containing urinary calculi in dogs: 102 cases (1981–1989). *J. Am. Vet. Med. Assoc.* **201:** 129–133.

Cornelius C.E., Bishop J.A., and Schaffer M.H. 1967. A quantitative study of amino aciduria in dachshunds with a history of cystine urolithiasis. *Cornell Vet.* **57:** 177–183.

Crane C.W. and Turner A.W. 1956. Amino acid patterns of urine and blood plasma in a cystinuric Labrador dog. *Nature* **177:** 237–238.

Dello Strologo L., Pras E., Pontesilli C., Beccia E., Ricci-Barbini V., de Sanctis L., Ponzone A., Gallucci M., Bisceglia L., Zelante L., et al. 2002. Comparison between *SLC3A1* and *SLC7A9* cystinuria patients and carriers: A need for a new classification. *J. Am. Soc. Nephrol.* **13:** 2547–2553.

DiBartola S.P., Chew D.J., and Horton M.L. 1991. Cystinuria in a cat. *J. Am. Vet. Med. Assoc.* **198:** 102–104.

Fernandez E., Carrascal M., Rousaud F., Abian J., Zorzano A., Palacin M., and Chillaron J. 2002. Rbat-b$^{0,+}$ at heterodimer is the main apical reabsorption system for cystine in the kidney. *Am. J. Physiol. Renal Physiol.* **283:** F540–F548.

Font-Llitjós M., Jimenez-Vidal M., Bisceglia L., Di Perna M., de Sanctis L., Rousaud F., Zelante L., Palacin M., and Nunes V. 2005. New insights into cystinuria: 40 new mutations, genotype-phenotype correlation, and digenic inheritance causing partial phenotype. *J. Med. Genet.* **42:** 58–68.

Garrod A.E. 1908. The croonina lectures on inborn errors of metabolism. *Lancet* **2:** 1, 73, 142, 214.

Green D.F., Morris M.L., Cahill G.F., and Brand E. 1936. Canine cystinuria ii. Analysis of cystine calculi and sulfur distribution in the urine. *J. Biol. Chem.* **114:** 91–94.

Harris H., Mittwoch U., Robson E.B., and Warren F.L. 1955. Phenotypes and genotypes in cystinuria. *Ann. Hum. Genet.* **20:** 57–91.

Henthorn P.S., Liu J., Gidalevich T., Fang J., Casal M.L., Patterson D.F., and Giger U. 2000. Canine cystinuria: Polymorphism in the canine *SLC3A1* gene and identification of a nonsense mutation in cystinuric Newfoundland dogs. *Hum. Genet.* **107:** 295–303.

Holtzapple P.G., Rea C., Bovee K., and Segal S. 1971. Characteristics of cystine and lysine transport in renal jejunal tissue from cystinuric dogs. *Metab. Clin. Exp.* **20:** 1016–1022.

Hoppe A., and Denneberg T. 2001. Cystinuria in the dog: Clinical studies during 14 years of medical treatment (erratum *J. Vet. Intern. Med.* [2001] **15:** 594). *J. Vet. Intern. Med.* **15:** 361–367.

Hoppe A., Denneberg T., Jeppsson J.O., and Kagedal B. 1993a. Canine cystinuria: An extended study on the effects of 2-mercaptopropionylglycine on cystine urolithiasis and urinary cystine excretion. *Br. Vet. J.* **149:** 235–251.

———. 1993b. Urinary excretion of amino acids in normal and cystinuric dogs [see comment]. *Br. Vet. J.* **149:** 253–268.

Lassaigne J.L. 1823. Observation sur l'existence de l'oxide cystique dans un calcul vesical du chien, et essai analytique sur la composition elementaire de cette substance particuliere. *Ann. Chim. Phys.* **23:** 328–334.

Lindell A., Denneberg T., Jeppsson J.O., and Tiselius H.G. 1995. Measurement of diurnal variations in urinary cystine saturation. *Urol. Res.* **23:** 215–220.

Ling G.V., Franti C.E., Ruby A.L., and Johnson D.L. 1998a. Urolithiasis in dogs. II. Breed prevalence, and interrelations of breed, sex, age, and mineral composition. *Am. J. Vet. Res.* **59:** 630–642.

Ling G.V., Franti C.E., Ruby A.L., Johnson D.L., and Thurmond M. 1998b. Urolithiasis in dogs. I. Mineral prevalence and interrelations of mineral composition, age, and sex. *Am. J. Vet. Res.* **59:** 624–629.

Morris M.L., Green D.F., Dinkel J.H., and Brand E. 1935. Canine cystinuria. An unusual case of urinary calculi in the dog. *N. Am. Vet.* **16:** 16–19.

Osborne C.A., O'Brien T.D., Ghobrial H.K., Meihak L., and Stevens J.B. 1986. Crystalluria. Observations, interpretations, and misinterpretations. *Vet. Clin. N. Am. Small Anim. Pract.* **16:** 45–65.

Osborne C.A., Sanderson S.L., Lulich J.P., Bartges J.W., Ulrich L.K., Koehler L.A., Bird K.A., and Swanson L.L. 1999a. Canine cystine urolithiasis. Cause, detection, treatment, and prevention. *Vet. Clin. N. Am. Small Anim. Pract.* **29:** 193–211.

Osborne C.A., Lulich J.P., Polzin D.J., Sanderson S.L., Koehler L.A., Ulrich L.K., Bird K.A., Swanson L.L., Pederson L.A., and Sudo S.Z. 1999b. Analysis of 77,000 canine uroliths. *Vet. Clin. N. Am. Small Anim. Pract.* **29:** 17–38.

Palacin M., Estevez R., Bertran J., and Zorzano A. 1998. Molecular biology of mammalian plasma membrane amino acid transporters. *Physiol. Rev.* **78:** 969–1054.

Palacin M., Goodyer P., Nunes V., and Gasparini P. 2001. Cystinuria. In *The metabolic and molecular bases of inherited disease* (ed. C.R. Scriver et al.), pp. 4909–4932. McGraw-Hill, New York.

Palacin M., Nunes V., Font-Llitjos M., Jimenez-Vidal M., Fort J., Gasol E., Pineda M., Feliubadalo L., Chillaron J., and Zorzano A. 2005. The genetics of heteromeric amino acid transporters. *Physiology* **20:** 112–124.

Peters T., Thaete C., Wolf S., Popp A., Sedlmeier R., Grosse J., Nehls M.C., Russ A., and Schlueter V. 2003. A mouse model for cystinuria type I. *Hum. Mol. Genet.* **12:** 2109–2120.

Rosenberg L.E., Downing S., Durant J.L., and Segal S. 1966. Cystinuria: Biochemical evidence for three genetically distinct diseases. *J. Clin. Invest.* **45:** 365–371.

Schmidt C., Vester U., Wagner C.A., Lahme S., Hesse A., Hoyer P., Lang F., Zerres K., and Eggermann T. 2003. Significant contribution of genomic rearrangements in *SLC3A1* and *SLC7A9* to the etiology of cystinuria. *Kidney Int.* **64:** 1564–1572.

Segal S. and Thier S.O. 1995. Cystinuria. In *Metabolic and molecular bases of inherited disease* (ed C. Scriver et al.), pp. 3581–3601, McGraw-Hill, New York.

Treacher R.J. 1963. The amino-aciduria of canine cystine-stone disease. *Res. Vet. Sci.* **4:** 556–567.

———. 1964a. The aetiology of canine cystinuria. *Biochem. J.* **90:** 494–498.

———. 1964b. Quantitative studies on the excretion of the basic amino acids in canine cystinuria. *Br. Vet. J.* **120:** 178–185.

———. 1966. Urolithiasis in the dog. II. Biochemical aspects. *J. Small Anim. Pract.* **7:** 537–547.

Tsan M.F., Jones T.C., Thornton G.W., Levy H.L., Gilmore C., and Wilson T.H. 1972. Canine cystinuria: Its urinary amino acid pattern and genetic analysis. *Am. J. Vet. Res.* **33:** 2455–2461.

Wallerstrom B.I. and Wagberg T.I. 1992. Canine urolithiasis in Sweden and Norway: Retrospective survey of prevalence and epidemiology. *J. Small Anim. Pract.* **33:** 534–539.

Wallerstrom B.I., Wagberg T.I., and Lagergren C.H. 1992. Cystine calculi in the dog: An epidemiologic retrospective study. *J. Small Anim. Pract.* **33:** 78–84.

Weaver A.D. 1970. Canine urolithiasis: Incidence, chemical composition and outcome of 100 cases. *J. Small. Anim. Pract.* **11:** 93–107.

Wollaston W.H. 1810. On cystic oxide: A new species of urinary calculus. *Trans. R. Soc. Lond. B Biol. Sci.* **100:** 223.

19

Molecular Genetics of Dilated Cardiomyopathy

Polona Stabej

Department of Clinical Sciences of Companion Animals
Faculty of Veterinary Medicine, Utrecht University
3508TD Utrecht, The Netherlands

Kathryn M. Meurs

Department of Veterinary Clinical Science
The Ohio State University College of Veterinary Medicine
Columbus, Ohio 43210-1089

Bernard A. van Oost

Department of Animals, Science, and Society
Faculty of Veterinary Medicine, Utrecht University
3508TD Utrecht, The Netherlands

CARDIOMYOPATHIES ARE DEFINED AS DISEASES of the myocardium associated with cardiac dysfunction. According to the WHO/ISFC criteria, cardiomyopathies in human beings are classified by the major pathophysiology as dilated cardiomyopathy, hypertrophic cardiomyopathy, restrictive cardiomyopathy, and right ventricular cardiomyopathy (Richardson et al. 1996). In dogs, dilated cardiomyopathy and arrhythmogenic right ventricular cardiomyopathy are the most common inherited forms of myocardial disease.

CANINE FAMILIAL CARDIOMYOPATHIES

Familial Dilated Cardiomyopathy

Dilated cardiomyopathy (DCM) is a primary myocardial disease characterized by cardiac enlargement and impaired contractile function of one or both ventricles. Although canine DCM is traditionally discussed as one disease, significant variation in the age of onset, presenting complaint, clinical features, and disease progression have been observed among and within breeds (Freeman et al. 1996; Tidholm and Jonsson 1996; Calvert et al. 1997; Kittleson et al. 1997; Brownlie and Cobb 1999; Meurs et al. 2001a,b; Sleeper et al. 2002), indicating the genetic diversity of DCM. The metabolic basis remains unknown, but deficiencies in myocardial L-carnitine in boxers (Keene et al. 1991) and taurine in American cocker spaniels (Kittleson et al. 1997) have been observed.

DCM is generally seen in large and medium-sized adult dog breeds. Some breeds are clearly overrepresented, and this seems to vary somewhat with geographical location. In North America, the Doberman pinscher, Irish wolfhound, Great Dane, and American cocker spaniels are among the most commonly reported breeds (Monnet et al. 1995; Sisson et al. 2000). Some European references suggest an increased incidence in the Airedale terrier, Doberman pinscher, Newfoundland, and English cocker spaniel (Tidholm and Jonsson 1997). Although familial, the mode of inheritance of DCM seems to vary between breeds and has not been clearly defined. An autosomal recessive inheritance appears likely in the Portuguese water dog and an autosomal dominant mode in Doberman pinschers (Hammer et al. 1996; Alroy et al. 2000), whereas a male predominance suggests an X-chromosomal recessive trait in dalmatians and Great Danes (Freeman et al. 1996; Meurs et al. 2001a). DCM is predominantly one of left ventricular dysfunction, but biventricular involvement and heart failure may be noted. Arrhythmias, such as ventricular premature complexes, tachycardia, and atrial fibrillation, occur commonly. The observed variation among breeds is illustrated below with some examples (Table 1).

DCM has been reported in both American and English cocker spaniels. Dilation of the left ventricle and a reduction in left ventricular contractility were observed in 5- to 13-year-old American cocker spaniels. It is not known whether this is a familial problem in the American cocker, although it should be noted that the strong breed specificity of decreased taurine levels may suggest a familial problem related to taurine metabolism or absorption (Kittleson et al. 1997). A relationship of DCM to taurine has not been well studied in the English cocker spaniels. Many reported dogs were from the same kennel, which suggests a heritable component. Profound left ventricular enlargement is frequently observed, some dogs die suddenly, and others show a prolonged, fairly asymptomatic course of disease (years) (Gooding et al. 1986; Wotton 1998).

Table 1. Breed characteristics of DCM in dogs

Breed	Mean age at diagnosis	DCM phenotype			Survival after diagnosis	Mode of inheritance	Histopathology	References
		congestive heart failure	arrhythmias	sudden death				
Doberman pinscher	6.7 years	left-sided	VT, (S)VPCs, AF	yes	45.5 days	autosomal dominant	fatty infiltration	Calvert et al. (1982, 1997); Hammer et al. (1996); Tidholm and Jonsson (2005)
Boxer	6 years	left-sided	(S)VPCs, AF	yes	3–6 months	autosomal dominant	fatty infiltration	Harpster (1983, 1991); Baumwart et al. (2005)
Great Dane	4.8 years	left-sided or biventricular	AF	no	N.A.	X-linked	N.A.	Meurs et al. (2001a)
Portuguese water dog	13 weeks	left-sided	no	no	1–5 days	autosomal recessive	attenuated wavy fiber type	Alroy et al. (2000); Dambach et al. (1999)
Irish wolfhound	4.2 years	left-sided or biventricular	AF, (S)VPCs	yes	5.1(0,07–14) months	autosomal dominant	N.A.	Vollmar (2000)
English cocker spaniel	Adulthood	left-sided	no	yes	N.A.	N.A.	Necrosis, foci of collagen	Staaden (1981)
Newfoundland	5 years	left followed by right-sided	AF	yes	5 months	autosomal dominant	Attenuated wavy fiber type	Tidholm and Jonsson (1996)
Presa canario	7.5 years	left-sided	AF, VPCs, VT	no	8.5 and 15 months	N.A.	N.A.	Morales et al. (2001)
Dalmatian	6.8 years	left-sided	SVT	no	10 months	N.A.	N.A.	Freeman et al. (1996)

(S)VPCs: (Supra)ventricular premature contractions; AF: atrial fibrillation; (S)VT: (supra)ventricular tachycardia; N.A.: data not available.

The Doberman pinscher is one of the most commonly reported breeds affected with DCM in North America and Europe (Monnet et al. 1995; Calvert and Meurs 2000; Domanjko-Petrič et al. 2002) and is characterized by left and/or biventricular failure, often with atrial and/or ventricular tachyarrhythmias. DCM in the adult Doberman pinscher appears to be very rapidly progressive, with the median survival time of only 10 weeks once heart failure has developed. Some Doberman pinschers suffer from collapsing episodes or die suddenly before left ventricular dilation or contractile dysfunction is ever evident (Calvert et al. 1982; O'Grady and Horne 1992; Domanjko-Petrič et al. 2002).

In contrast to other DCMs, affected Portuguese water dogs (Dambach et al. 1999; Sleeper et al. 2002) develop congestive heart failure with an average age of 13 weeks and die within days thereafter. An autosomal recessive mode of inheritance is suspected.

Arrhythmogenic Right Ventricular Cardiomyopathy in the Boxer

Since the early 1980s, the term boxer cardiomyopathy has been used to describe adult boxers with ventricular arrhythmias which can be asymptomatic and sometimes causes collapse, sudden death, or systolic dysfunction with congestive heart failure (Basso et al. 2004). Although few boxers have left ventricular dilatation with left or biventricular heart failure, the majority suffer primarily from arrhythmias as a result of right ventricular myocardial disease. Recent studies have demonstrated that the disease has many similarities to arrhythmogenic right ventricular cardiomyopathy (ARVC) in humans (Basso et al. 2004). ARVC in boxers appears to be inherited as an autosomal dominant trait (Meurs et al. 1999) with variable penetrance.

The cardiomyopathy phenotype varies between breeds, whereas it is more or less homogeneous within a particular breed. We suggest two possible explanations for inter- and intra-breed phenotypic heterogeneity: (1) the existence of diverse DCM mutations and (2) the presence of the same mutations resulting in various phenotypes due to different genetic backgrounds. This suggestion is based on the extensive body of knowledge obtained recently for mostly human familial DCM that is discussed in the following sections.

GENOTYPES AND PHENOTYPES OF FAMILIAL DCM IN HUMAN

DCM is a leading cause of cardiovascular morbidity and mortality in human medicine. No single clinical parameter has been found to distinguish familial from nonfamilial DCM. Based on family history, the cause is predicted to be genetic in at least 35% of cases (Grunig et al. 1998). Familial DCM is a heterogeneous group of disorders with autosomal dominant, autosomal recessive, X-chromosomal recessive, or mitochondrial modes of transmission (Fatkin and Graham 2002).

In past years, recognition of numerous DCM-causing mutations boosted the understanding of molecular mechanisms leading to DCM. Moreover, it contributed considerably toward understanding of the physiology of the heart. Until recently, DCM was described as idiopathic, indicating that the disease arises from an obscure or unknown cause (Dec and Fuster 1994). One of the earliest reports on familial autosomal dominant DCM with variable penetrance dates back to 1978 and suggests a mutant gene as a cause (Ross et al. 1978). Ten years later, Berko and Swift (1987) described a family with X-linked DCM, which was succeeded by identification of the first DCM-causing mutation—a deletion in the dystrophin promoter (Muntoni et al. 1993). The genes causing autosomal dominant DCM were somewhat harder to decipher. Mutations in the actin gene were the first to be revealed by the candidate gene approach (Olson et al. 1998). In the years between 1999 and 2005, the availability of a number of polymorphic markers in the human genome and high-throughput sequencing/genotyping methods enabled identification of mutations in several genes either by the candidate gene approach or by genome-wide linkage analysis.

Autosomal Dominant and Recessive Trait

Most frequently, DCM in human beings is inherited by an autosomal dominant trait with incomplete penetrance. The phenotypes of patients vary from mild, late-onset DCM to severe DCM beginning in infancy. Death may result from severe heart failure and/or ventricular arrhythmias (Fatkin and Graham 2002), and hence pure DCM is distinguished clinically from DCM with conduction defects. Thus far, 16 genes have been causally implicated in autosomal dominant DCM; in some cases, different mutations in the same gene are responsible for either pure DCM or DCM with abnormal conduction (Olson et al. 2001). In addition, a mutation in the cardiac troponin I gene (*TNNI3*) was determined to be the cause of an autosomal recessive form of DCM (Murphy et al. 2004).

X-linked DCM

Two genes have so far been associated with X-chromosomal recessive forms of DCM. Mutations of tafazzin *(TAZ)* cause severe infantile DCM, a disorder allelic to the Barth syndrome (D'Adamo et al. 1997). The tafazzin gene codes for an acyltransferase, which affects mitochondrial cardiolipin levels and mitochondrial dysfunction forms, likely the basis of the cardiomyopathy (Barth et al. 2004). X-linked DCM can be caused by various dystrophin (*DYS*) mutations and is allelic to Duchenne and Becker muscular dystrophies (DMD and BMD). These DCM patients lack the skeletal muscle signs (Nigro et al. 1994; Muntoni et al. 1997).

Matrilineal Inheritance

Deletions and other mutations in various mitochondrial transfer RNAs have been observed in association with DCM (Suomalainen et al. 1992; Marin-Garcia et al. 2001; http://www.neuro.wustl.edu/neuromuscular/ mitosyn.html#cardiac).

PATHOPHYSIOLOGY OF DCM

Cardiomyocyte Structure

Proteins of the Sarcomere—Contractile Unit of the Heart

The interior of the cardiac myocyte (see Fig. 1) contains bundles of longitudinally arranged myofibrils that have a characteristic striated appearance formed by repeating sarcomeres. The sarcomere is the structural and functional unit of cardiac muscle comprising thin (α-cardiac actin, α-tropomyosin, troponins C, I, T) and thick (myosin and myosin-binding proteins C, H, X) filaments. Each sarcomere has an I-band (composed of thin filaments only), A-band (composed of overlapping thin and thick filaments), and M-band (composed of thick filaments only). The scaffolding for the thick and thin filaments is provided by the giant protein titin that spans from the Z-line to the M-line and myomesins (titin-associated proteins) (Fatkin and Graham 2002). Z-discs demarcate the sarcomeres and cross-link the myofilaments into a highly ordered 3-dimensional lattice. Z-discs are linked with one another, to the sarcolemma, and to the nuclear envelope by the desmin network of the cytoskeleton. The major connection of the Z-disc to the rest of the sarcomere occurs through the titin. Titin's amino-terminal domains insert into the Z-disc, where the titin molecules from opposing sarcomeres overlap and titin's carboxyl terminus attaches to the M-line of the sarcomere and makes contact with the head–neck interface of crossbridges through its interaction with the myosin-binding protein C (MyBP-C)

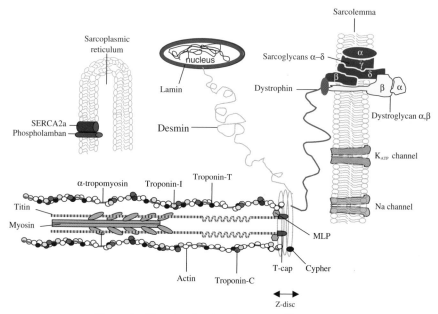

Figure 1. Structural network in a cardiomyocyte.

(Pyle and Solaro 2004). These Z-disc proteins function as a physical anchor for myofilaments and the cytoskeleton and have a major role in reception, transduction, and transmission of mechanical and biochemical signals.

Extrasarcomeric Proteins of the Cytoskeleton— The Dystrophin Glycoprotein Complex

The dystrophin glycoprotein complex (DGC) is a multicomponent complex that provides a mechanical link between the intracellular and extracellular matrix. It has structural and signal transduction properties and consists of dystrophin, syntrophins, dystroglycans, sarcoglycans, caveolin-3, NO-synthase, and sarcospan (Lapidos et al. 2004). Dystrophin links the F-actin and β-dystroglycan that connects to α-dystroglycan. The latter has an extracellular region attached to laminin 2 (a component of the extracellular matrix), which on its turn binds to collagen IV. The sarcoglycans are connected to α-dystroglycan through α-sarcoglycan. This is how a continuous connection is formed between the cytoskeletal actin network and the extracellular matrix (Rybakova et al. 2000). Mutations in genes coding for DGC components have been identified to cause DCM in experimental animals and humans.

DCM Genes and Molecular Mechanisms of Inherited DCM

Over the past years, several molecular pathways involved in DCM were elucidated by genetic studies, and we discuss below three main pathways that can be distinguished.

Disturbed Integrity of the Cytoskeleton

At the molecular level, the loss or structural damage of dystrophin results in destabilization of the rest of the DGC complex and impaired mechanical link between the sarcolemma and the extracellular matrix (Cohen and Muntoni 2004). Aside from the 16 *DYS* mutations, a missense mutation and a 3-bp deletion found in the sarcoglycan delta gene of DCM patients were also predicted to affect the secondary structure of the complex and to disturb the integrity of the cytoskeleton (Tsubata et al. 2000).

Actin is a sarcomeric thin filament, a major component of the cell cytoskeleton that participates in more protein–protein interactions than any other known protein (Dominguez 2004). One end forms cross bridges with the myosin and the other end is anchored in the Z-disc. Mutations in actin in patients with DCM were found to impair the cytoskeletal connection between actin and Z-disc proteins (Olson et al. 1998).

Two DCM-causing mutations in the β-myosin heavy chain gene (*MYH7*) are located in the part of the myosin heavy chain that interacts with the actin thin filament. They likely alter the stereospecific interaction between actin and myosin, which is essential for the power of cardiac contraction (Kamisago et al. 2000; Daehmlow et al. 2002). A third *MYH7* mutation resides in the converter region that transmits the movement from the head of the myosin to the neck, thereby hindering the propelling of the thick filament (Kamisago et al. 2000).

Lamin A and C proteins, encoded in the lamin A/C gene (*LMNA*), are intermediate filaments forming a two-dimensional matrix at the nuclear lamina. Lamins of the A-type maintain stability and regulation of the transcription factors required for the differentiation of adult stem cells. Mutations in the *LMNA* have been shown to cause a host of diseases, including cardiomyopathies (Arbustini et al. 2002; Burke and Stewart 2002; Hutchinson and Worman 2004).

Disturbed Calcium Kinetics and Sensitivity

Mutations of the cardiac troponin T gene (*TNNT2*) can cause dilated, hypertrophic, and restrictive cardiomyopathy (Gomes and Potter 2004).

DCM-causing *TNNT2* mutations cause diminished calcium sensitivity of force generation and in some cases display gender-specific phenotypic differences (Kamisago et al. 2000; Li et al. 2001; Lu et al. 2003; Stefanelli et al. 2004).

Impaired Intracellular Signaling Mechanisms

The research of Knoll et al. (2002) on muscle LIM protein gene (*MLP*) and titin-cap (*TCAP*) mutations provides functional evidence that the chamber dilatation in DCM is related to an inability of the cardiomyocytes to sense the mechanical stretch stimulus and to generate a primary effect on muscle tension. This molecular model suggests that the elastic segments within the titin-I-band domains serve as the intrinsic cardiac mechanical stress sensor that requires the binding of MLP and TCAP. In a DCM heart, the mutated MLP causes impaired anchoring of the Z-disc to the T-cap/titin complex and leads to a loss of elasticity of the titin elastic spring. The primary result of the latter is defective cardiomyocyte stretch sensing that leads to overstretching of the myocytes and activation of the cell death pathways (Knoll et al. 2002).

The giant molecule titin is the third most abundant of the cardiac proteins, after myosin and actin. It belongs to the proteins of the sarcomeric cytoskeleton and is the largest natural protein currently known. One molecule spans half a sarcomere from the Z-disc to the M-line (Labeit et al. 1997). The titin is dynamic in structure and function. It is responsible for the structural integrity of the sarcomere by acting as a scaffold, gives rise to passive muscle stiffness, influences active force development, and is essential for sarcomerogenesis; titin-based protein complexes have a role in signaling as biomechanical sensors (Gregorio et al. 1999; for review, see Granzier and Labeit 2004). Because titin is such a large, multifunctional protein, it is a prominent target for mutations that give rise to muscle diseases. It was suggested that the DCM-causing *TTN* mutations either affect the Z-disc–theletonin–titin component of the cardiac stretch sensor or give rise to a truncated, nonfunctional protein (Gerull et al. 2002; Itoh-Satoh et al. 2002; Knoll et al. 2002).

Phospholamban (PLN) is a small transmembrane phosphoprotein of 52 amino acids that plays an important role in cardiac contraction (occurs with elevated cytoplasmic $[Ca^{2+}]$) and relaxation (triggered by Ca^{2+} uptake into the sarcoplasmic reticulum through the Ca^{2+}-ATP (SERCA2a) pump, plasma membrane Ca^{2+}-ATPases (PMCAs), and Na^+/Ca^{2+} exchangers). Phospholamban in its dephosphorylated state binds to and inhibits the SERCA2a pump activity, whereas the phosphorylated PLN reverses the Ca^{2+} pump inhibition (MacLennan and Kranias 2003). The mutated PLNR9C found in a family with DCM shows enhanced affinity for protein kinase A (PKA), which becomes trapped in a mutant PLN–PKA complex and cannot dissociate and phosphorylate wild-type PLN molecules. DCM patients with a PLN mutation therefore have a chronically inhibited SERCA2a pump leading to DCM (MacLennan and Kranias 2003; Schmitt et al. 2003).

Cypher/ZASP is a gene that expresses a Z-disc-associated cytoskeletal protein. A missense mutation located in the third LIM domain of Cypher causes an increased binding affinity of mutated Cypher for protein kinase C. Consistent with the findings of mutations in MLP and PLN, Cypher/ZASP mutations cause abnormal recruitment of molecules participating in intracellular signaling (Arimura et al. 2004).

Mutations in two ion channel genes, SCN5A and ABCC9 (Bienengraeber et al. 2004; McNair et al. 2004), have been found in DCM. Additionally, ryanodine receptor (ion channel) mutations have been found in arrhythmogenic right ventricular cardiomyopathy, a phenotype related to DCM (Tiso et al. 2001). The phenotypic commonalities of mutations in the ion channel genes are the conduction disturbances associated with DCM.

In conclusion, the three pathways described certainly interlink, as altered connections between the proteins of the cytoskeleton can result in impaired signaling mechanisms (mutations in MLP) or cause disturbed calcium kinetics (mutations in TNNT2). Another example is mutated phospholamban (PLNR9C), which causes disturbed Ca^{2+} kinetics by hampering a signaling molecule protein kinase (PKA). Considering the incomplete knowledge of the physiology of the heart, the precise description of the pathways leading from a mutated gene to various forms of DCM are yet to be discovered.

THE DOG AS A MODEL OF HUMAN DCM

Parallels between the myocardial disease in man and dog were already drawn in the 1960s (Wagner 1968), and the dog has been suggested as a naturally occurring large animal model of human DCM (Smucker et al. 1990). The genetic basis of canine DCM has been recognized for the past 20 years (Staaden 1981; Calvert et al. 1982). Researchers have been gathering family material from dog breeds with frequently occurring DCM, and genome-wide linkage scans and candidate gene approaches have been used to identify the mutations causing DCM in various breeds (Meurs et al. 2001a,b; Spier et al. 2001; Dukes-McEwan and Jackson 2002; Jakobs et al. 2004; Stabej et al. 2005b). In families of Newfoundlands and Irish wolfhounds, genome-wide linkage studies with over 200 markers failed to detect linkage (Dukes-McEwan and Jackson 2002; Jakobs et al. 2004). In the Doberman pinscher, disease-causing mutations in the desmin, δ-sarcoglycan, and α-tropomyosin were excluded as a cause of DCM by typing of the microsatellite markers and/or single nucleotide polymorphisms in the DCM and DCM nonaffected dogs (Stabej et al. 2004, 2005b; Stabej 2005). Sequencing of the phospholamban gene excluded mutations in Doberman pinschers, Newfoundlands, and Great Danes with DCM (Stabej et al. 2005a).

Genotyping of the titin gene (*TTN*) with marker REN252E18 (Guyon et al. 2003), however, revealed a significant difference in the allele frequencies of REN252E18 between the DCM and DCM nonaffected groups (Chi-square test; $p = 0.002$). Analysis of additional markers revealed a striking difference of haplotype distribution. Unexpectedly, the DCM group displays a variety of haplotypes, whereas the nonaffected group is strongly enriched for one haplotype. These results suggest that the titin allele which is common to the nonaffected dogs confers protection against DCM and that the breed as a whole is susceptible to DCM due to another genetic cause (Stabej 2005). This genotype difference in the titin gene provides the first molecular genetic basis of canine DCM.

Despite the extensive knowledge of human cardiomyopathies, the molecular genetic background of this intricate heart disease remains ambiguous. Linkage studies in multiple families with DCM have suggested loci in which genes and the DCM-causing mutations remain to be identified. Another unsolved DCM riddle is the observed variability in phenotype of DCM among individuals with identical mutations (Olson et al. 2001). Furthermore, the male gender predominance in some families remains unexplained (Stefanelli et al. 2004). Finally, the variability of outcome of DCM patients highlights the importance of genetic background (modifier genes) and environmental factors in phenotype determination. Modifier genes are genes that are not involved in the genesis of the disease but modify the severity of the phenotypic expression once the disease has developed (Le Corvoisier et al. 2003a). The heterogeneity of the human population coupled with the generally modest effect of the modifier genes makes identification of modifier genes a complex task requiring complementary approaches. Mouse models generated to develop DCM circumvent the problem of heterogeneity to some degree. Quantitative trait locus (QTL) mapping in an experimental mouse model of DCM induced by cardiac-specific overexpression of calsequestrin (CSQ), and characterized by a strong strain-specific variability in phenotype, resulted in identification of several QTLs that differentially modify the cardiac phenotype. Although experimental models like the CSQ mouse recapitulate a number of human/dog DCM key features, the modifier loci identified in such a model are likely to be specific for DCM induced by CSQ overexpression (Le Corvoisier et al. 2003b). To overcome the limitation of experimental models and high heterogeneity of the human, dogs seem to be a perfect naturally occurring animal model that can help untangle the complex effects of the genetic background on the development of DCM. Studies of linkage disequilibrium (LD) in dogs demonstrated that the LD varies between breeds and extends up to 100 times farther than does LD in isolated human populations. In addition, the haplotype diversity in dog breeds is relatively low, with two and three haplotypes accounting for 80% of the chromosomes in each breed. Extensive LD, coupled with a high degree of haplotype sharing, makes whole-genome association scans more amenable in dog breeds than in human populations (Sutter et al. 2004). Given the knowledge about the genetic structure of dog breeds (Parker et al. 2004), as well as the availability of important genetic resources such as the 7.8× dog sequence (http://www.genome.ucsc.edu) and a high-resolution genetic map (Breen et al. 2004), the cloning of DCM-causing genes and modifier genes will be expedited.

Another disadvantage of human DCM is generally the small families available for study. Several DCM genes have been identified by screening hundreds of unrelated patients for mutations in DCM candidate genes. For example, Olson et al. (2001) screened 350 unrelated patients for mutations in *TPM1* and identified unique missense mutations in two cases, and Itoh-Satoh et al. (2002) screened 120 unrelated patients, identifying three *TTN* mutations. The family material obtained with such an approach is usually limited, and expanding the material to make it sufficient for linkage analysis often proves to be a daunting task (Daehmlow et al. 2002). In contrast, dog families are often much larger and generations are shorter, simplifying the approach.

Studies in specific canine breeds can bypass the extensive phenotypic heterogeneity seen in human patients, as the phenotype of DCM in a breed is relatively homogeneous. Therefore, either linkage analyses performed in closely related dogs or association studies performed in unrelated DCM patients of specific breeds are warranted. Several groups have been working on DCM in dogs and have gathered valuable clinicopathologic and DNA material. A good connection of the cardiologists and geneticists is essential in the genetic studies of DCM. Establishing good worldwide collaborations between DCM groups and making a common DNA bank of DCM dogs, accompanied with detailed phenotypic descriptions, would provide an invaluable resource for future DCM studies.

REFERENCES

Alroy J., Rush J.E., Freeman L., Amarendhra Kumar M.S., Karuri A., Chase K., and Sarkar S. 2000. Inherited infantile dilated cardiomyopathy in dogs: Genetic, clinical, biochemical, and morphologic findings. *Am. J. Med. Genet.* **95:** 57–66.

Arbustini E., Pilotto A., Repetto A., Grasso M., Negri A., Diegoli M., Campana C., Scelsi L., Baldini E., Gavazzi A., and Tavazzi L. 2002. Autosomal dominant dilated cardiomyopathy with atrioventricular block: A lamin A/C defect-related disease. *J. Am. Coll. Cardiol.* **39:** 981–990.

Arimura T., Hayashi T., Terada H., Lee S.Y., Zhou Q., Takahashi M., Ueda K., Nouchi T., Hohda S., Shibutani M., et al. 2004. A Cypher/ZASP mutation associated with dilated cardiomyopathy alters the binding affinity to protein kinase C. *J. Biol. Chem.* **279:** 6746–6752.

Barth P.G., Valianpour F., Bowen V.M., Lam J., Duran M., Vaz F.M., and Wanders R.J. 2004. X-linked cardioskeletal myopathy and neutropenia (Barth syndrome): An update. *Am. J. Med. Genet.* **126:** 349–354.

Basso C., Fox P.R., Meurs K.M., Towbin J.A., Spier A.W., Calabrese F., Maron B.J., and Thiene G. 2004. Arrhythmogenic right ventricular cardiomyopathy causing sudden cardiac death in boxer dogs: A new animal model of human disease. *Circulation* **109:** 1180–1185.

Baumwart R.D., Meurs K.M., Atkins C.E., Bonagura J.D., DeFrancesco T.C., Keene B.W., Koplitz S., Luis Fuentes V., Miller M.W., Rausch W., and Spier A.W. 2005. Clinical, echocardiographic, and electrocardiographic abnormalities in Boxers with cardiomyopathy and left ventricular systolic dysfunction: 48 cases (1985–2003). *J. Am. Vet. Med. Assoc.* **226:** 1102–1104.

Berko B.A. and Swift M. 1987. X-linked dilated cardiomyopathy. *N. Engl. J. Med.* **316:** 1186–1191.

Bienengraeber M., Olson T.M., Selivanov V.A., Kathmann E.C., O'Cochlain F., Gao F., Karger A.B., Ballew J.D., Hodgson D.M., Zingman L.V., et al. 2004. ABCC9 mutations identified in human dilated cardiomyopathy disrupt catalytic KATP channel gating. *Nat. Genet.* **36:** 382–387.

Breen M., Hitte C., Lorentzen T.D., Thomas R., Cadieu E., Sabacan L., Scott A., Evanno G., Parker H.G., Kirkness E.F., et al. 2004. An integrated 4249 marker FISH/RH map of the canine genome. *BMC Genomics* **13:** 65.

Brownlie S.E. and Cobb M.A. 1999. Observations on the development of congestive heart failure in Irish wolfhounds with dilated cardiomyopathy. *J. Small Anim. Pract.* **40:** 371–377.

Burke B. And Stewart C.L. 2002. Life at the edge: The nuclear envelope and human disease. *Nat. Rev. Mol. Cell. Biol.* **3:** 575–585.

Calvert C.A. and Meurs K.M. 2000. CVT Update: Doberman pinscher occult cardiomyopathy. In *Kirk's current veterinary therapy XIII* (ed. J.D. Bonagura), pp. 756–760. W.B. Saunders, Philadelphia.

Calvert C.A., Chapman W.L., Jr., and Toal R.L. 1982. Congestive cardiomyopathy in Doberman pinscher dogs. *J. Am. Vet. Med. Assoc.* **181:** 598–602.

Calvert C.A., Pickus C.W., Jacobs G.J., and Brown J. 1997. Signalment, survival, and prognostic factors in Doberman pinschers with end-stage cardiomyopathy. *J. Vet. Intern. Med.* **11**: 323–326.

Cohen N. and Muntoni F. 2004. Multiple pathogenetic mechanisms in X linked dilated cardiomyopathy. *Heart* **90**: 835–841.

D'Adamo P., Fassone L., Gedeon A., Janssen E.A., Bione S., Bolhuis P.A., Barth P.G., Wilson M., Haan E., Orstavik K.H., et al. 1997. The X-linked gene G4.5 is responsible for different infantile dilated cardiomyopathies. *Am. J. Hum. Genet.* **61**: 862–867.

Daehmlow S., Erdmann J., Knueppel T., Gille C., Froemmel C., Hummel M., Hetzer R., and Regitz-Zagrosek V. 2002. Novel mutations in sarcomeric protein genes in dilated cardiomyopathy. *Biochem. Biophys. Res. Commun.* **298**: 116–120.

Dambach D.M., Lannon A., Sleeper M.M., and Buchanan J. 1999. Familial dilated cardiomyopathy of young Portuguese water dogs. *J. Vet. Intern. Med.* **13**: 65–71.

Dec G.W. and Fuster V. 1994. Idiopathic dilated cardiomyopathy. *N. Engl. J. Med.* **331**: 1564–1575.

Domanjko-Petrič A., Stabej P., and Žemva A. 2002. Dilated cardiomyopathy in Doberman pinschers, survival, causes of death and pedigree review in a related line. *J. Vet. Cardiol.* **4**: 17–24.

Dominguez R. 2004. Actin-binding proteins—A unifying hypothesis. *Trends Biochem. Sci.* **29**: 572–578.

Dukes-McEwan J. and Jackson I.J. 2002. The promises and problems of linkage analysis by using the current canine genome map. *Mamm. Genome* **13**: 667–672.

Fatkin D. and Graham R.M. 2002. Molecular mechanisms of inherited cardiomyopathies. *Physiol. Rev.* **82**: 945–980.

Freeman L.M., Michel K.E., Brown D.J., Kaplan P.M., Stamoulis M.E., Rosenthal S.L., Keene B.W., and Rush J.E. 1996. Idiopathic dilated cardiomyopathy in Dalmatians: Nine cases (1990–1995). *J. Am. Vet. Med. Assoc.* **209**: 1592–1596.

Gerull B., Gramlich M., Atherton J., McNabb M., Trombitas K., Sasse-Klaassen S., Seidman J.G., Seidman C., Granzier H., Labeit S., et al. 2002. Mutations of TTN, encoding the giant muscle filament titin, cause familial dilated cardiomyopathy. *Nat. Genet.* **30**: 201–204.

Gomes A.V. and Potter J.D. 2004. Molecular and cellular aspects of troponin cardiomyopathies. *Ann. N.Y. Acad. Sci.* **1015**: 214–224.

Gooding J.P., Robinson W.F., and Mews G.C. 1986. Echocardiographic characterization of dilatation cardiomyopathy in the English cocker spaniel. *Am. J. Vet. Res.* **47**: 1978–1983.

Granzier H.L. and Labeit S. 2004. The giant protein titin: A major player in myocardial mechanics, signaling, and disease. *Circ. Res.* **94**: 284–295.

Gregorio C.C., Granzier H., Sorimachi H., and Labeit S. 1999. Muscle assembly: A titanic achievement? *Curr. Opin. Cell Biol.* **11**: 18–25.

Grunig E., Tasman J.A., Kucherer H., Franz W., Kubler W., and Katus H.A. 1998. Frequency and phenotypes of familial dilated cardiomyopathy. *J. Am. Coll. Cardiol.* **31**: 186–194.

Guyon R., Lorentzen T.D., Hitte C., Kim L., Cadieu E., Parker H.G., Quignon P., Lowe J.K., Renier C., Gelfenbeyn B., et al. 2003. A 1-Mb resolution radiation hybrid map of the canine genome. *Proc. Natl. Acad. Sci.* **100**: 5296–5301.

Hammer T.A., Venta P.J., and Eyster G.E. 1996. The genetic basis of dilated cardiomyopathy in Doberman pinschers. *Anim. Genet.* **27**: 101.

Harpster N.K. 1983. Boxer cardiomyopathy. In *Current veterinary therapy VIII* (ed. R.W. Kirk), pp. 329–337. W.B. Saunders, Philadelphia.

———. 1991. Boxer cardiomyopathy. A review of the long-term benefits of antiarrhythmic therapy. *Vet. Clin. N. Am. Small Anim. Pract.* **21:** 989–1004.

Hutchison C.J. and Worman H.J. 2004. A-type lamins: Guardians of the soma? *Nat. Cell Biol.* **6:** 1062–1067.

Itoh-Satoh M., Hayashi T., Nishi H., Koga Y., Arimura T., Koyanagi T., Takahashi M., Hohda S., Ueda K., Nouchi T., et al. 2002. Titin mutations as the molecular basis for dilated cardiomyopathy. *Biochem. Biophys. Res. Commun.* **291:** 385–393.

Jakobs P., Bestwick M.L., Ludwigsen S.J., Nelsen S.M., Winther M.J., Hershberger R.E., and Litt M. 2004. Genetic linkage analysis of cardiomyopathy in Irish Wolfhounds. In *2nd International Conference Advances in Canine and Feline Genomics*, p. 77. Utrecht, The Netherlands, October 14–17.

Kamisago M., Sharma S.D., DePalma S.R., Solomon S., Sharma P., McDonough B., Smoot L., Mullen M.P., Woolf P.K., Wigle E.D., et al. 2000. Mutations in sarcomere protein genes as a cause of dilated cardiomyopathy. *N. Engl. J. Med.* **343:** 1688–1696.

Keene B.W., Panciera D.P., Atkins C.E., Regitz V., Schmidt M.J., and Shug A.L. 1991. Myocardial L-carnitine deficiency in a family of dogs with dilated cardiomyopathy. *J. Am. Vet. Med. Assoc.* **198:** 647–650.

Kittleson M.D., Keene B., and Pion P. 1997. Results of the multicenter spaniel trial (MUST): Taurine- and carnitine-responsive dilated cardiomyopathy in American cocker spaniels with decreased plasma taurine concentration. *J. Vet. Intern. Med.* **11:** 204–211.

Knoll R., Hoshijima M., Hoffman H.M., Person V., Lorenzen-Schmidt I., Bang M.L., Hayashi T., Shiga N., Yasukawa H., Schaper W., et al. 2002. The cardiac mechanical stretch sensor machinery involves a Z disc complex that is defective in a subset of human dilated cardiomyopathy. *Cell* **111:** 943–955.

Labeit S., Kolmerer B., and Linke W.A. 1997. The giant protein titin. Emerging roles in physiology and pathophysiology. *Circ. Res.* **80:** 290–294.

Lapidos K.A., Kakkar R., and McNally E.M. 2004. The dystrophin glycoprotein complex: Signaling strength and integrity for the sarcolemma. *Circ. Res.* **94:** 1023–1031.

Le Corvoisier P., Park H.Y., and Rockman H.A. 2003a. Modifier genes and heart failure. *Minerva Cardioangiol.* **51:** 107–120.

Le Corvoisier P., Park H.Y., Carlson K.M., Marchuk D.A., and Rockman H.A. 2003b. Multiple quantitative trait loci modify the heart failure phenotype in murine cardiomyopathy. *Hum. Mol. Genet.* **12:** 3097–3107.

Li D., Czernuszewicz G.Z., Gonzalez O., Tapscott T., Karibe A., Durand J.B., Brugada R., Hill R., Gregoritch J.M., Anderson J.L., et al. 2001. Novel cardiac troponin T mutation as a cause of familial dilated cardiomyopathy. *Circulation* **104:** 2188–2193.

Lu Q.W., Morimoto S., Harada K., Du C.K., Takahashi-Yanaga F., Miwa Y., Sasaguri T., and Ohtsuki I. 2003. Cardiac troponin T mutation R141W found in dilated cardiomyopathy stabilizes the troponin T-tropomyosin interaction and causes a Ca^{2+} desensitization. *J. Mol. Cell. Cardiol.* **35:** 1421–1427.

MacLennan D.H. and Kranias E.G. 2003. Phospholamban: A crucial regulator of cardiac contractility. *Nat. Rev. Mol. Cell. Biol.* **4:** 566–577.

Marin-Garcia J., Goldenthal M.J., and Moe G.W. 2001. Mitochondrial pathology in cardiac failure. *Cardiovasc. Res.* **49:** 17–26.

McNair W.P., Ku L., Taylor M.R., Fain P.R., Dao D., Wolfel E., Mestroni L., and Familial Cardiomyopathy Registry Research Group. 2004. SCN5A mutation associated with dilated cardiomyopathy, conduction disorder, and arrhythmia. *Circulation* **110:** 2163–2167.

Meurs K.M., Miller M.W., and Wright N.A. 2001a. Clinical features of dilated cardiomyopathy in Great Danes and results of a pedigree analysis: 17 cases (1990–2000). *J. Am. Vet. Med. Assoc.* **218:** 729–732.

Meurs K.M., Spier A.W., Miller M.W., Lehmkuhl L.B., and Towbin J.A. 1999. Familial ventricular dysrhythmias in Boxer dogs. *J. Vet. Intern. Med.* **13:** 437–439.

Meurs K.M., Magnon A.L., Spier A.W., Miller M.W., Lehmkuhl L.B., and Towbin J.A. 2001b. Evaluation of the cardiac actin gene in Doberman Pinschers with dilated cardiomyopathy. *Am. J. Vet. Res.* **62:** 33–36.

Monnet E., Orton E.C., Salman M., Salman M., and Boon J. 1995. Idiopathic dilated cardiomyopathy in dogs: Survival and prognostic indicators. *J. Vet. Intern. Med.* **9:** 12.

Morales M., Ynaraja E., and Montoya J.A. 2001. Dilated cardiomyopathy in Presa canario dogs: ECG findings. *J. Vet. Med. A Physiol. Pathol. Clin. Med.* **48:** 577–580.

Muntoni F., Cau M., Ganau A., Congiu R., Arvedi G., Mateddu A., Marrosu M.G., Cianchetti C., Realdi G., Cao A., et al. 1993. Brief report: Deletion of the dystrophin muscle-promoter region associated with X-linked dilated cardiomyopathy. *N. Engl. J. Med.* **329:** 921–925.

Muntoni F., Di Lenarda A., Porcu M., Sinagra G., Mateddu A., Marrosu G., Ferlini A., Cau M., Milasin J., Melis M.A., et al. 1997. Dystrophin gene abnormalities in two patients with idiopathic dilated cardiomyopathy. *Heart* **78:** 608–612.

Murphy R.T., Mogensen J., Shaw A., Kubo T., Hughes S., and McKenna W.J. 2004. Novel mutation in cardiac troponin I in recessive idiopathic dilated cardiomyopathy. *Lancet* **363:** 371–372.

Nigro G., Politano L., Nigro V., Petretta V.R., and Comi L.I. 1994. Mutation of dystrophin gene and cardiomyopathy. *Neuromuscul. Disord.* **4:** 371–379.

O'Grady M.R. and Horne R. 1992. Occult dilated cardiomyopathy: An echocardiographic and electrocardiographic study of 193 asymptomatic Doberman pinschers. *J. Vet. Intern. Med.* **6:** 131.

Olson T.M., Kishimoto N.Y., Whitby F.G., and Michels V.V. 2001. Mutations that alter the surface charge of alpha-tropomyosin are associated with dilated cardiomyopathy. *J. Mol. Cell. Cardiol.* **33:** 723–732.

Olson T.M., Michels V.V., Thibodeau S.N., Tai Y.S., and Keating M.T. 1998. Actin mutations in dilated cardiomyopathy, a heritable form of heart failure. *Science* **280:** 750–752.

Parker H.G., Kim L.V., Sutter N.B., Carlson S., Lorentzen T.D., Malek T.B., Johnson G.S., DeFrance H.B., Ostrander E.A., and Kruglyak L. 2004. Genetic structure of the purebred domestic dog. *Science* **304:** 1160–1164.

Pyle W.G. and Solaro R.J. 2004. At the crossroads of myocardial signaling: The role of Z-discs in intracellular signaling and cardiac function. *Circ. Res.* **94:** 296–305.

Richardson P., McKenna W., Bristow M., Maisch B., Mautner B., O'Connell J., Olsen E., Thiene G., Goodwin J., Gyarfas I., et al. 1996. Report of the 1995 World Health Organization/International Society and Federation of Cardiology Task Force on the definition and classification of cardiomyopathies. *Circulation* **93:** 841–842.

Ross R.S., Bulkley B.H., Hutchins G.M., Harshey J.S., Jones R.A., Kraus H., Liebman J., Thorne C.M., Weinberg S.B., Weech A.A., and Weech A.A., Jr. 1978. Idiopathic familial myocardiopathy in three generations: A clinical and pathologic study. *Am. Heart J.* **96:** 170–178.

Rybakova I.N., Patel J.R., and Ervasti J.M. 2000. The dystrophin complex forms a mechanically strong link between the sarcolemma and costameric actin. *J. Cell Biol.* **150:** 1209–1214.

Schmitt J.P., Kamisago M., Asahi M., Li G.H., Ahmad F., Mende U., Kranias E.G., MacLennan D.H., Seidman J.G., and Seidman C.E. 2003. Dilated cardiomyopathy and heart failure caused by a mutation in phospholamban. *Science* **299:** 1410–1413.

Sisson D.D., Thomas W.P., and Keene B.W. 2000. Primary myocardial disease in the dog. In *Textbook of veterinary internal medicine*, 5th edition (ed. S.J. Ettinger and E.C. Feldman), pp. 874–895. W.B. Saunders, Philadelphia.

Sleeper M.M., Henthorn P.S., Vijayasarathy C., Dambach D.M., Bowers T., Tijskens P., Armstrong C.F., and Lankford E.B. 2002. Dilated cardiomyopathy in juvenile Portuguese Water Dogs. *J. Vet. Intern. Med.* **16:** 52–62.

Smucker M.L., Kaul S., Woodfield J.A., Keith J.C., Manning S.A., and Gascho J.A. 1990. Naturally occurring cardiomyopathy in the Doberman pinscher: A possible large animal model of human cardiomyopathy? *J. Am. Coll. Cardiol.* **16:** 200–206.

Spier A.W., Meurs K.M., Coovert D.D., Lehmkuhl L.B., O'Grady M.R., Freeman L.M., Burghes A.H., and Towbin J.A. 2001. Use of western immunoblot for evaluation of myocardial dystrophin, alpha-sarcoglycan, and beta-dystroglycan in dogs with idiopathic dilated cardiomyopathy. *Am. J. Vet. Res.* **62:** 67–71.

Staaden R.V. 1981. Cardiomyopathy of English cocker spaniels. *J. Am. Vet. Med. Assoc.* **178:** 1289–1292.

Stabej P. 2005. "Molecular genetics of dilated cardiomyopathy in the Dobermann dog." Ph.D. thesis, Utrecht University, The Netherlands.

Stabej P., Leegwater P., Stokhof A.A., Domanjko-Petrič A., and van Oost B.A. 2005a. Evaluation of the phospholamban gene in purebred dogs with dilated cardiomyopathy. *Am. J. Vet. Res.* **66:** 432–436.

Stabej P., Imholz S., Versteeg S.A., Zijlstra C., Stokhof A.A., Domanjko-Petrič A., Leegwater P.A., and van Oost B.A. 2004. Characterization of the canine desmin (DES) gene and evaluation as a candidate gene for dilated cardiomyopathy in the Dobermann. *Gene* **340:** 241–249.

Stabej P., Leegwater P.A.J, Imholz S., Versteeg S.A., Zijlstra C., Stokhof A.A., Domanjko-Petrič A., and van Oost B.A. 2005b. The canine sarcoglycan delta gene: BAC clones isolation, chromosome assignment and exclusion as a candidate gene for dilated cardiomyopathy in Dobermanns. *Cytogenet. Genome Res.* (in press).

Stefanelli C.B., Rosenthal A., Borisov A.B., Ensing G.J., and Russell M.W. 2004. Novel troponin T mutation in familial dilated cardiomyopathy with gender-dependant severity. *Mol. Genet. Metab.* **83:** 188–196.

Suomalainen A., Paetau A., Leinonen H., Majander A., Peltonen L., and Somer H. 1992. Inherited idiopathic dilated cardiomyopathy with multiple deletions of mitochondrial DNA. *Lancet* **340:** 1319–1320.

Sutter N.B., Eberle M.A., Parker H.G., Pullar B.J., Kirkness E.F., Kruglyak L., and Ostrander E.A. 2004. Extensive and breed-specific linkage disequilibrium in *Canis familiaris*. *Genome Res.* **14:** 2388–2396.

Tidholm A. and Jonsson L. 1996. Dilated cardiomyopathy in the Newfoundland: A study of 37 cases. *J. Am. Anim. Hosp. Assoc.* **32:** 465–470.

———. 1997. A retrospective study of canine dilated cardiomyopathy (189 cases). *J. Am. Anim. Hosp. Assoc.* **33:** 544–550.

———. 2005. Histologic characterization of canine dilated cardiomyopathy. *Vet Pathol.* **42:** 1–8.

Tiso N., Stephan D.A., Nava A., Bagattin A., Devaney J.M., Stanchi F., Larderet G., Brahmbhatt B., Brown K., Bauce B., et al. 2001. Identification of mutations in the cardiac ryanodine receptor gene in families affected with arrhythmogenic right ventricular cardiomyopathy type 2 (ARVD2). *Hum. Mol. Genet.* **10:** 189–194.

Tsubata S., Bowles K.R., Vatta M., Zintz C., Titus J., Muhonen L., Bowles N.E., and Towbin J.A. 2000. Mutations in the human delta-sarcoglycan gene in familial and sporadic dilated cardiomyopathy. *J. Clin. Invest.* **106:** 655–662.

Vollmar A.C. 2000. The prevalence of cardiomyopathy in the Irish wolfhound: A clinical study of 500 dogs. *J. Am. Anim. Hosp. Assoc.* **36:** 125–132.

Wagner B.M. 1968. Myocardial disease in man and dog, some properties. *Ann. N.Y. Acad. Sci.* **147:** 354–362.

Wotton P.R. 1998. Cardiomyopathy in English cocker and springer spaniels: A review of 38 cases. *Proc. Br. Small Anim. Vet. Assoc.*, p. 316 [Abstr.].

20

Sex Chromosomes, Sexual Development, and Sex Reversal in the Dog

Vicki N. Meyers-Wallen

J.A. Baker Institute for Animal Health
College of Veterinary Medicine
Cornell University
Ithaca, New York 14853

CANINE REPRODUCTIVE BIOLOGY HAS MANY FEATURES that are species-specific; however, prenatal development of the canine reproductive system follows the same steps observed in other mammals. The molecular aspects of canine developmental biology can now be studied in detail, being accelerated by recent advances in canine genomics, such as the 7.6X canine genomic sequence, and the integrated radiation hybrid map and linkage map of all chromosomes, including the sex chromosomes. This chapter outlines the molecular pathways controlling the three major steps in sexual development, summarizing studies in several mammals. Sections discussing abnormalities in sexual development identify those reported in the dog. An example summarized in some detail is sex reversal, a disorder for which the dog may be a particularly useful model. Although transgenic mouse models have been useful in defining mammalian sexual development, differences in gene expression patterns have been observed between mice and other mammals, including humans. Thus, the dog could be a more representative molecular model for comparative studies of mammalian sexual development that will benefit both dogs and humans.

OVERVIEW OF SEX DETERMINATION IN MAMMALS

Normal mammalian sexual development occurs in three steps, with each step depending on successful completion of the previous step:

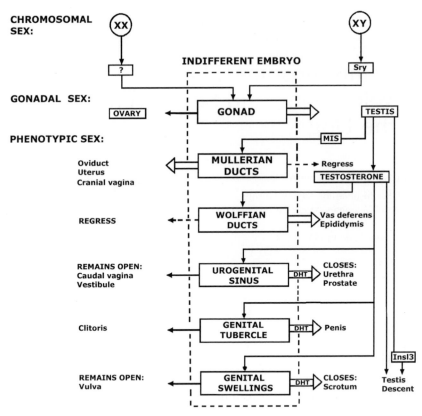

Figure 1. Overview of normal mammalian sexual development, showing the three basic steps: establishment of chromosomal sex complement, development of gonadal sex, and development of phenotypic sex (differentiation of internal and external genitalia). (Reprinted, with permission, from Meyers-Wallen 2000 [© Elsevier].)

(1) establishment of chromosomal sex complement at the time of fertilization, (2) development of gonadal sex, wherein either a testis or ovary emerges from the bipotential gonad, and (3) development of phenotypic sex by differentiation of either male or female internal and external genitalia (Fig. 1). At fertilization, normal canine zygotes obtain either a 78,XX or 78,XY chromosome complement, yet develop similarly during early embryonic stages, being sexually indifferent when the urogenital ridge first arises from the intermediate mesoderm. The morphology of the external genitalia at this time more closely resembles a female than male phenotype. Shortly after emergence of the bipotential gonad from the urogenital ridge, differential expression of genes on the sex chromosomes and autosomes determines whether the gonad will become a testis

or an ovary. The end of the sexually indifferent stage is marked by testis induction in males, which occurs prior to ovarian induction in female littermates. Since development of secondary sex characteristics is dependent on whether the testis or ovary develops, it is gonadal differentiation that determines phenotypic sex. The testis induction step is critical to further male development, as testis secretions are directly or indirectly required for extensive remodeling of the indifferent embryo that is necessary for masculinization of the genitalia. In the absence of a testis and its secretions, the indifferent embryo is free to develop along the female pathway, the genetic control of which is poorly understood at present.

Establishment of Chromosomal Sex

If the chromosomes are normal in number, structure, and genetic composition, it is the presence or absence of the Y chromosome that determines gonadal sex in most mammals. The *Sry* gene (Table 1) encodes a testis-determining factor, and is the only Y-linked gene that is necessary and sufficient to initiate testis development (Koopman et al. 1991). Thus, in normal dogs, 78,XX embryos develop ovaries whereas 78,XY embryos develop testes (Fig. 1). However, *Sry* is only the beginning switch for the testis pathway, as X-linked and autosomal genes also play important roles in sexual differentiation (Table 1).

The Canine Sex Chromosomes

The sex chromosomes are the only metacentric chromosomes in the canine karyotype, and the Y is the smallest chromosome (Fig. 2). Mammalian sex chromosomes are not homologous for most of their length. Thus, during meiosis, the canine X and Y chromosomes form a synaptonemal complex along only one segment (Pathak et al. 1982). In the mouse it has been shown that meiotic recombination occurs regularly between X and Y sequences in this segment, such that inheritance of genes here appears to be autosomal. Thus, the segments of X and Y that pair during meiosis are termed the pseudoautosomal regions of the X and Y chromosomes.

The pseudoautosomal region of CFAX has not been studied extensively, but present evidence indicates that it maps to Xp22.3-22.2 (Spriggs et al. 2003). This region includes two markers (AHTx21 and AHTx13) that also map to the Y chromosome. Thus the CFAX pseudoautosomal region extends from the telomere to at least Xp22.3, but does not include

Table 1. Canine homologs of genes known to have a role in mammalian sexual development

Gene Name	Gene symbol or other identifier	Chromosome location from CanFam 1.0	Human homolog GenBank#	Canine marker/gene GenBank#
Bipotential gonad				
Wilms' tumor-1	Wt1	chr18:45,690,705 to 45,740,153	NM_000378	U00687
Steroidogenic factor 1	Sf1 (Nr5a1)	chr9:51,023,812 to 51,048,896	NM_004959	
Lim homeobox 1	Lhx1/Lim1	chr9:29,356,422 to 29,362,415	NM_005568	
Lim homeobox 9	Lhx9	chr7:7,534,797 to 7,547,420	NM_020204	
GATA binding protein 4	Gata4	chr25:29,191,296 to 29,193,681	AF318320	
Empty spiracles homolog 2	Emx2	chr28:31,041,743 to 31,048,745	NM_004098	
Polycomb homolog	M33	chr7:28,910,433 to 28,912,424	NM_032647	
Insulin receptor	Ir(chr20_11.3)	chr20:54,888,404 to 55,002,612	NM_000208	
Insulin receptor-related receptor	Irr(chr7_9.10)	chr7:43,766,104 to 43,780,758	NM_014215	
Insulin-like growth factor 1 receptor	Igf1r	chr3:45,028,871 to 45,033,554		AY541499
Gonadal sex determination or differentiation				
Sex determining region Y	Sry	not in database		AF107021
Sry-box containing gene 9	Sox9	chr18:22,946,800 to 22,949,969		AY237827
DSS-AHC critical region on the X chromosome 1	Dax1 (Nr0b-1)	chrX:25,220,581 to 25,221,055	NM_000475	
Wingless-type MMTV Integration site family (member 4)	Wnt4 (chr2_15.79)	chr2:77,658,741 to 77,791,170	AY009398	
Double sex-and mab3-related transcription factor 1	Dmrt1	chr1:92,328,812 to 92,442,279	AF130728	
GATA binding protein 4	Gata4	chr25:29,191,296 to 29,193,681	AF318320	

Zinc finger protein multitype 2 or Friend of GATA protein 2	*Fog2*	chr13:9,443,281 to 9,804,852	AF119334
Desert hedgehog	*Dhh (chr27_2.10)*	chr27:8,495,916 to 8,500,822	NM_021044
Platelet derived growth factor alpha	*Pdgfr alpha (chr13_10.23)*	chr13:49,861,572 to 49,894,907	AY525124
Forkhead box L2	*FoxL2 (chr23_7.112)*	chr23:38,178,266 to 38,204,052	NM_023067
Polled intersex syndrome-regulated transcript 1	*Pisrt1 (ch23_7.128)*	ch23:38,443,527 to 38,445,377	NT_086641, AY319655
Fibroblast growth factor 9	*Fgf9*	chr25:19,622,295 to 19,652,705	NM_002010

Internal and external genitalia differentiation

Anti-Mullerian hormone/Mullerian Inhibiting Substance	*Amh/MIS*	chr20:59,804,832 to 59,807,674	NM_000479
Anti-Mullerian hormone/Mullerian Inhibiting Substance Receptor, TypeII	*Amh/MIS R, type II*	chr27:4,800,306 to 4,806,022	NM_020547
Steroidogenic factor 1	*Sf1 (Nr5a1)*	chr9:51,023,812 to 51,048,896	NM_004959
Androgen receptor	*AR*	chrX:54,755,862 to 54,939,471	NM_000044
5 alpha reductase	*5aR*	chr34:10,055,970 to 10,070,969	NM_001047
Insulin-like 3 factor	*Insl3*	chr20:47,930,454 to 47,931,836	NM_005543
Insulin-like 3 factor recepor	*Insl3R/Great*	chr25:11,289,099 to 11,349,712	AY749634
Wingless-type MMTV integration site family (member 4)	*Wnt4 (chr2_15.79)*	chr2:77,658,741 to 77,791,170	AY009398
Homeobox a 13	*Hoxa13*	chr14:42,929,942 to 42,933,163	NM_000522

Locations on canine chromosomes were derived by on-line search of the canine genome sequence with genes listed by GenBank accession number.

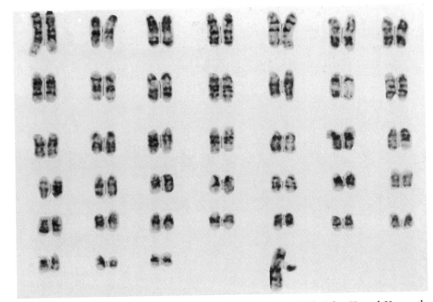

Figure 2. Giemsa banded karyotype of a male dog (78,XY). The X and Y are the only metacentric chromosomes. (Courtesy of Dr. Donald F. Patterson and Ulana Prociuk, Section of Medical Genetics, Veterinary School, University of Pennsylvania.)

Pdha1 (Table 2). Other investigators also mapped *Ant3*, *Csf2ra*, and *Sts* to the pseudoautosomal region (Toder et al. 1997). Some of these CFAX loci can be confirmed by on-line search of the canine genome sequence, using the RefSeq or human homologs deposited in GenBank, or on-line PCR with published primers for canine markers. Two X-linked genes with significance to sexual development, *Dax1* and the androgen receptor, are located far from the pseudoautosomal region (Table 2). Due to X chromosome conservation, there are several X-linked genes and diseases in common among mammals (for summary, see Spriggs et al. 2003). Interestingly, the gene order along CFAX is reported to be similar to that of humans, with less rearrangement of loci between the human and canine X than observed between the human and mouse X chromosome (Spriggs et al. 2003).

In comparison to the canine X chromosome, CFAY is poorly mapped (Guyon et al. 2003). The gene of significance to sex determination, canine *Sry* (Table 2), is located near the centromere (Breen et al. 2001). This differs from the *SRY* location in humans, which is just below the pseudoautosomal region (Yp11.3). *Sry* is included in the on-line genome sequence derived from the male poodle (GenBank AACN010454830), but CFAY sequences are absent from canine genome sequence derived from

Table 2. Genes, markers, and regions on the canine X chromosome that are of interest to sexual development

Canis familiaris X chromosome Gene name	Gene symbol or other identifier	Location from CanFam1.0 on chrX	Cytogenetic location	Reference
Colony stimulating factor 2 receptor α	Csf2ra (chrX_1.36)	740,741 to 749,280		Toder
Adenine nucleotide translocase	Ant3 (chrX_19.30)	94,594,932 to 94,596,284		Toder
Protein kinase	PrkX (chrX_1.76)	1,826,329 to 1,871,237		Toder
	AHTx13	no PCR matches found		Spriggs
	AHTx21	3,371,752 to 3,371,853	Xp22.3	Spriggs
Steroid sulfatase precursor	Sts	4,287,918 to 4,360,669	Xp22.2-22.3	Toder
Pyruvate dehydrogenase-α (lipoamide) 1	Pdha1	19,121,755 to 19,137,619		Spriggs
DSS-AHC critical region on the X chromosome 1	Dax1 (Nr0b-1)	25,220,581 to 25,221,055		
Androgen receptor	AR	54,755,862 to 54,939,471		Guyon

Shaded box indicates loci reported within the pseudoautosomal region.

the female boxer (http://www.ncbi.nlm.nih.gov/genome/guide/dog/). If the canine Y chromosome is similar to that of humans, it may contain many repetitive sequences that can present a challenge to mapping (Skaletsky et al. 2003).

Development of Gonadal Sex

Gonadal sex determination is the process whereby the bipotential gonad is induced to become either a testis or an ovary (Fig. 1). Thus, cells within the gonadal anlagen have the potential to organize and differentiate along two different pathways. The genetic control of gonadal sex determination has been elucidated primarily through studies of transgenic mice and human patients. Much has been learned regarding testis determination, but less is known concerning ovarian determination.

Several autosomal genes are involved in the emergence and mainte-nance of the bipotential gonad in XX and XY embryos, some of which promote cell growth or prevent apoptosis (Table 1). For example, null mutations of *Wt1* and *Sf1* result in a failure to develop gonads in both XX and XY embryos. Thus, both of these genes are involved in mainte-nance of the urogenital ridge near the end of the indifferent stage. Similar phenotypes have been described to result from transgenic manipulation of genes expressed in the intermediate mesoderm, urogenital ridge, or bipotential gonad. These include *Lhx9, Gata4, Fog2, Lhx1, Emx, M33, Dmrt1*, and members of the insulin receptor family (*Ir, Irr, Igf1r*) (for review, see Brennan and Capel 2004; MacLaughlin and Donahoe 2004). Complex interactions of such genes during gonadal sex determination are becoming evident. For example, at the time of testis induction, *Wt1, Gata4*, and *Fog2* may be involved in activating *Sry* expression (Brennan and Capel 2004).

Testis Determination

In most mammals, testis differentiation begins prior to that of the ovary. In the dog, seminiferous tubules are first identifiable in histologic sec-tions at day 34 of gestation (Meyers-Wallen et al. 1989, 1991, 1993), which is equivalent to Carnegie Stage (CS) 18.5–19 (O'Rahilly and Muller 1987). Between 34 and 36 days gestation, the canine ovary retains characteristics of the bipotential gonad. Ovarian features become evident at day 36 (CS 19.5–20). Thus, it has been hypothesized that the gonad is programmed to become an ovary, and that genes essential to testis induction and determination must act prior to expression of genes in the ovarian pathway. The mechanism of *Sry* action remains elusive, in that it is unclear whether *Sry* directly activates genes in the testis pathway. Furthermore, the autosomal gene *Sox9* (Table 1) also has the ability to induce the bipotential gonad to follow the testis determination pathway. Both *Sry* and *Sox9* are members of the high mobility group (HMG) non-histone proteins that associate with DNA. Transcriptional activation of *Sox9* is critical to testis induction in all vertebrates, whether activated genetically as in mammals, or environmentally as in reptiles. It is likely that the function of *Sry* in mammals is to directly or indirectly up-regulate *Sox9*. Unlike *Sry*, which is expressed exclusively in pre-Sertoli cells, *Sox9* is initially expressed in the mesenchyme of bipotential XX and XY gonads. Later, after *Sry* is expressed, *Sox9* is expressed only in pre-Sertoli cells within the XY gonad (Moreno-Mendoza et al. 2003). This is significant, because the first cellular step in testis induction is likely to be Sertoli cell differentiation, which is critical to further testis organization and differentiation.

Other autosomal genes (Table 1) have been shown to play a role in testis differentiation (for review, see Brennan and Capel 2004). In early testes, in response to *Sox9* expression, Sertoli cells secrete *Fgf9*, which induces migration of mesonephric cells necessary for testis cord formation. Other autosomal genes encode transcription factors important to the function of testis-specific cell types. For example, *Sf1*, *Wt1*, and *Sox9* up-regulate *MIS/AMh* expression in Sertoli cells. MIS is the first hormone product of the fetal testis, and its secretion marks the beginning of testis function. Transcription factors and growth factor signaling are also involved in differentiation of testis cell types. For example, *Dhh* and *Pdgfrα* are critical to early steroidogenic cell differentiation that culminates in Leydig cells (Yao et al. 2002; Brennan et al. 2003). *Sf1* also transcriptionally regulates enzymes controlling steroid hormone synthesis, such as testosterone in the Leydig cells.

Ovary Determination

In comparison to the testis, the ovarian pathway is poorly understood. At present, no gene has been identified to play an ovary-determining role analogous to that of *Sry* in testis determination. The X-linked *Dax1* gene was initially hypothesized to play such a role, as *Dax1* overexpression can interfere with testis determination in mice. However, ovarian development is unaffected in *Dax1* null mice. More recent studies suggest that *Dax1* may have multiple roles, both in the testis and in ovarian pathways (for review, see Park and Jameson 2004).

Autosomal genes have also been reported to contribute to ovarian differentiation, based on studies in transgenic mice and human patients. For example, *Wnt4* is normally up-regulated in developing ovaries and down-regulated in the testis. Studies in *Wnt4* knockout mice suggest that in the ovary, *Wnt4* inhibits formation of a testicular vascular pattern and suppresses the development of cells capable of androgen production (for review, see Park and Jameson 2004). Other autosomal genes include *FoxL2* and the untranslated gene *Pisrt1* (Table 1), which are normally expressed in the differentiating ovary and not in the testis. These genes were identified through studies of XX sex reversal in the goat (PIS, polled intersex syndrome). In this model, a large deletion causes a long-range effect on *FoxL2* and *Pisrt1*, which inhibits their expression. The fetal gonad of PIS-affected goats (XX) is initially similar to a normal fetal ovary. However, some ovarian characteristics soon regress as testicular characteristics appear in the supporting cell lineage, and at birth, most affected goats have bilateral testes (Pailhoux et al. 2002). This suggests that *FoxL2* and/or *Pisrt1* may normally suppress a branch of the testis pathway. Interestingly, XX *FoxL2* null mice are not sex-reversed, indicating that *FoxL2* apparently has no role in ovarian determination in that species (for review, see Brennan and Capel 2004). This finding may be due to species-specific differences.

Development of Phenotypic Sex

In this phase of sexual development, hormones mediate the remodeling of the indifferent internal and external genitalia (Fig. 1). Target organs express the type 2 receptor for *MIS/Amh*, the androgen receptor, the enzyme 5α-reductase, or the insulin-like factor 3 receptor (Table 1). In the presence of testis secretions, MIS, testosterone, and Insl3, the internal and external genitalia are masculinized. Mullerian ducts regress in response to MIS secreted by the Sertoli cells. In response to testosterone secreted by the Leydig cells, the canine Wolffian ducts differentiate into the vasa deferentia and epididymides. Cells within the urogenital sinus express 5α-reductase, an enzyme that converts testosterone to dihydrotesterone (DHT). In response to this androgen, the canine urogenital sinus forms the prostate and penile urethra, the genital tubercle forms the penis, and the genital folds close to form the scrotum.

Testis descent is the last step in masculinization of the external genitalia. Canine testes descend from the cranial abdomen and pass through the inguinal ring and inguinal canal into the scrotum. In mammals, testosterone prevents the outgrowth of the cranial suspensory ligament, which frees the testis from its cranial attachment. Growth and differentiation of the gubernaculum, attached to the caudal pole of the testis, is induced by Insl3 (Table 1), which is produced by Leydig cells. These changes in the gubernaculum move the testis caudally to the inguinal ring. Canine testes descend into the scrotum approximately 10 days after birth, completing the male phenotype (Gier and Marion 1969).

Thus far, no essential role for ovarian secretions has been found in the early differentiation of female internal and external genitalia. The receptors discussed above are expressed in the female genital target organs, but in the absence of their ligands, these fail to masculinize. The Mullerian ducts persist and differentiate into the oviducts, uterus, and cranial vagina, and the Wolffian ducts regress. The urogenital sinus remains open, forming the caudal vagina. The genital tubercle forms the clitoris, and the genital folds form the vulva. The ovary remains in the cranial abdomen since the cranial suspensory ligament persists and the gubernaculum fails to differentiate.

Abnormalities in Sexual Development

An abnormal sexual phenotype can arise from a defect at any level in the process of sexual development: in establishment of chromosomal sex, in development of gonadal sex, or in development of the internal or external genitalia (Fig. 1). Defects arising early in this hierarchy are generally more severe, as they have the potential to disrupt all subsequent dependent steps in sexual development.

Abnormalities in Chromosomal Sex

Defects arising at this level result from abnormalities in chromosome number or structure. The most common examples are monosomy or trisomy of the sex chromosomes. Individuals with these abnormalities usually develop unambiguously male or female genitalia, but gonadal function is abnormal: An individual with an XXY sex chromosome complement develops bilateral testes and, consequently, a male phenotype; however, the germ cells degenerate. Similarly, an individual with an X0 sex chromosome complement develops as a phenotypic female, but germ cells in the ovary degenerate. In the absence of germ cells, follicular cells degenerate, and the ovary loses its typical architecture early in development and fails to produce gametes or hormones. Many other types of sex chromosome abnormalities are possible, such as translocations and deletions, with phenotypic variations dependent on the specific genes affected.

Abnormalities in Gonadal Sex Determination

Abnormalities in this category include the *sex reversal* syndromes, in which there is disagreement between the chromosomal and gonadal sex in the affected individual. For example, XX sex-reversed individuals have a normal female karyotype (XX) yet have some degree of testicular differentiation in the gonads. This syndrome has been described in several breeds of dogs (for review, see Meyers-Wallen 2001). The second category of sex reversal syndromes, XY sex reversal, is a somewhat misleading term. Although XY sex-reversed individuals have a normal male karyotype, they usually do not develop normal ovaries. Instead, such individuals (XY females) commonly develop *gonadal dysgenesis*, wherein neither true testicular nor ovarian differentiation occurs. For example, approximately 15% of XY female patients have *SRY* mutations or deletions and usually have complete failure of testis development and testicular hormones yet do not develop ovaries (Hawkins et al. 1992). Mutations in several other genes have also been reported to cause XY sex reversal (for review, see Brennan and Capel 2004; MacLaughlin and Donahoe 2004); however, XY sex reversal has not been described in the dog.

XX Sex Reversal

Most human patients with XX sex reversal have *SRY*-positive XX sex reversal, which results from an *SRY* translocation to either the X chromosome or an autosome. These patients are usually unambiguously male. The associated absence of spermatogenesis may be due to the additional X chromosome in the germ cells combined with the absence of Y sequences that normally contribute to spermatogenesis. The location of *SRY*, which is just below the pseudoautosomal region of the human Y, may be a predisposing factor to *SRY* translocation to the X chromosome during meiosis. Interestingly, canine *Sry* is located near the centromere of CFAY (Breen et al. 2001), and *Sry* translocation leading to XX sex reversal has not been reported in the dog.

Naturally occurring *Sry*-positive sex reversal in the mouse (Sxr) has been studied extensively. However, in that case, a duplication of Y sequences containing *Sry* was attached to the pseudoautosomal region of the Y chromosome (Singh and Jones 1982). As a result of meiotic recombination in the pseudoautosomal region, *Sry* was transferred from the Y^{Sxr} to the X such that approximately 50% of the XX offspring inherited an X with a translocated *Sry* (X^{Sxr}). This type of XX sex reversal has not been reported in the dog.

The discovery of *Sry*-negative XX sex reversal indicated that testis induction could occur in the absence of *Sry*. Patients with this disorder develop testes in the absence of the Y chromosome and *SRY*. These 46,XX individuals are either XX males, having bilateral testes, or XX true hermaphrodites, having both testicular and ovarian tissue in one or both gonads. As many as 10% of XX male patients studied have this type of XX sex reversal, and in the majority of these patients, the causative mutation is unknown. Although this disorder has also been reported in dogs (Meyers-Wallen et al. 1995), goats (Pailhoux et al. 2001b), pigs (Pailhoux et al. 2001a), and horses (Buoen et al. 2000), it has not been described as a naturally occurring disorder in the mouse. Nevertheless, transgenic mouse models have been useful in dissecting the role of known genes that could be causative. For example, the ability of *Sox9* to induce testis differentiation in the absence of *Sry* was demonstrated in XX mice transgenically expressing *Sox9* in the gonad (Bishop et al. 2000). Although this mechanism has been verified in a human patient (Huang et al. 1999), it appears be an infrequent cause of this disorder.

In the PIS goat model of *Sry*-negative XX sex reversal, a large deletion reduces *Pisrt1* and *FoxL2* expression in XX gonads. This deletion is also responsible for the lack of horns (polled phenotype). In humans, blepharophimosis and epicanthal inversus syndrome (BPES) and premature ovarian failure (POF) have been reported in association with *FOXL2* mutations, but sex reversal has not. The molecular mechanism of *Sry*-negative XXSR remains unknown in dogs, pigs, horses, and most human patients. Thus, domestic animal models may provide novel causative genes and a useful model of the human disorder.

In the several canine breeds in which XX sex reversal has been investigated, only the *Sry*-negative form has been found (Meyers-Wallen et al. 1999). In those breeds in which pedigree information is available, an autosomal mode of inheritance has been hypothesized. Both XX males with bilateral testes and XX true hermaphrodites with ovotestes have been described as siblings in the American cocker spaniel model, as in some human families (Sarafoglou and Ostrer 2000). The basis for phenotypic variation, from testes to ovotestes in affected animals having apparently the same genotype, is unknown. Most affected dogs develop bilateral ovotestes (XX true hermaphrodites), wherein the cortical portion of the gonad differentiates as ovarian cortex and the medullary portion develops

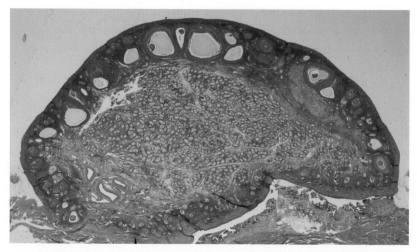

Figure 3. Histologic section of an ovotestis from a true hermaphrodite (78,XX). This affected dog is a member of the *Sry*-negative XX sex reversal pedigree generated from the American cocker spaniel. Testicular development is present in the medullary region and ovarian development is present in the cortical region (hematoxylin and eosin stain).

as testis (Fig. 3). These affected dogs develop a complete uterus and oviducts. Clitoral enlargement, with bone formation characteristic of canine penile differentiation (Fig. 4), occurs in response to androgens produced by the ovotestes. Only a small proportion of affected dogs develop as XX males with bilateral testes and incompletely masculinized genitalia. Testicular tissue in both XX males and XX true hermaphrodites contains all cell types of the normal testis except germ cells. The molecular mechanism that results in development of both testis and ovary from the bipotential gonad, and in the absence of *Sry*, is of great interest.

Our studies of *Sry*-negative XX sex reversal in the American cocker spaniel model began by defining the mode of inheritance and range of phenotypic expression. Current goals are to identify the mutation that causes this disorder and to determine how it alters the molecular pathways responsible for gonadal sex determination. For the first goal, two methods are being pursued in parallel: a genome-wide screen to identify novel genes and marker analysis of candidate genes. Linkage analysis is being performed on the multigenerational pedigree generated from a proven carrier male American cocker spaniel, which was bred to beagle females (Meyers-Wallen and Patterson 1988). A genome-wide screen, performed on a large subset of informative dogs, identified regions of interest on 5 chromosomes. However, the integrated canine linkage and radiation hybrid map was incomplete at the time these markers were

Figure 4. External genitalia of a true hermaphrodite (78,XX), showing clitoral enlargement in response to androgen production from ovotestes. This affected dog is a member of the *Sry*-negative XX sex reversal pedigree generated from the American cocker spaniel.

chosen. With the advent of the canine genome sequence, we were able to confirm changes in marker locations and marker densities on several chromosomes. Further screening is under way with additional markers to fine-map in regions of interest and cover each chromosome at a marker density of 8–10 cM.

A second method, to evaluate known genes as candidates, is being pursued in parallel with the genome-wide screen. Microsatellite markers within genes are designed to identify alleles that cosegregate with the affected phenotype in the sex reversal pedigree. Candidate genes include those identified to have a role in sex determination in other animal models or in humans. Prior to the availability of the canine genome sequence, each canine homolog was cloned and sequenced in the laboratory. The on-line genome sequence has expedited progress, as microsatellite sequences are now identified in silico. A small subset of informative dogs in the pedigree are screened with intragenic microsatellites that are polymorphic in our pedigree. Candidate genes, and other genes nearby, can be excluded if the alleles segregate independently of the affected phenotype. Eight canine homologs of genes in the sex determination pathway have thus far been excluded as candidates (Kothapalli et al. 2003, 2004, 2005 and in prep.; Pujar et al. 2005).

A second goal of our studies is to determine how the mutation causing canine *Sry*-negative XX sex reversal alters the molecular pathways responsible for gonadal sex determination or differentiation. One method is to characterize and compare gene expression in affected gonads with that of normal gonads of the same developmental stage. Methods were developed to accurately predict canine gestational age by objective criteria (Kutzler et al. 2003a,b) and to define canine embryonic developmental stages by morphologic criteria (V. Meyers-Wallen, unpubl.) equivalent to Carnegie Stages (CS) used for human embryos (O'Rahilly and Muller 1987). Since normal gonadal gene expression had not been reported for the dog, we began by evaluating expression through in situ hybridization with digoxygenin-labeled riboprobes and quantitative RT-PCR. Currently we obtain canine gene sequences from the on-line genome (Table 1). Studies of normal gonads indicate that canine sex determination begins at CS 15 with *Sf1* expression in the emerging bipotential gonad. This expression becomes pronounced at CS 17 in gonads of both sexes, which is consistent with *Sf1* function in regulating gene pathways in both sexes (Meyers-Wallen 2005). The timing of *Sry* and *Sox9* expression is consistent with a role in testis determination in the dog: In the testes, *Sry* expression begins at CS 16 and is followed by *Sox9* up-regulation at CS 17 (Meyers-Wallen 2003). Furthermore, canine gonadal *Sox9* expression is rapidly up-regulated in testes, whereas it becomes absent in female gonads at these stages (Meyers-Wallen 2003). The onset of *MIS/AMh* expression occurs at CS 18 in canine testes (Meyers-Wallen 2005). Taken together, these data indicate that canine testis determination occurs between CS 15 and CS 18. The temporal and spatial gene expression patterns in canine embryonic gonads are similar to those reported in humans (Hanley et al. 1999, 2000), goats (Pailhoux et al. 2002), and pigs (Pailhoux et al. 2001a), which is consistent with the view that the dog is a representative model of mammalian sex determination.

Abnormalities in Development of Phenotypic Sex

Abnormalities in development of the internal and external genitalia can occur in at least three ways. In the first, should the anlagen of internal or external genitalia fail to develop, then the genitalia fail to develop in both sexes. In the second, functional gonads fail to develop due to errors in chromosomal sex or gonadal sex, which then affects development of the genitalia. In the third, phenotypic abnormalities arise from disorders in hormone production or reception.

First, recent studies have revealed genes controlling the emergence of primordial internal duct systems during the indifferent stage. For example, *Wnt4* (Table 1), which is first expressed in the mesonephric mesenchyme, participates in gonadal differentiation but is also necessary for Mullerian duct formation (for review, see Park and Jameson 2004). Absence of Mullerian duct derivatives is observed in *Wnt4* null mice as well as in human patients with a similar defect, Mayer-Rokitansky-Kuster-Hauser syndrome. *Wnt7a* may mediate later Mullerian duct differentiation, as its expression begins at the indifferent stage and persists thereafter in murine Mullerian duct derivatives (for review, see Heikkila et al. 2001). Finally, homeobox genes are involved primarily in formation and patterning of caudal structures destined to contribute to the external genitalia. For example, both male and female patients and mice with *Hoxa13* mutations have malformations of external genitalia derived from the urogenital sinus and genital tubercle (for review, see MacLaughlin and Donahoe 2004).

Second, the failure to produce a functional testis has a severe effect on male phenotypic sex differentiation, since MIS, testosterone, and Insl3 are required for normal masculinization of the genitalia. Thus, genetic defects causing XY sex reversal or gonadal dysgenesis, in which the testis is nonfunctional or absent, result in development of female genitalia (see page 394, Abnormalities in Gonadal Sex Determination). In contrast, failure to develop a functional ovary does not affect fetal development of female genitalia. Human patients with these disorders (testis absent, gonadal dysgenesis, or ovary absent) are usually diagnosed at the time of puberty due to primary amenorrhea. Dogs affected with such disorders would fail to initiate estrous cycles at puberty.

Pseudohermaphroditism

Finally are the abnormalities of phenotypic sexual development that are classified as *pseudohermaphroditism*. In these individuals, the chromosomal and gonadal sex of the individual are in agreement, yet the internal or external genitalia have some features of the opposite sex due to abnormalities in hormone production or reception. Thus, these disorders are *not* sex reversal syndromes.

Female pseudohermaphrodites have an XX chromosome complement and normal ovaries, yet the internal or external genitalia undergo partial or complete androgen-dependent masculinization. The androgen source can be endogenous or exogenous. In humans, congenital adrenal hyperplasia due to an enzyme defect leads to excess adrenal androgen production and masculinization in XX females (for review, see MacLaughlin and Donahoe 2004). This disorder has not been reported in the dog. The most commonly reported cause of canine female pseudo-hermaphroditism is exogenous steroid hormone administration during gestation (for review, see Meyers-Wallen 2001).

Male pseudohermaphrodites have an XY chromosome complement and normal testes, yet have defects in Mullerian duct regression or androgen-dependent masculinization. The first type of male pseudoher-maphroditism is exemplified by persistent Mullerian duct syndrome, in which Mullerian ducts fail to regress in XY males having bilateral testes. In affected dogs, the uterus, oviducts, and cranial vagina are present alongside the Wolffian duct derivatives of epididymides and vasa defer-entia. In approximately 50% of affected dogs, one or both testes fail to descend into the scrotum, possibly due to the physical constraints of uterine horn attachment near the testis caudal pole (Meyers-Wallen et al. 1993). In some human patients, this disorder results from defects in MIS/AMh or its receptor, and in others, the cause remains unknown. The causative mutation in miniature schnauzer and basset hound dogs is unknown (Nickel et al. 1992). However, biologically active MIS is produced by testes of affected miniature schnauzers, which suggests a defect at the target organ level (Meyers-Wallen et al. 1993).

Several types of male pseudohermaphroditism due to defects in androgen-dependent masculinization have been described in humans and other mammals. These include defects in production of testosterone or 5 α-reductase, and androgen receptor defects. Defects in testosterone production or in the androgen receptor have the potential to affect development of all structures derived from the Wolffian ducts, urogenital sinus, genital tubercle, and genital folds. The genitalia in such patients vary between female, ambiguous, and partially masculinized. The degree of masculinization is related to the timing and amount of testosterone production, the degree of functionality of the androgen receptor, and the critical period for masculinization of each target organ. In the one reported canine case of this type, nonfunctional androgen receptors were found in an XY mixed breed dog. The testes were undescended, Mullerian and Wolffian duct derivatives were absent, and the external genitalia were female (Peter et al. 1993). Defects in 5 α-reductase have not been reported in the dog. However, human patients with these disorders are XY males with bilateral testes and partially masculinized or female external genitalia.

Other Defects in Male Development

The next two disorders result in an abnormal, but not ambiguous, male external phenotype. They are likely to be abnormalities in hormone production or reception, at least in part, but their genetic etiology is incompletely understood at present.

The first of these is hypospadias, a midline fusion defect in the wall of the urethra, such that the urethral canal is open as it courses through the ventral surface of the penis. The urethral orifice is ventrally displaced, commonly to the shaft of the penis or the periscrotal region. In the dog, the prepuce is also usually affected, being incompletely fused ventrally. Hypospadias can occur by itself or in association with other abnormalities of sexual development, as in disorders of androgen-dependent masculinization (above). Hypospadias in human patients has been attributed to both genetic and environmental factors. Canine hypospadias is likely to be an inherited trait in some breeds, as it has a high prevalence in the Boston terrier and terrier breeds (Hayes and Wilson 1986).

A second disorder that results in abnormal, but not ambiguous, development of the male external genitalia is cryptorchidism: failure of one or both testes to descend into the scrotum. This can occur in association with other abnormalities of sexual development, such as in XX males (see page 394, XX Sex Reversal). However, canine cryptorchidism occurs most frequently as the only defect in the sexual phenotype; that is, as isolated cryptorchidism. This is the most common disorder of the reproductive tract reported in dogs (Cox 1986), and a high prevalence is reported in some breeds (for review, see Meyers-Wallen 2001). In at least one family with a high incidence of cryptorchidism, delayed testicular descent was also observed, indicating that these phenotypes may be genetically related (Cox 1986). Inheritance of isolated canine cryptorchidism as a sex-limited autosomal trait is consistent with available data. It is likely that more than one gene is involved, judging from other species. For example, homozyous deletion of *Insl3* or its receptor leads to failure of gubernacular differentiation and bilateral intra-abdominal cryptorchidism in mice (Zimmerman et al. 1999). Heterozygous *Insl3* deletions in humans have been associated with late testicular descent, but mutations in this gene account for only a small percentage of cryptorchid human patients (Tomboc et al. 2002).

Defects in *Insl3* and its receptor are currently under investigation as a cause of canine cryptorchidism (Truong et al. 2003).

SUMMARY

Mammalian sexual development is a multistep process mediated by complex, genetically controlled pathways. Progress in this research has been largely dependent on studies of human patients and mouse models. Although transgenic models remain valuable, the rodent model may not be the most representative of mammalian sex determination. Some genes have different expression patterns and functions in the mouse than in humans and other mammals. It is important to study other mammals to define pathways that are common to all. Present data indicate that the canine model could be a useful and representative model of mammalian sexual development. Furthermore, it is important to increase our understanding of developmental pathways in this species to improve canine health and well-being. The increasing availability of canine genomic tools and clinical recognition of abnormal sexual development provide excellent opportunities to determine the genetic basis of canine sexual abnormalities, to the benefit of dogs and other mammals.

ACKNOWLEDGMENTS

The author thanks her collaborators, Harald Goring of the Southwest Foundation for Biomedical Research; Ewen Kirkness of The Institute for Genomic Research; and members of the laboratory, Kumar Kothapalli, Shashikant Pujar, and Roxanne Van Wormer. These studies were supported by National Institutes of Health grants (R01 HD 40351 and R03 HD 35896) and by the National Heart, Lung, and Blood Institute Mammalian Genotyping Service in Marshfield, Wisconsin.

REFERENCES

Bishop C.E., Whitworth D.J., Qin Y., Agoulnik A.I., Agoulnik I.U., Harrison W.R., Behringer R.R., and Overbeek P.A. 2000. A transgenic insertion upstream of Sox9 is associated with dominant XX sex reversal in the mouse. *Nat. Genet.* **26:** 490–494.

Breen M., Jouquand S., Renier C., Mellersh C.S., Hitte C., Holmes N.G., Cheron A., Suter N., Vignaux F., Bristow A.E., et al. 2001. Chromosome-specific single-locus FISH probes allow anchorage of an 1800-marker integrated radiation-hybrid/linkage map of the domestic dog genome to all chromosomes. *Genome Res.* **11:** 1784–1795.

Brennan J. and Capel B. 2004. One tissue, two fates: Molecular genetic events that underlie testis versus ovary development. *Nat. Rev. Genet.* **5:** 509–521.

Brennan J., Tilmann C., and Capel B. 2003. *Pdgfr-a* mediates testis cord organization and fetal Leydig cell development in the XY gonad. *Genes Dev.* **17:** 800–810.

Buoen L.C., Zhang T.Q., Weber A.F., and Ruth G.R. 2000. SRY-negative, XX intersex horses: The need for pedigree studies to examine the mode of inheritance of the condition. *Equine Vet. J.* **32:** 78–81.

Cox V.S. 1986. Cryptorchidism in the dog. In *Current therapy in theriogenology: Diagnosis, treatment, and prevention of reproductive diseases in small and large animals,* 2nd edition (ed. D.A. Morrow), pp. 541–544. Saunders, Philadelphia.

Gier H.T. and Marion G.B. 1969. Development of mammalian testes and genital ducts. *Biol. Reprod.* **1:** 1–23.

Guyon R., Lorentzen T.D., Hitte C., Kim L., Cadieu E., Parker H.G., Quignon P., Lowe J.K., Renier C., Gelfenbeyn B., et al. 2003. A 1-Mb resolution radiation hybrid map of the canine genome. *Proc. Natl. Acad. Sci.* **100:** 5296–5301.

Hanley N.A., Ball S.G., Clement-Jones M., Hagan D.M., Strachan T., Lindsay S., Robson S., Ostrer H., Parker K.L., and Wilson D.I. 1999. Expression of steroidogenic factor 1 and Wilms' tumour 1 during early human gonadal development and sex determination. *Mech. Dev.* **87:** 175–180.

Hanley N.A., Hagan D.M., Clement-Jones M., Ball S.G., Strachan T., Salas-Cortés L., McElreavey K., Lindsay S., Robson S., Bullen P., et al. 2000. SRY, SOX9 and DAX1 expression patterns during human sex determination and gonadal development. *Mech. Dev.* **91:** 403–407.

Hawkins J.R., Taylor A., Berta P., Levilliers J., Van der Auwera B., and Goodfellow P.N. 1992. Mutational analysis of SRY: Nonsense and missense mutations in XY sex reversal. *Hum. Genet.* **88:** 471–474.

Hayes H.M. and Wilson G.P. 1986. Hospital incidence of hypospadias in dogs in North America. *Vet. Rec.* **118:** 605–607.

Heikkila M., Peltoketo H., and Vainio S. 2001. Wnts and the female reproductive system. *J. Exp. Zool.* **290:** 616–623.

Huang B., Wang B., Ning Y., Lamb A.N., and Bartley J. 1999. Autosomal XX sex reversal caused by duplication of SOX9. *Am. J. Med. Genet.* **87:** 349–353.

Koopman P., Gubbay J., Vivian N., Goodfellow P., and Lovell-Badge R. 1991. Male development of chromosomally female mice transgenic for Sry. *Nature* **351:** 117–121.

Kothapalli D.S.D., Kirkness E., Natale L.J., and Meyers-Wallen V.N. 2003. Exclusion of *PISRT1* as a candidate locus for canine *Sry*-negative XX sex reversal. *Anim. Genet.* **34:** 465–476.

Kothapalli K., Kirkness E.F., Pujar S., and Meyers-Wallen V.N. 2004. Exclusion of WT1 as a candidate gene for canine *Sry*-negative XX sex reversal. *Anim. Genet.* **35:** 462–504.

Kothapalli K., Kirkness E.F., VanWormer R., and Meyers-Wallen V.N. 2005. Exclusion of *DMRT1* as a candidate gene for canine *SRY*-negative XX sex reversal. *Vet. J.* (in press).

Kutzler M.A., Yeager A.E., Mohammed H.O., and Meyers-Wallen V.N. 2003a. Accuracy of canine parturition date prediction using fetal measurements obtained by ultrasonography. *Theriogenology* **60:** 1309–1317.

Kutzler M.A., Mohammed H.O., Lamb S.V., and Meyers-Wallen V.N. 2003b. Accuracy of canine parturition date prediction from the initial rise in preovulatory progesterone concentration. *Theriogenology* **60:** 1187–1196.

MacLaughlin D.T. and Donahoe P.K. 2004. Sex determination and differentiation. *N. Engl. J. Med.* **350:** 367–378.

Meyers-Wallen V.N. 2000. CVT update: Inherited disorders of the reproductive tract in dogs and cats. In *Current veterinary therapy XIII* (ed. R.W. Kirk and J.D. Bonagura) pp. 904–909. Saunders, Philadelphia.

——. 2001. Inherited abnormalities of sexual development in dogs and cats. In *Recent advances in small animal reproduction* (ed. P.W. Concannon et al.). International Veterinary Information Service (www.ivis.org), document no. A1217.0901, Ithaca, New York.

——. 2003. *Sry* and *Sox9* expression during canine gonadal sex determination assayed by quantitative reverse transcription-polymerase chain reaction. *Mol. Reprod. Dev.* **65:** 373–381.

——. 2005. *Sf1* and *Mis* expression: Molecular milestones in the canine sex determination pathway. *Mol. Reprod. Dev.* (in press).

Meyers-Wallen V.N. and Patterson D.F. 1988. XX sex reversal in the American cocker spaniel dog: Phenotypic expression and inheritance. *Hum. Genet.* **80:** 23–30.

Meyers-Wallen V.N., Palmer V.L., Acland G.M., and Hershfield B. 1995. Sry-negative XX sex reversal in the American cocker spaniel dog. *Mol. Reprod. Dev.* **41:** 300–305.

Meyers-Wallen V.N., Donahoe P.K., Ueno S., Manganaro T.F., and Patterson D.F. 1989. Mullerian inhibiting substance is present in testes of dogs with persistent Mullerian duct syndrome. *Biol. Reprod.* **41:** 881–888.

Meyers-Wallen V.N., Schlafer D., Barr I., Lovell-Badge R., and Keyzner A. 1999. Sry-negative XX sex reversal in purebred dogs. *Mol. Reprod. Dev.* **53:** 266–273.

Meyers-Wallen V.N., Lee M.M., Manganaro T.F., Kuroda T., MacLaughlin D., and Donahoe P.K. 1993. Mullerian inhibiting substance is present in embryonic testes of dogs with Persistent Mullerian Duct Syndrome. *Biol. Reprod.* **48:** 1410–1418.

Meyers-Wallen V.N., Manganaro T.F., Kuroda T., Concannon P.W., MacLaughlin D.T., and Donahoe P.K. 1991. The critical period for Mullerian duct regression in the dog embryo. *Biol. Reprod.* **45:** 626–633.

Moreno-Mendoza N., Harley V., and Merchant-Larios H. 2003. Cell aggregation precedes the onset of Sox9-expressing preSertoli cells in the genital ridge of the mouse. *Cytogenet. Genome Res.* **101:** 219–223.

Nickel R.F., Ubbink G., van der Gaag I., and van Sluijs F.J. 1992. Persistent Mullerian duct syndrome in the basset hound. *Tijdschr. Diergeneeskd.* **117:** 31S.

O'Rahilly R. and Muller F. 1987. Developmental stages of human embryos. Carnegie Institution of Washington, Publication no. 637.

Pailhoux E., Mandon-Pepin B., and Cotinot C. 2001a. Mammalian gonadal differentiation: The pig model. *Reprod. Suppl.* **58:** 65–80.

Pailhoux E., Vigier B., Vaiman D., Servel N., Chaffaux S., Cribiu E.P., and Cotinot C. 2002. Ontogenesis of female-to-male sex-reversal in XX polled goats. *Dev. Dyn.* **224:** 39–50.

Pailhoux E., Vigier B., Chaffaux S., Servel N., Taourit S., Furet J.-P., Fellous M., Grosclaude F., Cribiu E.P., Cotinot C., and Vaiman D. 2001b. A 11.7-kb deletion triggers intersexuality and polledness in goats. *Nat. Genet.* **29:** 453–458.

Park S.Y. and Jameson J.L. 2004. Minireview: Transcriptional regulation of gonadal development and differentiation. *Endocrinology* **146:** 1035–1042.

Pathak S., Van Tuinen P., and Merry D.E. 1982. Heterochromatin, synaptonemal complex, and NOR activity in the somatic and germ cells of a male domestic dog, *Canis familiaris* (Mammalia, Canidae). *Cytogenet. Cell Genet.* **34:** 112–118.

Peter A.T., Markwelder D., and Asem E.K. 1993. Phenotypic feminization in a genetic male dog caused by nonfunctional androgen receptors. *Theriogenology* **40:** 1093–1105.

Pujar S., Kothapalli K.S.D., Kirkness E., Van Wormer R.H., and Meyers-Wallen V.N. 2005. Exclusion of *Lhx9* as a candidate gene for *Sry*-negative XX sex reversal in the American Cocker Spaniel model. J. Hered. (in press).

Sarafoglou K. and Ostrer H. 2000. Familial sex reversal: A review. *J. Clin. Endocrinol. Metab.* **85:** 483–493.

Singh L. and Jones K.W. 1982. Sex reversal in the mouse (*Mus musculus*) is caused by a recurrent nonreciprocal cross-over involving the X and an aberrant Y chromosome. *Cell* **28:** 205–216.

Skaletsky H., Kuroda-Kawaguchi T., Minx P.J., Cordum H.S., Hillier L., Brown L.G., Repping S., Pyntikova T., Ali J., Bieri T., et al. 2003. The male-specific region of the human Y chromosome is a mosaic of discrete sequence classes. *Nature* **423:** 825–837.

Spriggs H.F., Holmes N.G., Breen M.G., Deloukas P.G., Langford C.F., Ross M.T., Carter N.P., Davis M.E., Knights C.E., Smith A.E., et al. 2003. Construction and integration of radiation-hybrid and cytogenetic maps of dog Chromosome X. *Mamm. Genome* **14:** 214–221.

Toder R., Glaser B., Schiebel K., Wilcox S.A., Rappold G., Graves J.A., and Schempp W. 1997. Genes located in and near the human pseudoautosomal region are located in the X-Y pairing region in dog and sheep. *Chromosome Res.* **5:** 301–306.

Tomboc M., Lee P.A., Mitwally M.F., Schneck F.X., Bellinger M., and Witchel S.F. 2002. Insulin-like 3/relaxin-like factor gene mutations are associated with cryptorchidism. *Clin. Endocrinol. Metab.* **85:** 4013–4018.

Truong A., Bogatcheva N.V., Schelling C., Dolf G., and Agoulnik A.I. 2003. Isolation and expression analysis of the canine *insulin-like factor 3* gene. *Biol. Reprod.* **69:** 1658–1664.

Yao H.H., Whoriskey W., and Capel B. 2002. Desert Hedgehog/Patched 1 signaling specifies fetal Leydig cell fate in testis organogenesis. *Genes Dev.* **16:** 1433–1440.

Zimmerman S., Steding G., Emmen J.M.A., Brinkmann A.O., Nayernia K., Holstein A.F., Engel W., and Adham I.M. 1999. Targeted disruption of the Insl3 gene causes bilateral cryptorchidism. *Mol. Endocrinol.* **13:** 681–691.

21

QTL Mapping Using Crossbreed Pedigrees: Strategies for Canine Hip Dysplasia

Raluca G. Mateescu, Nathan L. Dykes, and Rory J. Todhunter
Department of Clinical Sciences
College of Veterinary Medicine
Cornell University, Ithaca, New York 14853

Gregory M. Acland, Nancy I. Burton-Wurster, and George Lust
Baker Institute for Animal Health
College of Veterinary Medicine
Cornell University, Ithaca, New York 14853

Zhiwu Zhang and Richard L. Quaas
Department of Animal Breeding
College of Agriculture and Life Sciences
Cornell University, Ithaca, New York 14853

Kate Tsai and Keith Murphy
Department of Veterinary Pathobiology
College of Veterinary Medicine
Texas A&M University, College Station, Texas 77843

Cᴀɴɪɴᴇ ʜɪᴘ ᴅʏsᴘʟᴀsɪᴀ (CHD) is a common inherited trait in dogs and the secondary debilitating hip osteoarthritis (OA) results in pain, lameness, and physical disability. Because CHD is a complex trait, the genotype of a dog cannot be judged from its phenotype. Furthermore, dysplastic dogs also have abnormalities in other joints (Olsewski et al. 1983; Kealy et al. 1992, 1997, 2000; Farquhar et al. 1997; Morgan et al. 1999), indicating that the primary defect is systemic, rather than restricted to the hip. We developed an informative crossbreed pedigree for mapping the quantitative trait loci (QTL) contributing to CHD expression. In the context of this unique pedigree, we review here recent advances in CHD and our first attempts in defining the molecular genetic basis of this common complex disease trait, although the trigger event for CHD remains elusive. The dysplastic dog presents itself as an important natural large animal model of a complex human trait.

BREED PREVALENCE

Hip dysplasia occurs in any pure or mixed breed dog but more commonly affects large breed dogs (Hedhammar et al. 1974; Leighton et al. 1977; Cardinet et al. 1983; Willis 1989; Kaneene et al. 1997; Breur et al. 2002). Breed prevalence, as estimated by the Orthopedic Foundation for Animals (OFA), varies from 1% to 75% (http://www.ofa.org/hipstatbreed.html).

Phenotype

Traditional

Mature dogs are positioned in dorsal recumbency with the hips extended, and a ventrodorsal radiograph is taken (Fig. 1). In North America, radiographs are scored based on the degree of subluxation and secondary OA according to the 7-point scale described by the OFA (Henry 1992), although other scales are used in Europe and Australia (http://www.ofa.org/hipstatbreed.html). The OFA (extended-hip) scale ranges from excellent, good, and fair hip conformation through borderline to mild, moderate, and severe hip dysplasia. Radiographic signs of hip OA include osteophytes on the acetabular rim and femoral neck, flattening of the acetabulum and femoral head, and subchondral bone sclerosis (Fig. 1).

Norberg Angle

The Norberg angle is measured from the ventrodorsal, extended-hip radiograph that is also used to derive the OFA subjective hip score (Fig. 1) (Olsson 1961). A line is drawn between the geometric center of each femoral head and connected to a line that just contacts the craniodorsal acetabular rim. The included angle is the Norberg angle, which ranges from about 75° (a completely subluxated hip) to 115° (an unaffected hip).

Distraction Index

The maximum amount of passive lateral hip laxity can be measured on a radiograph taken in the distraction, dorsal position (PennHIP, hip

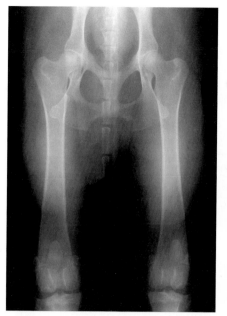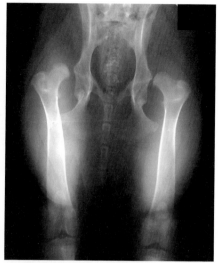

Figure 1. Images of radiographs (unaffected, *left panel*; dysplastic, *right panel*) of the extended-hip method (*top*) from which a score of hip conformation is derived that ranges from excellent (1) to the poorest (7) hip conformation. The Norberg angle is measured from this radiograph by extending a line connecting the geometric centers of each femoral head to the cranial (proximal) and dorsal (posterior) acetabulum and measuring this included angle (Olsson 1961).

registry, Malvern, PA) (Fig. 2) (Smith 1997). Data on CHD gathered by PennHIP is less biased as every dog evaluated will be registered, thereby also permitting an evaluation of any breeding program over time. Labrador retrievers with distraction indexes (DIs) less than 0.3 at 8 months of age had a probability greater than 80% of not developing secondary hip OA as defined below. Those with DIs greater than 0.7 had a high probability of developing hip OA (Lust et al. 1993). Some breeds appear to display different susceptibilities to CHD based on their DIs, and some breeds may tolerate more passive hip laxity than other breeds (Popovitch et al. 1995; Smith et al. 1995).

Dorsolateral Subluxation Score

The dorsolateral subluxation (DLS) test elicits dorsal subluxation of the femoral head from the acetabulum with the hips in a natural weight-bearing position (Fig. 3), as they would be when the dog is walking.

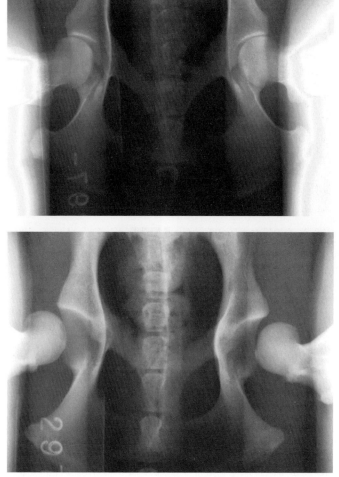

Figure 2. The distraction index is measured on the distraction view as the maximum lateral passive hip laxity (Smith 1997). The distance between the geometric centers of the femoral head and acetabulum is divided by the diameter of the femoral head. Unaffected hips (*top*) and dysplastic hips (*bottom*).

The stifles are flexed and the dog is placed in a kneeling position, so that natural load-bearing forces are transmitted to the hips by flexing and adducting the stifles that must be in contact with the tabletop. On the dorsoventral radiograph, the DLS score is measured as the percentage of the femoral head covered by the craniodorsal acetabulum (Fig. 3). The DLS score ranges from over ~55% for tight-hipped Labrador retrievers to 80% for greyhound hips. For dysplastic hips, the DLS score drops to as low as 20%.

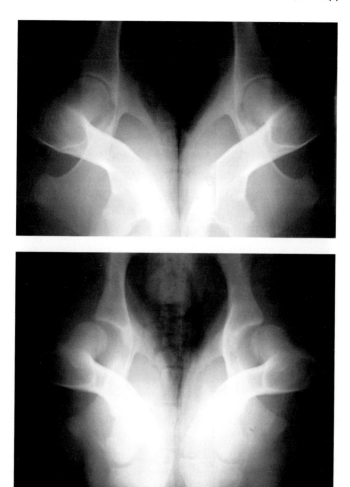

Figure 3. The dorsolateral subluxation score is measured on this radiographic image (Farese et al. 1998) and is equivalent to the percent of femoral head covered by the acetabulum. The distance of the femoral head within the acetabulum is expressed as a percentage of the femoral head diameter. Unaffected hip (*top*) and dysplastic (subluxated) hip (*bottom*).

Ascertainment Reliability and Repeatability

Ascertainment reliability is critical for mapping of genetic traits. The DLS score (Farese et al. 1998; Burton-Wurster et al. 1999) has proven the most accurate (highest overall sensitivity and specificity) method for early diagnosis of hip dysplasia in 8-month-old dogs belonging to our crossbreed greyhound/Labrador retriever pedigree and Labrador retrievers (Lust et al. 1993, 2001a,b). The intraclass correlation coefficient for the DLS score method was 0.85 and for remeasuring the same radiographs was 0.97 (Farese et al. 1998). The DI method was reported to have an intraclass correlation coefficient of 0.85–0.94 (Smith et al. 1997). The traits are correlated, but the correlations are modest (Lust et al. 2001b). The DLS score combined with the Norberg angle provided the best prediction of subsequent development of hip OA (Todhunter et al. 2003c).

Secondary Hip OA

The traditional standard that marks a hip as dysplastic has been the subsequent development of secondary OA. Synovitis, joint capsule thickening, articular cartilage injury, and resulting hip pain and disability are characteristic of secondary OA. The initial cartilage lesion occurs perifoveally, suggesting that abnormal magnitude or direction of load results in increased focal stress in this area (Farese et al. 1998; Burton-Wurster et al. 1999). Abnormal weight bearing continues to cause excess wear on the articular cartilage and damages the underlying bone, causing painful microfractures and sclerosis. Eventually, the hip laxity decreases as the capsule fibroses and the synovial effusion resolves.

Crossbreed Pedigree

Seven greyhounds (2 males and 5 females) from racing stock with excellent hip conformation (Beling et al. 1975) (PennHIP™ hip registry, Malvern, PA) (Cardinet et al. 1983) and 8 Labrador retrievers (4 males and 4 females) with hip dysplasia and secondary hip OA or from dysplastic lineages were selected as founders for development of a crossbreed pedigree for QTL mapping (Fig. 4) (Todhunter et al. 1999). A research colony of dysplastic Labrador retrievers has been maintained for eight generations at the Baker Institute, Cornell University, for more than 30 years (Lust et al. 1973; Lust 1997). These Labrador retrievers are both affected and unaffected with a wide range in phenotypes as measured by the distraction index (DI ranging from 0.2 to 1.2) and the DLS score (DLS ranging from 28% to 75%). Because of the breeding pressure exerted for racing performance, the greyhound founders were assumed

Figure 4. Diagram of a crossbreed pedigree founded on trait-free greyhounds and dysplastic Labrador retrievers. Squares and circles represent males and females, respectively. Filled and open portions of each symbol represent the proportion of greyhound and Labrador retriever alleles, respectively, possessed by that dog.

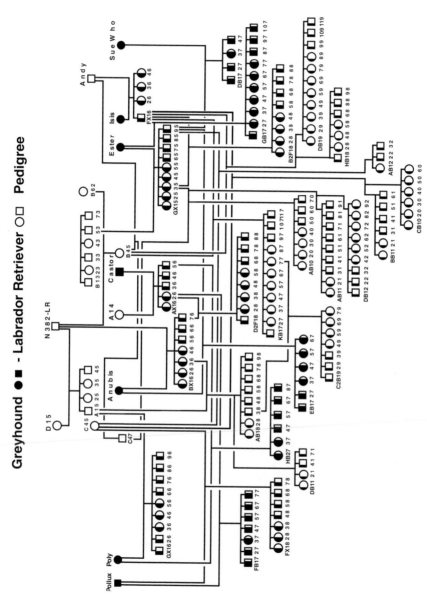

Figure 4. (*See facing page for legend.*)

to be homozygous for the alleles protective against hip dysplasia. The dysplastic Labrador retrievers were likely to be homozygous at the loci contributing to hip dysplasia because of selection pressure to produce families with dysplastic hips. The crossbreed pedigree consists of four generations (F_1 × both greyhound and Labrador retriever founder, F_2, and ¾ × ¾ Labrador retriever litters were bred). One hundred and fifty-nine dogs comprise the greyhound/Labrador retriever pedigree (Fig. 4), and analysis implies resegregation of the alleles underlying the hip traits in the backcross to the Labrador retriever (Table 1), while all but two dogs with greater than or equal to 50% greyhound genes had unaffected hips at 8 months of age (Todhunter et al. 1999).

GENETICS

Dysplasia of the canine hip is a complex trait (Henricson et al. 1966; Leighton et al. 1977; Hedhammar et al. 1979), and several major and many minor QTL interacting with environmental factors are probably involved in phenotypic expression. A major locus may influence this trait based on biometric methods outlined by Leighton (1997). A recent study based on Gibbs sampling and a Markov chain Monte Carlo algorithm showed that major genes contributed to hip dysplasia in four Finnish dog populations (Mäki et al. 2004).

The alleles at these QTL have putative positive and negative influences on the dysplastic traits. Based on human studies, there may be inductive or protective QTL which control expression of hip OA that are independent of those controlling the dysplastic phenotype (Vingsbo-Lundberg et al. 1998; Chapman et al. 1999). Some breeds appear to display different susceptibilities to CHD based on their distraction indices, and some breeds may tolerate more passive hip laxity than other breeds (Popovitch et al. 1995; Smith et al. 1995). Some breeds develop radiographically detectable hip OA as a result of antecedent CHD faster than other breeds (Popovitch et al. 1995), a factor also influenced by nongenetic factors like the nutritional plane and physical activity level (Willis 1989).

The crossbreed offspring from German shepherd–greyhound breedings had close to 50% CHD (Cardinet et al. 1983) and from normal or dysplastic German shepherds and golden retrievers and normal beagles had over 50% CHD (Lust et al. 1973). In contrast, only 2 dogs out of 34 were dysplastic in the F_1 generation of our greyhound/dysplastic Labrador retriever pedigree (Todhunter et al. 1999), similar to the findings of Gustafsson et al. (1972). These data suggest that the QTL which control CHD may be expressed differently in different breeds and their crossbreed offspring.

Table 1. Mean (and standard deviation) according to breeds in a crossbreed pedigree founded on greyhounds and dysplastic Labrador retrievers

Breed	No. of Dogs	DIL	DIR	DLSL	DLSR	NAL	NAR
GH	7	0.21(0)	0.21(0)	74(0)	74(0)	105(3.7)	110.4(4.5)
75% GH/25% Lab	33	0.31(0.15)	0.33(0.15)	66.5(5.6)	66.2(4.2)	108.9(3.0)	110.7(3.9)
50% GH/50% Lab F_1	7	0.45(0.14)	0.48(0.12)	63.8(0)	67.8(0)	104.03(4.1)	108.9(4.2)
50% GH/50% Lab F_2	16	0.37(0.1)	0.35(0.09)	67.3(5.5)	66.0(4.7)	110.0(3.5)	110.1(3.9)
25% GH/75% Lab	80	0.51(0.17)	0.57(0.17)	50.6(13.3)	50.8(12.7)	107.2(6.1)	107.7(5.6)
Lab	7	0.52(0.21)	0.58(0.18)	45.0(19.6)	49.1(8.2)	102.8(4.1)	102.5(5.4)

(L) Left; (R) right; (DI) hip distraction index; (DLS) dorsolateral subluxation score; (NA) Norberg angle; (GH) greyhound; (Lab) Labrador retriever.

Based on the hip-extended, ventrodorsal, OFA-type radiographic studies, heritability estimates for CHD range from 0.1 to 0.68 (Lingaas and Heim 1987; Willis 1989; Lingaas and Klementsdal 1990; Distl et al. 1991; Leighton 1997; Swenson et al. 1997; Breur et al. 2002). In both German shepherds and Labrador retrievers, CHD heritability for the extended-hip score was about 0.35, but for the distraction index was 0.45 (Leighton 1997).

Mode of Inheritance and Heritability

Both the DI and the DLS score have a significant additive genetic component, and the DLS score has a significant nonadditive (dominance) mode of inheritance (Bliss et al. 2002). The "narrow sense" heritability (the proportion of additive genetic variance of the total phenotypic variance) for the DI and DLS score was estimated at 0.5 and 0.6, respectively, in the greyhound/Labrador retriever pedigree and the Labrador retrievers related to the founders (Todhunter et al. 2003a). Based on the DLS scores and the DIs of the founder and F_1 breeding groups, Monte Carlo simulations of 5000 crossings of the founder and F_1 breeders to produce a backcross generation showed a unimodal probability density distribution function indicative of polygenic inheritance controlling the DLS score and a bimodal distribution for the DI, suggesting the presence of a major gene locus superimposed on a polygenic background (Todhunter et al. 2003a).

Comparison between Developmental Dysplasia of the Human Hip and Canine Hip Dysplasia

Developmental dysplasia of the human hip (congenital dislocation of the hip, hip dysplasia) occurs with a frequency between 1 and 10 per 1000 live births (Henricson et al. 1966; Czeizel et al. 1975; Cilliers and Beighton 1990; Weinstein 1996; Brier 1999; Evans 2001). Females appear to be overrepresented, similar to some canine studies (Swenson et al. 1997). The crucial phenotypic characteristics of developmental dysplasia of the human hip (delayed femoral capital ossification, hip joint laxity, and hip subluxation) are also present during the natural development of the dysplastic canine hip. Delayed endochondral ossification of the dysplastic femoral head occurs in human and canine neonates (Todhunter et al. 1997; Harcke and Grissom 1999). Chondro-osseous or acetabular dysplasia has also been described as a component of developmental dysplasia of the human hip and recently as the primary defect in a 12-member kindred (Beals 2003).

Because human hip dysplasia is often detectable at birth (Harcke and Grissom 1999), whereas the canine condition is not observable until after pups begin walking (Henricson et al. 1966; Gustafsson et al. 1972; Riser 1987), the two conditions are not temporally analogous. Because large-breed dogs are skeletally mature at 8 months of age, the development of the hip can be studied from incipience to maturity in a short period compared to the evolution of human hip dysplasia. Interestingly, the left hip joint can be more affected to dysplasia than the right hip in humans (Smith et al. 1963) as well as in dogs (Chase et al. 2004), although we did not find a predilection for left-sidedness in our crossbred population (Todhunter et al. 2003c).

The joint laxity and subluxation characteristic of hip dysplasia may reflect abnormalities in the joint capsule or the developing chondroepiphyses of the hip. Both primary inherited acetabular dysplasia and hip joint laxity may lead to hip subluxation (dislocation) in humans (Beals 2003). Trait expression in humans follows a continuous distribution, as it does in dogs, as clinicians have difficulty in defining a cutoff between affected and unaffected hips based on standard radiographic methods like the Center edge angle of Wiberg (Pompe et al. 2003), a measure of femoral head subluxation in humans.

The Genetic Component of Developmental Dysplasia of the Human Hip

Familial aggregation of human hip dysplasia was reported in sibling and twin studies (Holderbaum et al. 1999). The concordance of hip dislocation in identical twins was 41% and only 2.8 % for dizygotic twins (James and Wynne-Davies 1969), suggesting a complex condition affected by individual genetic background and environmental factors. Genetic predisposition to developmental dysplasia of the human hip and severe hip OA ("Beukes" familial hip dysplasia) was found in a family in South Africa (Beighton et al. 1994) which mapped to an 11-cM region on 4q35 (Roby et al. 1999). The incidence of developmental dysplasia of the hip is high in Italy, ranging between 10 and 18.5 cases per 1000 live births. Complex segregation analyses indicated that a two-locus, recessive-gene model was the most parsimonious hereditary transmission model in this population (Sollazzo et al. 2000).

In children with developmental dysplasia of the hip, increased joint laxity is present in 29% of females and in 73% of males (James and

Wynne-Davies 1969). An interstitial duplication of human chromosome 15q24-26 (DUP25) was associated with joint laxity in humans; stratification of these patients according to joint laxity enabled identification of the DUP25 locus which was also associated with social phobias (Gratacos et al. 2001). This region is a recombination hot spot (Ventura et al. 2003).

Loci for Human Hip OA as Candidates for Hip Dysplasia

Several authors have discussed the relationship between developmental dysplasia of the hip and early-onset hip OA in humans (Harris 1986; Hoaglund and Healy 1990; Treble et al. 1990; Beighton et al. 1994; Weinstein 1996). The biochemical and clinical features of hip OA that result from hip dysplasia in dogs are the same as those of human hip OA (Burton-Wurster et al. 1993; Long and Langley 1999). Secondary OA is a slowly progressive, insidious disease that may take decades to show clinical expression. Because no loci for human hip dysplasia have been mapped, it seems prudent to consider loci associated with human hip OA as an important positional cloning strategy as we proceed with fine mapping and candidate gene approaches that will involve comparative human–dog mapping. The 2000 NIH Conference report on OA stated that OA genetic factors account for at least 50% of the variance in cases of OA of the hands and hip and that increased contact stress precipitated OA (Spector et al. 1996; Felson et al. 2000). Asians, Hispanics, and African-Americans have a lower incidence of hip OA than Caucasians (Hoaglund et al. 1995), whereas Icelanders are predisposed to hip OA (Ingvarsson et al. 1999, 2001).

Mutations in minor cartilage collagens and cartilage matrix proteins may lead to OA. The genes for COL5A2 (encoding the $\alpha 2$ chain of collagen type V) and fibronectin (found in abundance in early osteoarthritic articular cartilage) lie in the region of CFA37 in which we have detected putative QTL for CHD in the form of the Norberg angle in our initial coarse genome-wide screen. Both cartilage oligomeric matrix protein and insulin-like growth factor-1 are on human chromosome 19, which is syntenic to the telomeric region of CFA01 in which a QTL locus for CHD in Portuguese water dogs was reported (Chase et al. 2004). Susceptibility loci for human hip OA have also been mapped to human chromosomes 6, 11, and 16 with various levels of statistical support (Chapman et al. 1999; Mustafa et al. 2000; Chapman et al. 2003). In general, the earlier the trait is expressed, the more likely it is to be genetically driven, whereas the later it is expressed, the more likely it is to be environmentally regulated or affected by modifying loci (Lander and Schork 1994).

ENVIRONMENTAL FACTORS

Milk-borne factors may influence hip development, as hormones and growth-promoting molecules such as insulin, cortisol, epidermal growth factor, insulin-like growth factors, parathyroid-hormone related peptide, relaxin, estrogen, and estrogen precursors are present in colostrum and milk. These substances, which are absorbed from the gastrointestinal tract in the postnatal period, can influence connective tissue metabolism of genetically susceptible pelvic tissues.

Relaxin affects the structure and metabolism of joint capsule and ligaments. Relaxin persists longer in the serum of dysplastic Labrador retriever bitches throughout lactation, whereas it is detectable for only the first 1–2 weeks of lactation in nondysplastic bitches. Total serum estrogens were similar in pups born of dysplastic and normal matings, but testosterone was detected only in the milk of dysplastic Labrador retriever bitches, and estradiol-17β appeared only in the serum of pups born of dysplastic matings. An aromatase inhibitor (prevents conversion of testosterone to 17β-estradiol) injected into pups from birth through lactation significantly reduced hip joint laxity at maturity in dysplastic-bred Labrador retrievers. Repetitive estrogen dosing in the growing period can induce a condition that appears similar to CHD.

Influence of seasonality is suggested by the fact that dogs born in the spring and summer had less severe hip dysplasia than did dogs born in the winter (Wood and Lakhani 2003).

Abundant food consumption was shown to shorten the time to first appearance, and to increase the severity, of CHD. Ad libitum feeding of Great Danes induced the expression of several unwanted orthopedic traits, including CHD, when compared to dogs on restricted feeding. Although overfeeding itself does not cause CHD, it maximizes trait expression in genetically susceptible individuals. For dogs between 6 weeks and 1 year of age, the frequency and severity of CHD and concomitant OA in affected hips were reduced markedly in Labrador retrievers by limiting food consumption to 25% of ad libitum-fed control littermates (Larson et al. 2003). The associated hip OA is also more severe in dogs with rapid weight gain and growth than in dogs on restricted feeding (Kealy et al. 2000). Similarly, human hip OA is exacerbated by obesity (Karlson et al. 2003; Manek et al. 2003). The mechanism of operation of these nutritional effects remains unknown. It may be explained simply by mechanics, as maximizing growth allows maximum load to be placed on genetically susceptible hips.

Besides reduced food intake, the only treatment that has been shown to significantly reduce the expression of hip OA in CHD is systemic poly-sulfated glycosaminoglycan (Adequan®) given to dysplasia-prone Labrador retrievers from 6 weeks to 8 months of age. Polysulfated glycosamino-glycan is an inhibitor of neutral matrix metalloproteinase activity and has effects on transcriptional response of isolated chondrocytes (Mertens et al. 2003).

For genetic mapping studies, the diet and other environmental factors should be standardized in order to reduce the contribution of nongenetic factors to complex trait variance. This environmental control is a critically important attribute of our research animal colony.

POWER ANALYSIS FOR GENOME-WIDE SCREENING

Until the report by Chase et al. (2004), no locus or mutation had been described for hip dysplasia in any species. Candidate gene approaches might have resulted in isolation of polymorphisms or mutations in susceptibility genes, but that approach carries a high risk. Under the pre-vailing circumstances and in the absence of a strong candidate locus, the logical approach was to undertake a genome-wide screen to localize the contributing chromosomal regions, and then to proceed with a combi-nation of fine mapping and candidate gene screening.

The power to find linkage of a single marker to a QTL contributing to these dysplastic traits in the backcross generations was determined (Todhunter et al. 2003b). Sample size calculations were performed assuming (1) that one or more QTL controlled trait expression, (2) a 10-cM, genome-wide microsatellite scan was undertaken, (3) a power of 0.8 for recombination fractions (θ) of 0, 0.05, and 0.1, (4) an α level of 0.05, and (5) three configurations of the trait means (high, observed, and low) in the founder and first filial (F_1) generations from which the back-cross generation was produced. Mean heterozygosity of 22 microsatellites in the F_1's was 0.78. For the DI controlled by one QTL, at the observed effect size, $\theta = 0.05$, and a heterozygosity of 0.75, 35 dogs in the backcross of the F_1 to the greyhound generation would yield linkage at a power of 0.8. For the DLS score controlled by one QTL, 35 dogs in the backcross to the Labrador retriever generation would be required for linkage at the same power. Fewer dogs were projected to be necessary to find linkage to hip OA. Testing for linkage at four loci simultaneously, with 0.25 power at each of the four loci, yielded an overall power of 0.7. On the basis of this conservative single-marker estimate, we concluded that this pedigree has the requisite power to find microsatellites linked to susceptibility loci for CHD and OA by breeding a reasonable number of backcross dogs.

Genotype Fidelity

Genotype data can be accumulated in large volumes as a result of high-throughput genotyping with multiplex PCR reactions and sequencing on hundreds of individuals. In genome-wide screens accumulated on the Cornell pedigrees, for example, over 400 genotypes on 320 dogs (128,000 genotypes) must be evaluated. If genotypes are accumulated at an external agency or by collaborators at another institution and genotypes from several sources are merged, the data must be checked for allele call errors. This can be based on replicates run in the same screen and on inheritance patterns in the pedigrees. We analyze genotypic data to detect Mendelian errors using an algorithm that checks for genotype inconsistencies between parents and offspring (http://www.people.cornell.edu/pages/zz19/research/genoped). To check the flow of marker alleles through the pedigree for consistency, a program, developed in SAS (Statistical Analysis System, Carey, NC) starts the analysis at the top of the pedigree, assumes that the marker allele size in grandparents is correct, follows the flow of each allele through the pedigree, checks for inconsistencies in allele size within a narrow interval, and corrects the errors. The majority of errors appear to be due to small differences in called allele sizes (1–2 bp difference), but not all errors can be attributable to allele call errors. There was on average about a 5% error rate in the genotypes from the Mammalian Genotyping Service in our first genome-wide screen, and the error rate was 2.8% in the second genome-wide screen undertaken there on our pedigrees.

Marker Informativeness

Microsatellites are based on short repeat units (1–5 bp) that are randomly distributed throughout vertebrate genomes. In comparison with human populations, purebreed dogs express higher levels of genetic homogeneity (Jeffreys and Morton 1987), which is caused mainly by the inbreeding and line breeding that is required to maintain the individual type and character of different breeds. Despite the lower degree of variation of microsatellite loci compared to minisatellites, canine microsatellites have proved their usefulness in dog mapping (Ostrander et al. 1993; Lynch and Walsh 1998).

One requirement for successful QTL mapping is individuals with a sufficient number of informative genetic markers. Marker informativeness is a function of both the number of alleles per locus and allele frequencies and is most accurately represented by polymorphism information content (PIC) and degree of heterozygosity (Het) (Botstein et al. 1980; Lynch and Walsh 1998).

To assess the quality of marker genotype data in our crossbred pedigree with CHD, 247 markers spanning 80% of the canine genome were analyzed on 9 Labrador retriever and 7 greyhound founders, 7 F_1 and 136 backcross individuals. The total number of alleles and the mean number of alleles per locus were determined for each marker for the two founder groups, the F_1 breeders and the backcross population. The heterozygosity for each marker was determined separately for the four groups of animals using the equation

$$Het = 1 - \sum_{i=1}^{n} p_i^2$$

The PIC for each marker was calculated using the equation

$$PIC = 1 - \sum_{i=1}^{n} p_i^2 - 2\left[\sum_{i=1}^{n-1} \sum_{j=i+1}^{n} p_i^2 p_j^2\right]$$

where p_i is the frequency of the ith allele and n is the number of alleles (Botstein et al. 1980).

Summary statistics for PIC and Het for the two founder breeds, F_1 and backcross (BC) generations, are listed in Table 2. The increase in average marker informativeness in a crossbreed pedigree is associated with an increase in proportion of moderately and highly informative markers in F_1 and BC generations relative to the founders (Fig. 5).

Table 2. Descriptive statistics from 247 microsatellite markers screened on 159 dogs in an experimental canine pedigree representing the Labrador retriever and greyhound founders, F_1, and backcross generations

	n	No. of alleles mean/marker	maximum	Median PIC	Het
L	9	3.61	8	0.54	0.60
G	7	4.1	9	0.58	0.64
F_1	7	4.3	10	0.62	0.67
BC	136	5.58	16	0.63	0.69

(n) Number of individuals; (PIC) polymorphism information content; (Het) heterozygosities; (L) Labrador retriever; (G) greyhound; (BC) backcross.

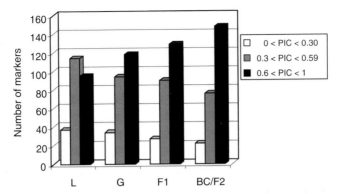

Figure 5. Proportion of uninformative (PIC < 0.30), moderately informative (0.3 > PIC > 0.59), and highly informative (PIC > 0.60) microsatellite markers in the Labrador retriever (L), greyhound (G), F_1, and backcross/F_2 (BC/F_2) generations of a crossbreed pedigree.

Approximately 60% of the markers in the BC generation demonstrated high information content compared to only 39% in the Labrador retriever and 48% in the greyhound founders. The increase in PIC is a direct consequence of the increase in number of alleles for many markers.

MAPPING STRATEGIES

There are two main strategies for finding trait loci: association tests using candidate genes and genome scans based on linkage mapping with anonymous DNA markers. The candidate gene approach can be powerful and can detect loci with small effect, provided that the candidate gene represents a true causative gene. However, there are often many candidate genes for the trait of interest and it is more time-consuming to evaluate all candidates than it is to do a genome scan. Furthermore, the candidate gene approach might fail to identify a major trait locus simply because of the gaps in our knowledge regarding gene function.

Often in candidate gene studies a nominal significance threshold is applied (i.e., as appropriate when a single study is performed). As Lander and Kruglyak (1995) argued, researchers may often perform several or many candidate gene studies; hence significance thresholds should be as stringent as those for a genome scan. On these grounds, a number of the reported associations may be erroneous. Another potential pitfall is population stratification, where the population is composed of several subpopulations which may differ in the frequencies of many loci. A test for association between any two of these loci may show spurious association if the stratification is not taken into account. Another potential problem with candidate gene association studies may be the prevalence of linkage disequilibrium. The candidate gene approach relies on the presence of localized linkage disequilibrium to be successful. A detected polymorphism must be in disequilibrium with the causative polymorphism in order to have an association with phenotypic variation. However, if linkage disequilibrium is widespread, associations may be found with a locus or loci some distance away.

Methods based on candidate genes are not appropriate for poorly understood traits. Even for well-understood traits there may be advantages in genome-wide scans, because they may reveal previously unsuspected loci. In principle, if a sufficiently large number of animals are phenotyped for a quantitative trait, and genotyped with a set of polymorphic markers that provides full genome coverage, a genome scan will always find the map location of a trait locus with a major effect. However, a genome scan will fail to detect trait loci with smaller effects if they do not reach the stringent significance thresholds that must be applied when doing a large number of tests in a full genome scan.

Genome-wide QTL analysis was first applied to fruit characteristics in the tomato (Paterson et al. 1990), but it was soon used in mammals to study epilepsy in mice and hypertension in rats (Hilbert et al. 1991; Jacob et al. 1991; Rise et al. 1991). In only a short time, there has been an explosion of interest in QTL mapping in both agriculture and biomedicine.

Microsatellite Density and Progeny Number Considerations

An essential component for an efficient genome-wide scan is the availability of a high-density genetic map. Richman et al. (2001) published canine Minimal Screening Set 1 containing 172 markers, and Breen et al. (2001) published an integrated linkage-radiation hybrid canine map of 1800 markers. A 3300-marker canine map at 1 Mb resolution was recently made available (Guyon et al. 2003).

Finding the optimal balance between the number of marker loci to test and individuals to sample is an important question in QTL mapping. Although no clear-cut answer exists, the following suggestion seems reasonable: For an initial QTL study, a "framework" DNA marker map with loci sampled every 20–30 cM (so that no QTL is farther than 15 cM from a marker locus) might be sufficient to detect the most important QTL. If one or more regions of particular interest are identified based on the initial analysis, more markers known to map to the intervals most likely to contain the QTL can be used to develop a fine map of the regions.

The number of individuals needed for mapping varies with the heritability of the trait (if only a few loci with large effects are expected to control the trait, fewer progeny need to be screened). The power of t-tests comparing homozygous genotypic means depends on both the sample size and the amount of recombination between the marker and QTL. Lynch and Walsh (1998) concluded that when markers are already closely spaced, there is a point of diminishing returns. For example, for $r = 20\%$ recombination between markers and QTL (corresponding to markers spaced 26 cM apart, using the Haldane mapping function), moving to an infinitely dense map ($r = 0$) requires only 36% as many individuals to be scored to give the same power. However, for marker–QTL recombination frequencies of 10%, and 5%, these become 81%, and 90%. Thus, for markers spaced at 10 cM or closer, there is little advantage in increasing marker density when the goal is simple detection of a linked QTL (Lynch and Walsh 1998).

Interval Mapping

The approach we adopted for mapping trait loci was to use intercrosses between divergent populations. The F_1 individuals in such a cross tend to show a high heterozygosity at marker loci, particularly at those loci that account for phenotypic differences between the two populations. Several intercrosses have been generated in farm animals: the European wild boar and large white domestic pig cross (Andersson et al. 1994), the Asian and European pig crosses (Rohrer and Keele 1998; de Koning et al. 1999), and Bos taurus and Bos indicus cattle crosses (Brenneman et al. 1996).

Association between a quantitative trait and genetic markers can be evaluated using single markers or multiple markers. When using single markers, it is possible to make inference about the segregation of a QTL linked to that marker. However, it is not possible to distinguish between the size of a QTL effect and its position (relative to the marker). In addition, single-marker analyses have less power if the markers are far apart. If two (or more) markers are jointly used in an analysis, there is less confounding between the position and size of the QTL effect and more power in detecting a QTL, even if the markers are far apart. Inference about the QTL effect as well as the recombination rate between QTL and markers (i.e., position of QTL) is possible. The recombination rate between markers is usually assumed known.

There are two different approaches to perform interval mapping: regression (Haley and Knott 1992) and maximum likelihood-based methods (Lander and Botstein 1989). In the maximum likelihood methods, information on the presence of a QTL is derived both from the mean differences between the flanking marker genotype classes and from the distribution of the trait within each marker genotype class. Although these methods can provide accurate parameter estimates and hypothesis testing, custom-written software is required for their implementation. Furthermore, iterative numerical methods (algorithms) are required to maximize the likelihood, and thus, these methods become increasingly troublesome as the models become more complex.

Haley and Knott (1992) demonstrated that ordinary least squares can be used for interval mapping, providing similar estimates and test statistics to those obtained from maximum likelihood approaches. This allows relatively complex (and potentially more realistic) models to be used without placing severe demands on computational resources. For incremental positions along the marker map, the identity-by-descent (IBD) probabilities are estimated from flanking marker data. Subsequently, a statistical model is fitted to trait observations and the IBD probabilities. The general linear model allows for fixed effects and covariates to be included in the analysis. Every centiMorgan, the offspring phenotypes are regressed against the coefficients of additive (a) and dominance (d) variables. F-test statistics calculated as the ratio of the residual sums of squares in a model with the QTL ("full") and a model without it ("reduced") are performed at each location along the chromosomes. Estimates are obtained for the additive and dominance effect of the putative QTL at that location in the backcross/F_2 population. The location giving the highest F ratio is considered to be the best estimate for the position of the QTL.

Interval and Regression Mapping in Our Crossbreed Hip Dysplasia Pedigree

We genotyped 159 dogs from three generations of the crossbreed pedigree described above with 240 microsatellite markers distributed in 38 autosomes and the X chromosome. Hip traits measured and analyzed included the DI, the DLS score, and the Norberg angle. Permutation testing was used to derive the chromosome-wide level of significance at $p < 0.05$ for each QTL. Chromosomes 4, 9, 10, 11 ($p < 0.01$), 16, 20, 22, 25, 29 ($p < 0.01$), 30, 35, and 37 harbored putative QTL for one or more traits (Todhunter et al. 2005). Measuring multiple traits for hip dysplasia allowed us to detect the maximum number of QTL with this coarse genome-wide screen. Significant QTL on four chromosomes were identified for at least two traits, whereas significant QTL on eight chromosomes were unique to different traits. QTL for the DLS score were detected on five, for the NA on five, for the DI on two, and for different principal components on six chromosomes. However, on most of the chromosomes harboring QTL for only one trait, many other traits exhibited a high F-test statistic without reaching statistical significance. QTL for several traits (on CFA10, 11, 16, and 22) had alleles with a protective effect, whereas QTL identified on CFA4, 20, 29, 30, and 35 had alleles conferring a worsening influence on hip conformation. Our conclusions must be tempered by the knowledge that mapping in modest-sized pedigrees inflates parameter estimates, compared to that which might be observed in large populations. In addition, the limited number of recombinations and markers used reduced mapping resolution and accuracy.

RANDOM MODEL AND VARIANCE COMPONENT APPROACH

Literature about QTL mapping in populations with complex pedigrees can be classified according to the allelic assumptions associated with the QTL. Mapping methods assume the QTL is either a fixed effect with a finite number of alleles or a random effect with an infinite number of alleles. Analysis of statistical models that treat the QTL as a fixed effect range from the simple regression-based methodologies (Knott et al. 1996) described to complex statistical analyses involving Markov chain Monte Carlo methods within frequentist (Heath 1997; Jansen et al. 1998) and Bayesian (George et al. 2000; Uimari and Hoeschele 1997) paradigms. The statistical models are based on mixture distributions, where the number of component densities is determined by the number of QTL genotypes, and assumptions regarding the total number of segregating alleles have a profound effect on the formulation of the statistical model.

The IBD-based variance component analyses are derived from the assumption that individuals of similar phenotype are more likely to share alleles that are IBD. The construction of IBD matrices for alleles at each position tested along the genome and the fitting of random-effect models (which assumes that QTL effects are normally distributed) offer an appropriate method to map QTL if the mapping population is large enough and if the progenies are connected. In addition, these models do not assume a known, finite set of alleles at each putative QTL and, therefore, a less parameterized statistical environment in which to map QTL is created. This is achieved by assuming that the QTL effects are normally distributed—an assumption that circumvents the estimation of QTL allele frequencies and is robust to violation (Hoeschele et al. 1997). Random effects models have long been utilized by human geneticists interested in partitioning the genetic variance of quantitative traits into effects due to specific chromosomal regions. As early as the 1970s, variance component approaches (i.e., analytical methods that estimate the parameters of random effects models) were being used to detect QTL in phase-known pedigrees (Jayakar 1970). Since then, the development of increasingly sophisticated variance component methods has enabled QTL to be mapped in increasingly general pedigrees (Amos 1994; Almasy and Blangero 1998).

DETECTION OF THE MAXIMUM NUMBER OF LOCI PRIOR TO FINER MAPPING

The principal challenge with quantitative traits lies not in detecting QTL, but in unraveling the genes that underlie them. Despite intense efforts to identify the genes that affect quantitative traits, in particular those that are involved in common human diseases, there are few success stories (Darvasi and Pisante-Shalom 2002; Glazier et al. 2002). The identification of genes and mutations that underlie QTL is problematic for several reasons. First, it remains difficult to determine the exact chromosomal location of a QTL, as the marker and crossover density in the region of interest limits the mapping precision. The lack of a direct relationship between genotype and phenotype, as exists for monogenic traits, prohibits the unambiguous identification of recombinant individuals that is required for high-resolution mapping. This is due to the fact that individual QTL only account for part of the phenotypic variance, the rest being due to environmental factors as well as other QTL. The situation is particularly difficult in the case of numerous loosely linked QTL, which seem to account for a significant proportion of the largest QTL effects that are detected in livestock. Although composite interval mapping and multiple QTL mapping might help to unravel such situations in experimental crosses between inbred lines, the development of a suitable statistical framework to address such situations in outbred designs is only in its infancy (Meuwissen and Goddard 2004). Epistatic interactions might also add to the challenge of dissecting the genetic basis of complex traits. Certainly, data from model organisms and from coat-color inheritance in mammals indicate that epistasis between QTL might be an important factor for consideration. However, studies of epistasis among QTL in vertebrates have been rare (Brockmann et al. 2000; Shimomura et al. 2001; Carlborg et al. 2003). For these reasons, QTL are often mapped to chromosomal regions that are over 20 cM long [~20 megabase pairs (Mb)] and that might contain several hundred genes.

Most QTL have a mild phenotypic effect, and the mutations that cause them are difficult to distinguish from neutral polymorphisms. In contrast, mutations that cause monogenic disorders generally knock out gene expression or lead to an altered protein function. Another complicating factor in the identification of QTL mutations is that a good proportion of these are likely to be regulatory mutations. Our ability to map and evaluate functionally important mutations in noncoding regions is still poorly developed. For example, the identification and annotation of regulatory regions in sequenced genomes is still very rudimentary compared with that of coding sequences, although our ability to identify important regulatory elements through sequence comparison will improve as more mammalian genomes are sequenced. The situation might be even more complicated if QTL effects reflect the combined action of clusters of tightly linked mutations (Steinmetz et al. 2002), or if epigenetic inheritance contributes to quantitative genetic variation.

QTL MAPPING LIMITATIONS

High levels of intrabreed genetic homogeneity and marked genetic heterogeneity across breeds characterize the domestic dog (Parker et al. 2004). The stratification of breed lines can complicate linkage analysis in crossbreed dog populations. Between breeds, different alleles may segregate with the same QTL, or different QTL may exist for a given trait. These differences result in decreased ability to detect weak QTL (Lynch and Walsh 1998). It is possible to increase the probability of identical QTL segregating within two breeds through selection of populations with common ancestry. However, in order to get the largest array of phenotypic traits, it is often more efficient to use breeds that are evolutionarily distant. For the previous reasons, linkage disequilibrium mapping between breeds may be a more efficient method to fine-map complex traits in dogs than linkage mapping. Nevertheless, all of the above constraints aside, our initial approach to undertake QTL mapping in a crossbreed pedigree based on phenotypically divergent breeds has been rewarding.

Limitations to successful QTL mapping can be a function of both the power of the pedigree and the tools available for statistical analysis. Determining which software is best for a given analysis is much more challenging now because the number of programs available is rapidly increasing. Most programs are tailored for a specific purpose, such as sib-pair, nuclear family, inbred pedigree, or crossbred pedigree analyses. Complicating this further is the fact that most software is designed for human genetic linkage studies and generally focuses on sib-pair and nuclear family analytical approaches because of the nature of human pedigrees. Few programs can handle pedigrees with high levels of inbreeding and/or breeding loops characteristic of canine pedigrees without breaking up the loops, thereby reducing power (http://linkage.rockefeller.edu/soft/list.html). An excellent program for analysis of animal pedigrees that works well in our 3-generation cross breed is QTL Express™ (Seaton et al. 2002). The module we use is specifically designed to handle crossbred pedigrees with both backcross and intercross families. Other programs that are useful include GENEHUNTER and LINKAGE (Ott 1999).

Frequently in complex trait mapping, several traits, often correlated, are collected or measured on each individual. Principal component analysis (PCA) combines correlated phenotypic traits into independent components. An excellent example is provided by the recent work of Chase et al. (2002), where PCA was used to define phenotypes that correspond to skeletal structure traits and the resulting groups of phenotypic traits were then used to identify QTL. The same group of investigators also searched for QTL that contributed to hip dysplasia in the Portuguese water dog population. Because hip dysplasia is a bilateral trait, they found that analyzing the phenotype on each side in a separate analysis enabled the detection of different QTL contributing to the right and the left hip on each end of CFA01. Other options are to analyze the best or the worst hip of each animal (Lust et al. 2001b). In our analysis of QTL contributing to hip dysplasia in an experimental pedigree of greyhounds and affected Labrador retrievers, we concluded that the analysis of the left and right side separately, the best and the worst hip, and principal components are all useful in the mapping effort (Todhunter et al. 2005). As outlined above, it was clear from our analysis that measuring several hip traits held more power for mapping than a single trait, as we detected contributing QTL for one trait on some chromosomes and for another trait on other chromosomes.

Statistical programs generally assume that the QTL is normally distributed; however, programs and statistical tools that account for violations of normality do so differently. For example, the regression mapping method is robust to modest departures from normality (Feingold 2002). However, if assumptions are incorrect when using variance components, the power may be significantly decreased and type I errors may increase (Feingold 2002). Conversely, nonparametric measurements do not initially assume any particular form for the underlying distributions, allowing a model-free statistical analysis and making it a more appropriate choice for analysis of complex traits (Lynch and Walsh 1998). However, because variables are not defined in nonparametric analyses, confidence in the accuracy of the parameter estimates is lowered.

MAPPING RESOLUTION

Classic linkage is first undertaken to identify regions of interest because it can localize QTL based on coarse screens across large chromosomal regions. To initially locate a QTL, one must be able to detect recombination events across generations (Doerge et al. 1997). With more recombination events, the region of interest can be narrowed. Thus, adding to the sample size may be more beneficial than increasing marker density (Doerge et al. 1997). However, as we outlined, there is a limit to QTL mapping resolution using linkage, probably to a region of 10–40 cM. Traditionally, linkage disequilibrium (LD) mapping, and association mapping, are used to fine-map areas in which linkage has already been identified. Nevertheless, the genetic homogeneity within a breed increases the extent of LD in the dog compared to human populations, making it less useful for fine-mapping efforts than in humans (Sutter and Ostrander 2004). Therefore, even upon identification of QTL, it is challenging to narrow the region of interest.

In conclusion, a crossbreed pedigree increases the power to initially detect QTL, but further complicates attempts to narrow regions harboring loci of interest. Linkage mapping uses only the recombinations that occurred within the data set, which typically contains three generations, and the putative QTL could be mapped only to regions of 5–10 cM. LD mapping uses all recombinations since the mutation occurred, which increases the precision of the estimate of the position. Therefore, linkage mapping using a crossbreed pedigree is more useful for a genome-wide scan for QTL, whereas LD mapping is more useful for precise estimation of QTL position. A crossbreed dog population, although still possessing genetic limitations, increases marker informativeness and may demonstrate maximal phenotypic variation. The utility of such a pedigree is not restricted to the study of the genetic mechanisms underlying complex traits of the dog alone. Rather, the dog has emerged as an excellent model for study of many human hereditary diseases, including developmental dislocation of the hip, a counterpart of canine hip dysplasia. This is amplified by the fact that the dog is genetically and anatomically more similar to the human than is the mouse, a traditional mammalian model of human traits, and shares many of the most frequently occurring human diseases. Finally, there is no shortage of canine pedigrees to study diseases. The facts that approximately 36% of households in the U.S. own at least one dog and that medical surveillance of the dog is second only to that of the human (Sutter and Ostrander 2004) make the dog an increasingly important subject of biomedical research and a popular choice as a disease model system.

REFERENCES

Almasy L. and Blangero J. 1998. Multipoint quantitative-trait linkage analysis in general pedigrees. *Am. J. Hum. Genet.* **62:** 1198–1211.

Amos C.I. 1994. Robust variance-components approach for assessing genetic linkage in pedigrees. *Am. J. Hum. Genet.* **54:** 535–543.

Andersson L., Haley C.S., Ellegren H., Knott S.A., Johansson M., Andersson K., Andersson-Eklund L., Edfors-Lilja I., Fredholm M., Hansson I, et al. 1994. Genetic mapping of quantitative trait loci for growth and fatness in pigs. *Science* **263:** 1771–1774.

Beals R.K. 2003. Familial primary acetabular dysplasia and dislocation of the hip. *Clin. Orthop. Relat. Res.* **406:** 109–115.

Beighton P., Cilliers H.J., and Ramesar R. 1994. Autosomal dominant (Beukes) premature degenerative osteoarthropathy of the hip joint unlinked to COL2A1. *Am. J. Med. Genet.* **53:** 348–351.

Beling C.G., Gustafsson P.O., and Kasstrom H. 1975. Metabolism of estradiol in greyhounds and German shepherd dogs. An investigation with special reference to hip dysplasia. *Acta Radiol. Suppl.* **344:** 109–120.

Bliss S., Todhunter R.J., Quaas R., Casella G., Wu R., Lust G., Williams A.J., Hamilton S., Dykes N.L., Yeager A., Gilbert R.O., Burton-Wurster N.I., and Acland G.M. 2002. Quantitative genetics of traits associated with hip dysplasia in a canine pedigree constructed by mating dysplastic Labrador Retrievers with unaffected Greyhounds. *Am. J. Vet. Res.* **63:** 1029–1035.

Botstein D., White R., Skolnick M., and David R.W. 1980. Construction of a genetic linkage map using restriction fragment length polymorphisms. *Am. J. Hum. Genet.* **32:** 314–331.

Breen M., Jouquand S., Renier C., Mellersh C.S., Hitte C., Holmes N.G., Cheron A., Suter N., Vignaux, Bristow A.E., et al. 2001. Chromosome-specific single-locus FISH probes allow anchorage of an 1800-marker integrated radiation-hybrid/linkage map of the domestic dog genome to all chromosomes. *Genome Res.* **11:** 1784–1795.

Brenneman R.A., Davis S.K., Sanders J.O., Burns B.M., Wheeler T.C., Turner J.W., and Taylor J.F. 1996. The polled locus maps to BTA1 in a *Bos indicus* x *Bos taurus* cross. *J. Hered.* **87:** 156.

Breur G.J., Lust G., and Todhunter R.J., eds. 2002. Genetics of hip dysplasia and other orthopedic traits. In *The genetics of the dog*, pp. 267–298. CAB International, Wallingford, Oxon, United Kingdom.

Brier S.R. 1999. *Primary care orthopedics*. Mosby, St. Louis, Missouri.

Brockmann G.A., Kratzsch J., Haley C.S., Renne U., Schwerin M., and Karle S. 2000. Single QTL effects, epistasis, and pleiotropy account for two-thirds of the phenotypic F(2) variance of growth and obesity in DU6i x DBA/2 mice. *Genome Res.* **10:** 1941–1957.

Burton-Wurster N., Todhunter R.J., and Lust G. 1993. Animal models of osteoarthritis. In *Joint cartilage degradation: Basic and clinical aspects* (ed. J.F. Woessner and D.S. Howell), pp. 347–384. Marcel Dekker, New York.

Burton-Wurster N., Farese J.P., Todhunter R.J., and Lust G. 1999. Site-specific variation in femoral head cartilage composition in dogs at high and low risk for development of osteoarthritis: Insights into cartilage degeneration. *Osteoarthritis Cartilage* **7:** 486–497.

Cardinet G.H., III, Guffy M.M., Wallace L.J., and Laben R.C. 1983. Canine hip dysplasia in German Shepherd dog-Greyhound crossbreeds. *J. Am. Vet. Med. Assoc.* **182:** 393–395.

Carlborg O., Kerje S., Schutz K., Jacobsson L., Jensen P., and Andersson L. 2003. A global search reveals epistatic interaction between QTL for early growth in the chicken. *Genome Res.* **13:** 413–421.

Chapman J.M., Cooper J.D., Todd J.A., and Clayton D.G. 2003. Detecting disease associations due to linkage disequilibrium using haplotype tags: A class of tests and the determinants of statistical power. *Hum. Hered.* **56:** 18–31.

Chapman K., Mustafa Z., Irven C., Carr A.J., Clipsham K., Smith A., Chitnavis J., Sinsheimer J.S., Bloomfield V.A., McCartney M., et al. 1999. Osteoarthritis-susceptibility locus on chromosome 11q, detected by linkage. *Am. J. Hum. Genet.* **65:** 167–174.

Chase K., Lawler D.F., Adler F.R., Ostrander E.A., and Lark K.G. 2004. Bilaterally asymmetric effects of quantitative trait loci (QTLs): QTLs that affect laxity in the right versus left coxofemoral (hip) joints of the dog (*Canis familiaris*). *Am. J. Med. Genet.* **124A:** 239–247.

Chase K., Carrier D.R., Adler F.R., Jarvik T., Ostrander E.A., Lorentzen T.D., and Lark K.G. 2002. Genetic basis for systems of skeletal quantitative traits: Principal component analysis of the canid skeleton. *Proc. Natl. Acad. Sci.* **99:** 9930–9935.

Cilliers H.J. and Beighton P. 1990. Beukes familial hip dysplasia: An autosomal dominant entity. *Am. J. Med. Genet.* **36:** 386–390.

Czeizel A., Tusnady G., Vaczo G., and Vizkelety T. 1975. The mechanism of genetic predisposition in congenital dislocation of the hip. *J. Med. Genet.* **12:** 121–124.

Darvasi A. and Pisante-Shalom A. 2002. Complexities in the genetic dissection of quantitative trait loci. *Trends Genet.* **18:** 489–491.

de Koning D.J., Janss L.L., Rattink A.P., van Oers P.A., de Vries B.J., Groenen M.A., van der Poel J.J., de Groot P.N., Brascamp E.W., and van Arendonk J.A. 1999. Detection of quantitative trait loci for backfat thickness and intramuscular fat content in pigs (*Sus scrofa*). *Genetics* **152:** 1679–1790.

Distl O., Grussler W., Schwarz J., and Krausslich H. 1991. Analysis of environmentally-conditioned and genetic influences on the frequency of hip joint dysplasia in German Shepherd dogs (in German). *Zentbl. Vetmed. A.* **38:** 460–471.

Doerge R.W., Zeng Z.B., and Weir G.S. 1997. Statistical issues in the search for genes affecting quantitative traits in experimental populations. *Stat. Sci.* **12:** 295–219.

Evans R.C. 2001. *Illustrated orthopedic physical assessment.* Mosby, St. Louis, Missouri.

Farese J.P., Todhunter R.J., Lust G., Williams A.J., and Dykes N.L. 1998. Dorsolateral subluxation of hip joints in dogs measured in a weight-bearing position with radiography and computed tomography. *Vet. Surg.* **27:** 393–405.

Farquhar T., Bertram J., Todhunter R.J., Burton-Wurster N., and Lust G. 1997. Variations in composition of cartilage from the shoulder joints of young adult dogs at risk for developing canine hip dysplasia. *J. Am. Vet. Med. Assoc.* **210:** 1483–1485.

Feingold E. 2002. Regression-based quantitative-trait-locus mapping in the 21st century. *Am. J. Hum. Genet.* **71:** 217–222.

Felson D.T., Lawrence R.C., Dieppe P.A., Hirsch R., Helmick C.G., Jordan J.M., Kington R.S., Lane N.E., Nevitt M.C., Zhang Y., et al. 2000. Osteoarthritis: New insights. 1. The disease and its risk factors. *Ann. Intern. Med.* **133:** 635–646.

George A.W., Mengersen K.L., and Davis G.P. 2000. Localization of a quantitative trait locus via a Bayesian approach. *Biometrics* **56:** 40–51.

Glazier A.M., Nadeau J.H., and Aitman T.J. 2002. Finding genes that underlie complex traits. *Science* **298:** 2345–2349.

Gratacos M., Nadal M., Martin-Santos R., Pujana M.A., Gago J., Peral B., Armengol L., Ponsa I., Miro R., Bulbena A., and Estivill X. 2001. A polymorphic genomic duplication on human chromosome 15 is a susceptibility factor for panic and phobic disorders. *Cell* **106:** 367–379.

Guyon R., Lorentzen T.D., Hitte C., Kim L., Cadieu E., Parker H.G., Quignon P., Lowe J.K., Renier C., Gelfenbeyn B., et al. 2003. A 1-Mb resolution radiation hybrid map of the canine genome. *Proc. Natl. Acad. Sci.* **100:** 5296–5301.

Gustafsson P.O., Kasstrom H., Olsson S.E., and Wennman B. 1972. Skeletal development and sexual maturation in German Shepherds, Greyhounds and their crossbreed offspring. An investigation with special reference to hip dysplasia. *Acta Radiol. Suppl.* **319:** 187–190.

Haley C.S. and Knott S.A. 1992. A simple regression method for mapping quantitative trait loci in line crosses using flanking markers. *Heredity:* 315–324.

Harcke H.T. and Grissom L.E. 1999. Pediatric hip sonography. Diagnosis and differential diagnosis. *Radiol. Clin. N. Am.* **37:** 787–796.

Harris W.H. 1986. Etiology of osteoarthritis of the hip. *Clin. Orthop. Relat. Res.* **213:** 20–33.

Heath S.C. 1997. Markov chain Monte Carlo segregation and linkage analysis for oligogenic models. *Am. J. Hum. Genet.* **61:** 748–760.

Hedhammar A., Olsson S.E., Andersson S.A., Persson L., Pettersson L., Olausson A., and Sundgren P.E. 1979. Canine hip dysplasia: Study of heritability in 401 litters of German Shepherd dogs. *J. Am. Vet. Med. Assoc.* **174:** 1012–1016.

Hedhammar A., Wu F.M., Krook L., Schryver H.F., De Lahunta A., Whalen J.P., Kallfelz F.A., Nunez E.A., Hintz H.F., Sheffy B.E., and Ryan G.D. 1974. Overnutrition and skeletal disease. An experimental study in growing Great Dane dogs. *Cornell Vet.* **64:** 5–160.

Henricson B., Norberg I., and Olsson S.E. 1966. On the etiology and pathogenesis of hip dysplasia: A comparative review. *J. Small Anim. Pract.* **7:** 673–688.

Henry G.A. 1992. Radiographic development of canine hip dysplasia. *Vet. Clin. N. Am. Small Anim. Pract.* **22:** 559–578.

Hilbert P., Lindpaintner K., Beckmann J.S., Serikawa T., Soubrier F., Dubay C., Cartwright P., De Gouyon B., Julier C., and Takahasi S. 1991. Chromosomal mapping of two genetic loci associated with blood-pressure regulation in hereditary hypertensive rats. *Nature* **353:** 521–529.

Hoaglund F.T. and Healy J.H. 1990. Osteoarthrosis and congenital dysplasia of the hip in family members of children who have congenital dysplasia of the hip. *J. Bone Joint Surg. Am.* **72:** 1510–1518.

Hoaglund F.T., Oishi C.S., and Gialamas G.G. 1995. Extreme variations in racial rates of total hip arthroplasty for primary coxarthrosis: A population-based study in San Francisco. *Ann. Rheum. Dis.* **54:** 107–110.

Hoeschele I., Uimari P., Grignola F.E., Zhang Q., and Gage K.M. 1997. Advances in statistical methods to map quantitative trait loci in outbred populations. *Genetics* **147:** 1445–1457.

Holderbaum D., Haqqi T.M., and Moskowitz R.W. 1999. Genetics and osteoarthritis: Exposing the iceberg. *Arthritis Rheum.* **42:** 397–405.

Ingvarsson T., Hagglund G., Jonsson H., Jr., and Lohmander L.S. 1999. Incidence of total hip replacement for primary osteoarthrosis in Iceland 1982–1996. *Acta Orthop. Scand.* **70:** 229–233.

Ingvarsson T., Stefansson S.E., Gulcher J.R., Jonsson H.H., Jonsson H., Frigge M.L., Palsdottir E., Olafsdottir G., Jonsdottir T., Walters G.B., et al. 2001. A large Icelandic family with early osteoarthritis of the hip associated with a susceptibility locus on chromosome 16p. *Arthritis Rheum.* **44:** 2548–2555.

Jacob H.J., Lindpaintner K., Lincoln S.E., Kusumi K., Bunker R.K., Mao Y.P., Ganten D., Dzau V.J., and Lander E.S. 1991. Genetic mapping of a gene causing hypertension in the stroke-prone spontaneously hypertensive rat. *Cell* **67:** 213–224.

James J.I.P. and Wynne-Davies R. 1969. Factors in orthopedics. In *Recent advances in orthopedics* (ed. A.G. Apley), pp. 1–35. J & A Churchill, London.

Jansen R.C., Johnson D.L., and Van Arendonk J.A. 1998. A mixture model approach to the mapping of quantitative trait loci in complex populations with an application to multiple cattle families. *Genetics* **148:** 391–399.

Jayakar S.D. 1970. On the detection and estimation of linkage between a locus influencing a quantitative character and a marker locus. *Biometrics* **26:** 451–464.

Jeffreys A.J. and Morton D.B. 1987. DNA fingerprints of dogs and cats. *Anim. Genet.* **18:** 1–15.

Kaneene J.B., Mostosky U.V., and Padgett G.A. 1997. Retrospective cohort study of changes in hip joint phenotype of dogs in the United States. *J. Am. Vet. Med. Assoc.* **211:** 1542–1544.

Karlson E.W., Mandl L.A., Aweh G.N., Sangha O., Liang M.H., and Grodstein F. 2003. Total hip replacement due to osteoarthritis: The importance of age, obesity, and other modifiable risk factors. *Am. J. Med.* **114:** 93–98.

Kealy R.D., Lawler D.F., Ballam J.M., Lust G., Biery D.N., Smith G.K., and Mantz S.L. 2000. Evaluation of the effect of limited food consumption on radiographic evidence of osteoarthritis in dogs. *J. Am. Vet. Med. Assoc.* **217:** 1678–1680.

Kealy R.D., Lawler D.F., Ballam J.M., Lust G., Smith G.K., Biery D.N., and Olsson S.E. 1997. Five-year longitudinal study on limited food consumption and development of osteoarthritis in coxofemoral joints of dogs. *J. Am. Vet. Med. Assoc.* **210:** 222–225.

Kealy R.D., Olsson S.E., Monti K.L., Lawler D.F., Biery D.N., Helms R.W., Lust G., and Smith G.K. 1992. Effects of limited food consumption on the incidence of hip dysplasia in growing dogs. *J. Am Vet. Med. Assoc.* **201:** 857–863.

Knott S.A., Elsen J.M., and Haley C.S. 1996. Methods for multiple-marker mapping of quantitative trait loci in half-sib populations. *Theor. Appl. Genet.* **93:** 71–80.

Lander E.S. and Botstein D. 1989. Mapping mendelian factors underlying quantitative traits using RFLP linkage maps. *Genetics* **121:** 185–199.

Lander E.S. and Kruglyak L. 1995. Genetic dissection of complex traits: Guidelines for interpreting and reporting linkage results. *Nat. Genet.* **11:** 241–247.

Lander E.S. and Schork N.J. 1994. Genetic dissection of complex traits. *Science* **265:** 2037–2048.

Larson B.T., Lawler D.F., Spitznagel E.L.J., and Kealy R.D. 2003. Improved glucose tolerance with lifetime diet restriction favorably affects disease and survival in dogs. *J. Nutr.* **133:** 2887–2892.

Leighton E.A. 1997. Genetics of canine hip dysplasia. *J. Am. Vet. Med. Assoc.* **210:** 1474–1479.

Leighton E.A., Linn J.M., Willham R.L., and Castleberry M.W. 1977. A genetic study of canine hip dysplasia. *Am. J. Vet. Res.* **38:** 241–244.

Lingaas F. and Heim P. 1987. En genetisk undersokelse av hoftleddsdysplasi I norski hunderaser. *Nor. Veterinaertidsskr.* **99:** 617–623.

Lingaas F. and Klementsdal G. 1990. Breeding values and genetic trend for hip dysplasia in the Norwegian Golden Retriever population. *J. Anim. Breed. Genet.* **107:** 437–443.

Long A.D. and Langley C.H. 1999. The power of association studies to detect the contribution of candidate genetic loci to variation in complex traits. *Genome Res.* **9:** 720–731.

Lust G. 1997. An overview of the pathogenesis of canine hip dysplasia. *J. Am. Vet. Med. Assoc.* **210:** 1443–1445.

Lust G., Geary J.C., and Sheffy B.E. 1973. Development of hip dysplasia in dogs. *Am. J. Vet. Res.* **34:** 87–91.

Lust G., Todhunter R.J., Erb H.N., Dykes N.L., Williams A.J., Burton-Wurster N.I., and Farese J.P. 2001a. Repeatability of dorsolateral subluxation scores in dogs and correlation with macroscopic appearance of hip osteoarthritis. *Am. J. Vet. Res.* **62:** 1711–1715.

———. 2001b. Comparison of three radiographic methods for diagnosis of hip dysplasia in eight-month-old dogs. *J. Am. Vet. Med. Assoc.* **219:** 1242–1246.

Lust G., Williams A.J., Burton-Wurster N., Pijanowski G.J., Beck K.A., Rubin G., and Smith G.K. 1993. Joint laxity and its association with hip dysplasia in Labrador retrievers. *Am. J. Vet. Res.* **54:** 1990–1999.

Lynch M. and Walsh B. 1998. *Genetics and analysis of quantitative traits.* Sinauer, Sunderland, Massachusetts.

Manek N.J., Hart D., Spector T.D., and MacGregor A.J. 2003. The association of body mass index and osteoarthritis of the knee joint: An examination of genetic and environmental influences. *Arthritis Rheum.* **48:** 1024–1029.

Mäki K., Janss L.L.G., Groen A.F., Liinamo A.-E., and Ojala M. 2004. An indication of major genes affecting hip and elbow dysplasia in four Finnish dog populations. *Heredity* **92:** 402–408.

Mertens W.D., MacLeod J.N., Fubini S.L., Vernier-Singer M., Nixon A.J., and Todhunter R.J. 2003. Polysulfated glycosaminoglycans modulate transcription of interleukin-treated chondrocytes in monolayer culture. *Vet. Comp. Orthop. Traumatol.* **16:** 93–98.

Meuwissen T.H. and Goddard M.E. 2004. Mapping multiple QTL using linkage disequilibrium and linkage analysis information and multitrait data. *Genet. Sel. Evol.* **36:** 261–279.

Morgan J., Wind A., and Davidson A. 1999. Bone dysplasias in the Labrador Retriever: A radiographic study. *J. Am. Anim. Hosp. Assoc.* **35:** 332–340.

Mustafa Z., Chapman K., Irven C., Carr A.J., Clipsham K., Chitnavis J., Sinsheimer J.S., Bloomfield V.A., McCartney M., Cox O., et al. 2000. Linkage analysis of candidate genes as susceptibility loci for osteoarthritis-suggestive linkage of COL9A1 to female hip osteoarthritis. *Rheumatology* **39:** 299–306.

Olsewski J.M., Lust G., Rendano V.T., and Summers B.A. 1983. Degenerative joint disease: Multiple joint involvement in young and mature dogs. *Am. J. Vet. Res.* **44:** 1300–1308.

Olsson S.E. 1961. Roentgen examination of the hip joints of German Shepherd dogs. *Adv. Small Anim. Pract.* **3:** 117–118.

Ostrander E.A., Sprague G.F.J., and Rine J. 1993. Identification and characterization of dinucleotide repeat (CA)n markers for genetic mapping in dog. *Genomics* **16:** 207–213.

Ott J. 1999. Methods of analysis and resources available for genetic trait mapping. *J. Hered.* **90:** 68–70.

Parker H.G., Kim L.V., Sutter N.B., Carlson S., Lorentzen T.D., Malek T.B., Johnson G.S., DeFrance H.B., Ostrander E.A., and Kruglyak L. 2004. Genetic structure of the pure-bred domestic dog. *Science* **304:** 1160–1164.

Paterson A.H., DeVerna J.W., Lanini B., and Tanksley S.D. 1990. Fine mapping of quantitative trait loci using selected overlapping recombinant chromosomes, in an interspecies cross of tomato. *Genetics.* **47:** 137–149.

Pompe B., Daniel M., Sochor M., Vengust R., Kralj-Iglic V., and Iglic A. 2003. Gradient of contact stress in normal and dysplastic human hips. *Med. Eng. Phys.* **25:** 379–385.

Popovitch C.A., Smith G.K., Gregor T.P., and Shofer F.S. 1995. Comparison of suscepti-bility for hip dysplasia between Rottweilers and German shepherd dogs. *J. Am. Vet. Med. Assoc.* **206:** 648–650.

Richman M., Mellersh C.S., Andre C., Galibert F., and Ostrander E.A. 2001. Characterization of a minimal screening set of 172 microsatellite markers for genome-wide screens of the canine genome. *J. Biochem. Biophys. Methods* **47:** 137–149.

Rise M.L., Frankel W.N., Coffin J.M., and Seyfried T.N. 1991. Genes for epilepsy mapped in the mouse. *Science* **253:** 669–673.

Riser W.H. 1987. A half century of canine hip dysplasia. *Semin. Vet. Med. Surg. (Small Anim.)* **2:** 87–91.

Roby P., Eyre S., Worthington J., Ramesar R., Cilliers H., Beighton P., Grant M., and Wallis G. 1999. Autosomal dominant (Beukes) premature degenerative osteoarthropathy of the hip joint maps to an 11-cM region on chromosome 4q35. *Am. J. Hum. Genet.* **64:** 904–908.

Rohrer G.A. and Keele J.W. 1998. Identification of quantitative trait loci affecting carcass composition in swine. I. Fat deposition traits. *J. Anim. Sci.* **76:** 2247–2254.

Seaton G., Haley C.S., Knott S.A., Kearsey M., and Visscher P.M. 2002. QTL Express: Mapping quantitative trait loci in simple and complex pedigrees. *Bioinformatics* **18:** 339–340.

Shimomura K., Low-Zeddies S.S., King D.P., Steeves T.D., Whiteley A., Kushla J., Zemenides P.D., Lin A., Vitaterna M.H., Churchill G.A., and Takahashi J.S. 2001. Genome-wide epistatic interaction analysis reveals complex genetic determinants of circadian behavior in mice. *Genome Res.* **11:** 959–980.

Smith G.K. 1997. Advances in diagnosing canine hip dysplasia. *J. Am. Vet. Med. Assoc.* **210:** 1451–1457.

Smith G.K., Popovitch C.A., Gregor T.P., and Shofer F.S. 1995. Evaluation of risk factors for degenerative joint disease associated with hip dysplasia in dogs. *J. Am. Vet. Med. Assoc.* **206:** 642–647.

Smith G.K., LaFond E., Gregor T.P., Lawler D.F., and Nie R.C. 1997. Within- and between-examiner repeatability of distraction indices of the hip joints in dogs. *Am. J. Vet. Res.* **58:** 1076–1077.

Smith W.S., Cjoleman C.R., Olix M.L., and Slager R.F. 1963. Etiology of congenital dislocation of the hip. *J. Bone Joint Surg. Am.* **45-A:** 491–500.

Sollazzo V., Bertolani G., Calzolari E., Atti G., and Scapoli C. 2000. A two-locus model for non-syndromic congenital dysplasia of the hip (CDH). *Ann. Hum. Genet.* **64:** 51–59.

Spector T.D., Cicuttini F., Baker J., Loughlin J., and Hart D. 1996. Genetic influences on osteoarthritis in women: A twin study. *BMJ* **312:** 940–943.

Steinmetz L.M., Sinha H., Richards D.R., Spiegelman J.I., Oefner P.J., McCusker J.H., and Davis R.W. 2002. Dissecting the architecture of a quantitative trait locus in yeast. *Nature* **416:** 326–330.

Sutter N.B. and Ostrander E.A. 2004. Dog star rising: The canine genetic system. *Nat. Rev. Genet.* **5**: 900–910.

Swenson L., Audell L., and Hedhammar A. 1997. Prevalence and inheritance of and selection for hip dysplasia in seven breeds of dogs in Sweden and benefit: Cost analysis of a screening and control program. *J. Am. Vet. Med. Assoc.* **210**: 207–214.

Todhunter R.J., Zachos T.A., Gilbert R.O., Erb H.N., Williams A.J., Burton-Wurster N., and Lust G. 1997. Onset of epiphyseal mineralization and growth plate closure in radiographically normal and dysplastic Labrador retrievers. *J. Am. Vet. Med. Assoc.* **210**: 1458–1462.

Todhunter R.J., Bliss S.P., Casella G., Wu R., Lust G., Burton-Wurster N.I., Williams A.J., Gilbert R.O., and Acland G.M. 2003a. Genetic structure of susceptibility traits for hip dysplasia and microsatellite informativeness of an outcrossed canine pedigree. *J. Hered.* **94**: 39–48.

Todhunter R.J., Acland G.M., Olivier M., Williams A.J., Vernier-Singer M., Burton-Wurster N., Farese J.P., Grohn Y.T., Gilbert R.O., Dykes N.L., and Lust G. 1999. An outcrossed canine pedigree for linkage analysis of hip dysplasia. *J. Hered.* **90**: 83–92.

Todhunter R.J., Casella G., Bliss S.P., Lust G., Williams A.J., Hamilton S., Dykes N.L., Yeager A.E., Gilbert R.O., Burton-Wurster N.I., Mellersh C.C., and Acland G.M. 2003b. Power of a Laborador Retriever-Greyhound pedigree for linkage analysis of hip dysplasia and osteoarthritis. *Am. J. Vet. Res.* **64**: 418–424.

Todhunter R.J., Grohn Y.T., Bliss S.P., Wilfand A., Williams A.J., Vernier-Singer M., Burton-Wurster N.I., Dykes N.L., Wu R., Casella G., Acland G.M., and Lust G. 2003c. Evaluation of multiple radiographic predictors of cartilage lesions in the hip joints of eight-month-old dogs. *Am. J. Vet. Res.* **64**: 1472–1478.

Todhunter R.J., Mateescu R., Lust G., Burton-Wurster N.I., Dykes N.L., Bliss S.P., Williams A.J., Vernier-Singer M., Corey E., Harjes C., et al. 2005. Quantitative trait loci for hip dysplasia in a crossbreed canine pedigree. *Mamm. Genome.* (in press).

Treble N.J., Jensen F.O., Bankier A., Rogers J.G., and Cole W.G. 1990. Development of the hip in multiple epiphyseal dysplasia. Natural history and susceptibility to premature osteoarthritis. *J. Bone Jt. Surg. Br. Vol.* **72**: 1061–1064.

Uimari P. and Hoeschele I. 1997. Mapping-linked quantitative trait loci using Bayesian analysis and Markov chain Monte Carlo algorithms. *Genetics* **146**: 735–743.

Ventura M., Mudge J.M., Palumbo V., Burn S., Blennow E., Pierluigi M., Giorda R., Zuffardi O., Archidiacono N., Jackson M.S., and Rocchi M. 2003. Neocentromeres in 15q24-26 map to duplicons which flanked an ancestral centromere in 15q25. *Genome Res.* **13**: 2059–2068.

Vingsbo-Lundberg C., Nordquist N., Olofsson P., Sundvall M., Saxne T., Pettersson U., and Holmdahl R. 1998. Genetic control of arthritis onset, severity and chronicity in a model for rheumatoid arthritis in rats. *Nat. Genet.* **20**: 401–404.

Weinstein S. 1996. Developmental hip dysplasia and dislocation. In *Lovell and Winter's pediatric orthopaedics*, 4th edition (ed. R.T. Morrissy and S.L. Weinstein), pp. 903–943. Lippincott-Raven, Philadelphia, Pennsylvania.

Willis M.B. 1989. Hip dysplasia. In *Genetics of the dog*, pp. 144–179. Howell Book House, New York.

Wood J.L. and Lakhani K.H. 2003. Effect of month of birth on hip dysplasia in Labrador retrievers and Gordon setters. *Vet. Rec.* **152**: 69–72.

22

Breed-specific Canine Lymphoproliferative Diseases

Jaime F. Modiano
Integrated Department of Immunology and AMC Cancer Center
School of Medicine
University of Colorado Health Sciences Center
Denver, Colorado 80214

Matthew Breen
Department of Molecular Biomedical Sciences
College of Veterinary Medicine
North Carolina State University
Raleigh, North Carolina 27606

Anne C. Avery
Department of Microbiology, Immunology, and Pathology
College of Veterinary Medicine and Biomedical Sciences
Colorado State University, Fort Collins, Colorado 80523

Cheryl A. London
Department of Surgical and Radiological Sciences
School of Veterinary Medicine
University of California, Davis
Davis, California 95616

THE LYMPHATIC SYSTEM CONSISTS OF T CELLS, B cells, NK cells, and other less common subsets of small mononuclear cells that arise from pluripotent hematopoietic stem cells. These cells differentiate primarily in the bone marrow or in the thymus into long-lived cells that are responsible for adaptive immune responses. During a normal immune response to a pathogenic challenge, these cells undergo frequent rounds of cell division to expand their numbers. T cells and B cells, in particular, follow this process of clonal expansion with a carefully orchestrated apoptotic

The Dog and Its Genome ©2006 Cold Spring Harbor Laboratory Press 0-87969-742-3

program that contracts their populations to physiological levels while allowing survival of memory cells.

Considering the high rate of proliferation that is required for self-renewal and to support immune responses, as well as the absolute requirement for extensive apoptosis that restores the balance of lymphocyte populations, it is not surprising that cells of the lymphoid system are highly susceptible to neoplastic transformation. Lymphoproliferative diseases (LPDs) encompass a constellation of tumors that originate from lympho-cytes. LPDs that manifest as solid tumors are called lymphomas, whereas LPDs that arise from the bone marrow or spleen and (generally) present with malignant cells in the peripheral circulation are called leukemias (Vernau and Moore 1999; Valli et al. 2002). A group of tumors called non-Hodgkin's lymphoma (NHL) comprises the most common LPDs in people. As a group, NHL is the fifth most common cancer in the U.S. with a life-time risk of ~1 in 50 and an overall 5-year survival rate of approximately 55% (Jemal et al. 2005). The incidence rates of NHL in people have almost doubled over the past 30 years, making the disease the third fastest growing cancer in the world. The lifetime risk and incidence of leukemias are only slightly lower than are those of NHL. Combined, LPDs are the most com-mon childhood cancers, and they account for 7–10% of all adult human cancers and cancer-related deaths (Jemal et al. 2005).

COMPARATIVE ASPECTS AND SUBTYPES OF HUMAN AND CANINE LPD

The full spectrum of LPDs also occurs in dogs with features of clinical presentation, histology, and biology that closely parallel those of human malignancies (Carter et al. 1986; Hahn et al. 1994; Fournel-Fleury et al. 1997; Fernandes et al. 2002; Valli et al. 2002). Although the lifetime risk of cancer in dogs is similar to that of humans (~1 in 3 [Peterson et al. 2000; Jemal et al. 2005]), LPDs in general, and NHL in particular, occur more frequently in dogs than in people. NHL is reported to account for as many as 20–25% of all canine malignancies (MacEwen 1990), suggest-ing the average lifetime risk for a dog to develop this disease is about 1 in 15. Although LPDs affect dogs of all ages, and some types of leukemia and lymphoma are seen in puppies and in young adults, these are largely diseases of middle-aged to older dogs (Priester and McKay 1980; Modiano et al. 2005). There are important similarities and differences in the distribution of NHL subtypes between dogs and humans. Diffuse large-B-cell lymphoma (DLBCL) is the most common type of NHL in both species. On the other hand, indolent B-cell lymphomas such as follicular lymphoma and mantle cell lymphoma are quite common in humans, but are not commonly recognized in dogs. The endemic form of Burkitt's lymphoma (BL) that is associated with Epstein-Barr virus (EBV) infection is common in people, whereas a counterpart to EBV has not been described for dogs. Conversely, sporadic BL (not associated with EBV infection) occurs rarely in humans and dogs (Valli et al. 2002; Ponce et al. 2004). In people, T-cell-derived NHL, which is almost exclusively seen outside the lymph nodes, is rare. NHL that occurs outside the lymph nodes is also mainly a disease of T cells in dogs, but unlike humans, T-cell NHL of the lymph nodes occurs commonly in dogs. Finally, chronic lymphocytic leukemia/small lymphocytic lymphoma complex (CLL/SLL) is a slowly progressive disease that occurs in a similar segment of the population for both species (adults and elderly patients). Intriguingly, the disease in people overwhelmingly originates from B cells, whereas in dogs, the majority of cases arise from T cells (Vernau and Moore 1999). Yet, the biological behavior and outcomes of patients with these tumors are remarkably similar in humans and dogs.

CLINICAL PRESENTATION AND STAGING OF CANINE NHL

The most common presentation for canine NHL is a dog with no clinical signs other than generalized lymph node enlargement (lymphadenopathy) with or without infiltration of abdominal organs such as the spleen and liver. Clinical signs, when present, are usually nonspecific (decreased appetite, lethargy, weight loss, cough, vomiting, diarrhea, fever, bleeding tendencies). Anemia, low platelet counts, and high white blood cell counts may be evident in some cases, although circulating malignant cells are detectable using polymerase-chain-reaction-based assays in many or most cases of canine NHL (Keller et al. 2004). Other signs or symptoms are usually dependent on the site of origin (within the lymph nodes or outside the lymph nodes). The most common sites outside the lymph nodes are the alimentary tract and skin, which are normally inhabited by specialized types of lymphocytes.

Determination of the clinical stage of the disease is based on the World Health Organization (WHO) system, "Clinical Stages of Lymphoma and Lymphatic Leukemia in Domestic Animals." Staging includes the morphologic appearance of the tumor and the anatomic sites involved, and it is useful to characterize the disease and predict its response to therapy. The stages range from I, which indicates the least severe involvement and that which is most likely to respond to treatment, to stages IV and V, which include involvement of liver and spleen or bone marrow or sites outside the lymph nodes, respectively.

In dogs, untreated cases of diffuse NHL (excluding CLL/SLL) rarely survive beyond 3 months after diagnosis. However, canine NHL is generally considered a treatable disease that is chemoresponsive, and treatment increases both the length and quality of an affected dog's life (Mellanby et al. 2003). The standard of care is based on chemotherapy protocols that use multiple drugs. Median survival time for all dogs is between 10 and 14 months. Significant differences in survival are reported depending on whether the dog shows signs of illness or not, as well as depending on the phenotype and histologic subtype (Carter et al. 1986; Teske et al. 1994c; Ponce et al. 2004). Nevertheless, there remains considerable variation (i.e., wide confidence intervals) in the response to therapy and overall survival time in cases with the same clinical stage. A limited number of diagnostic and prognostic indicators are useful in the clinical management of these cases (Greenlee et al. 1990; Thomas et al. 2003c; Ponce et al. 2004). However, additional criteria are required to generate more refined schemes to predict responses, such as those that have been derived for various types of human lymphoma based on gene expression profiles (Alizadeh et al. 2000; Siebert et al. 2001; Husson et al. 2002; Rosenwald et al. 2002; Shipp et al. 2002; Tracey et al. 2002; Evans et al. 2003; Lee 2003; Dave et al. 2004; Lossos et al. 2004). A first-generation genomic array has been developed to analyze numerical aberrations of cancer-related genes in canine lymphoma (Thomas et al. 2003a,c), and various studies are under way to assess the predictive ability of gene expression profiles for LPD.

GENETIC CHANGES ASSOCIATED WITH LPD SUBTYPES

Recent work has defined recurrent genetic changes that are associated with canine LPD. First, improved methods of diagnosis allow the identification of clonally expanded malignant populations with more precision. Second, contemporary technologies and the completion of the canine genome sequence offer tools to classify these diseases on the basis of their molecular signatures that are pathologically significant. For example, comparative genomic hybridization studies identified recurrent numerical abnormalities in various types of canine LPD, including amplification of canine chromosome (CFA) 13 in diffuse lymphomas, deletion of CFA 11 in lymphoblastic T-cell lymphoma (LBT), and deletion of CFA 22 in CLL/SLL complex (Thomas et al. 2003c; M. Breen and J.F. Modiano, in prep.). Among genes that are pathogenetically important for LPDs, the c-*myc* proto-oncogene resides in CFA 13; the *INK4* tumor suppressor locus that contains *ink4a*, *ink4b*, and *arf* resides in CFA 11; and the *RB1* tumor suppressor locus resides in CFA 22 (Thomas et al. 2003a). It was recently shown that c-*myc* is overexpressed in DLBCL, and that the p16 and p110 products of *ink4a* and *rb* are undetectable in LBT and CLL, respectively (M. Breen and J.F. Modiano, in prep.; J.F. Modiano et al., unpubl.). In addition to structural abnormalities, canine LPDs also harbor pathognomonic structural cytogenetic changes. For example, the molecular signature of human BL is a reciprocal translocation of *MYC* from HSA 8q24 to HSA 14q32, which places *MYC* under the control of the immunoglobulin heavy-chain enhancer (*IGH*). The same translocation was recently documented to occur in the canine form of BL (M. Breen and J.F. Modiano, in prep.). This suggests that, as is true in humans (Taub et al. 1984; Hecht and Aster 2000), the pathogenesis of BL in dogs seems to require temporally regulated overexpression of c-*myc*. The same may be true for reciprocal translocations that involve the *ABL* locus in CFA 9 in canine acute lymphoid leukemia (M. Breen and J.F. Modiano, in prep.). Finally, reproducible epigenetic changes may also contribute to the origin and progression of canine LPDs. Together, these results indicate that recurrent chromosomal aberrations in naturally occurring LPD are evolutionarily conserved in dogs and humans, and they support the idea of a fundamental association between cytogenetic abnormalities and tumor phenotype. Furthermore, careful investigation of recurrent genetic changes in canine LPD offers the potential to define regions of the canine genome—and thus the human genome—containing genes that have so far not been associated with tumorigenesis. This is especially so for cytogenetic lesions that are key features of the canine tumor genome but that, at least with current technologies, remain undetectable in corresponding human tumors.

BREED-SPECIFIC RISK FOR LPD IN DOGS

There is a distinct, significant, and reproducible breed predisposition for LPDs in dogs, suggesting that heritable risk factors for the disease have segregated with breed-specific traits. Previous studies documented that specific dog breeds, including boxers, golden retrievers, Labrador retrievers, cocker spaniels, basset hounds, Saint Bernards, Scottish terriers, Doberman pinschers, bullmastiffs, and German shepherd dogs, had a statistically significantly increased risk for lymphomas, whereas others, such as the Chihuahua, dachshund, and Pomeranian showed a significantly lower risk (Dorn et al. 1968; Priester and McKay 1980; Onions 1984). When compared to the average risk of any dog to develop lymphoma, the risk for boxers is more than four times higher (Dorn et al. 1968;

Priester and McKay 1980; Lurie et al. 2004), the lifetime risk for the disease in golden retrievers is approximately 1 in 8 (Glickman et al. 2000), and familial clustering has been reported for rottweilers and Scottish terriers (Teske et al. 1994a). This suggests that breed barrier rules, which perpetuate inbreeding and line breeding, may have increased the homogeneity of alleles that contribute to risk, and possibly may have even influenced the penetrance of these factors in the population. The observation that breed type also influences response to therapy (Garrett et al. 2002) suggests that genetic factors modulate disease progression and are thus prognostically significant.

When all breeds are combined, LPDs of B cells and T cells occur at similar frequencies in dogs (Carter et al. 1986; Greenlee et al. 1990; Teske et al. 1994b; Fournel-Fleury et al. 1997; Ruslander et al. 1997; Modiano 1998) as they do in some human populations: specifically, non-Caucasians in the U.S. (Groves et al. 2000), indigenous Japanese (in HTLV-I non-endemic areas) (Shih and Liang 1991; Lymphoma Study Group 2000), Indians (Bhargava et al. 1988; Naresh et al. 2000), and Chinese (Shih and Liang 1991). Across 1,226 dogs from 87 pure breeds and a group of mixed-breed dogs, the distribution of ~61% B-cell tumors and 39% T-cell tumors reported recently (Lurie et al. 2004; Modiano et al. 2005) was not significantly different from the combined prevalence reported in 642 dogs from 10 other contemporary studies (Appelbaum et al. 1984; Greenlee et al. 1990; Teske et al. 1994b,c; Fisher et al. 1995; Caniatti et al. 1996; Teske and van Heerde 1996; Vail et al. 1996; Fournel-Fleury et al. 1997; Ruslander et al. 1997). However, when breeds were examined separately, there was distinct breed-specific variability in the prevalence of B- or T-cell lymphomas. Three independent reference populations were used, including dogs previously reported in the literature, all other dogs in the same data set, and mixed-breed dogs in the same data set. The latter reference group was included not only because it contained the largest number of individuals, providing high power for comparisons, but also because it represented an "experiment of nature" that showed how interbreeding might affect disease predisposition across genetically restricted populations. These results documented excess occurrences of T-cell LPD in Irish wolfhounds, shih tzus, Airedale terriers, cavalier King Charles spaniels, Yorkshire terriers, Siberian huskies, Chinese shar-pei, Australian shepherds, boxers, and golden retrievers, whereas there were excess occurrences of B-cell LPD in cocker spaniels and basset hounds (Fig. 1). The prevalence of B-cell and T-cell LPDs in other popular breeds such as rottweilers and Labrador retrievers was not significantly different from that of the reference populations. The significance of the associations between breed and LPD phenotype was

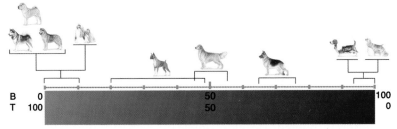

Figure 1. Breed-specific distribution of T and B lymphoproliferative diseases. Between 60% and 80% of LPDs in dogs are B-cell-derived tumors, and most of the remaining 20–40% are T-cell-derived tumors. When analyzed independently, the prevalence of B-cell-derived and T-cell-derived tumors in most dog breeds (represented here by the German shepherd dog) and in mixed breed dogs fall within these ranges. However, the occurrence of excess T-cell or B-cell tumors has been shown in certain breeds. Specifically, the oldest domestic breeds, including spitz-type dogs (e.g., Alaskan malamute, Siberian husky, Chinese shar-pei) and small Asian "lap dogs" (e.g., shih tzu) share a predisposition for excess T-cell-derived LPD, suggesting they retain ancestrally inherited risk factors. In contrast, some European breeds of recent derivation are predisposed to excess B-cell tumors. Two popular breeds, boxers and golden retrievers, show an increased predilection for T-cell-derived tumors, albeit less pronounced than the ancient breeds described above. Efforts are currently under way to define the precise factors that account for these unique risk patterns. (Dog illustrations by Chet Jezierski, reprinted, with permission, © American Kennel Club.)

further illustrated by analyzing the phenotypes of CLL. As noted above, CLL is almost exclusively a B-cell disease in people (Tefferi and Phyliky 1992; Guipaud et al. 2003), but more commonly (73%) a T-cell disease in dogs (Vernau and Moore 1999). In this study, six dogs had B-cell CLL, and five of these were from breeds that show a preponderance of B-cell tumors (Australian cattle dog, chow chow, Doberman pinscher, poodle, standard schnauzer).

Sex and gonadal status (neutering) did not influence the distribution of LPD phenotypes, and although there was a significant association between tumor type or location and molecular phenotype (i.e., extranodal lymphomas and most leukemias were of T-cell origin and all plasma cell tumors were of B-cell origin), this was not sufficient to explain the observed breed predilections. The frequency of T-cell tumors was also greater than the average in dogs <3 years of age, and the frequency of B-cell tumors was greater than expected in dogs >14, but this similarly did not explain the observed breed predilections for B- and T-cell LPD.

The associations between LPD phenotypes and breed relatedness were also analyzed. An excess of T-cell LPD was identified in spitz breeds belonging to the oldest domestic dog group that includes Akita, basenji, Siberian husky, Alaskan malamute, and Chinese shar-pei, as well as wolves, and in the shih tzu group that includes shih tzu, Lhasa apso, and other Asian "lap" dogs (Fig. 1) (Parker et al 2004; Modiano et al. 2005).

HERITABILITY AND CANINE LPD

These recent observations documented a significant difference in the frequency of B-cell and T-cell LPD between boxers or golden retrievers and appropriate reference groups. Unlike other mastiff breeds, boxers had a higher frequency of T-cell LPD, as high as 82% in one study (Lurie et al. 2004). Similarly, the frequency of T-cell LPD in golden retrievers, which was approximately 50% in both studies, differed from most other European breeds of recent derivation (Modiano et al. 2005). Environmental factors cannot explain these associations (nor can they account for the increased incidence of LPD in people over the past 10 years [Jemal et al. 2005]). Thus, given the similarity of the canine and human genomes and the fact that both also have similar exposure to environmental factors which contribute to the origin and progression of LPD, we propose that identification of genes which mediate heritable risk for LPD in dogs will be useful to find cancer-associated genes which have thus far been elusive in studies of human families, populations, and tumors.

There is, therefore, a basis for the hypothesis that heritable risk factors predispose dogs (and probably humans) to develop B-cell or T-cell malignancies. These risk factors are presumably mutations or epigenetic changes in genes that regulate lymphocyte development, although the involvement of genes that regulate the fidelity of the genome cannot be ruled out, as familial LPDs in people are largely associated with conditions such as the p53 mutation in Li-Fraumeni syndrome (Segel and Lichtman 2004; Siddiqui et al. 2004). As noted above, shared ancestral risk factors may account for the LPD phenotype of "excess" T-cell tumors in spitz breeds and Asian lap dogs. Conversely, risk factors that predispose European breeds to B-cell tumors may be ancestral or may have developed independently during the process of breed derivation. Nevertheless, these factors are now embedded in the genome, and their identification will provide information that will be valuable for prevention and treatment. Although this task may seem difficult, data suggest there are specific genetic and epigenetic abnormalities associated not only with the LPD immunophenotype, but also with its morphology and clinical presentation. Furthermore, analyses of pedigrees from golden retriever families also suggest there is shared susceptibility to specific forms of LPD among family members. Thus, optimized marker sets can now be used to search for genes that define familial predisposition. Genome scans, expression arrays, and proteomic analysis will offer additional tools to refine and stratify familial data by tumor type and clinical response, providing greater insight and statistical power to find genes important in disease susceptibility and progression.

SUMMARY

Here, we describe comparative features of LPDs in dogs and humans. Not only are these diseases phenotypically similar in both species, but also their genetic signatures are evolutionarily conserved. In dogs, selective breeding has led to segregation of risk factors that influence the lifetime risk for LPDs as well as the prevalence of specific disease phenotypes. Advances in molecular profiling, along with the unique genetic structure of modern dog breeds and the availability of the canine genome sequence, provide necessary tools to begin to identify these factors. Ongoing work to clarify heritable influences that contribute to the origin and progression of LPDs will allow us to design better strategies for prevention and treatment of these diseases in both dogs and people.

ACKNOWLEDGMENTS

This work was supported in part by grants from the AKC Canine Health Foundation, AVMA Foundation, and Morris Animal Foundation.

REFERENCES

Alizadeh A.A., Eisen M.B., Davis R.E., Ma C., Lossos I.S., Rosenwald A., Boldrick J.C., Sabet H., Tran T., Yu X., et al. 2000. Distinct types of diffuse large B-cell lymphoma identified by gene expression profiling. *Nature* **403:** 503–511.

Appelbaum F.R., Sale G.E., Storb R., Charrier K., Deeg H.J., Graham T., and Wulff J.C. 1984. Phenotyping of canine lymphoma with monoclonal antibodies directed at cell surface antigens: Classification, morphology, clinical presentation and response to chemotherapy. *Hematol. Oncol.* **2:** 151–168.

Bhargava M., Kumar R., Karak A., Kochupillai V., Arya L.S., and Mohanakumar T. 1988. Immunological subtypes of acute lymphoblastic leukemia in north India. *Leuk. Res.* **12:** 673–678.

Caniatti M., Roccabianca P., Scanziani E., Paltrinieri S., and Moore P.F. 1996. Canine lymphoma: Immunocytochemical analysis of fine-needle aspiration biopsy. *Vet. Pathol.* **33:** 204–212.

Carter R.F., Valli V.E., and Lumsden J.H. 1986. The cytology, histology and prevalence of cell types in canine lymphoma classified according to the National Cancer Institute Working Formulation. *Can. J. Vet. Res.* **50:** 154–164.

Dave S.S., Wright G., Tan B., Rosenwald A., Gascoyne R.D., Chan W.C., Fisher R.I., Braziel

R.M., Rimsza L.M., Grogan T.M., et al. 2004. Prediction of survival in follicular lymphoma based on molecular features of tumor-infiltrating immune cells. *N. Engl. J. Med.* **351:** 2159–2169.

Dorn C.R., Taylor D.O., Schneider R., Hibbard H.H., and Klauber M.R. 1968. Survey of animal neoplasms in Alameda and Contra Costa Counties, California. II. Cancer morbidity in dogs and cats from Alameda County. *J. Natl. Cancer Inst.* **40:** 307–318.

Evans W.E., Cheok M., Yang W., and Sherr C.J. 2003. Expression arrays illuminate a way forward for mantle cell lymphoma. *Cancer Cell* **3:** 100–102.

Fernandes P.J., Modiano J.F., Wojcieszyn J., Thomas J.S., Benson P.A., Smith R., 3rd, Avery A.C., Burnett R.C., Boone L.I., Johnson M.C., and Pierce K.R. 2002. Use of the Cell-Dyn 3500 to predict leukemic cell lineage in peripheral blood of dogs and cats. *Vet. Clin. Pathol.* **31:** 167–182.

Fisher D.J., Naydan N., Werner L.L., and Moore P.F. 1995. Immunophenotyping lymphomas in dogs: A comparison of results from fine needle aspirate and needle biopsy samples. *Vet. Clin. Pathol.* **24:** 118–123.

Fournel-Fleury C., Magnol J.P., Bricaire P., Marchal T., Chabanne L., Delverdier A., Bryon P.A., and Felman P. 1997. Cytohistological and immunological classification of canine malignant lymphomas: Comparison with human non-Hodgkin's lymphomas. *J. Comp. Pathol.* **117:** 35–59.

Garrett L.D., Thamm D.H., Chun R., Dudley R., and Vail D.M. 2002. Evaluation of a 6-month chemotherapy protocol with no maintenance therapy for dogs with lymphoma. *J. Vet. Intern. Med.* **16:** 704–709.

Glickman L., Glickman N., and Thorpe R. 2000. The Golden Retriever Club of America National Health Survey (GRCA, available at http://www.grca.org/healthsurvey.pdf).

Greenlee P.G., Filippa D.A., Quimby F.W., Patnaik A.K., Calvano S.E., Matus R.E., Kimmel M., Hurvitz A.I., and Lieberman P.H. 1990. Lymphomas in dogs. A morphologic, immunologic, and clinical study. *Cancer* **66:** 480–490.

Groves F.D., Linet M.S., Travis L.B., and Devesa S.S. 2000. Cancer surveillance series: Non-Hodgkin's lymphoma incidence by histologic subtype in the United States from 1978 through 1995. *J. Natl. Cancer Inst.* **92:** 1240–1251.

Guipaud O., Deriano L., Salin H., Vallat L., Sabatier L., Merle-Beral H., and Delic J. 2003. B-cell chronic lymphocytic leukaemia: A polymorphic family unified by genomic features. *Lancet Oncol.* **4:** 505–514.

Hahn K.A., Bravo L., Adams W.H., and Frazier D.L. 1994. Naturally occurring tumors in dogs as comparative models for cancer therapy research. *In Vivo* **8:** 133–143.

Hecht J.L. and Aster J.C. 2000. Molecular biology of Burkitt's lymphoma. *J. Clin. Oncol.* **18:** 3707–3721.

Husson H., Carideo E.G., Neuberg D., Schultze J., Munoz O., Marks P.W., Donovan J.W., Chillemi A.C., O'Connell P., and Freedman A.S. 2002. Gene expression profiling of follicular lymphoma and normal germinal center B cells using cDNA arrays. *Blood* **99:** 282–289.

Jemal A., Murray T., Ward E., Samuels A., Tiwari R.C., Ghafoor A., Feuer E.J., and Thun M.J. 2005. Cancer statistics, 2005. *CA Cancer J. Clin.* **55:** 10–30.

Keller R.L., Avery A.C., Burnett R.C., Walton J.A., and Olver C.S. 2004. Detection of neoplastic lymphocytes in peripheral blood of dogs with lymphoma by polymerase chain reaction for antigen receptor gene rearrangement. *Vet. Clin. Pathol.* **33:** 145–149.

Lee D. 2003. Factors predicting the response to rituximab in indolent lymphoma. *Clin. Lymphoma* **4:** 19–21.

Lossos I.S., Czerwinski D.K., Alizadeh A.A., Wechser M.A., Tibshirani R., Botstein D., and Levy R. 2004. Prediction of survival in diffuse large-B-cell lymphoma based on the expression of six genes. *N. Engl. J. Med.* **350:** 1828–1837.

Lurie D.M., Lucroy M.D., Griffey S.M., Simonson E., and Madewell B.R. 2004. T-cell-derived malignant lymphoma in the boxer breed. *Vet. Comp. Oncol.* **2:** 171–175.

Lymphoma Study Group. 2000. The World Health Organization classification of malignant lymphomas in Japan: Incidence of recently recognized entities. Lymphoma Study Group of Japanese Pathologists. *Pathol. Int.* **50:** 696–702.

MacEwen E.G. 1990. Spontaneous tumors in dogs and cats: Models for the study of cancer biology and treatment. *Cancer Metastasis Rev.* **9:** 125–136.

Mellanby R.J., Herrtage M.E., and Dobson J.M. 2003. Owners' assessments of their dog's quality of life during palliative chemotherapy for lymphoma. *J. Small Anim. Pract.* **44:** 100–103.

Modiano J.F. 1998. Prognostic significance of malignant cell phenotypes in canine lymphoma. *Adv. Vet. Med. Surg.* **11:** 1–2.

Modiano J.F., Breen M., Burnett R.C., Parker H.G., Inusah S., Thomas R., Avery P.R., Lindblad-Toh K., Ostrander E.A., Cutter G.C., and Avery A.C. 2005. Distinct B-cell and T-cell lymphoproliferative disease prevalence among dog breeds indicates heritable risk. *Cancer Res.* **65:** 5654–5661.

Naresh K.N., Srinivas V., and Soman C.S. 2000. Distribution of various subtypes of non-Hodgkin's lymphoma in India: A study of 2773 lymphomas using R.E.A.L. and WHO classifications. *Ann. Oncol.* (suppl. 1) **11:** 63–67.

Onions D.E. 1984. A prospective survey of familial canine lymphosarcoma. *J. Natl. Cancer Inst.* **72:** 909–912.

Parker H.G., Kim L.V., Sutter N.B., Carlson S., Lorentzen T.D., Malek T.B., Johnson G.S., DeFrance H.B., Ostrander E.A., and Kruglyak L. 2004. Genetic structure of the purebred domestic dog. *Science* **304:** 1160–1164.

Peterson M.R., Frommelt R.A., and Dunn D.G. 2000. A study of the lifetime occurrence of neoplasia and breed differences in a cohort of German shepherd dogs and Belgian Malinois military working dogs that died in 1992. *J. Vet. Intern. Med.* **14:** 140–145.

Ponce F., Magnol J.P., Ledieu D., Marchal T., Turinelli V., Chalvet-Monfray K., and Fournel-Fleury C. 2004. Prognostic significance of morphological subtypes in canine malignant lymphomas during chemotherapy. *Vet. J.* **167:** 158–166.

Priester W.A. and McKay F.W. 1980. The occurrence of tumors in domestic animals. *Natl. Cancer Inst. Monogr.* **54:** 1–210.

Rosenwald A., Wright G., Chan W.C., Connors J.M., Campo E., Fisher R.I., Gascoyne R.D., Muller-Hermelink H.K., Smeland E.B., Giltnane J.M., et al. 2002. The use of molecular profiling to predict survival after chemotherapy for diffuse large-B-cell lymphoma. *N. Engl. J. Med.* **346:** 1937–1947.

Ruslander D.A., Gebhard D.H., Tompkins M.B., Grindem C.B., and Page R.L. 1997. Immunophenotypic characterization of canine lymphoproliferative disorders. *In Vivo* **11:** 169–172.

Segel G.B. and Lichtman M.A. 2004. Familial (inherited) leukemia, lymphoma, and myeloma: An overview. *Blood Cells Mol. Dis.* **32:** 246–261.

Shih L.Y. and Liang D.C. 1991. Non-Hodgkin's lymphomas in Asia. *Hematol. Oncol. Clin. N. Am.* **5:** 983–1001.

Shipp M.A., Ross K.N., Tamayo P., Weng A.P., Kutok J.L., Aguiar R.C., Gaasenbeek M., Angelo M., Reich M., Pinkus G.S., et al. 2002. Diffuse large B-cell lymphoma outcome prediction by gene-expression profiling and supervised machine learning. *Nat. Med.*

8: 68–74.

Siddiqui R., Onel K., Facio F., and Offit K. 2004. The genetics of familial lymphomas. *Curr. Oncol. Rep.* **6:** 380–387.

Siebert R., Rosenwald A., Staudt L.M., and Morris S.W. 2001. Molecular features of B-cell lymphoma. *Curr. Opin. Oncol.* **13:** 316–324.

Taub R., Moulding C., Battey J., Murphy W., Vasicek T., Lenoir G.M., and Leder P. 1984. Activation and somatic mutation of the translocated c-myc gene in Burkitt lymphoma cells. *Cell* **36:** 339–348.

Tefferi A. and Phyliky R.L. 1992. A clinical update on chronic lymphocytic leukemia. I. Diagnosis and prognosis. *Mayo Clinic Proc.* **67:** 349–353.

Teske E. and van Heerde P. 1996. Diagnostic value and reproducibility of fine-needle aspiration cytology in canine malignant lymphoma. *Vet. Q.* **18:** 112–115.

Teske E., de Vos J.P., Egberink H.F., and Vos J.H. 1994a. Clustering in canine malignant lymphoma. *Vet. Q.* **16:** 134–136.

Teske E., Wisman P., Moore P.F., and van Heerde P. 1994b. Histologic classification and immunophenotyping of canine non-Hodgkin's lymphomas: Unexpected high frequency of T cell lymphomas with B cell morphology. *Exp. Hematol.* **22:** 1179–1187.

Teske E., van Heerde P., Rutteman G.R., Kurzman I.D., Moore P.F., and MacEwen E.G. 1994c. Prognostic factors for treatment of malignant lymphoma in dogs. *J. Am. Vet. Med. Assoc.* **205:** 1722–1728.

Thomas R., Bridge W., Benke K., and Breen M. 2003a. Isolation and chromosomal assignment of canine genomic BAC clones representing 25 cancer-related genes. *Cytogenet. Genome Res.* **102:** 249–253.

Thomas R., Smith K.C., Ostrander E.A., Galibert F., and Breen M. 2003b. Chromosome aberrations in canine multicentric lymphomas detected with comparative genomic hybridisation and a panel of single locus probes. *Br. J. Cancer* **89:** 1530–1537.

Thomas R., Fiegler H., Ostrander E.A., Galibert F., Carter N.P., and Breen M. 2003c. A canine cancer-gene microarray for CGH analysis of tumors. *Cytogenet. Genome Res.* **102:** 254–260.

Tracey L., Villuendas R., Ortiz P., Dopazo A., Spiteri I., Lombardia L., Rodriguez-Peralto J.L., Fernandez-Herrera J., Hernandez A., Fraga J., et al. 2002. Identification of genes involved in resistance to interferon-α in cutaneous T-cell lymphoma. *Am. J. Pathol.* **161:** 1825–1837.

Vail D.M., Kisseberth W.C., Obradovich J.E., Moore F.M., London C.A., MacEwen E.G., and Ritter M.A. 1996. Assessment of potential doubling time (Tpot), argyrophilic nucleolar organizer regions (AgNOR), and proliferating cell nuclear antigen (PCNA) as predictors of therapy response in canine non-Hodgkin's lymphoma. *Exp. Hematol.* **24:** 807–815.

Valli V.E., Jacobs R.M., Parodi A.L., Vernau W., and Moore P.F. 2002. *Histological classification of hematopoietic tumors of domestic animals, Series II.* American Registry of Pathology, Armed Forces Institute of Pathology, Washington, D.C.

Vernau W. and Moore P.F. 1999. An immunophenotypic study of canine leukemias and preliminary assessment of clonality by polymerase chain reaction. *Vet. Immunol. Immunopathol.* **69:** 145–164.

23

Cancer Biology in Dogs

Chand Khanna and Melissa C. Paoloni

Center for Cancer Research, National Cancer Institute
National Institutes of Health
Rockville, Maryland 20850

THE RECENT AVAILABILITY OF THE DOG GENOME SEQUENCE has contributed, and will continue to contribute, to our understanding of canine physiology, pathophysiology, and disease. This may be most true for the problem of cancer in dogs. It has been estimated that there are approximately 65 million pet dogs at risk of developing cancer in the U.S. (Vail and MacEwen 2000). Twenty-three percent of all dogs, regardless of age, and 45% of dogs 10 years of age or older die of cancer (Withrow and MacEwan 2001). Estimates of age-adjusted cancer incidence range from 243 to 381 per 100,000 dogs/year at risk (Bronson 1982). These rates are comparable to those reported by the National Cancer Institute SEER program for human beings (~300 per 100,000) (Dorn and Taylor 1968; Priester and McKay 1980; Bronson 1982; Khanna and Vail 2003). Cancers in dogs share many of the basic biological features of human cancers and cancers of other species. Interestingly, the spectrum of cancer histologies seen in the dog does not completely mirror those seen in humans. Whereas humans most commonly develop carcinomas (breast, prostate, and lung cancer), dogs more commonly develop lymphoma, sarcomas, and then carcinomas.

Cancer in all species is strictly defined as an uncontrolled growth (or proliferation) of cells that results in an abnormal accumulation of cells in a particular region of the body. A tumor is the mass that results from this abnormal proliferation of cells within a tissue. Transformation of a normal cell into a phenotypic cancer cell involves a change in the morphological, biochemical, or growth parameters of the cell. This process involves the loss of anchorage dependence, contact inhibition, decreased requirement for growth factors, and colony formation in a semi-solid medium.

Progress in the field of molecular genetics has led to our understanding of cancer as a genetic disease. Alterations in the genes that control cell growth and death, as well as those that mediate the cell cycle, are the gatekeepers of key signaling pathways which are responsible for tumor growth. The identification of these key genes as oncogenes and tumor suppressor genes, respectively, has helped to elucidate many of the biological pathways of cancer (Tannock and Hill 1998; DeVita et al. 2001).

Cancer is no longer believed to be the result of a single random event, but rather a series of related and unrelated steps that lead to the development of a primary tumor from a previously normal cell and then the growth, progression, and spread of this tumor to distant parts of the body (metastasis). The steps associated with the development and progression of cancer are both genetic and epigenetic, occurring not only in the tumor cell but also in the "normal" cells located in the microenvironment of the tumor (Fig. 1). The types and numbers of genetic changes that are

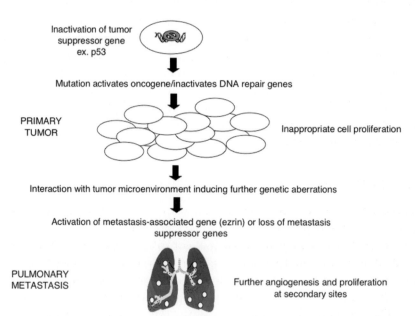

Figure 1. Multistep pathogenesis of cancer. Multiple genetic changes lead to the development of a primary tumor, and others contribute to the progression of the cancer and ultimately the acquisition of the metastatic phenotype. This entire process is duly influenced by the interaction between the tumor and the tumor microenvironment.

acquired during the process of cancer development can be distinctive to a specific cancer; however, the result of these genetic changes is the acquisition of a common set of "credentials" of cancer suggested first by Hanahan and Weinberg and modified herein (Hanahan and Weinberg 2000). Each of these cancer-associated credentials distinguishes normal cells from cancer cells and may be considered as the requisites of the cancer phenotype:

- Manipulation of growth signaling
- Loss of growth inhibition
- Evading programmed cell death (apoptosis)
- Inhibited cellular aging and senescence
- Acquisition of the metastatic phenotype

Within this chapter, the cellular and genetic mechanisms that make up the cancer credentials are described using specific examples taken from canine cancers. Detailed analysis of each credential in the biology of cancer is beyond the scope of this chapter. Readers are referred to recent reviews (Weiner and Cance 1994; Steele et al. 1998; Tsao et al. 1998; Gemmill and Idell 2003; Lawen 2003; Muehlbauer 2003). As suggested above, the use of genomic information that is now available for the dog will allow the biology of canine cancers to be as completely resolved and understood as in other species with a genome sequence. This improved understanding of cancer biology will be translated into novel management strategies for dogs with cancer. It is expected that the treatment of companion dogs with novel cancer drugs will allow the dog to "take the lead" in the delivery of new cancer treatments for both humans and dogs (Hansen and Khanna 2004).

ONCOGENES AND TUMOR SUPPRESSOR GENES

Two classes of genes have been broadly defined as contributing to cancer development and progression: (1) oncogenes and (2) tumor suppressor genes. These two classes of genes may contribute to one or many of the credentials of cancer, listed above, within a given tumor. As such, a brief review of tumor oncogene and tumor suppressor gene biology precedes our discussion of the distinguishing functional credentials of cancer.

The activation of a tumor oncogene results from the activation of a single allele, which acts dominantly over the wild-type allele. Therefore, activation of one of the two alleles of a proto-oncogene is sufficient for its activation to an oncogene. Tumor suppressor genes are recessively regulated. As such, the loss of function of both alleles is required for the

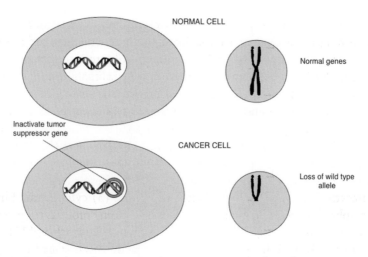

Figure 2. Inactivation of tumor suppressor genes. Inactivation of a tumor suppressor gene occurs due to the loss or mutation of both of its alleles. Tumor suppressor genes normally regulate cell proliferation, DNA repair, and cell signaling pathways. The loss of heterozygosity at a tumor suppressor gene allele is the hallmark of tumor suppressor gene dysfunction.

abnormal function of a tumor suppressor gene (Fig. 2). Mutations in one allele of a tumor suppressor gene can be carried asymptomatically as a germ-line mutation and passed silently into successive generations (Tannock and Hill 1998; DeVita et al. 2001; Withrow and MacEwan 2001). This explains the association of tumor suppressor genes with familial or inherited cancers. According to Knudson's two-hit model of carcinogenesis, a "second hit" or subsequent somatic mutation in the remaining wild-type (normal) tumor suppressor allele is necessary for loss of tumor suppressor gene function and for increased cancer risk. This second hit is more often categorized as a duplication of the mutant allele and subsequent loss of heterozygosity, rather than as a somatic mutation in the wild-type allele. Regardless, it is this dual loss in a tumor suppressor gene that contributes to tumor formation (Tannock and Hill 1998; DeVita et al. 2001). The activation of proto-oncogenes or loss of tumor suppressor gene function may contribute to many of the credentials of cancers suggested in the Hanahan and Weinberg model. For most cancers, additional genetic or nongenetic changes are needed to complete the requisite list of credentials required for cancer formation and progression.

Oncogenes, first recognized in the study of RNA viruses, are integral genes which, when inappropriately regulated, can contribute to the development of cancers (Tannock and Hill 1998). When expressed under the control of intact and physiological regulatory pathways, these genes are referred to as proto-oncogenes. Proto-oncogenes are involved in several physiological functions, including cellular proliferation, differentiation, modulation of gene expression, progression through the cell cycle, cell adhesion, and motility, as well as the coordination of DNA synthesis (DeVita et al. 2001). Beyond their role as effectors in these diverse cellular processes, their expression also facilitates signal transduction and therein is responsible for a cell's ability to respond to external signals. The protein products of proto-oncogenes can be grouped into distinct classes based on their cellular localization and functions. These categories include (1) growth factors, (2) growth factor receptors, (3) cytoplasmic kinases, (4) membrane-associated guanine-nucleotide-binding proteins (G-proteins), (5) nuclear proteins, and (6) cytoplasmic proteins (Tannock and Hill 1998; DeVita et al. 2001; Withrow and MacEwan 2001). Examples of oncogenes that are activated in human and canine cancers are listed in Table 1.

Proto-oncogene activation, resulting in a cancer-causing oncogene, may occur through several mechanisms, including insertional mutagenesis, amplification, chromosome translocation, or gene mutation (Fig. 3)

Table 1. Examples of oncogenes activated in canine and human tumors

Category	Mechanism	Associated human tumors	Associated canine tumors
Growth factors			
FGF	amplification	bladder, breast CA	bladder CA (TCC)?
PDGF	overexpression	glioma	unknown
IGF-1	overexpression	osteosarcoma	osteosarcoma
Growth factor receptors			
c-kit	mutation	GIST, mastocytosis	mast cell tumors, GIST
EGF-R	mutation	breast, lung CA	osteosarcoma, mammary CA
c-met	overexpression	many	osteosarcoma
Cytoplasmic kinases			
abl	translocation	CML	unknown
G proteins			
ras	mutation	lung carcinoma	lung carcinoma
Nuclear proteins			
myc	translocation	Burkitt's lymphoma	unknown
MDM2	overexpression	soft tissue sarcoma	soft tissue sarcoma

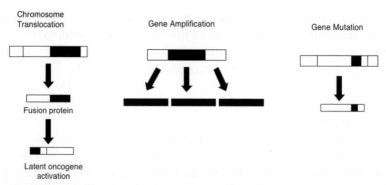

Figure 3. Proto-oncogene activation. The activation of a proto-oncogene to an oncogene can occur by gene mutation, gene amplification, or chromosome translocation. Gene mutation results in constitutive activation of a gene, gene amplification leads to overexpression of an otherwise normal gene product, and chromosome translocation either causes expression of an aberrant fusion protein or activates a latent oncogene in a new location.

(Tannock and Hill 1998; DeVita et al. 2001; Withrow and MacEwan 2001). Retrovirus infection and integration is a classic example of proto-oncogene activation by insertional mutagenesis. Retroviral transcripts directly introduce a retroviral oncogene, inappropriately activate a host proto-oncogene, or cause a fusion protein between viral and host elements which results in an oncogenic protein product (Withrow and MacEwan 2001). Cancer-causing retroviruses have been described in humans, cats, cows, chickens, and many other species; however, none has been definitively identified in the dog. The curious absence of natural retroviruses may suggest an unusual species-specific mechanism that resists retroviral infection or may simply be the result of a lack of detection.

Chromosomal translocations may also result in proto-oncogene activation. Translocations occur when part of one chromosome is exchanged with part of another during cell division. This can be oncogenic by creating either a novel protein capable of malignant transformation (e.g., Philadelphia chromosome in chronic myelogenous leukemia in humans) or through the transcriptional activation of a latent oncogene in an aberrant location (c-myc in Burkitt's lymphoma) (Tannock and Hill 1998; DeVita et al. 2001). Recent work by M. Breen et al. (pers. comm.) has identified chromosomal translocations in canine leukemias that resemble those seen in the Philadelphia chromosome in humans. It is likely that work with newly available technologies, including FISH and CGH array, will identify additional canine cancer-associated translocations.

Gene amplification may occur from a number of mechanisms leading to high expression of a gene product. This overexpression overwhelms normal cellular control mechanisms and activates further downstream signaling pathways, resulting in oncogenic transformation. The aberrant expression of c-met in canine osteosarcoma may be due to this mechanism and may influence its malignant potential (Ferracini et al. 2000; C.A. London, unpubl.).

Gene mutation can result in the constitutive activation of a gene. Mutations in c-kit, a tyrosine kinase growth factor receptor for stem cell factor, have been identified in a number of cancers, including gastrointestinal stromal tumors (human and dog) and mast cell cancers (humans and dogs) (London et al. 1996, 1999; Ma et al. 1999). Kit mutations are thought to be important mechanisms in the development and progression of these tumor types.

Tumor suppressor genes act to control the progress of cells through the cell cycle, control messenger pathways, survey damage to the genome, activate DNA repair, and initiate cell death when irreversible damage occurs (Tannock and Hill 1998; DeVita et al. 2001; Withrow and MacEwan 2001). Their key role in tumorigenesis was first recognized through the studies of familial cancers in humans. The predispositions of Bernese mountain dogs to malignant histiocytosis and the German shepherd to renal cystadenocarcinoma and nodular dermatofibrosis are heritable (Padgett et al. 1995; Lingaas et al. 2003) cancers that have been linked to loss of tumor suppressor gene function. Interestingly, the gene responsible for renal cystadenocarcinoma in dogs is thought to be a mutated tumor suppressor gene that is similar to a recently identified gene in humans responsible for the Britt-Hogge Dubay syndrome, a cancer phenotype in humans that resembles this disease in dogs (Schmidt 2004). Examples of tumor suppressor genes that are mutated or lost in human and canine cancers are listed in Table 2.

Retinoblastoma (Rb) was one of the first tumor suppressor genes discovered. It is a nuclear phosphoprotein that plays a key role in regulating the cell cycle. In its active state, Rb serves as a "brake" for cells advancing from G_1 to S phase, where it acts to sequester E2F family transcription factors, preventing replication (Fig. 4) (Tannock and Hill 1998; DeVita et al. 2001). When healthy cells are stimulated by growth factors, Rb protein is inactivated by phosphorylation, and the brake is released, with cells then advancing beyond this checkpoint of the cell cycle. When the Rb protein is absent, due to gene mutations/deletions, the brakes on the cell cycle are unregulated, and cells move unchecked into S phase. Germ-line loss or mutation of the Rb gene predisposes children to

Table 2. Examples of tumor suppressor genes in human and canine cancer

Subcellular location	Function	Associated human tumors	Associated canine tumors
Cell surface			
E-cadherin	cell adhesion	gastric, breast CA	unknown
Cytoskeleton			
NF-2	unknown	meningioma, schwanoma	unknown
Cytosol			
APC	inhibits signal transduction	colon, pancreas CA	unknown
Nucleus			
p53	regulates cell cycle and apoptosis	many	osteosarcoma, lymphoma, mammary CA, soft tissue and hemangiosarcoma
Rb	regulates cell cycle	retinoblastoma, osteosarcoma	osteosarcoma?
p21	regulates cell cycle	melanoma	melanoma
p16			
PTEN	controls survival signaling	breast, prostate CA	osteosarcoma

retinoblastoma and, less commonly, osteosarcoma (Tannock and Hill 1998; DeVita et al. 2001). Rb status has been examined in both canine osteosarcoma cell lines and tumors, as well as canine melanomas. Although the Rb family protein mutations were found in three of four osteosarcoma cell lines, these mutations are not common in tumor tissues themselves (Mendoza et al. 1998; Levine and Fleischli 2000; Koenig et al. 2002). Additional studies in larger, well-characterized tissue sets are necessary.

The most frequent tumor suppressor gene mutated in human cancers is p53. Mutations and dysregulation of p53 have been recognized in canine cancers in varying degrees, including osteosarcoma, lymphoma, soft tissue sarcomas, hemangiosarcoma, testicular tumors, and mammary carcinoma (Johnson et al. 1998; Mayr et al. 1999, 2002; Nasir et al. 2001; Setoguchi et al. 2001). p53 is a nuclear phosphoprotein that controls both transcriptional activity and repression of several genes (Fig. 5). p53 or its regulated genes control an important cell cycle checkpoint that allows cells to recognize DNA damage and initiate repair or entry to a program of cell death (apoptosis) if repair is impossible (Tannock and Hill 1998; DeVita et al. 2001). The loss of p53 also contributes to the aggressive

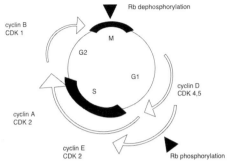

Figure 4. The cell cycle consists of four phases: M (mitosis), G_1 (protein synthesis to prepare for S phase), S (synthesis), and G_2 (protein synthesis to prepare for mitosis). Movement through the cell cycle is tightly regulated by cyclins and CDKs (cyclin dependent kinases). Rb (retinoblastoma protein) in its active phosphorylated state prevents movement of cells from G_1 to S phase, controlling individual cell division.

behavior of cancers and their natural resistance to many types of chemotherapy and radiation therapy (Withrow and MacEwan 2001). These mutations are most commonly mis-sense mutations that affect the ability of p53 to bind to DNA. Accordingly, p53 overexpression, related to mutation or loss of normal p53 function, has been identified as a negative prognostic indicator in many human cancers. Similarly, the over-expression of p53 in dogs has been associated with poor outcome in dogs with mammary tumors, squamous cell carcinomas of the skin, and mast cell tumors. (Ginn et al. 2000; Jaffe et al. 2000; Haga et al. 2001; Wakui

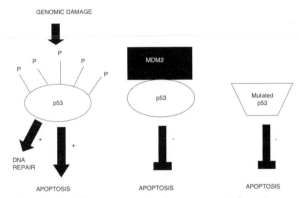

Figure 5. p53 function. p53 or its regulated genes control an important cell cycle checkpoint that allows cells to recognize DNA damage and initiate repair, or if irreparable, direct the cell into apoptosis. MDM2 is a binding protein of p53 that down-regulates its activity; it can be overexpressed in certain cancers as a mechanism to avoid apoptosis. p53 mutations contribute to the aggressive behavior of certain tumor types and to their resistance to standard therapies that require intact apoptotic pathways.

et al. 2001; Lee et al. 2004) Interestingly, in one study p53 overexpression was found in appendicular but not axial osteosarcoma, possibly accounting for the more aggressive phenotype of these tumors compared to their non-appendicular counterparts (Sagartz et al. 1996). p53 function can also be modulated by mutations in its many regulatory proteins, including its cellular binding protein, MDM2 (Tannock and Hill 1998; DeVita et al. 2001). In humans and in dogs, MDM2 has been shown to be overexpressed by gene amplification in a subset of soft-tissue sarcomas (Taubert et al. 2000). In most of these cancers, there was a reciprocal relationship between MDM2 overexpression and p53 down-regulation (Nasir et al. 2001).

PTEN is a well-recognized tumor suppressor gene that is deleted in a variety of human cancers, namely prostate, breast, and endometrial carcinomas. PTEN is a phosphatase that acts in part to control the "life signal" provided by the AKT pathway (DeVita et al. 2001; Steelman et al. 2004). PTEN was found to be mutated in 10 of 15 canine osteosarcoma tumor samples and may play a role in the pathogenesis of osteosarcoma in the dog (Levine et al. 2002). PTEN mutations have also been found in primary canine melanomas. The inbred nature of dog breeds and their well-known pedigrees have resulted in the identification of familial cancer inheritance patterns that are reasonably associated with loss of tumor suppressor gene function. It is well known that large and giant breed dogs have a high incidence of appendicular osteosarcoma (MacEwen and Kurzman 1996). Within these large and giant breeds, work is ongoing to identify genetic risk factors of osteosarcoma. The breeds under closest examination include the rottweiler, greyhound, golden retriever, and boxer. Other cancers are also seen at higher rates in some breeds and include mast cell tumors in boxers, beagles, and Boston terriers (Withrow and MacEwan 2001; Ozaki et al. 2002); lymphoma in the golden retriever; both testicular tumors and hemangiosarcoma in the German shepherd dog (Mays and Bergeron 1986; Peterson et al. 2000); and transitional cell carcinoma of the urinary bladder in Scottish terriers.

MANIPULATION OF GROWTH SIGNALING

Malignant cells proliferate in an exponential fashion. The hypothetical growth of tumors, known as Gompertzian, suggests that the initial and rapid proliferation of tumor cells occurs before the disease is clinically detectable. Once the tumor reaches 1 gram in weight (10^9 cells), believed to be the limit of gross clinical detection, its growth rate begins to plateau. A balance between the rate of cell growth and cell death determines the eventual size of a tumor (Tannock and Hill 1998). Tumor cells can shift this balance toward growth by becoming independent of external signals or by amplifying available growth signals (Tannock and Hill 1998). Mutations in growth factor receptors may result in their sustained activation, independent of ligand (signal). Tumor cells may become independent of growth factor signals by producing both the signal (ligand) and receptor for a growth factor. In this situation, the tumor cell is considered to have autocrine activation of its receptor. Both mechanisms result in tumor growth and progression through the manipulation of growth signaling. Many examples of manipulation of growth signaling have been defined in human and canine cancers. Insulin-growth factor (IGF-1) and its receptor IGF-R represent a signaling pathway that tumor cells can use to become self-sufficient. IGF-1 activation can result in tumor cell proliferation and anti-apoptotic or pro-metastatic signaling. Osteosarcomas in both children and dogs have been found to express IGF-1 and IGF-1 receptors and to proliferate in response to this growth factor (Withrow and MacEwan 2001; Khanna et al. 2002). This mechanism may help drive the constitutive propagation of these malignant cells. Similarly, mutation or autocrine activation of the epidermal growth factor receptor (EGF-R) family has been implicated in the development and metastatic potential of several human cancers. Subsets of canine osteosarcoma and mammary adenocarcinoma have been shown to express the putative mutated form of the EGF-R (i.e., the erb-2 oncogene) (Flint et al. 2004). As discussed above, c-kit mutations have been identified in dog cancers, including mast cell tumors. Mutations in c-kit are thought to result in ligand-independent activation of the receptor. Dogs with mutations in this receptor have been shown to have a more aggressive course of disease than dogs with mast cell tumors where no c-kit mutations are seen (Reguera et al. 2000; Kiupel et al. 2004).

The fact that cancers manipulate growth signals through growth factor receptors suggests that the wild-type or mutated receptor may be a valuable target for new cancer therapies (Fig. 6). Several agents targeting growth factor receptors have entered clinical development and have received approval as new cancer drugs for human cancer patients. Studies with Gefitinib, a recently FDA-approved small-molecule inhibitor of EGF-R, in patients with lung cancer have suggested that specific mutations in EGF-R are necessary for Gefitinib to be most active (Lynch et al. 2004; Paez et al. 2004). Patients with these mutations had nearly 100% response rates, whereas responses in the remaining patients were infrequently observed. Small-molecule inhibitors of the c-kit receptor tyrosine kinase (SU11654) have entered clinical trials in dogs with mast cell tumors. Similar to the data seen in human patients, the presence of an activating mutation in the c-kit receptor in dogs with mast cell tumors yielded more

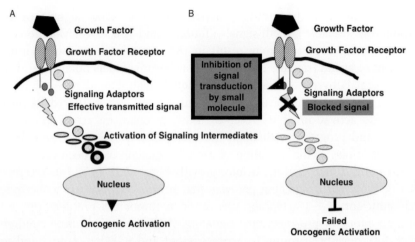

Figure 6. Signal transduction and the inhibition of signaling by small molecule inhibitors. (**A**) Oncogenic signals may be received at the nucleus by a series of activation steps that lead from growth factor interaction with a growth factor receptor at the cell membrane, leading to the transfer of activation signals (most often including the sequential transfer of phosphate groups between signaling intermediates eventually leading to the oncogenic signals being received and transmitted at the nucleus. In cases of dysregulated signaling, cancer cells may not require growth factor/ growth factor receptor interaction to yield an effective signal. This ligand-independent signaling may be the result of mutations in the growth factor or the autocrine production of both growth factor and growth factor receptor by the same tumor cell. (**B**) The generation of small molecule inhibitors of signal transduction, specifically cancer-associated growth factor receptors that are dysregulated (autoactivated), have become available as anticancer drugs. Examples of such drugs include SU11546, a small molecule inhibitor of the c-kit growth factor receptor. Blockage of signaling through this growth factor receptor has been shown to be an effective treatment in dogs with mast cell tumors and other cancers.

than 90% response rates, whereas the absence of activating mutations resulted in benefits in approximately 50% of dogs. These clinical trial data suggest that although cancers may amplify growth signals by several mechanisms, it is the activating mutations of growth signals that will yield the most productive therapeutic targets (London et al. 2003; Pryer et al. 2003).

LOSS OF GROWTH INHIBITION

The ability to resist the mechanisms that control or suppress cell proliferation is an essential credential of the cancer cell. These controls are most finely tuned in the regulation of the cell cycle (Fig. 4). The cell cycle is divided into four distinct phases: M (mitosis), S (synthesis), G_1, and G_2.

The time that a cell spends in the phases of the cell cycle is based on the type of cell, its function, and stage of differentiation. Nonproliferating cells retain the capacity to divide and arrest in G_0 (rest phase between M and G_1) until stimulated to continue through the cell cycle (Tannock and Hill 1998). The movement of cells through the cell cycle is regulated at a variety of checkpoints, each controlled by one or several genes. Cells may receive signals to enter the cell cycle either in G_0 or G_1. There is an important stop point known as the restriction site (R point) in G_1, beyond which a cell is committed to enter S phase and DNA synthesis occurs. It is the inhibition of the Rb (retinoblastoma gene product) that is required for the movement of cells through the R point from G_1 to S phase. Although this biology is well defined in tissue culture, in vivo experiments suggest that the situation is more complex and that multiple restriction points are likely to exist within the cell cycle (Tannock and Hill 1998).

The transitions through phases of the cell cycle are governed by interactions between cyclins and cyclin-dependent kinases (cdks). It is the cyclins/cdks and their inhibitors which regulate the inhibition (by phosphorylation) of Rb and the cell cycle (Tannock and Hill 1998). Cyclins bind to their respective cdks, initiating phosphorylation events that regulate cell cycle transitions. The kinases are at consistent levels throughout the cell cycle, while the cyclin levels fluctuate, making them the regulatory units of the cell cycle (Withrow and MacEwan 2001). Cdk inhibitory proteins further control the binding of cdks to cyclins. These inhibitors, which are made up of two distinct families, KIP-kinase inhibitory proteins (e.g., p21, p27) and INK-inhibitors of kinases (e.g., p15, p16) bind to their respective cdks and hinder their activity (Tannock and Hill 1998). The KIP molecules bind to the cyclin/cdk complexes, inactivating them, while the INK family leads to the dissociation of the cyclin/cdk complex and prevents their reassociation (Tannock and Hill 1998). The control mechanisms that regulate the cell cycle, cyclins, cdks, and their inhibitors, may be mutated or inhibited in the process of cancer development. p21 is a cdk inhibitor that controls the G_1/S transition and the phosphorylation of Rb. Therefore, it acts to impede progression through the cell cycle. p21 and p16 mutations have been recognized in canine and human malignant melanoma and may play a role in their pathogenesis (Koenig et al. 2002).

EVADING PROGRAMMED CELL DEATH

The ability to evade programmed cell death, apoptosis, is a key feature of the cancer phenotype. Apoptosis is distinguished from necrosis, the other mechanism of cell death, by being an active process and by the absence of inflammation that is usually associated with necrosis. Apoptosis occurs through a series of molecular events that result in the activation of proteolytic enzymes known as caspases, which are ultimately responsible for the death of cells (Tannock and Hill 1998; DeVita et al. 2001). Many of the regulatory genes of apoptosis are co-opted by malignant cells to resist this programmed death process. Normal apoptotic pathway function is required to survey and detect DNA damage. If the detected damage is significant, the initiation of the apoptosis pathway occurs. Defects in the detection of DNA damage (i.e., p53), defects in the apoptosis initiation pathway (i.e., bcl-2 family proteins) or apoptosis execution pathway (i.e., caspases) have been shown to contribute to cancer development and progression (Tannock and Hill 1998). Defects in apoptosis initiation by Bcl-2 overexpression have been documented in a variety of human and canine tumor cell lines and are thought to be a mechanism of chemotherapy resistance (Reed 1995; Lin et al. 1999; Sano et al. 2003). The bcl-2 superfamily of proteins includes both pro-apoptotic (bax, bad, bcl-xS) and anti-apoptotic (bcl-2, bcl-xL) elements (DeVita et al. 2001) and was one of the first identified apoptosis-associated gene families.

INHIBITED CELLULAR AGING AND SENESCENCE

The failure of malignant cells to follow appropriate signals of aging (senescence) is an increasingly recognized feature of the cancer phenotype (Tannock and Hill 1998). To a large extent, the cellular age of a cell is linked to the number of cellular divisions an individual cell will undergo through its life. Telomeres are 6-nucleotide repeats found at the 3′ ends of chromosomes. Telomeres act as the internal clock of cells by protecting the chromosome ends from degradation during cellular division. Normal telomere attrition occurs during successive replications and will continue until a "critical" telomere length is reached. At this point, the cell initiates signals for cellular senescence and eventual apoptosis (Tannock and Hill 1998; Argyle and Nasir 2003).

Telomerase, a ribonucleoprotein, is the enzyme responsible for maintaining telomere length. It is made up of two components, a RNA subunit, which acts as a template for the construction of the growing telomere, and an enzymatic subunit, which is the catalytic component of telomerase activity. Increased telomerase activity is a mechanism by which tumor cells evade cell death and maintain replicative potential (Ahmed and Tollefsbol 2003). This confers cellular immortality to malignant cells. Telomerase stands as both a marker of malignancy and a potential target for interventional therapeutics due to its specificity for malignant cells.

Telomere biology is similar in dogs and humans. Normal human telomeres are about 15 kb, whereas average dog telomeres are 22–23 kb in size. This contrasts to murine telomeres, which are much larger, approximately 60 kb in size. Telomerase activity is normally absent in somatic tissues, save germinal tissues, but is very high in 85–90% of both human and canine tumors (Nasir et al. 2001; Yazawa et al. 2001; Argyle and Nasir 2003). A number of different canine cancer cell lines and tumor tissues, including mammary gland adenocarcinoma, osteosarcoma, hemangiosarcoma, and lymphoma, have high telomerase activity (Argyle et al. 2003). Up-regulation of telomerase is thought to be an influential means by which cancer cells can maintain their telomeres and hence avoid senescence. Although another mechanism for maintaining telomere length known as ALT (alternative mechanism for telomere lengthening) is believed to exist, the exact interplay between these two mechanisms within cancer is still unclear (Ulaner 2004). A variety of methods have been utilized to attempt to abrogate telomerase activity in canine tumor cells; these have included pharmacological agents (AZT), dominant negative mutants, oligonucleotides, and, most recently, RNA interference. Which is the best method for telomerase inhibition is under study, although a recent report using RNA interference in human cancer cell lines and a viral vector delivery system was very successful in causing acute apoptosis. (Li et al. 2004). Future studies in dogs are pursuant on these same mechanisms (D.J. Argyle and M.C. Paoloni, pers. comm.).

ACQUISITION OF THE METASTATIC PHENOTYPE

The spread of a tumor from its primary site to distant locations is the most significant problem associated with cancer. Most deaths associated with cancer in animals and humans are the result of cancer metastasis. Based on the work of several groups, it is increasingly understood that for metastasis to occur, a cancer cell must be able to complete a diverse set of complex tasks (Poste and Fidler 1980; Fidler 2003; Pantel and Brakenhoff 2004). They must leave the site of the primary tumor and enter the circulation (**intravasation**) (Muraoka et al. 2002). While in the circulation, tumor cells must **resist** programmed cell death associated with loss of cellular contact (**anoikis**), **evade immune recognition and the damage of blood flow-induced shearing,** and eventually **arrest at distant organs** (Chambers et al. 2000; Wong et al. 2002; Wang et al. 2004). At this distant site the cell must **survive** in a foreign tissue microenvironment (Khanna and Hunter 2004), **proliferate**, create new blood vessels (**angiogenesis**) or co-opt existing blood vessels, and then successfully grow into a metastatic lesion (Chambers et al. 1992; Al-Mehdi et al. 2000). A more detailed understanding of each of the steps associated with the metastatic process is emerging from work in human and veterinary oncology and has been recently reviewed elsewhere (Khanna and Hunter 2004; Pantel and Brakenhoff 2004).

The processes required for a metastatic cancer cell to be successful (listed in bold text above) may be the functional building blocks of the metastatic phenotype. A metastatic cancer may achieve its necessary complement of building blocks through a number of genetic events in a similar way to developing the credentials that define the cancer phenotype (discussed earlier). This set of genetic events may include a single gene that can contribute to several functional metastasis building blocks, or a set of multiple events that contribute to a single building block. Two classes of genes have been broadly defined as contributing to the metastatic phenotype: metastasis-promoting genes (Clark et al. 2000; Khanna et al. 2001; Ramaswamy et al. 2003) and metastasis suppressors (Shevde and Welch 2003; Steeg 2004). The function of these genes is not limited to the promotion or control of metastasis, as these genes may also function in normal development or physiology (i.e., cell migration, tissue invasion, and angiogenesis) that are subverted by the cancer cell in the acquisition of the metastatic phenotype. The use of high-throughput and genome-wide investigations has uncovered many putative metastasis-associated genes in human and veterinary cancers, including osteosarcoma, mammary carcinoma, melanoma, hemangiosarcoma, and hematopoietic neoplasias (Khan et al. 1998; Cohen et al. 2001; Jiang et al. 2001; Khanna et al. 2001; Boon et al. 2003; Squire et al. 2003; Stam et al. 2003; Tsutsumi et al. 2003; Yagi et al. 2003; Bicciato et al. 2004; Khanna and Hunter 2004; Kohlmann et al. 2004; Wang et al. 2004; Yu et al. 2004). Although they are likely to exist, metastasis-suppressor genes have not yet been identified in veterinary cancers.

A key factor in tumor growth and metastasis is the creation of new blood vessels, i.e., angiogenesis/vasculogenesis, or the subversion of existing blood vessels to the cancer. For both processes, endothelial cells or endothelial progenitors are activated by tumor-derived growth factors and result in new capillaries at the tumor site (Kerbel 2000). In both normal tissue and tumor tissue, a balance between proteins that activate or antagonize endothelial cells controls endothelial cell proliferation. Malignant tumors provide signals that result in endothelial cell survival, motility, invasion, differentiation, and organization. These steps are required to create a supportive vasculature for the tumor. In many ways, these required endothelial processes share parallel features with the processes required for the success of a metastatic cancer cell. The creation of new blood vessels requires the recruitment of circulating endothelial cells to the site of the primary tumor, presumably through the release of growth factors, e.g., vascular endothelial growth factor (VEGF). Circulating endothelial cells must survive at their new site and then form vascular tubes that are reorganized to sustain blood flow.

The importance of angiogenesis in the biology of metastasis is supported by many lines of evidence: The vascularity of a primary tumor (measured by microvessel density) has been correlated with metastatic behavior for many human and many veterinary tumors; the expression of angiogenesis-associated growth or survival factors and their receptors (i.e., VEGF-R, FGF-R, PDGF-R) in serum and in tumors, respectively, has also been correlated with outcome; and more recently, functional imaging studies using MRI and other means have provided correlates of vascularity with poor outcome. The validation of angiogenesis as a critical biology associated with cancer progression has supported the development of a number of novel therapeutic agents with anti-angiogenic activities. These agents have moved through discovery and development and are now appearing as approved drugs (e.g., VEGF-R antibody, Bevacizumab). These same agents, as well as novel drugs developed for veterinary patients, are likely to be included in the management of cancer patients in the future.

CONCLUSIONS AND FUTURE AIMS

To a large extent our understanding of cancer biology in dogs and in other species, discussed above, is "candidate" based. The candidate approach rests on the understanding of a single gene or protein in a biological process. With the advent of novel technology platforms, including gene and tissue microarrays and genomic and proteomic databases, it is increasingly possible and essential that connections between candidates are identified and understood. This connected "understanding of systems" approach to understanding cancer biology is likely to yield new candidates, define the vital candidates in a cancer, and most importantly, demonstrate the interconnectedness of abnormalities that drive cells toward cancer and metastatic cancer phenotype.

Several factors contribute to the value of spontaneous cancers in dogs as relevant models for human cancer. These animals share many environmental risk factors with their human caregivers, suggesting their value as sentinels of disease (Kelsey et al. 1998). In most instances, cancers in dogs display similar biological behaviors as their human counterparts, exhibiting the same tumor histologies and response rates to conventional chemotherapeutics, although the time it takes to complete a clinical trial in dogs is much shorter. Perhaps most importantly, cancers that develop spontaneously in dogs often possess identical genetic mutations when compared to those found in corresponding human cancers, indicating common molecular mechanisms of malignant transformation.

The focused study of canine cancer has lead to successful development of new therapeutic techniques as well as informative testing of new anticancer agents in dogs with cancer. We predict that this trend will continue with the availability of all the same genomic and molecular tools as for human and mouse, and that many of the biological conclusions and resulting treatments will be applicable to humans as well as dogs.

REFERENCES

Ahmed A. and Tollefsbol T.O. 2003. Telomerase, telomerase inhibition, and cancer. *J. Anti Aging Med.* **6:** 315–325.

Al-Mehdi A.B., Tozawa K., Fisher A.B., Shientag L., Lee A., and Muschel R.J. 2000. Intravascular origin of metastasis from the proliferation of endothelium-attached tumor cells: A new model for metastasis. *Nat. Med.* **6:** 100–102.

Argyle D.J. and Nasir L. 2003. Telomerase: A potential diagnostic and therapeutic tool in canine oncology. *Vet. Pathol.* **40:** 1–7.

Bicciato S., Luchini A., and Di Bello C. 2004. Marker identification and classification of cancer types using gene expression data and SIMCA. *Methods Inf. Med.* **43:** 4–8.

Boon K., Edwards J.B., Siu I.M., Olschner D., Eberhart C.G., Marra M.A., Strausberg R.L., and Riggins G.J. 2003. Comparison of medulloblastoma and normal neural transcriptomes identifies a restricted set of activated genes. *Oncogene* **22:** 7687–7694.

Bronson R.T. 1982. Variation in age at death of dogs of different sexes and breeds. *Am. J. Vet. Res.* **43:** 2057–2059.

Chambers A.F., Naumov G.N., Vantyghem S.A., and Tuck A.B. 2000. Molecular biology of breast cancer metastasis. Clinical implications of experimental studies on metastatic inefficiency. *Breast Cancer Res.* **2:** 400–407.

Chambers A.F., Schmidt E.E., MacDonald I.C., Morris V.L., and Groom A.C. 1992. Early steps in hematogenous metastasis of B16F1 melanoma cells in chick embryos studied by high-resolution intravital videomicroscopy. *J. Natl. Cancer Inst.* **84:** 797–803.

Clark E.A., Golub T.R., Lander E.S., and Hynes R.O. 2000. Genomic analysis of metastasis reveals an essential role for RhoC. *Nature* **406:** 532–535.

Cohen N., Rozenfeld-Granot G., Hardan I., Brok-Simoni F., Amariglio N., Rechavi G., and Trakhtenbrot L. 2001. Subgroup of patients with Philadelphia-positive chronic myelogenous leukemia characterized by a deletion of 9q proximal to ABL gene: Expression profiling, resistance to interferon therapy, and poor prognosis. *Cancer Genet. Cytogenet.* **128:** 114–119.

DeVita V.T., Hellman S., and Rosenberg S.A. 2001. *Cancer, principles and practice of oncology.* Lippincott Williams & Wilkins, Philadelphia.

Dorn C. and Taylor D. 1968. Survey of animal neoplasms in Alameda and Contra Costa Counties, California. II. *J. Natl. Cancer Inst.* **40:** 307–318.

Ferracini R., Angelini P., Cagliero E., Linari A., Martano M., Wunder J., and Buracco P. 2000. MET oncogene aberrant expression in canine osteosarcoma. *J. Orthop. Res.* **18:** 253–256.

Fidler I.J. 2003. The pathogenesis of cancer metastasis: The 'seed and soil' hypothesis revisited. *Nat. Rev. Cancer* **3**: 453–458.

Flint A.F., U'Ren L., Legare M.E., Withrow S.J., Dernell W., and Hanneman W.H. 2004. Overexpression of the erbB-2 proto-oncogene in canine osteosarcoma cell lines and tumors. *Vet. Pathol.* **41**: 291–296.

Gemmill R. and Idell C.S. 2003. Biological advances for new treatment approaches. *Semin. Oncol. Nurs.* **19**: 162–168.

Ginn P.E., Fox L.E., Brower J.C., Gaskin A., Kurzman I.D., and Kubilis P.S. 2000. Immunohistochemical detection of p53 tumor-suppressor protein is a poor indicator of prognosis for canine cutaneous mast cell tumors. *Vet. Pathol.* **37**: 33–39.

Haga S., Nakayama M., Tatsumi K., Maeda M., Imai S., Umesako S., Yamamoto H., Hilgers J., and Sarkar N.H. 2001. Overexpression of the p53 gene product in canine mammary tumors. *Oncol. Rep.* **8**: 1215–1219.

Hanahan D. and Weinberg R.A. 2000. The hallmarks of cancer. *Cell* **100**: 57–70.

Hansen K. and Khanna C. 2004. Spontaneous and genetically engineered animal models; use in preclinical cancer drug development. *Eur. J. Cancer* **40**: 858–880.

Jaffe M.H., Hosgood G., Taylor H.W., Kerwin S.C., Hedlund C.S., Lopez M.K., Davidson J.R., Miller D.M., and Paranjpe M. 2000. Immunohistochemical and clinical evaluation of p53 in canine cutaneous mast cell tumors. *Vet. Pathol.* **37**: 40–46.

Jiang Z., Woda B.A., Rock K.L., Xu Y., Savas L., Khan A., Pihan G., Cai F., Babcook J.S., Rathanaswami P., et al. 2001. P504S: A new molecular marker for the detection of prostate carcinoma. *Am. J. Surg. Pathol.* **25**: 1397–1404.

Johnson A.S., Couto C.G., and Weghorst C.M. 1998. Mutation of the p53 tumor suppressor gene in spontaneously occurring osteosarcomas of the dog. *Carcinogenesis* **19**: 213–217.

Kelsey J.L., Moore A.S., and Glickman L.T. 1998. Epidemiologic studies of risk factors for cancer in pet dogs. *Epidemiol. Rev.* **20**: 204–217.

Kerbel R.S. 2000. Tumor angiogenesis: Past, present and the near future. *Carcinogenesis* **21**: 505–515.

Khan J., Simon R., Bittner M., Chen Y., Leighton S.B., Pohida T., Smith P.D., Jiang Y., Gooden G.C., Trent J.M., and Meltzer P.S. 1998. Gene expression profiling of alveolar rhabdomyosarcoma with cDNA microarrays. *Cancer Res.* **58**: 5009–5013.

Khanna C. and Hunter K. 2004. Modeling metastasis in vivo. *Carcinogenesis* **26**: 513–523.

Khanna C. and Vail D.M. 2003. Targeting the lung: Preclinical and comparative evaluation of anticancer aerosols in dogs with naturally occurring cancers. *Curr. Cancer Drug Targets* **3**: 265–273.

Khanna C., Khan J., Nguyen P., Prehn J., Caylor J., Yeung C., Trepel J., Meltzer P., and Helman L. 2001. Metastasis-associated differences in gene expression in a murine model of osteosarcoma. *Cancer Res.* **61**: 3750–3759.

Khanna C., Prehn J., Hayden D., Cassaday R.D., Caylor J., Jacob S., Bose S.M., Hong S.H., Hewitt S.M., and Helman L.J. 2002. A randomized controlled trial of octreotide pamoate long-acting release and carboplatin versus carboplatin alone in dogs with naturally occurring osteosarcoma: Evaluation of insulin-like growth factor suppression and chemotherapy. *Clin. Cancer Res.* **8**: 2406–2412.

Kiupel M., Webster J.D., Kaneene J.B., Miller R., and Yuzbasiyan-Gurkan V. 2004. The use of KIT and tryptase expression patterns as prognostic tools for canine cutaneous mast cell tumors. *Vet. Pathol.* **41**: 371–377.

Koenig A., Bianco S.R., Fosmire S., Wojcieszyn J., and Modiano J.F. 2002. Expression and significance of p53, rb, p21/waf-1, p16/ink-4a, and PTEN tumor suppressors in canine melanoma. *Vet. Pathol.* **39**: 458–472.

Kohlmann A., Schoch C., Schnittger S., Dugas M., Hiddemann W., Kern W., and Haferlach T. 2004. Pediatric acute lymphoblastic leukemia (ALL) gene expression signatures classify an independent cohort of adult ALL patients. *Leukemia* **18:** 63–71.

Lawen A. 2003. Apoptosis—An introduction. *Bioessays* **25:** 888–896.

Lee C.H., Kim W.H., Lim J.H., Kang M.S., Kim D.Y., and Kweon O.K. 2004. Mutation and overexpression of p53 as a prognostic factor in canine mammary tumors. *J. Vet. Sci.* **5:** 63–69.

Levine R.A. and Fleischli M.A. 2000. Inactivation of p53 and retinoblastoma family pathways in canine osteosarcoma cell lines. *Vet. Pathol.* **37:** 54–61.

Levine R.A., Forest T., and Smith C. 2002. Tumor suppressor PTEN is mutated in canine osteosarcoma cell lines and tumors. *Vet. Pathol.* **39:** 372–378.

Li S., Rosenberg J.E., Donjacour A.A., Botchkina I.L., Hom Y.K., Cunha G.R., and Blackburn E.H. 2004. Rapid inhibition of cancer cell growth induced by lentiviral delivery and expression of mutant template telomerase RNA and anti-telomerase short interfering RNA. *Cancer Res.* **64:** 4833–4840.

Lin H.H., Yang T.P., Jiang S.T., Yang H.Y., and Tang M.J. 1999. Bcl-2 overexpression prevents apoptosis-induced Madin-Darby canine kidney simple epithelial cyst formation. *Kidney Int.* **55:** 168–178.

Lingaas F., Comstock K.E., Kirkness E.F., Sorensen A., Aarskaug T., Hitte C., Nickerson M.L., Moe L., Schmidt L.S., Thomas R., et al. 2003. A mutation in the canine BHD gene is associated with hereditary multifocal renal cystadenocarcinoma and nodular dermatofibrosis in the German Shepherd dog. *Hum. Mol. Genet.* **12:** 3043–3053.

London C.A., Kisseberth W.C., Galli S.J., Geissler E.N., and Helfand S.C. 1996. Expression of stem cell factor receptor (c-kit) by the malignant mast cells from spontaneous canine mast cell tumors. *J. Comp. Pathol.* **155:** 399–414.

London C.A., Galli S.J., Yuuki T., Hu Z.Q., Helfand S.C., and Geissler E.N. 1999. Spontaneous canine mast cell tumors express tandem duplications in the proto-oncogene c-kit. *Exp. Hematol.* **27:** 689–697.

London C.A., Hannah A.L., Zadovoskaya R., Chien M.B., Kollias-Baker C., Rosenberg M., Downing S., Post G., Boucher J., Shenoy N., et al. 2003. Phase I dose-escalating study of SU11654, a small molecule receptor tyrosine kinase inhibitor, in dogs with spontaneous malignancies. *Clin. Cancer Res.* **9:** 2755–2768.

Lynch T.J., Bell D.W., Sordella R., Gurubhagavatula S., Okimoto R.A., Brannigan B.W., Harris P.L., Haserlat S.M., Supko J.G., Haluska F.G., et al. 2004. Activating mutations in the epidermal growth factor receptor underlying responsiveness of non-small-cell lung cancer to gefitinib. *N. Engl. J. Med.* **350:** 2129–2139.

Ma Y., Longley B.J., Wang X., Blount J.L., Langley K., and Caughey G.H. 1999. Clustering of activating mutations in c-KIT's juxtamembrane coding region in canine mast cell neoplasms. *J. Invest. Dermatol.* **112:** 165–170.

MacEwen E.G. and Kurzman I.D. 1996. Canine osteosarcoma: Amputation and chemoimmunotherapy. *Vet. Clin. N. Am. Small Anim. Pract.* **26:** 123–133.

Mayr B., Reifinger M., and Alton K. 1999. Novel canine tumour suppressor gene p53 mutations in cases of skin and mammary neoplasms. *Vet. Res. Commun.* **23:** 285–291.

Mayr B., Zwetkoff S., Schaffner G., and Reifinger M. 2002. Tumour suppressor gene p53 mutation in a case of haemangiosarcoma of a dog. *Acta Vet. Hung.* **50:** 157–160.

Mays M.B. and Bergeron J.A. 1986. Cutaneous histiocytosis in dogs. *J. Am. Vet. Med. Assoc.* **188:** 377–381.

Mendoza S., Konishi T., Dernell W.S., Withrow S.J., and Miller C.W. 1998. Status of the p53, Rb and MDM2 genes in canine osteosarcoma. *Anticancer Res.* **18:** 4449–4453.

Muehlbauer P.M. 2003. Anti-angiogenesis in cancer therapy. *Semin. Oncol. Nurs.* **19:** 180–192.

Muraoka R.S., Dumont N., Ritter C.A., Dugger T.C., Brantley D.M., Chen J., Easterly E., Roebuck L.R., Ryan S., Gotwals P.J., et al. 2002. Blockade of TGF-beta inhibits mammary tumor cell viability, migration, and metastases. *J. Clin. Invest.* **109:** 1551–1559.

Nasir L., Rutteman G.R., Reid S.W., Schulze C., and Argyle D.J. 2001. Analysis of p53 mutational events and MDM2 amplification in canine soft-tissue sarcomas. *Cancer Lett.* **174:** 83–89.

Ozaki K., Yamagami T., Nomura K., and Narama I. 2002. Mast cell tumors of the gastrointestinal tract in 39 dogs. *Vet. Pathol.* **39:** 557–564.

Padgett G.A., Madewell B.R., Keller E.T., Jodar L., and Packard M. 1995. Inheritance of histiocytosis in Bernese mountain dogs. *J. Small Anim. Pract.* **36:** 93–98.

Paez J.G., Janne P.A., Lee J.C., Tracy S., Greulich H., Gabriel S., Herman P., Kaye F.J., Lindeman N., Boggon T.J., et al. 2004. EGFR mutations in lung cancer: Correlation with clinical response to gefitinib therapy. *Science* **304:** 1497–1500.

Pantel K. and Brakenhoff R.H. 2004. Dissecting the metastatic cascade. *Nat. Rev. Cancer* **4:** 448–456.

Peterson M.R., Frommelt R.A., and Dunn D.G. 2000. A study of the lifetime occurrence of neoplasia and breed differences in a cohort of German Shepherd Dogs and Belgian Malinois military working dogs that died in 1992. *J. Vet. Intern. Med.* **14:** 140–145.

Poste G. and Fidler I.J. 1980. The pathogenesis of cancer metastasis. *Nature* **283:** 139–146.

Priester W.A. and McKay F.W. 1980. The occurrence of tumors in domestic animals. *Natl. Cancer Inst. Monogr.* **54:** 1–210.

Pryer N.K., Lee L.B., Zadovaskaya R., Yu X., Sukbuntherng J., Cherrington J.M., and London C.A. 2003. Proof of target for SU11654: Inhibition of KIT phosphorylation in canine mast cell tumors. *Clin. Cancer Res.* **9:** 5729–5734.

Ramaswamy S., Ross K.N., Lander E.S., and Golub T.R. 2003. A molecular signature of metastasis in primary solid tumors. *Nat. Genet.* **33:** 49–54.

Reed J.C. 1995. Bcl-2: Prevention of apoptosis as a mechanism of drug resistance. *Hematol. Oncol. Clin. N. Am.* **9:** 451–473.

Reguera M.J., Rabanal R.M., Puigdemont A., and Ferrer L. 2000. Canine mast cell tumors express stem cell factor receptor. *Am. J. Dermatopathol.* **22:** 49–54.

Sagartz J.E., Bodley W.L., Gamblin R.M., Couto C.G., Tierney L.A., and Capen C.C. 1996. p53 tumor suppressor protein overexpression in osteogenic tumors of dogs. *Vet. Pathol.* **33:** 213–221.

Sano J., Oguma K., Kano R., and Hasegawa A. 2003. Canine Bcl-xL gene and its expression in tumor cell lines. *J. Vet. Med. Sci.* **65:** 149–151.

Schmidt L.S. 2004. Birt-Hogg-Dube syndrome, a genodermatosis that increases risk for renal carcinoma. *Curr. Mol. Med.* **4:** 877–885.

Setoguchi A., Okuda M., Nishida E., Yazawa M., Ishizaka T., Hong S.H., Hisasue M., Nishimura R., Sasaki N., Yoshikawa Y., et al. 2001. Results of hyperamplification of centrosomes in naturally developing tumors of dogs. *Am. J. Vet. Res.* **62:** 1134–1141.

Shevde L.A. and Welch D.R. 2003. Metastasis suppressor pathways—An evolving paradigm. *Cancer Lett.* **198:** 1–20.

Squire J.A., Pei J., Marrano P., Beheshti B., Bayani J., Lim G., Moldovan L., and Zielenska M. 2003. High-resolution mapping of amplifications and deletions in pediatric osteosarcoma by use of CGH analysis of cDNA microarrays. *Genes Chromosomes Cancer* **38:** 215–225.

Stam R.W., den Boer M.L., Meijerink J.P., Ebus M.E., Peters G.J., Noordhuis P., Janka-Schaub G.E., Armstrong S.A., Korsmeyer S.J., and Pieters R. 2003. Differential mRNA expression of Ara-C-metabolizing enzymes explains Ara-C sensitivity in MLL gene-rearranged infant acute lymphoblastic leukemia. *Blood* **101:** 1270–1276.

Steeg P.S. 2004. Perspectives on classic article: Metastasis suppressor genes. *J. Natl. Cancer Inst.* **96:** E4.

Steele R.J., Thompson A.M., Hall P.A., and Lane D.P. 1998. The p53 tumour suppressor gene. *Br. J. Surg.* **85:** 1460–1467.

Steelman L.S., Bertrand F.E., and McCubrey J.A. 2004. The complexity of PTEN: Mutation, marker and potential target for therapeutic intervention. *Expert Opin. Ther. Targets* **8:** 537–550.

Tannock I. and Hill R.P. 1998. *The basic science of oncology.* McGraw-Hill Health Professions Division, New York.

Taubert H., Koehler T., Meye A., Bartel F., Lautenschlager C., Borchert S., Bache M., Schmidt H., and Wurl P. 2000. mdm2 mRNA level is a prognostic factor in soft tissue sarcoma. *Mol. Med.* **6:** 50–59.

Tsao S.W., Zhang D.K., Cheng R.Y., and Wan T.S. 1998. Telomerase activation in human cancers. *Chin. Med. J. (Engl.)* **111:** 745–750.

Tsutsumi S., Taketani T., Nishimura K., Ge X., Taki T., Sugita K., Ishii E., Hanada R., Ohki M., Aburatani H., and Hayashi Y. 2003. Two distinct gene expression signatures in pediatric acute lymphoblastic leukemia with MLL rearrangements. *Cancer Res.* **63:** 4882–4887.

Ulaner G.A. 2004. Telomere maintenance in clinical medicine. *Am. J. Med.* **117:** 262–269.

Vail D.M. and MacEwen E.G. 2000. Spontaneously occurring tumors of companion animals as models for human cancer. *Cancer Invest.* **18:** 781–792.

Wakui S., Muto T., Yokoo K., Yokoo R., Takahashi H., Masaoka T., Hano H., and Furusato M. 2001. Prognostic status of p53 gene mutation in canine mammary carcinoma. *Anticancer Res.* **21:** 611–616.

Wang H., Fu W., Im J.H., Zhou Z., Santoro S.A., Iyer V., DiPersio C.M., Yu Q.C., Quaranta V., Al-Mehdi A., and Muschel R.J. 2004. Tumor cell α3 β1 integrin and vascular laminin-5 mediate pulmonary arrest and metastasis. *J. Cell Biol.* **164:** 935–941.

Weiner T. and Cance W.G. 1994. Molecular mechanisms involved in tumorigenesis and their surgical implications. *Am. J. Surg.* **167:** 428–434.

Withrow S.J. and MacEwan E.G. 2001. *Small animal clinical oncology.* Saunders, Philadelphia, Pennsylvania.

Wong C.W., Song C., Grimes M.M., Fu W., Dewhirst M.W., Muschel R.J., and Al-Mehdi A.B. 2002. Intravascular location of breast cancer cells after spontaneous metastasis to the lung. *Am. J. Pathol.* **161:** 749–753.

Yagi T., Morimoto A., Eguchi M., Hibi S., Sako M., Ishii E., Mizutani S., Imashuku S., Ohki M., and Ichikawa H. 2003. Identification of a gene expression signature associated with pediatric AML prognosis. *Blood* **102:** 1849–1856.

Yazawa M., Okuda M., Setoguchi A., Iwabuchi S., Nishimura R., Sasaki N., Masuda K., Ohno K., and Tsujimoto H. 2001. Telomere length and telomerase activity in canine mammary gland tumors. *Am. J. Vet. Res.* **62:** 1539–1543.

Yu Y., Khan J., Khanna C., Helman L., Meltzer P.S., and Merlino G. 2004. Expression profiling identifies the cytoskeletal organizer ezrin and the developmental homeoprotein Six-1 as key metastatic regulators. *Nat. Med.* **10:** 175–181.

24

Gene Therapy for Metabolic Inherited Diseases in Dogs

N. Matthew Ellinwood
Department of Animal Science
Iowa State University
Ames, Iowa 50011

Margret L. Casal
Department of Clinical Studies, School of Veterinary Medicine
University of Pennsylvania
Philadelphia, Pennsylvania 19104

Mark E. Haskins
Departments of Pathobiology and Clinical Studies
School of Veterinary Medicine, University of Pennsylvania
Philadelphia, Pennsylvania 19104

MOST OF THE GENETIC DISEASES FOR WHICH gene therapy is a good candidate for potential therapy are considered rare disorders, and gene transfer is still a somewhat uncertain proposition in terms of safety, as has been borne out by recent human trials (Hacein-Bey-Abina et al. 2003; Raper et al. 2003). Driven by the need for safety, and the paucity of patients for many of these disorders, and with the potential difficulty of conducting studies with statistically meaningful outcomes, evaluating potential gene transfer protocols in animal models is an essential preliminary step. Although mouse models are the most practical for initial assessments, larger animal models are needed because mouse models may fail to faithfully mirror the human disease (Phaneuf et al. 1996; Ohshima et al. 1997), or because the larger animal models are superior for preclinical work, as their size is closer to that of humans, and they are longer-lived and less inbred, as are humans.

The benefits of using large animal models represent an important step in the preclinical evaluation of human-directed gene transfer protocols. For practical reasons, large animals have usually been limited to dogs and cats. Although some larger species, including ovine, porcine, caprine, and bovine, have been used in the past to better understand genetic diseases, considerations of size, housing, and reproduction have limited their use in treatment protocols. Dogs are particularly well represented as models for genetic diseases, because the wide diversity of breeds and breeding practices has led to a large number of distinct and isolated genetic populations. This, combined with the medical scrutiny from the veterinary profession that has been focused on dogs and the need to develop animal models of human disease, has led to the identification and characterization of numerous canine genetic disorders, many of which exist in research colonies.

Generally, there are two approaches using gene transfer to treat genetic diseases. The most common is to provide a normal copy of the cDNA of the gene that is defective. This approach usually involves a viral vector, which either may integrate into the chromosomal DNA of the patient's cells or may remain episomal. Another approach involves the use of an oligonucleotide of the normal DNA sequence surrounding a single base pair mutation, which will allow repair of the mutant sequence. Below we describe those disorders in dogs which have been the focus of gene transfer experiments that are not discussed in other chapters. The genetic diseases are grouped into six categories: disorders of coagulation, immunological disorders, lysosomal storage diseases, retinal disorders, disorders of connective tissue and the musculoskeletal system, and genodermatoses.

COAGULOPATHIES

Hemophilia A and B are X-linked inherited bleeding disorders caused by a deficiency of the blood-clotting proteins Factors VIII (FVIII) and IX (FIX), respectively. These factors are normally synthesized in the liver and secreted into blood. Current treatment of hemophilia involves infusions of plasma-derived or recombinant clotting factor in response to bleeding crises. Hemophilias are particularly attractive disorders for the development of gene-based therapies, as conventional therapy is lifelong and expensive, and patients remain at risk for emerging blood-borne infectious diseases. Finally, therapeutic control of hemostasis can be accomplished by supplying only a small percentage of normal factor levels, thus making gene transfer-based therapies achievable. The canine models of hemophilia are useful for developing and evaluating such gene therapies because the canine proteins are very well characterized, the genes are cloned (Evans et al. 1989a,b; Mauser et al. 1996; Brooks et al. 1997, 2003; Cameron et al. 1998; Gallo-Penn et al. 1999; Gu et al. 1999; Hough et al. 2002; Lozier et al. 2002), and cDNAs are available. Moreover, in contrast to many diseases, there is no requirement for a specific target tissue for gene delivery, and the gene product itself does not require precise regulation of expression (Lynch 1999).

Several reports review the preclinical data underlying current strategies and the design of the ongoing and proposed clinical trials in humans to identify a safe dose of gene transfer vector that reliably yields circulating levels of FVIII or FIX (High 2001a,b, 2002, 2003).

Hemophilia A

Hemophilia A is due to deficient FVIII, a trace plasma glycoprotein that is a cofactor in the activation of Factor X by FIXa. Using adenovirus vectors administered intravenously with either the canine or human FVIII cDNA has resulted in transient therapeutic plasma levels of FVIII and phenotypic correction of the coagulation defect, but was often accompanied by liver toxicity and the development of inhibitory antibodies (Connelly et al. 1996; Zhang et al. 1999; Gallo-Penn et al. 2001; Chuah et al. 2003; Brown et al. 2004).

Adeno-associated virus (AAV) vectors have a small packaging capacity, which limits their use in treating hemophilia A caused by mutations in a large gene. An approach to overcome this limitation involved using small regulatory elements to drive expression of a B-domain-deleted form of FVIII, which produced sustained (>14 months) expression of biologically active FVIII (2–4%), correlated with a partial correction in the whole-blood clotting time and cuticle bleeding time (Scallan et al. 2003).

Hemophilia B due to Deficient FIX

Administration of an adenovirus vector with the canine FIX cDNA via the portal vein directly to the liver produced plasma FIX concentrations in hemophilia B dogs ranging from 0 to 300% of the level present in normal dogs (Kay et al. 1994). This resulted in amelioration of the disease demonstrated by normal blood coagulation and hemostatic measurements. However, as is typical with adenoviral vectors, the activity then declined

(Kay et al. 1994). An approach using a vector with a temperature-sensitive mutation (ts125) was also evaluated in hemophilia B dogs (Fang et al. 1996). However, no significant difference in the duration of transgene expression was observed. In another adenovirus vector experiment, FIX was in the therapeutic range for up to 2.5 months, followed by a decrease in FIX levels and a stabilized partial correction of clotting time for 4.5 months, with no detectable toxicity (Ehrhardt et al. 2003).

The most encouraging results were from experiments using serotype 2 AAV vectors. The most promising approach has been delivery of the vector by percutaneous injections into skeletal muscle. Five hemophilia B dogs with a missense mutation were treated with the canine cDNA and showed stable (>17 months), vector dose-dependent partial correction of the whole-blood clotting time and, at higher doses, of the activated partial thromboplastin time (Herzog et al. 1999). Using hemophilia B dogs carrying a null mutation of FIX resulted in the induction of inhibitory anti-canine FIX antibodies that could be blocked in one animal by transient immune suppression with cyclophosphamide at the time of vector administration. Treatment with a combination of gene transfer and transient immune modulation resulted in sustained expression (>8 months) of canine FIX at levels sufficient for partial correction of coagulation parameters (Herzog et al. 2001). The intramuscular approach was shown to be dose-dependent, with the dog treated at the highest dose showing prolonged expression (>3 years and still under observation) of 1.4% of normal circulating levels of FIX (High 2001a). However, large doses of vector in single sites produced anti-FIX antibodies, which could be avoided by limiting the vector dose or by transient immune modulation (Herzog et al. 2002). Reports suggesting that vectors based on other serotypes transduced murine skeletal muscle much more efficiently than serotype 2 led to experiments to determine whether this increased efficacy could be observed in hemophilic dogs (Arruda et al. 2004). The intramuscular approach using serotype 1 yielded an approximately 50-fold higher level of expression due partly to higher gene copy number and a larger number of cells transduced at each injection site. In all of the dogs, inhibitory antibodies to FIX developed, emphasizing that the increased efficacy of the serotype 1 vectors carried the risk of inhibitor formation to FIX (Arruda et al. 2004). A novel alternate approach used an intravascular isolated limb delivery technique to achieve extensive transduction of most of the muscle in the dog limb, resulting in long-term (>3 years, with observation ongoing), robust, FIX expression (circulating levels of 4–14% of normal), and essentially complete correction of the bleeding disorder (Arruda et al. 2005).

Intravenous and portal vein administration of AAV vectors producing transduction of hepatocytes resulted in from 1% to 14% of normal canine FIX levels, the absence of inhibitors, and a sustained partial correction of the coagulation defect (Snyder et al. 1999; High 2002; Harding et al. 2004). Long-term (>17 months) substantial correction of canine hemophilia B in three of four animals, including two dogs with a FIX null mutation, was seen with a comparatively low dose of vector (Mount et al. 2002). This approach produced 5–12% of normal canine FIX levels, and normalization of activated and whole-blood clotting times. Activated partial thromboplastin times were substantially reduced, and anti-canine FIX antibodies were not detected. The fourth animal, also with a null mutation, showed transient expression (4 weeks) and the development of neutralizing anti-canine FIX inhibitory antibodies (Mount et al. 2002). In a third approach, a retroviral vector (RV) was administered intravenously to three newborn hemophilia B dogs, resulting in hepatocyte transduction, 12–36% of normal canine FIX antigen levels, and improved coagulation tests. Most animals failed to generate antibodies to canine FIX, demonstrating that neonatal gene transfer may have induced tolerance (Xu et al. 2003).

LYSOSOMAL STORAGE DISORDERS

Lysosomal storage disorders (LSDs) are a group of inherited diseases characterized by the disruption of normal lysosomal catabolism and subsequent accumulation of incompletely degraded substrates. There are more than 40 well-characterized genetic conditions that are classified as LSDs (Scriver et al. 2001). Most, but not all, LSDs are caused by the loss of the normal function of a specific lysosomal acid hydrolase, which acts to degrade large complex substrates after endocytosis or autophagy. The clinical signs associated with LSDs are often neurological and/or skeletal, but can also be multisystemic, with cardiovascular, visceral, muscle, and ocular system involvement in addition to the skeletal and central nervous systems.

A basis for therapy for most LSDs depends on providing normal enzyme to the lysosomes of abnormal cells. Efficient transfer of enzyme between cells and trafficking within cells usually involves a critical posttranslational modification of the lysosomal enzyme, which takes place in the Golgi as part of normal enzyme maturation. The posttranslational modification adds a mannose 6-phosphate (M6P) moiety to an enzyme. Critically important to gene therapy, a proportion of the M6P-modified enzyme is secreted by the cell and enters the extracellular fluid, where it can be taken up by other cells using the M6P receptor that is present in the plasma membrane of most cells (Kaplan et al. 1977; Distler et al. 1979; Natowicz et al. 1979). This aspect of normal cell biology allows gene transfer to produce normal active enzyme that can be delivered to distant affected cells, leading to a correction of intracellular storage. As with hemophilia, only a small percentage of normal lysosomal enzyme activity is needed to effect a substantial therapeutic response in a cell.

Canine Mucopolysaccharidosis I

Canine mucopolysaccharidosis I (MPS I) occurs in the plott hound dog (Shull et al. 1984; Stoltzfus et al. 1992). The disease results from a complete null mutation in α L-iduronidase (IDUA) by a G > A transition in the donor splice site of intron 1, creating a premature termination codon (Menon et al. 1992). Six MPS I dogs treated with in vitro RV-transduced bone marrow cells had a strong humoral immune response to the congenitally absent canine protein, and there was a disappointing in vivo survival of transduced cells. Two dogs that were re-treated had an anamnestic response, producing neutralizing antibodies and not even a short-lived appearance of enzyme-positive cells (Shull et al. 1996b). Three MPS I dogs were also treated by gene therapy targeted to muscle using direct injection of a plasmid encoding the canine IDUA cDNA, but produced no detectable enzyme. Myoblasts grown from MPS I dog skeletal muscle biopsies and transduced with a RV containing the normal canine cDNA produced a several hundredfold overexpression of enzyme in vitro. However, following reintroduction of the cultured cells into the dogs, enzyme production declined rapidly with the production of specific antibodies and a cellular infiltration at injection sites (Shull et al. 1996a). Autologous marrow cells modified in vitro by a RV expressed enzymatically active canine IDUA 10–200 times the levels seen in normal dogs (Lutzko et al. 1999a). The gene-modified cells were then infused into non-myeloablated recipients. Neither enzyme nor transcripts were detected in any dog. Humoral responses to the enzyme and serum components of the culture media were identified that apparently interfered with the therapeutic potential of hematopoietic gene therapy in these experiments (Lutzko et al. 1999a). To try to avoid the immune response, in vitro canine IDUA-transduced MPS I marrow cells were transplanted into pre-immune, mid-gestation fetal pups. Neither enzyme nor proviral-specific transcripts were detected in blood or marrow leukocytes of any MPS I dog, and all died at 8–11 months of age from complications of MPS I disease. There was no evidence of an immune response to the enzyme (Lutzko et al. 1999b).

In experiments to extend treatment strategies from mice to dogs, as a preclinical feasibility study, eight MPS I dogs were treated either with AAV2 or AAV5 gene therapy vectors containing the human IDUA cDNA (Desmaris et al. 2004). Dogs received either a single or four intrastriatal injections (two in each hemisphere). Various immunosuppression regimens were employed to prevent immune reactions against what was, for the affected dogs, a novel protein. Analysis was performed 7–17 weeks post-injection. Detection of vector genome by Q-PCR revealed spreading to the entire encephalon after four AAV5 injections. Animals receiving mild immunosuppression showed local immune reaction. Inflammation and the presence of anti-IDUA antibodies in brain tissue extracts were observed in vector-positive areas. Animals receiving a more intense immunosuppression regimen showed IDUA activity in the entire brain. Mild lysosomal distensions were observed in the brains of these young animals in areas where enzyme activity was not detected. Lesions were not observed in enzyme positive areas.

Mucopolysaccharidosis VII

Mucopolysaccharidosis VII is caused by deficient activity of β-glucuronidase (GUSB) and has been described in German shepherd dogs (Haskins et al. 1984; Silverstein et al. 2004). The normal cDNA is known, and the mutation, an arginine-to-histidine substitution at amino acid 166, has been identified (Ray et al. 1998). The first gene transfer approach used RV-transduced fibroblasts in artificial matrices called "organoids" that were placed in the omentum (Wolfe et al. 2000). This led to GUSB secretion sufficient to reduce liver storage, but insufficient to affect other tissues. This preliminary experiment was followed by neonatal intravenous RV gene therapy (Ponder et al. 2002; Xu et al. 2002; Mango et al. 2004; Sleeper et al. 2004), producing stable serum GUSB activity of between 40% and 6,000% of normal for over 4 years (the current time). Important clinical signs of disease, such as cardiac abnormalities, were absent or minimal (Sleeper et al. 2004), there was a marked improvement in the growth of treated dogs, and the dysostosis multiplex was improved in many long bones (Ponder et al. 2002; Mango et al. 2004). The dogs have remained ambulatory to beyond 4 years, versus most affected dogs that are unable to stand or walk by 6 months of age (Fig. 1). Corneal clouding has been absent or very mild in all treated dogs. Recently, an AAV vector encoding GUSB was injected into the vitreous humor of

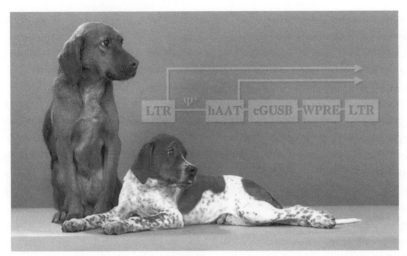

Figure 1. Two dogs with MPS VII. The dog on the left was treated with an intravenous retrovirus vector (illustrated above the dogs) containing the canine β-glucuronidase cDNA. The treated dog has been able to stand and walk for 4 years. The untreated dog on the right cannot stand and is 6 months old.

young MPS VII dogs in experiments similar to those in mice, which reduced lysosomal distension in regions of the thalamus, tectum, visual cortex, and hippocampus contralateral to the injected eye (Hennig et al. 2003).

α-Fucosidosis

α-Fucosidosis (Kelly et al. 1983; Healy et al. 1984) is a glycoproteinosis resulting from a 14-bp deletion at the end of exon one, leading to a frameshift and premature stop codon in the transcript of the canine gene for α-fucosidase (Occhiodoro and Anson 1996; Skelly et al. 1996). Dogs with α-fucosidosis have been treated by heterologous bone marrow transplantation, which was efficacious for the CNS disease, provided therapy was accomplished at an early age (Ferrara et al. 1992; Taylor et al. 1992). However, transplantation of affected dogs with in vitro RV-transduced bone marrow resulted in early graft failure (Ferrara et al. 1997).

Glycogen Storage Disease Ia

Glycogen storage disease Ia (GSD Ia) (Brix et al. 1995; Kishnani et al. 1997, 2001) in the Maltese dog is due to defective glucose-6-phosphatase (G6Pase) activity and, similar to the human disease, has clinical signs of tremors, weakness, and other neurological manifestations during hypo-glycemia, postnatal growth retardation, and progressive hepatomegaly. Histologically, hepatocellular and proximal convoluted tubular epithelium vacuolation has been described (Kishnani et al. 2001). The cause is a point mutation with a substitution of a methionine by isoleucine at codon 121 (Kishnani et al. 1997). Intravenous administration of an AAV vector containing the canine cDNA to three neonatal affected dogs resulted in reduced glycogen and lipid accumulation in the liver. At 2 months post-therapy, one affected dog had normalization of fasting glucose, choles-terol, triglycerides, and lactic acid. Six weeks after vector administration, the level of vector DNA signal in each dog varied from one to five copies per cell, consistent with variation in the efficiency of transduction within the liver. Thus, AAV vector administration in canine GSD Ia resulted in sustained G6Pase expression and improvement in liver histology and biochemical parameters (Beaty et al. 2002).

RETINAL DISORDERS

A canine model of Leber congenital amaurosis has been identified in briards, and is characterized genetically by a mutation in the gene for the retinal pigmented epithelial-specific 65-kD protein (RPE65) (Acland et al. 2001). Affected dogs have early and severe visual impairment similar to that seen in human infants with Leber congenital amaurosis. Visual impairment in affected dogs is caused by a 4-bp deletion in the gene, which results in a frameshift, a premature stop codon, and truncation of the normal protein. The effect of this mutation is that the visual pigment rhodopsin is undetectable. Histopathology reveals prominent inclusions in the RPE with slightly abnormal rod photoreceptor morphology early in life, and slowly progressive photoreceptor degeneration, resulting in early and severe visual impairment.

The normal RPE65 cDNA was incorporated into a recombinant AAV vector. The AAV-RPE65 vector was injected subretinally, into one eye of affected dogs. Thus, the contralateral eye served as a control. Retinal function measured by electroretinograms (ERGs) was improved in treated eyes compared with the same eyes before treatment, and with the untreated contralateral eye. Since only a regional area of the retina was treated, as expected, the ERGs did not normalize completely, but clinical improvement, which allowed the treated animals to remain visual, was demonstrated. Qualitative assessments of visual function were made of the three treated animals at 4 months after injection. Treated dogs, which were scored under room lighting, consistently avoided objects directly in front and on the treated side, but failed to avoid objects on the uninjected side. Injection of the vector into the vitreous was not sufficient to improve the ERG.

IMMUNE SYSTEM DISORDERS

Cyclic Neutropenia

Cyclic neutropenia in gray collie dogs, characterized by recurrent severe neutropenia leading to bacterial infections and shortened life expectancy, is caused by a mutation of the gene encoding the dog adapter protein complex 3 (AP3) β-subunit, directing trans-Golgi export of transmembrane cargo proteins to lysosomes (Benson et al. 2003). A lentiviral vector pseudotyped with vesicular stomatitis virus G protein expressing canine granulocyte colony-stimulating factor was administered intramuscularly to affected dogs, resulting in elevated neutrophil production for longer than 17 months (Yanay et al. 2003). The neutrophil counts continued to cycle, however, with a range at nadirs well above what was seen prior to therapy, and an absence of clinical signs of infection and fever (Yanay et al. 2003).

X-Linked Severe Combined Immunodeficiency

X-linked severe combined immunodeficiency (XSCID) is characterized by growth retardation and increased susceptibility to bacterial and viral infections in young pups beginning when maternal serum antibodies decline (Jezyk et al. 1989). Affected dogs rarely survive past 3–4 months of age. A pathologic feature of canine XSCID is a small, dysplastic thymus (Fig. 2), and grossly identifiable lymph nodes, tonsils, and Peyer's patches are absent. During the neonatal period, XSCID dogs have few, if any, peripheral T cells and an increased number of peripheral B cells (Jezyk et al. 1989; Felsburg et al. 1998, 1999). XSCID in the bassett hound and Cardigan Welsh corgi is due to different mutations in the common γ subunit of the interleukin-2 (IL-2), IL-4, IL-7, IL-9, IL-15, and IL-21 receptors (Henthorn et al. 1994; Somberg et al. 1995).

Bone marrow transplantation of XSCID dogs without pre-transplantation conditioning, using untreated bone marrow cells or purified canine CD34+ bone marrow cells from normal littermates, has been successful and demonstrated a selective advantage to the donor T cells (Felsburg et al. 1997; Hartnett et al. 2002). These studies form the foundation to develop and evaluate strategies for gene therapy protocols that target CD34 cells. Ex vivo treatment of XSCID bone marrow cells

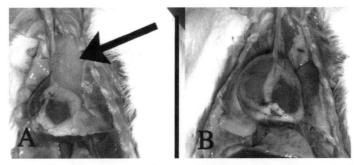

Figure 2. (A) The arrow points to the normal thymus above the heart in the chest of a normal 10-day-old puppy. (B) The same region in a 10-day-old littermate with XSCID showing the absence of a thymus.

from 3-week-old dogs with a RV vector containing the human common γ chain cDNA achieved 60–70% transduction. Following transplantation, a second round of marrow aspiration and transduction was performed a month later. Intravenous administration of the cells following 5 days of culture produced 4–10% CD3+ lymphocytes expressing human common γ chain at 6 and 11 weeks (Ting et al. 2002). In another experiment, RV coding for human common γ chain and green fluorescent protein was injected intravenously to three XSCID dogs at 3 days of age and showed GFP-marking of both lymphoid and myeloid compartments, associated with steadily improved lymphocyte count (Ting et al. 2003).

Canine Leukocyte Adhesion Defect

Canine leukocyte adhesion defect (CLAD) was originally described in juvenile Irish setter dogs with severe, recurrent bacterial infections (Giger et al. 1987; Bauer et al. 2004a). The disease was shown to result from a mutation in the leukocyte integrin CD18 subunit, which prevents leukocyte surface expression of the CD11/CD18 complex (Kijas et al. 1999; Foureman et al. 2002). Because very low levels of donor CD18+ neutrophils reversed the disease phenotype following allogeneic hematopoietic stem cell transplantation (Bauer et al. 2004b), gene therapy is currently under investigation in a colony of mixed-breed dogs (Fig. 3) (Creevy et al. 2003). CD34+ cells from bone marrow from four 3–4-month-old dogs with CLAD were pre-stimulated in vitro with cytokines and treated with a RV containing the normal canine CD18 subunit cDNA (Hai et al. 2004). The cells were infused back into the animal following conditioning with 200 cGy total body irradiation. Posttransplantation immunosuppression

Figure 3. The male heterozygote founder of a colony of dogs with canine leukocyte adhesion defect being used in gene therapy experiments. (Courtesy of Dr. Dennis Hickstein. NCI, NIH, Bethesda, Maryland.)

consisted of cyclosporine and mycophenolate mofetil. All four CLAD dogs that were treated were alive with marked improvement of their CLAD disease at 6–7 months of age (Hai et al. 2004).

MUSCLE AND COLLAGEN DISEASES

Muscular Dystrophy

Muscular dystrophy in the golden retriever (GRMD), the canine model of Duchenne muscular dystrophy, is an X-linked disease caused by a splice site mutation in the dystrophin gene, predicting a premature termination codon in exon 8. Western blot analysis of skeletal muscle from GRMD dogs revealed a slightly truncated 390-kD protein that was approximately 91% the size of normal dystrophin (Schatzberg et al. 1998). The first clinical sign is atrophy of the temporal muscles (Fig. 4) with progressive gait abnormality as the disease progresses. Using a replication-deficient adenovirus vector, high-level dystrophin expression was achieved in skeletal muscle of GRMD dogs (Howell et al. 1998). However, a humoral and cellular immune response against antigens of viral and transgene origin led to a decline of dystrophin expression over a 2-month period. Immunosuppression significantly prolonged transgene expression. In an

Figure 4. A golden retriever showing the temporal muscle atrophy, which is one of the first clinical signs of Duchenne muscular dystrophy. (Courtesy of Yan Cherel, Ecole Nationale Veterinaire, Nantes, France.)

alternate approach, DNA and RNA chimeric oligonucleotides designed to induce host cell mismatch repair mechanisms were injected into the cranial tibialis of a 6-week-old GRMD dog. Repair of the GRMD mutation was demonstrated by reverse transcription polymerase chain reaction (RT-PCR) to be sustained for 48 weeks, and an isolated exon-7-specific dystrophin antibody confirmed localization to the sarcolemma (Bartlett et al. 2000).

Myotonia Congenita

Myotonia congenita is an autosomal recessive disorder of sarcolemmal excitation in miniature schnauzers leading to delayed relaxation of skeletal muscle following contraction. The mutation results in replacement of a threonine residue with methionine in the D5 transmembrane segment (Rhodes et al. 1999; Bhalerao et al. 2002). Using an engineered ribozyme that targeted the mRNA of the mutant canine skeletal muscle chloride channel to replace the mutant-containing 3' portion by trans-splicing the corresponding 4-kb wild-type sequence, repair efficiency, assessed by quantitative RT-PCR, was $1.2 \pm 0.1\%$ in a population of treated cells. However, when chloride channel function was examined in single cells, a wide range of electrophysiological activity was observed (Rogers et al. 2002).

Alport Syndrome

Alport syndrome, an X-linked disease in dogs, is caused by a 10-bp deletion in exon 9 of the COL4A5 gene resulting in absence of type IV collagen α 5 chains from the glomerular basement membrane (Cox et al. 2003). The consequence of this defect is progressive renal failure, for which the only available treatments are dialysis and transplantation. An adenoviral vector containing the α 5(IV) transgene was injected into bladder smooth muscle that lacked both the α 5(IV) and α 6(IV) chains in affected dogs. At 5 weeks after injection, there was expression of both the α 5(IV) and α 6(IV) chains by smooth muscle cell basement membrane at the injection site. The recombinant α 5(IV) chain was capable of restoring expression of a second α (IV) chain that required the presence of the α 5(IV) chain for incorporation into collagen trimers, establishing the potential for gene therapy for this disease (Harvey et al. 2003).

GENODERMATOSES

The skin is an attractive organ for gene therapy. The epidermis can be easily biopsied, and both the keratinocytes and the fibroblasts can be separated and expanded in culture. Genes can be transferred into the appropriate cells, which can then be transferred back onto the patient (Meneguzzi and Vailly 2001). Treatment of recessive disorders is particularly appealing, as the expression or overexpression of the deficient protein product should result in correction of a defect. However, in dominant traits the defective protein product hinders the normal protein from functioning properly, usually by inhibiting formation of the proper quaternary structure. Current experiments aimed at correcting dominant disorders involve inactivation of the defective allele and correction of the mutant allele. However, these studies are currently extremely inefficient and have not been performed in any dog models (Meneguzzi and Vailly 2001). It is important to realize that some of the genetic skin diseases involve other organ systems as well, which must be taken into consideration when devising a therapeutic strategy.

Dystrophic Epidermolysis Bullosa

Dystrophic epidermolysis bullosa (RDEB) in golden retrievers is caused by a mutation (G1906S) in the 8.8-kb canine collagen type VII cDNA that leads to a glycine substitution which likely interferes with the assembly of the homotrimers and, thus, the formation of the anchoring fibrils (Baldeschi et al. 2003). Retroviral vectors were used to deliver the complete canine cDNA to primary cell cultures, resulting in sustained and permanent expression of the transgene product in canine RDEB collagen type VII-null keratinocytes. Typical triple helical homotrimers were formed by recombinant collagen type VII secreted from the canine keratinocytes and were indistinguishable from the wild-type homotrimers. The genetically modified keratinocytes were then used to make artificial skin in which high levels of recombinant collagen type VII were correctly expressed at the dermal–epidermal junction. Untreated RDEB keratinocytes were hypermotile in petri dishes coated with collagen type I compared to normal keratinocytes. Hypermotility was strongly reduced in the canine RDEB keratinocytes after retroviral correction (Baldeschi et al. 2003; Gache et al. 2004).

Junctional Epidermolysis Bullosa

Junctional epidermolysis bullosa (JEB) (see Fig. 5) in German shorthaired pointers appears to be caused by a mutation in the $\alpha3$ cDNA of laminin-5 (Meneguzzi 2002). The full-length wild-type canine $\alpha3$ cDNA was cloned into a RV vector, and canine JEB keratinocytes were transduced. The protein product was expressed at the basement membrane zone in the keratinocytes and enhanced the proliferative and clonogenic potential compared to the untransduced JEB cells (Meneguzzi 2002).

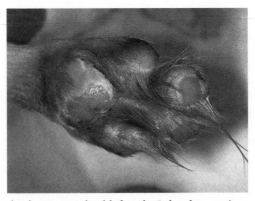

Figure 5. Footpad of a 3-month-old female Labrador retriever with junctional epidermolysis bullosa. Open sores were noted on all palmar and plantar footpad surfaces.

MODELS OF POTENTIAL USE IN GENE THERAPY PROTOCOLS

A large number of genetic diseases described in the dog can be found from searching the relevant literature. Only a small percentage of these potential models are well suited as targets for the evaluation of gene therapy. Many occur as single reports, remain uncharacterized at the genetic level, or have not been developed into colonies, all of which limits use of such models.

ACKNOWLEDGMENTS

The authors acknowledge the support of the National MPS Society, Inc., the Ryan Foundation, the National Foundation for Ectodermal Dysplasias, and the National Institutes of Health (grants 5P40RR002512, 5T32RR007063, DK54481, K01AR049817, and DK066448).

REFERENCES

Acland G.M., Aguirre G.D., Ray J., Zhang Q., Aleman T.S., Cideciyan A.V., Pearce-Kelling S.E., Anand V., Zeng Y., Maguire A.M., et al. 2001. Gene therapy restores vision in a canine model of childhood blindness. *Nat. Genet.* **28:** 92–95.

Arruda V.R., Stedman H.H., Nichols T.C., Haskins M.E., Nicholson M., Herzog R.W., Couto L.B., and High K.A. 2005. Regional intravascular delivery of AAV-2-F.IX to skeletal muscle achieves long-term correction of hemophilia B in a large animal model. *Blood.* **105:** 3458–3464.

Arruda V.R., Schuettrumpf J., Herzog R.W., Nichols T.C., Robinson N., Lotfi Y., Mingozzi F., Xiao W., Couto L.B., and High K.A. 2004. Safety and efficacy of factor IX gene transfer to skeletal muscle in murine and canine hemophilia B models by adeno-associated viral vector serotype 1. *Blood* **103:** 85–92.

Baldeschi C., Gache Y., Rattenholl A., Bouille P., Danos O., Ortonne J.P., Bruckner-Tuderman L., and Meneguzzi G. 2003. Genetic correction of canine dystrophic epidermolysis bullosa mediated by retroviral vectors. *Hum. Mol. Genet.* **12:** 1897–1905.

Bartlett R.J., Stockinger S., Denis M.M., Bartlett W.T., Inverardi L., Le T.T., thi Man N., Morris G.E., Bogan D.J., Metcalf-Bogan J., and Kornegay J.N. 2000. In vivo targeted repair of a point mutation in the canine dystrophin gene by a chimeric RNA/DNA oligonucleotide. *Nat. Biotechnol.* **18:** 615–622.

Bauer T.R., Jr., Gu Y.C., Creevy K.E., Tuschong L.M., Embree L., Holland S.M., Sokolic R.A., and Hickstein D.D. 2004a. Leukocyte adhesion deficiency in children and Irish setter dogs. *Pediatr. Res.* **55:** 363–367.

Bauer T.R., Jr., Creevy K.E., Gu Y.C., Tuschong L.M., Donahue R.E., Metzger M.E., Embree L.J., Burkholder T., Bacher J.D., Romines C., et al. 2004b. Very low levels of donor CD18+ neutrophils following allogeneic hematopoietic stem cell transplantation reverse the disease phenotype in canine leukocyte adhesion deficiency. *Blood* **103:** 3582–3589.

Beaty R.M., Jackson M., Peterson D., Bird A., Brown T., Benjamin D.K., Jr., Juopperi T., Kishnani P., Boney A., Chen Y.T., and Koeberl D.D. 2002. Delivery of glucose-6-phosphatase in a canine model for glycogen storage disease, type Ia, with adeno-associated virus (AAV) vectors. *Gene Ther.* **9:** 1015–1022.

Benson K.F., Li F.Q., Person R.E., Albani D., Duan Z., Wechsler J., Meade-White K., Williams K., Acland G.M., Niemeyer G., Lothrop C.D., and Horwitz M. 2003. Mutations associated with neutropenia in dogs and humans disrupt intracellular transport of neutrophil elastase. *Nat. Genet.* **35:** 90–96.

Bhalerao D.P., Rajpurohit Y., Vite C.H., and Giger U. 2002. Detection of a genetic mutation for myotonia congenita among miniature Schnauzers and identification of a common carrier ancestor. *Am. J. Vet. Res.* **63:** 1443–1447.

Brix A.E., Howerth E.W., McConkie-Rosell A., Peterson D., Egnor D., Wells M.R., and Chen Y.T. 1995. Glycogen storage disease type Ia in two littermate Maltese puppies. *Vet. Pathol.* **32:** 460–465.

Brooks M.B., Gu W., and Ray K. 1997. Complete deletion of factor IX gene and inhibition of factor IX activity in a labrador retriever with hemophilia B. *J. Am. Vet. Med. Assoc.* **211:** 1418–1421.

Brooks M.B., Gu W., Barnas J.L., Ray J., and Ray K. 2003. A Line 1 insertion in the Factor IX gene segregates with mild hemophilia B in dogs. *Mamm. Genome* **14:** 788–795.

Brown B.D., Shi C.X., Powell S., Hurlbut D., Graham F.L., and Lillicrap D. 2004. Helper-dependent adenoviral vectors mediate therapeutic factor VIII expression for several months with minimal accompanying toxicity in a canine model of severe hemophilia A. *Blood* **103:** 804–810.

Cameron C., Notley C., Hoyle S., McGlynn L., Hough C., Kamisue S., Giles A., and Lillicrap D. 1998. The canine factor VIII cDNA and 5′ flanking sequence. *Thromb. Haemostasis* **79:** 317–322.

Chuah M.K., Schiedner G., Thorrez L., Brown B., Johnston M., Gillijns V., Hertel S., Van Rooijen N., Lillicrap D., Collen D., VandenDriessche T., and Kochanek S. 2003. Therapeutic factor VIII levels and negligible toxicity in mouse and dog models of hemophilia A following gene therapy with high-capacity adenoviral vectors. *Blood* **101:** 1734–1743.

Connelly S., Mount J., Mauser A., Gardner J.M., Kaleko M., McClelland A., and Lothrop C.D., Jr. 1996. Complete short-term correction of canine hemophilia A by in vivo gene therapy. *Blood* **88:** 3846–3853.

Cox M.L., Lees G.E., Kashtan C.E., and Murphy K.E. 2003. Genetic cause of X-linked Alport syndrome in a family of domestic dogs. *Mamm. Genome* **14:** 396–403.

Creevy K.E., Bauer T.R., Jr., Tuschong L.M., Embree L.J., Colenda L., Cogan K., Starost M.F., Haskins M.E., and Hickstein D.D. 2003. Canine leukocyte adhesion deficiency colony for investigation of novel hematopoietic therapies. *Vet. Immunol. Immunopathol.* **94:** 11–22.

Desmaris N., Verot L., Ciron C., Puech J.P., Chen F., Ferry N., Caillaud C., Cherel Y., Lajat Y., Vanier M.T., et al. 2004. Correction of neurodegeneration in mouse and dog models of Hurler disease. *Gene Ther.* **11:** S133–S134.

Distler J., Hieber V., Sahagian G., Schmickel R., and Jourdian G.W. 1979. Identification of mannose 6-phosphate in glycoproteins that inhibit the assimilation of beta-galactosidase by fibroblasts. *Proc. Natl. Acad. Sci.* **76:** 4235–4239.

Ehrhardt A., Xu H., Dillow A.M., Bellinger D.A., Nichols T.C., and Kay M.A. 2003. A gene-deleted adenoviral vector results in phenotypic correction of canine hemophilia B without liver toxicity or thrombocytopenia. *Blood* **102:** 2403–2411.

Evans J.P., Brinkhous K.M., Brayer G.D., Reisner H.M., and High K.A. 1989a. Canine hemophilia B resulting from a point mutation with unusual consequences. *Proc. Natl. Acad. Sci.* **86:** 10095–10099.

Evans J.P., Watzke H.H., Ware J.L., Stafford D.W., and High K.A. 1989b. Molecular cloning of a cDNA encoding canine factor IX. *Blood* **74:** 207–212.

Fang B., Wang H., Gordon G., Bellinger D.A., Read M.S., Brinkhous K.M., Woo S.L., and Eisensmith R.C. 1996. Lack of persistence of E1-recombinant adenoviral vectors containing a temperature-sensitive E2A mutation in immunocompetent mice and hemophilia B dogs. *Gene Ther.* **3:** 217–222.

Felsburg P.J., Somberg R.L., Hartnett B.J., Henthorn P.S., and Carding S.R. 1998. Canine X-linked severe combined immunodeficiency. A model for investigating the requirement for the common gamma chain (gamma c) in human lymphocyte development and function. *Immunol. Res.* **17:** 63–73.

Felsburg P.J., Hartnett B.J., Henthorn P.S., Moore P.F., Krakowka S., and Ochs H.D. 1999. Canine X-linked severe combined immunodeficiency. *Vet. Immunol. Immunopathol.* **69:** 127–135.

Felsburg P.J., Somberg R.L., Hartnett B.J., Suter S.F., Henthorn P.S., Moore P.F., Weinberg K.I., and Ochs H.D. 1997. Full immunologic reconstitution following nonconditioned bone marrow transplantation for canine X-linked severe combined immunodeficiency. *Blood* **90:** 3214–3221.

Ferrara M.L., Taylor R.M., Griffin A.D., Hawthorne W.J., Williamson P., and Stewart G.J. 1992. Optimal bone marrow dose for long-term engraftment after total lymphoid irradiation in dogs. *Transplant. Proc.* **24:** 2280–2281.

Ferrara M.L., Occhiodoro T., Fuller M., Hawthorne W.J., Teutsch S., Tucker V.E., Hopwood J.J., Stewart G.J., and Anson D.S. 1997. Canine fucosidosis: A model for retroviral gene transfer into haematopoietic stem cells. *Neuromuscul. Disord.* **7:** 361–366.

Foureman P., Whiteley M., and Giger U. 2002. Canine leukocyte adhesion deficiency: Presence of the Cys36Ser beta-2 integrin mutation in an affected U.S. Irish Setter cross-breed dog and in U.S. Irish Red and White Setters. *J. Vet. Intern. Med.* **16:** 518–523.

Gache Y., Baldeschi C., Del Rio M., Gagnoux-Palacios L., Larcher F., Lacour J.P., and Meneguzzi G. 2004. Construction of skin equivalents for gene therapy of recessive dystrophic epidermolysis bullosa. *Hum. Gene Ther.* **15:** 921–933.

Gallo-Penn A.M., Shirley P.S., Andrews J.L., Tinlin S., Webster S., Cameron C., Hough C., Notley C., Lillicrap D., Kaleko M., and Connelly S. 2001. Systemic delivery of an adenoviral vector encoding canine factor VIII results in short-term phenotypic correction, inhibitor development, and biphasic liver toxicity in hemophilia A dogs. *Blood* **97:** 107–113.

Gallo-Penn A.M., Shirley P.S., Andrews J.L., Kayda D.B., Pinkstaff A.M., Kaloss M., Tinlin S., Cameron C., Notley C., Hough C., Lillicrap D., et al. 1999. In vivo evaluation of an adenoviral vector encoding canine factor VIII: High-level, sustained expression in hemophiliac mice. *Hum. Gene Ther.* **10:** 1791–1802.

Giger U., Boxer L.A., Simpson P.J., Lucchesi B.R., and Todd R.F., III. 1987. Deficiency of leukocyte surface glycoproteins Mo1, LFA-1, and Leu M5 in a dog with recurrent bacterial infections: An animal model. *Blood* **69:** 1622–1630.

Gu W., Brooks M., Catalfamo J., Ray J., and Ray K. 1999. Two distinct mutations cause severe hemophilia B in two unrelated canine pedigrees. *Thromb. Haemostasis* **82:** 1270–1275.

Hacein-Bey-Abina S., von Kalle C., Schmidt M., Le Deist F., Wulffraat N., McIntyre E., Radford I., Villeval J.L., Fraser C.C., Cavazzana-Calvo M., and Fischer A. 2003. A serious adverse event after successful gene therapy for X-linked severe combined immunodeficiency. *N. Engl. J. Med.* **348:** 255–256.

Hai M., Bauer T.R., Gu Y.-C., Tuschong L.M., Sokolic R.A., and Hickstein D.D. 2004. Retroviral-mediated gene transfer of CD18 into hematopoietic stem cells in dogs with canine leukocyte adhesion deficiency reverses the severe deficiency phenotype. *Blood* **104:** 866a.

Harding T.C., Koprivnikar K.E., Tu G.H., Zayek N., Lew S., Subramanian A., Sivakumaran A., Frey D., Ho K., VanRoey M.J., et al. 2004. Intravenous administration of an AAV-2 vector for the expression of factor IX in mice and a dog model of hemophilia B. *Gene Ther.* **11:** 204–213.

Hartnett B.J., Yao D., Suter S.E., Ellinwood N.M., Henthorn P.S., Moore P.E., McSweeney P.A., Nash R.A., Brown J.D., Weinberg K.I., and Felsburg P.J. 2002. Transplantation of X-linked severe combined immunodeficient dogs with CD34+ bone marrow cells. *Biol. Blood Marrow Transplant.* **8:** 188–197.

Harvey S.J., Zheng K., Jefferson B., Moak P., Sado Y., Naito I., Ninomiya Y., Jacobs R., and Thorner P.S. 2003. Transfer of the alpha 5(IV) collagen chain gene to smooth muscle restores in vivo expression of the alpha 6(IV) collagen chain in a canine model of Alport syndrome. *Am. J. Pathol.* **162:** 873–885.

Haskins M.E., Desnick R.J., DiFerrante N., Jezyk P.F., and Patterson D.F. 1984. Beta-glucuronidase deficiency in a dog: A model of human mucopolysaccharidosis VII. *Pediatr. Res.* **18:** 980–984.

Healy P.J., Farrow B.R., Nicholas F.W., Hedberg K., and Ratcliffe R. 1984. Canine fucosidosis: A biochemical and genetic investigation. *Res. Vet. Sci.* **36:** 354–359.

Hennig A.K., Levy B., Ogilvie J.M., Vogler C.A., Galvin N., Bassnett S., and Sands M.S. 2003. Intravitreal gene therapy reduces lysosomal storage in specific areas of the CNS in mucopolysaccharidosis VII mice. *J. Neurosci.* **23:** 3302–3307.

Henthorn P.S., Somberg R.L., Fimiani V.M., Puck J.M., Patterson D.F., and Felsburg P.J. 1994. IL-2R gamma gene microdeletion demonstrates that canine X-linked severe combined immunodeficiency is a homologue of the human disease. *Genomics* **23:** 69–74.

Herzog R.W., Mount J.D., Arruda V.R., High K.A., and Lothrop C.D., Jr. 2001. Muscle-directed gene transfer and transient immune suppression result in sustained partial correction of canine hemophilia B caused by a null mutation. *Mol. Ther.* **4:** 192–200.

Herzog R.W., Fields P.A., Arruda V.R., Brubaker J.O., Armstrong E., McClintock D., Bellinger D.A., Couto L.B., Nichols T.C., and High K.A. 2002. Influence of vector dose on factor IX-specific T and B cell responses in muscle-directed gene therapy. *Hum. Gene Ther.* **13:** 1281–1291.

Herzog R.W., Yang E.Y., Couto L.B., Hagstrom J.N., Elwell D., Fields P.A., Burton M., Bellinger D.A., Read M.S., Brinkhous K.M., et al. 1999. Long-term correction of canine hemophilia B by gene transfer of blood coagulation factor IX mediated by adeno-associated viral vector. *Nat. Med.* **5:** 56–63.

High K.A. 2001a. AAV-mediated gene transfer for hemophilia. *Ann. N.Y. Acad. Sci.* **953:** 64–74.

———. 2001b. Gene transfer as an approach to treating hemophilia. *Circ. Res.* **88:** 137–144.

———. 2002. AAV-mediated gene transfer for hemophilia. *Genet. Med.* **4:** 56S–61S.

———. 2003. Theodore E. Woodward Award. AAV-mediated gene transfer for hemophilia. *Trans. Am. Clin. Climatol. Assoc.* **114:** 337–352.

Hough C., Kamisue S., Cameron C., Notley C., Tinlin S., Giles A., and Lillicrap D. 2002. Aberrant splicing and premature termination of transcription of the FVIII gene as a cause of severe canine hemophilia A: Similarities with the intron 22 inversion mutation in human hemophilia. *Thromb. Haemostasis* **87:** 659–665.

Howell J.M., Lochmuller H., O'Hara A., Fletcher S., Kakulas B.A., Massie B., Nalbantoglu J., and Karpati G. 1998. High-level dystrophin expression after adenovirus-mediated dystrophin minigene transfer to skeletal muscle of dystrophic dogs: Prolongation of expression with immunosuppression. *Hum. Gene Ther.* **9:** 629–634.

Jezyk P.F., Felsburg P.J., Haskins M.E., and Patterson D.F. 1989. X-linked severe combined immunodeficiency in the dog. *Clin. Immunol. Immunopathol.* **52:** 173–189.

Kaplan A., Achord D.T., and Sly W.S. 1977. Phosphohexosyl components of a lysosomal enzyme are recognized by pinocytosis receptors on human fibroblasts. *Proc. Natl. Acad. Sci.* **74:** 2026–2030.

Kay M.A., Landen C.N., Rothenberg S.R., Taylor L.A., Leland F., Wiehle S., Fang B., Bellinger D., Finegold M., Thompson A.R., et al. 1994. In vivo hepatic gene therapy: Complete albeit transient correction of factor IX deficiency in hemophilia B dogs. *Proc. Natl. Acad. Sci.* **91:** 2353–2357.

Kelly W.R., Clague A.E., Barns R.J., Bate M.J., and MacKay B.M. 1983. Canine alpha-L-fucosidosis: A storage disease of Springer spaniels. *Acta Neuropathol.* **60:** 9–13.

Kijas J.M., Bauer T.R., Jr., Gafvert S., Marklund S., Trowald-Wigh G., Johannisson A., Hedhammar A., Binns M., Juneja R.K., Hickstein D.D., and Andersson L. 1999. A missense mutation in the beta-2 integrin gene (ITGB2) causes canine leukocyte adhesion deficiency. *Genomics* **61:** 101–107.

Kishnani P.S., Bao Y., Wu J.Y., Brix A.E., Lin J.L., and Chen Y.T. 1997. Isolation and nucleotide sequence of canine glucose-6-phosphatase mRNA: Identification of mutation in puppies with glycogen storage disease type Ia. *Biochem. Mol. Med.* **61:** 168–177.

Kishnani P.S., Faulkner E., VanCamp S., Jackson M., Brown T., Boney A., Koeberl D., and Chen Y.T. 2001. Canine model and genomic structural organization of glycogen storage disease type Ia (GSD Ia). *Vet. Pathol.* **38:** 83–91.

Lozier J.N., Dutra A., Pak E., Zhou N., Zheng Z., Nichols T.C., Bellinger D.A., Read M., and Morgan R.A. 2002. The Chapel Hill hemophilia A dog colony exhibits a factor VIII gene inversion. *Proc. Natl. Acad. Sci.* **99:** 12991–12996.

Lutzko C., Kruth S., Abrams-Ogg A.C., Lau K., Li L., Clark B.R., Ruedy C., Nanji S., Foster R., Kohn D., Shull R., and Dube I.D. 1999a. Genetically corrected autologous stem cells engraft, but host immune responses limit their utility in canine alpha-L-iduronidase deficiency. *Blood* **93:** 1895–1905.

Lutzko C., Omori F., Abrams-Ogg A.C., Shull R., Li L., Lau K., Ruedy C., Nanji S., Gartley C., Dobson H., Foster R., Kruth S., and Dube I.D. 1999b. Gene therapy for canine alpha-L-iduronidase deficiency: In utero adoptive transfer of genetically corrected hematopoietic progenitors results in engraftment but not amelioration of disease. *Hum. Gene Ther.* **10:** 1521–1532.

Lynch C.M. 1999. Gene therapy for hemophilia. *Curr. Opin. Mol. Ther.* **1:** 493–499.

Mango R.L., Xu L., Sands M.S., Vogler C., Seiler G., Schwarz T., Haskins M.E., and Ponder K.P. 2004. Neonatal retroviral vector-mediated hepatic gene therapy reduces bone, joint, and cartilage disease in mucopolysaccharidosis VII mice and dogs. *Mol. Genet. Metab.* **82:** 4–19.

Mauser A.E., Whitlark J., Whitney K.M., and Lothrop C.D., Jr. 1996. A deletion mutation causes hemophilia B in Lhasa Apso dogs. *Blood* **88:** 3451–3455.

Meneguzzi G. 2002. Gene therapy: The canine models. In *18th Annual Congress of the ESVD/ECVD*, Nice, France.

Meneguzzi G. and Vailly J. 2001. Gene therapy of inherited skin diseases. In *The skin and gene therapy* (ed. U.R. Hengge and B. Volc-Platzer), pp. 97–116. Springer, New York.

Menon K.P., Tieu P.T., and Neufeld E.F. 1992. Architecture of the canine IDUA gene and mutation underlying canine mucopolysaccharidosis I. *Genomics* **14**: 763–768.

Mount J.D., Herzog R.W., Tillson D.M., Goodman S.A., Robinson N., McCleland M.L., Bellinger D., Nichols T.C., Arruda V.R., Lothrop C.D., Jr., and High K.A. 2002. Sustained phenotypic correction of hemophilia B dogs with a factor IX null mutation by liver-directed gene therapy. *Blood* **99**: 2670–2676.

Natowicz M.R., Chi M.M., Lowry O.H., and Sly W.S. 1979. Enzymatic identification of mannose 6-phosphate on the recognition marker for receptor-mediated pinocytosis of beta-glucuronidase by human fibroblasts. *Proc. Natl. Acad. Sci.* **76**: 4322–4326.

Occhiodoro T. and Anson D.S. 1996. Isolation of the canine alpha-L-fucosidase cDNA and definition of the fucosidosis mutation in English Springer spaniels. *Mamm. Genome* **7**: 271–274.

Ohshima T., Murray G.J., Swaim W.D., Longenecker G., Quirk J.M., Cardarelli C.O., Sugimoto Y., Pastan I., Gottesman M.M., Brady R.O., and Kulkarni A.B. 1997. α-Galactosidase A deficient mice: A model of Fabry disease. *Proc. Natl. Acad. Sci.* **94**: 2540–2544.

Phaneuf D., Wakamatsu N., Huang J.Q., Borowski A., Peterson A.C., Fortunato S.R., Ritter G., Igdoura S.A., Morales C.R., Benoit G., et al. 1996. Dramatically different phenotypes in mouse models of human Tay-Sachs and Sandhoff diseases. *Hum. Mol. Genet.* **5**: 1–14.

Ponder K.P., Melniczek J.R., Xu L., Weil M.A., O'Malley T.M., O'Donnell P.A., Knox V.W., Aguirre G.D., Mazrier H., Ellinwood N.M., et al. 2002. Therapeutic neonatal hepatic gene therapy in mucopolysaccharidosis VII dogs. *Proc. Natl. Acad. Sci.* **99**: 13102–13107.

Raper S.E., Chirmule N., Lee F.S., Wivel N.A., Bagg A., Gao G.P., Wilson J.M., and Batshaw M.L. 2003. Fatal systemic inflammatory response syndrome in a ornithine transcarbamylase deficient patient following adenoviral gene transfer. *Mol. Genet. Metab.* **80**: 148–158.

Ray J., Bouvet A., DeSanto C., Fyfe J.C., Xu D., Wolfe J.H., Aguirre G.D., Patterson D.F., Haskins M.E., and Henthorn P.S. 1998. Cloning of the canine beta-glucuronidase cDNA, mutation identification in canine MPS VII, and retroviral vector-mediated correction of MPS VII cells. *Genomics* **48**: 248–253.

Rhodes T.H., Vite C.H., Giger U., Patterson D.F., Fahlke C., and George A.L., Jr. 1999. A missense mutation in canine C1C-1 causes recessive myotonia congenita in the dog. *FEBS Lett.* **456**: 54–58.

Rogers C.S., Vanoye C.G., Sullenger B.A., and George A.L., Jr. 2002. Functional repair of a mutant chloride channel using a trans-splicing ribozyme. *J. Clin. Invest.* **110**: 1783–1789.

Scallan C.D., Lillicrap D., Jiang H., Qian X., Patarroyo-White S.L., Parker A.E., Liu T., Vargas J., Nagy D., Powell S.K., Wright J.F., et al. 2003. Sustained phenotypic correction of canine hemophilia A using an adeno-associated viral vector. *Blood* **102**: 2031–2037.

Schatzberg S.J., Anderson L.V., Wilton S.D., Kornegay J.N., Mann C.J., Solomon G.G., and Sharp N.J. 1998. Alternative dystrophin gene transcripts in golden retriever muscular dystrophy. *Muscle Nerve* **21**: 991–998.

Scriver C.R., Beaudet A.L., Sly W.S., and Valle D. 2001. *Lysosomal storage diseases.* McGraw Hill, New York.

Shull R.M., Helman R.G., Spellacy E., Constantopoulos G., Munger R.J., and Neufeld E.F. 1984. Morphologic and biochemical studies of canine mucopolysaccharidosis I. *Am. J. Pathol.* **114:** 487–495.

Shull R.M., Lu X., McEntee M.F., Bright R.M., Pepper K.A., and Kohn D.B. 1996a. Myoblast gene therapy in canine mucopolysaccharidosis. I: Abrogation by an immune response to alpha-L-iduronidase. *Hum. Gene Ther.* **7:** 1595–1603.

Shull R., Lu X., Dube I., Lutzko C., Kruth S., Abrams-Ogg A., Kiem H.P., Goehle S., Schuening F., Millan C., and Carter R. 1996b. Humoral immune response limits gene therapy in canine MPS I. *Blood* **88:** 377–379.

Silverstein D.C., Carmichael K.P., Wang P., O'Malley T.M., Haskins M.E., and Giger U. 2004. Mucopolysaccharidosis type VII in a German Shepherd dog. *J. Am. Vet. Med. Assoc.* **224:** 553–557.

Skelly B.J., Sargan D.R., Herrtage M.E., and Winchester B.G. 1996. The molecular defect underlying canine fucosidosis. *J. Med. Genet.* **33:** 284–288.

Sleeper M.M., Fornasari B., Ellinwood N.M., Weil M.A., Melniczek J., O'Malley T.M., Sammarco C.D., Xu L., Ponder K.P., and Haskins M.E. 2004. Gene therapy ameliorates cardiovascular disease in dogs with mucopolysaccharidosis VII. *Circulation* **110:** 815–820.

Snyder R.O., Miao C., Meuse L., Tubb J., Donahue B.A., Lin H.F., Stafford D.W., Patel S., Thompson A.R., Nichols T., et al. 1999. Correction of hemophilia B in canine and murine models using recombinant adeno-associated viral vectors. *Nat. Med.* **5:** 64–70.

Somberg R.L., Pullen R.P., Casal M.L., Patterson D.F., Felsburg P.J., and Henthorn P.S. 1995. A single nucleotide insertion in the canine interleukin-2 receptor gamma chain results in X-linked severe combined immunodeficiency disease. *Vet. Immunol. Immunopathol.* **47:** 203–213.

Stoltzfus L.J., Sosa-Pineda B., Moskowitz S.M., Menon K.P., Dlott B., Hooper L., Teplow D.B., Shull R.M., and Neufeld E.F. 1992. Cloning and characterization of cDNA encoding canine alpha-L-iduronidase. mRNA deficiency in mucopolysaccharidosis I dog. *J. Biol. Chem.* **267:** 6570–6575.

Taylor R.M., Farrow B.R., and Stewart G.J. 1992. Amelioration of clinical disease following bone marrow transplantation in fucosidase-deficient dogs. *Am. J. Med. Genet.* **42:** 628–632.

Ting S.S., Hartnett B.J., Linton G.F., Malech H., and Felsburg P.J. 2002. Gene therapy in canine X-linked severe combined immunodeficiency by RD114 pseudotyped oncoretroviral vector. *Blood* **100:** 116a.

Ting S.S., Christopher S., Choi U., DeLeon J., Linton G., Theobald-Whiting N., Malech H., and Felsburg P.J. 2003. In vivo oncoretroviral gene transfer by intravenous injection of RD114-pseudotyped oncoretroviral vector achieves early and significant lymphoid marking in XSCID dogs. *Blood* **102:** 249a.

Wolfe J.H., Sands M.S., Harel N., Weil M.A., Parente M.K., Polesky A.C., Reilly J.J., Hasson C., Weimelt S., and Haskins M.E. 2000. Gene transfer of low levels of beta-glucuronidase corrects hepatic lysosomal storage in a large animal model of mucopolysaccharidosis VII. *Mol. Ther.* **2:** 552–561.

Xu L., Mango R.L., Sands M.S., Haskins M.E., Ellinwood N.M., and Ponder K.P. 2002. Evaluation of pathological manifestations of disease in mucopolysaccharidosis VII mice after neonatal hepatic gene therapy. *Mol. Ther.* **6:** 745–758.

Xu L., Gao C., Sands M.S., Cai S.R., Nichols T.C., Bellinger D.A., Raymer R.A., McCorquodale S., and Ponder K.P. 2003. Neonatal or hepatocyte growth factor-potentiated adult gene therapy with a retroviral vector results in therapeutic levels of canine factor IX for hemophilia B. *Blood* **101:** 3924–3932.

Yanay O., Barry S.C., Katen L.J., Brzezinski M., Flint L.Y., Christensen J., Liggitt D., Dale D.C., and Osborne W.R. 2003. Treatment of canine cyclic neutropenia by lentivirus-mediated G-CSF delivery. *Blood* **102:** 2046–2052.

Zhang W.W., Josephs S.F., Zhou J., Fang X., Alemany R., Balague C., Dai Y., Ayares D., Prokopenko E., Lou Y.C., et al. 1999. Development and application of a minimal-adenoviral vector system for gene therapy of hemophilia A. *Thromb. Haemostasis* **82:** 562–571.

25

Behavioral Genetics of Dog Cognition: Human-like Social Skills in Dogs Are Heritable and Derived

Brian Hare and Michael Tomasello

Max Planck Institute for Evolutionary Anthropology
Leipzig D-04103, Germany

ALMOST EVERYWHERE THERE ARE PEOPLE THERE ARE DOGS, and although many people like dogs, very few behaviorists have found them of much scientific interest. A notable exception is Charles Darwin, who found them exceedingly interesting, and indeed launched *On the Origin of Species* (Darwin 1859) with a flurry of examples outlining variability in domestic animals, including dogs, since nowhere at the time could evidence for descent with modification be more clearly observed than in familiar domestic species (Ritvo 1998).

Recently, a number of cognitive scientists have begun to find dogs interesting as well (Cooper et al. 2003; Miklosi et al. 2004). The reason is that dogs seem to have some special skills for reading human social and communicative behavior. These skills appear to be more flexible—and possibly more human-like—than those of other animals more closely related to humans phylogenetically, such as chimpanzees. This raises the possibility that convergent evolution has occurred: Both *Canis familiaris* and *Homo sapiens* may have evolved some similar (though obviously not identical) social-communicative skills—in both cases adapted for certain kinds of social and communicative interactions with human beings. Cases of convergent evolution potentially provide a unique opportunity for making inferences regarding how heritable traits evolve. If two distantly related species share a similar trait(s), it is possible that this analogous trait(s) arose independently due to a similar evolutionary process. Moreover, it is possible that similar evolutionary processes may have affected similar ontogenetic pathways in order to produce analogous phenotypes. Therefore, if dogs' social skills represent a case of convergent evolution with humans and conclusions are drawn about the system(s) that is affected and the selection pressure(s) that drove such evolution in dogs, it might also be possible to devise ways of testing whether a similar process played a role in shaping human cognition. We review recent research with dogs and other nonhumans which suggest that these kinds of inferences regarding the evolution of human social intelligence are possible from recent comparative studies.

HUMAN-LIKE SOCIAL SKILLS IN DOGS?

The test is necessarily simple (it must be simple, if you are to test a wide range of species). Hide a reward in one of several opaque containers, and then look at or point to that location in an attempt to help the subject find the hidden object. Human infants find this type of task trivially easy from around 14 months of age, as they are just beginning to learn language (Behne et al. 2005). However, perhaps surprisingly, chimpanzees, impressive in solving so many other social problems, show little skill in using such social and communicative cues to solve the task (Tomasello et al. 2003). Meanwhile, give domestic dogs a crack at it and they show impressive flexibility in solving the same problem (Hare et al. 1998; Miklosi et al. 1998).

The majority of results, such as these, examining the use of social and communicative behaviors across species have been obtained using the so-called object-choice paradigm (originally developed for use with primates; Anderson et al. 1995). In the object-choice paradigm a piece of food or an attractive object acts as a reward and is hidden in one of several hiding locations (usually two opaque cups spread ~1 m apart) while the subject watches. However, while the reward is hidden in one container, the other container is sham-baited so that the subject only knows the reward has been hidden, but does know in which container it is hidden. It is only then that the human experimenter tries to help the subject by indicating the reward's location (e.g., pointing or looking at the baited cup). Once the experimenter indicates the reward's location, the subject is allowed to search in one of the two containers. If the subject chooses the baited container first, he is allowed to retrieve his reward, whereas, if he chooses the wrong container, he is only shown where the reward was hidden. This procedure is then repeated 10–20 times in order to compare the individual's performance against chance (e.g., chance is 50% when two hiding locations are used) and to give subjects an opportunity to learn to use the experimenter's indicative behavior if they do not do so spontaneously.

All primates tested in the object-choice task, including both species of chimpanzees (*Pan troglodytes* and *Pan paniscus*) are universally poor (with the exception of some individuals raised with humans; Itakura and Tanaka 1998; Call et al. 2000) at finding hidden food rewards using human social-communicative behaviors. It takes dozens of trials for primates to learn to use such information when provided by a helpful human or conspecific (Anderson et al. 1995; Call et al. 2000). Once one type of cue is mastered, even chimpanzees do not generalize these skills when slightly modified or novel cues are available that closely resemble those they previously learned (Povinelli et al. 1997; Tomasello et al. 1997). For example, a group of chimpanzees trained to choose a container that a human was pointing to by touching it were no longer able to choose the correct cup if a human again pointed to the correct cup without touching it (Povinelli et al. 1997).

The story with dogs is very different. In all six initial studies (Hare et al. 1998; Miklosi et al. 1998, 2000; Hare and Tomasello 1999; Agnetta et al. 2000; McKinley and Sambrook 2000) conducted in three different laboratories, the majority of dogs were spontaneously able to use several different cues to locate hidden food at above chance levels: (1) a human pointing to the target location (including "distal pointing" in which the experimenter stands over a meter away from the target and points in its direction using her cross-lateral hand); (2) a human gazing to the target location (dog either sees the head turn or only a static head looking to a location); (3) a human bowing or nodding to the target location; and (4) a human placing a marker in front of the target location (a totally novel communicative cue). Some dogs were even skillful when the human walked toward the wrong container while pointing in the opposite direction toward the correct container. Similarly, dogs were capable of finding hidden food if a conspecific was oriented toward the baited container. In all of these cases, the dogs used the cues effectively from their very first trials, showing that they came to the experiment already possessing the required skills. Typically in these tests, two-thirds of the dogs were above chance as individuals. Meanwhile, control procedures were always run (i.e., forcing dogs to choose one of the hiding locations without giving any visual cues to the food's location) to ensure that subjects were not simply using olfactory cues to locate the food in all the different tests. In all cases, dogs were unable to find hidden food above chance levels when a visual cue was absent.

Perhaps most surprisingly, when dogs are directly compared to chimpanzees on identical object-choice tasks, they are much more skilled at using a novel social cue to find hidden food (i.e., a human places a block on the correct container conspicuously while the subject watches) (Hare et al. 2002). In other studies, dogs have performed well when they see the block placed but then it is removed before their choice—demonstrating that they are not just attracted to the block itself (Reidel et al. 2005). With respect to gaze cues, dogs—like human infants and unlike chimpanzees (Povinelli et al. 1999)—only use the direction someone is looking (head and eye direction) to locate hidden food if the human is gazing directly at one of two possible hiding locations; they ignore a human's gaze if the human stares into space above the correct hiding location (Soproni et al. 2001). This suggests that dogs identify human communicative behaviors, as opposed to other behaviors, in ways more similarly to human infants than do chimpanzees.

Another line of research has also demonstrated that dogs can assess what a human can or cannot see in a number of other contexts. For example, if a human throws a ball for the dog to fetch and then turns his back, the dog almost always brings the ball back around his body to drop it in front of his face (Hare et al. 1998; Miklosi et al. 2000). In addition, when dogs are forced to choose between two humans with food, they prefer to beg from a human whose head and eyes are visible and are not covered with either a blindfold or a bucket (Gasci et al. 2004; Viranyi et al. 2004; Bishop and Young 2005), something that chimpanzees do not do spontaneously (Povinelli and Eddy 1996). Dogs are also more likely to avoid approaching forbidden food when a human's eyes are open than when they are closed (Call et al. 2003), again, something that chimpanzees do not do spontaneously (Povinelli and Eddy 1996; Kaminksi et al. 2004b). Moreover, dogs also avoid approaching forbidden food when it is visible to the forbidding human through a transparent window in a barrier between the dog and the human. That is, in this situation dogs make the decision not to approach the food at a moment when they cannot see the human and the human cannot see them, apparently projecting what the human can see through the small window (they do approach in the absence of the window) (Brauer et al. 2004).

Finally, it seems that dogs can learn about their environment from humans in a variety of surprising ways that can be either highly communicative or even noncommunicative. For example, when the owner of a particular Border collie tells him to fetch using a novel word ("The dax!"), he reliably retrieves a novel toy instead of a familiar one whose name he already knows (e.g., a ball)—a kind of reasoning by exclusion (if she had wanted the ball she would have asked for "The ball," so the dax must be the other one). Such social inferences have previously been demonstrated only for language-learning human children (Kaminski et al. 2004a; but see Markman and Abelev 2004). Finally, dogs who witness a human or conspecific demonstrator solving simple instrumental and detour tasks are quicker and more successful at those tasks than other dogs who have not witnessed a demonstration (Slabbart and Rasa 1997; Pongracz et al. 2001, 2003; Kubinyi et al. 2003a,b).

One thing that makes the human-like social skills of dogs so fascinating is that dogs do not seem to show unusual flexibility in other, nonsocial domains. For example, dogs are unable to infer the location of a hidden object based on the trajectory of a hider's movements and as the result of their own previous searches (Watson et al. 2001; Bräuer et al. 2004), whereas great apes are capable of such inferences in a number of situations (Call and Carpenter 2001). When dogs have been directly compared to primates in their ability to do such things as make inferences about the location of hidden food based on nonsocial cues—for example, seeing one board lying flat versus another board tilted up as if food could be under it—dogs fail miserably as compared with nonhuman great apes (J. Bräuer et al., unpubl.). Dogs also fail means–end tasks that require them to avoid pulling a string that is not connected to food in favor of one that is connected to the food (Osthaus et al. 2005), a task most primates easily solve (Tomasello and Call 1997). The fact that dogs are so skilled in the social domain but unremarkable in solving physical tasks leads one to wonder how and why dogs have acquired a social specialization for solving problems involving humans.

CONVERGENT COGNITIVE EVOLUTION IN DOGS AND HUMANS?

Hence, if man goes on selecting, and thus augmenting, any peculiarity, he will almost certainly unconsciously modify other parts of the structure, owing to the mysterious laws of the correlation of growth.

DARWIN 1859

The fact that domestic dogs, but not nonhuman apes, possess such human-like social-cognitive skills begs the question of the origin of these abilities. Could it be that the similarities between dogs and humans represent a case of convergent cognitive evolution? There are three obvious explanations for the origin of such specialized social skills in dogs, and they have been explored by comparing the use of basic human social-communicative behaviors (i.e., a point or gaze cues) in the object-choice task both within and between various canid species.

The most straightforward explanation for dogs' special social-cognitive skills with humans is that dogs grow up with unusual exposure to humans and so have an unusual opportunity to learn from them; that is, there is a kind of "enculturation" similar to the kind proposed to explain the unusual problem-solving abilities of apes raised by humans (Call and Tomasello 1996). For example, two different studies have found that two different pairs of adult chimpanzees, who were all raised as infants by humans, spontaneously use a number of social cues when tested in the object-choice task (Itakura and Tanaka 1998; Call et al. 2000). Given that all the tests described thus far were on adult dogs, it is entirely possible that, like human-reared apes, all the observed skill is acquired. This hypothesis predicts that the ability to read human social cues should improve over a dog's lifetime and should vary depending on the amount of exposure a dog has had to humans. However, a cross-sectional comparison of puppies found that different age groups did not differ in their ability to use a human pointing or gaze cue; even puppies as young as 9 weeks old were nearly perfect in the basic tests. In addition, when a group of puppies in an obedience class were compared in their ability to use pointing and gazing cues to a group of litter-reared puppies (who had relatively little exposure to humans), both groups were skilled at using the cues and there was no difference between them (Hare et al. 2002). This is, of course, not to say that learning cannot occur during a dog's life. For example, there is evidence for learning from a comparison between active and pet working dogs. Although both groups of dogs were quite skilled at using human social cues, active working dogs were more successful than pet working dogs at using a pointing cue in the object-choice task (McKinley and Sambrook 2000). However, taken together, these findings do not support the hypothesis that dogs require unusual amounts of exposure to humans (i.e., relative to most primates tested) in order to learn to read human social and communicative behavior, although, at the same time, they do not rule out that such exposure can shape such abilities.

A second candidate hypothesis invokes the canid ancestry of dogs. Amazingly, a convergence of genetic comparisons shows that all the phenotypic variance observed in dogs is inherited from a single species: the wolf (*Lupus lupus*). Comparisons between wolf and dog mitochondrial control regions suggest that dogs arose through multiple domestication events within multiple wolf populations (Vila et al. 1997; Savalainen et al. 2002). A comparison of mitochondrial DNA in a worldwide sample of dogs suggests that initial domestication events may have occurred in East Asia. In particular, Chinese wolves that are relatively small and docile (compared to their larger European and American relatives) may have been repeatedly domesticated in a number of locations throughout East Asia (Savolainen et al. 2002). Perhaps then dogs, as direct ancestors of wolves, simply inherited their skill for reading the behavior of others from wolves. Indeed, wolves are social pack hunters and likely need to read the social cues of their fellow hunters as well as prey in order to effectively cooperate while hunting, providing a plausible explanation for the skill observed in their dog descendants (Clutton-Brock 1999; Coppinger and Coppinger 2001). This hypothesis predicts that the social skills dogs exhibit should also be seen if the same types of tests are used to examine wolves (Hare and Tomasello 1999). However, the wolves tested thus far are not particularly skilled at reading human social cues; they are much more like chimpanzees than dogs in this respect. Two independent studies have found that captive wolves reared by humans (and tested by the humans who reared them) are not as skilled as dogs in using human social cues to find hidden food (Hare et al. 2002; Miklosi et al. 2003). Importantly, this difference is unlikely due to the wolves' being poor in all such human-led cognitive tasks. Indeed, studies comparing wolves and dogs in simple nonsocial problem-solving or memory tasks typically find that wolves perform as well if not better than dogs (in some cases these were the same wolves and dogs compared in the social tasks; Frank 1980, 1982; Frank et al. 1989; Hare et al. 2002). Instead, it seems that for some reason wolves just relate to people very differently than do dogs. For example, when human-reared dogs and wolves were both presented with an impossible task (opening a locked box with food inside) the dogs almost immediately gave up and gazed alternately at the human and the box, whereas the wolves continued to try to solve the task on their own until the test ended (Miklosi et al. 2003). Therefore, it seems that, when uncertain, dogs want or expect human help, whereas wolves behave as if they do not want human help or realize that a human might help them. Overall then, the comparisons made thus far do not provide any support for the idea that dogs inherited their social-cognitive skills directly from wolves.

This leaves the possibility that dogs' social-cognitive skills evolved during the process of domestication: that is, these skills evolved during the tens of thousands of years that our two species have lived together. In some sense, this seems an unlikely explanation. Is it really plausible that during the same period that dogs underwent a 25% reduction in cranial capacity (as compared to wolves) they also became more skilled at solving social problems involving humans? In fact, it has often been suggested that if domestication has any effect on the cognitive abilities of domesticates it is to dull their problem-solving abilities (for review, see Kruska 2004). Perhaps most problematic is that even if such evolution took place during domestication, how would one test for such a possibility and how could any test determine with any precision or certainty the selection pressure(s) that drove such evolution during domestication? Unfortunately, until very recently, there was no way to systematically compare different dog breeds in order to test such a hypothesis. We just knew too little about the breeding history of dogs over the past millennia to have any confidence in within-species comparisons. Luckily, dogs are not the only domesticated canids.

In 1959, Dr. Dmitry Belyaev and his colleagues in Siberia, Russia, began one of the longest-running experiments in history with the goal of studying the behavioral genetics of domestication (Belyaev 1979). Since the start of the experiment, two separate populations of foxes (*Vulpes vulpes*) have been maintained. An experimental population was selectively bred based on a single criterion—whether they fearlessly and nonaggressively approached a human. The second population was maintained as a control and has been bred randomly with respect to their behavior toward humans (Trut 1999, 2001).

Based on the single selection criterion the Russians used, behavioral, physiological, and morphological changes were observed in the experimental fox population that were either not found in control foxes or were expressed at a significantly higher frequency in the experimental population (a number of these differences appeared after only 20 generations of selection). First, not surprisingly, the individuals in the experimental population were observed from 1 month of age onward to show little fear or aggression toward humans. Subsequent experiments have demonstrated that selection for tameness retarded the development of fear responses to novelty while enhancing exploratory behavior in unfamiliar situations (Plyusnina et al. 1991). Second, physiological changes were observed after several generations of selection. The first change detected was attenuated activity of the pituitary–adrenal axis. The common pool of circulating glucocorticoids, their in vitro production, the basal level of the adrenocorticotrophic hormone (ACTH), and the adrenal response to stress were all reduced in the experimental foxes relative to controls (Oskina 1996). Subsequently, the activities of the serotonin, noradrenaline, and dopamine transmitter systems in specific brain regions that are implicated in the regulation of emotional-defensive responses have also been found to be altered in the experimental foxes (Popova et al. 1991; Trut et al. 2000). Finally, and perhaps most surprising, the domesticated foxes manifested a number of correlated changes that are commonly found in a number of other domesticated species. A higher frequency of domesticated foxes have floppy ears, short or curly tails, depigmentation of hair, extended reproductive seasons, and even changes in the size and shape of the crania and dentition (Trut 1999, 2001; Trut et al. 2000)—traits commonly found in other domesticated mammals (Kruska 2005). Thus, the results of the fox farm experiment demonstrate with unprecedented certainty that selection against aggressive and fearful behavior likely has been the driving force behind the heritable changes observed in domesticated mammals. In addition, the two populations of foxes provide a perfect test of the hypothesis that dogs evolved their unusual social skills not only during domestication, but also as a direct result of domestication. If the process of domesticating dogs is responsible for the evolution of their social-cognitive specialization, then the experimentally domesticated foxes should perform much like dogs when tested in the object-choice task, whereas the control foxes should perform more like wolves (and nonhuman primates).

Indeed, when fox kits from the domesticated population were compared with age-matched dog puppies on the basic pointing and gaze-following tests, the foxes were as skilled as the dogs in using the human social cues. In addition, to test whether the experimental foxes' skill was a result of the selection regime, a group of experimental fox kits were compared to an age-matched group of control foxes in two different social tasks. One of the tasks compared their ability to find hidden food using a human pointing cue in a standard object-choice task, and the other task compared their preferences for playing with a toy a human gestured toward and touched as opposed to a toy the human ignored (no food reward was given in the second test regardless of a subject's response). Not only were the experimental kits more skilled than control kits at using a human gesture to find hidden food, but they also were more likely to play with the toy a human had gestured toward than were control kits. As in the case of the comparisons between wolves and dogs, these differences are unlikely to be due to differences in rearing histories between the two populations or an inability of the control foxes to be skillful in any human-led task. First, both groups of foxes were reared in an identical manner in which they had negligible contact with humans before testing. Second, the control foxes were as skillful (if not more so) as the experimental foxes on a nonsocial task even though a human experimenter was present throughout the testing (Hare et al. 2005).

These findings with this special population of foxes seem to confirm the likelihood that dogs' unusual ability to read human social-communicative behaviors evolved not only during the process of domestication, but also as a result of domestication—a process of selecting against individuals with fear and aggression toward humans while selecting for those individuals with interest in humans. Perhaps most surprisingly, however, the research with domesticated foxes also suggests that dogs' skills for reading human social-communicative behavior may have initially evolved as an incidental by-product of selection for tame behavior. It is important to note that neither population of foxes was ever bred or tested on the basis of their ability to use human communicative gestures or behaviors; the single selection criterion was simply based on their approach behavior toward a human in a totally noncommunicative context (i.e., there was nothing to find or choose between and thus nothing about which to communicate during the selection trials).

In addition, the fox results raise the possibility that the unusual social skills of dogs may have evolved in stages. It may have been that dogs' special social-cognitive skills, like the foxes', first appeared after systems mediating fear and aggression were altered, systems which normally are not thought of as cognitive systems at all. Once this initial evolution occurred so that dogs were able to solve a new set of social problems involving humans (i.e., exapted), it may have been then that variance in social cognitive abilities came under direct selection. However, such evolution (if it indeed happened at all) would not have been possible without the initial selection on systems controlling emotional reactivity that allowed dogs to enter an entirely new adaptive space, one in which they interacted with humans in similar ways as they do with conspecifics. An exciting evolutionary scenario, indeed, and one that might even be implicated more broadly than the current case of observed cognitive evolution in canids, but once again, we are left with the task of coming up with a way to test this stage model. How are we to test such an idea using dogs, if we remain without any solid evidence about the selective history of the different dog breeds?

Timing is on our side. Thanks to a brand-new phylogenetics study that provides the first reliable data for mapping out the relationship among 85 breeds of domestic dogs, we have conducted an initial test of the two-stage model of cognitive evolution in canids (Parker et al. 2004). Indeed, these new phylogenetics data seem to provide tentative support for the stage theory of dog evolution in that nine dog breeds (chow chow, husky, malamute, shar-pei, Shibu-Inu, Akita, Afghan hound, saluki, and basenji) were identified as being more genetically wolf-like than the remaining breeds examined. This raises the specter that these breeds remain more wolf-like because they were "left behind" and did not undergo a second wave of selection as did the rest of the dog breeds sampled. Perhaps then, the first stage of dog domestication occurred due to natural selection as the least fearful East Asian wolves were at a selective advantage in exploiting a new niche created by garbage left near human settlements—a domestication event that would be highly similar to that observed in experimental populations of foxes. Then a second stage possibly occurred within the last millennium as humans began intensively and intentionally selecting individuals for their appearance or working ability—with those dogs best at reading human social-communicative behaviors having a fitness advantage when being bred to actively work with humans (Coppinger and Coppinger 2001). This stage hypothesis for selection during domestication predicts that whereas all domestic dogs will likely be skilled at reading social cues relative to wolves, dogs that are thought to have been bred for the ability to communicate with humans for the purpose of cooperating (after controlling for the degree to which they are genetically wolf-like) should be more flexible in their ability to use human communicative cues (i.e., selection on communicative ability leads to increases of skills and this ability segregates independently of other heritable traits under selection).

To test this hypothesis, we recently compared the ability of four groups of dog breeds for their ability to use human social-communicative behaviors in the ubiquitous object-choice task. We chose our breed groups based on whether they were considered working or nonworking breeds (according to the American Kennel Club) and whether they had been determined to be more or less wolf-like genetically: (1) nonworking more genetically wolf-like: basenji and New Guinea singing dogs, (2) working more genetically wolf-like: huskies, (3) nonworking less genetically wolf-like: a variety of toy dogs, and (4) working less genetically wolf-like: golden and Labrador retrievers. As the two-stage hypothesis predicts, although all the groups of dog breeds were skilled at using human social-communicative cues to find hidden food, the dogs—regardless of the genetic relation with wolves—thought to be bred in order to communicate in cooperative interactions with humans (retrievers and huskies) were the most skilled overall at using all the social cues when tested. These findings suggest that the ability to read human communicative behavior may have been under direct selection in these breeds. In addition, this finding is consistent with the idea that there is a heritable component to this phenotypic trait that may segregate independently of other traits that have been under selection. As increasingly sensitive phylogenetic comparisons are made between differing dog breeds, behaviorists will increasingly be able to compare breeds in a meaningful way to test various hypotheses for how genetic differences may code for meaningful behavioral and even cognitive differences.

In summary, comparative studies within and between canid species seem to suggest that dogs have evolved a social-cognitive specialization not only during domestication, but as a result of domestication. Initial comparisons between dog breeds seem to support the idea that an initial wave of selection on emotional reactivity may have been followed by further selection and heritable change resulting in a further specialized ability to use human communicative gestures in non-wolf-like working dogs. Future studies will be needed to test such an idea more fully. Finally, because domestication has resulted in the evolution of unusual social-cognitive skills in dogs, it now seems fair to characterize such skills as being convergent with those that seem to have evolved in our own lineage (i.e., skills that among primates are only robustly expressed in humans). Are there then lessons to be learned about human evolution from research on dog cognition?

IMPLICATIONS FOR HUMAN COGNITIVE EVOLUTION

These findings with dogs and their close relatives suggest that some aspects of human-like social intelligence may potentially evolve not as adaptations specifically, but rather as a result of selection on seemingly unrelated social-emotional systems—underlain not by the neocortex but by the limbic and endocrine systems. These findings suggest the further possibility that some of humans' social-cognitive and communicative abilities may have begun evolving in a similar way. That is, following this line of reasoning, one might seriously entertain the hypothesis that an important first step in the evolution of modern human societies was a kind of self-domestication (selection on systems controlling emotional reactivity) in which a human-like temperament was selected (e.g., individuals of a social group either killed or ostracized those who were overaggressive or despotic; [Boehm 1999; Leach 2003; Wrangham 2005]). Thus, much like domestic dogs, this selection for more tame forms of emotional reactivity put our hominid ancestors in a new adaptive space (where new human-like forms of social interaction and communication came into existence), enabling subsequent selection to favor variance that allowed for complex forms of these new social interactions and communication that characterize human cultural life today.

In support of the hypothesis that human temperament evolution preceded significant social cognitive evolution in our species is the fact that forms of potential cooperative behavior in chimpanzees are highly constrained by their levels of inter-individual tolerance (i.e., likely controlled by systems mediating levels of emotional reactivity). Whereas humans (including young infants) are quite helpful and tolerant of other unfamiliar humans when having to work together to accomplish a common goal (Tomasello et al. 2005; F. Warneken et al., in prep.), even familiar chimpanzees are unable to work together in the simplest of cooperative tasks except in a very constrained set of social situations. For example, chimpanzees are only able to work together successfully in an instrumental task (i.e., pulling two ropes simultaneously to retrieve a box full of food that is too heavy for any single individual to obtain) if they are (1) paired with another chimpanzee with whom they have previously shared food, (2) the food is sharable, and (3) there is enough working space so individuals are out of each other's reach (A. Melis et al., in prep.). If such social criteria are not met, then chimpanzees (subordinate individuals of any pair in particular) will refuse to ever cooperate even over a span of dozens of trials. It seems from such tests that subordinate chimpanzees are simply not willing to risk being physically attacked by intolerant dominants, and dominants are not able to control their aggression toward subordinates trying to obtain food—even if it means they will never receive any food. Chimpanzees, it seems, just do not have the temperament for human-like flexibility in their cooperative repertoire.

Such an emotional reactivity hypothesis should then also apply to a wide range of social problems. For example, such a hypothesis would seem to predict that chimpanzees should be more skillful at using social-communicative cues in the object-choice task if the task can be made more emotionally engaging. Indeed, chimpanzees demonstrate spontaneous skill at finding food when they see a human, who has previously established a competitive relation with them, reaching unsuccessfully toward a potential food location in an apparent attempt to obtain hidden food. However, if the same chimpanzees see a helpful human pointing (i.e., morphologically similar to reaching) toward the hiding location, they do not use the cue (Hare and Tomasello 2004). This finding suggests that chimpanzees only view humans as relevant to their success when competing—a context in which chimpanzees seem particularly engaged emotionally (i.e., the human might eat the food!). Otherwise it seems, in any other context, chimpanzees view their potential social partner as largely irrelevant to their success in the object-choice task. Therefore, it may be that chimpanzees do not demonstrate human-like skills at using social-communicative behaviors in the object-choice task because they lack a human-like temperament for sharing information—especially when a prized piece of food is on the line. Arguably then, the production and comprehension of cooperative communicative intentions—intended merely to inform others of things, with no direct benefit to oneself—is one of the species-specific social-cognitive abilities of human beings (the one possible exception being that of dogs). In addition, it would seem that the evolution of this unique ability must have been preceded by temperament evolution.

Therefore, taken together, comparisons between chimpanzees and humans seem to suggest that prerequisite for flexible forms of human cooperation and communication is a human-like temperament. In fact, the evolution of the human temperament may have necessarily preceded the evolution of more complex forms of social cognition that make us uniquely human (e.g., a more sophisticated theory of others' behavior or mental states would be of little use when cooperating if individuals are unable to share the rewards of their joint effort). It is then only once the human temperament evolved that variance in more complex forms of communicative and cooperative behaviors could have been shaped by evolution into unique forms of cognition present in our species today.

It seems then, based on these comparative findings between humans and chimpanzees, that careful behavioral and physiological studies are now needed to understand more precisely how humans and chimpanzees may differ in the systems that mediate their emotional reactivity, if we are to understand more precisely what it is about the human temperament that might have made such evolution possible. However, at this early stage of this research it is important to recognize that it was the comparative studies of canid social cognition which made this model of human cognitive evolution seem not only plausible, but worth pursuing further.

SUMMARY: DOGS AND THE FUTURE OF BEHAVIORAL GENETICS

In this chapter, we have reviewed a fast-growing literature on dog cognition and an area of research that promises to be an exciting avenue for gaining not only a better understanding about what the animal mind is capable of, but also more about the process by which intelligence evolves in mammals. The work thus far leads to the conclusion that dogs have evolved a social-cognitive specialization that allows them unusual skill in cooperating and communicating with humans. In addition, the latest research suggests that this specialization evolved not only during the process of domestication, but also as an incidental result of domestication (i.e., the specialization is an unintentional and unselected by-product of selection for tame behavior). However, there are now some data possibly implicating direct selection on communicative ability in dogs. Comparisons between different breed groups (determined by phylogenetic analysis) suggest that there may have been intentional selection by humans on social-communicative skills during the last millennia on certain breeds of working dogs. Taken together, this work with dogs has had important implications for thinking about how the cognitive abilities of our own species may have evolved.

Overall then, it would seem that our canine companions have come to join into the human conversation in some unusually human-like ways. It is hoped that further investigation of the ways in which dogs do and do not do this—and how they come to have their special skills—will provide us with further insights into the evolutionary processes leading to human-style cooperative interactions and communication. For example, future research with canids should potentially provide behavioral geneticists with the opportunity of identifying and exploring fully the ontogenetic pathway(s) that has been under selection and allow for social cognitive evolution—perhaps even those that are in some way involved in making our own species human.

ACKNOWLEDGMENT

The research of B.H. is supported by a Sofja Kovaleveskaja award received from the Alexander von Humboldt Foundation and the German Federal Ministry for Education and Research.

REFERENCES

Agnetta B., Hare B., and Tomasello M. 2000. Cues to food location that domestic dogs (*Canis familiaris*) of different ages do and do not use. *Anim. Cognition* **3:** 107–112.

Anderson J.R., Sallaberry P., and Barbier H. 1995. Use of experimenter-given cues during object-choice tasks by capuchin monkeys. *Anim. Behav.* **49:** 201–208.

Behne T., Carpenter M., and Tomasello M. 2005. One-year-olds comprehend the communicative intentions behind gestures in a hiding game. *Dev. Sci.* (in press).

Belyaev D. 1979. Destabilizing selection as a factor in domestication. *J. Hered.* **70:** 301–308.

Bishop S. and Young R. 2005. Do domestic dogs understand the importance of eye contact with humans? *Anim. Cognition* (in press).

Boehm C. 1999. *Hierarchy in the forest: The evolution of egalitarian behavior.* Harvard University Press, Cambridge, Massachusetts.

Bräuer J., Call J., and Tomasello M. 2004. Visual perspective taking in dogs (*Canis familiaris*) in the presence of barriers. *Appl. Anim. Behav. Sci.* **88:** 299–317.

Call J. and Carpenter M. 2001. Do apes and children know what they have seen? *Anim. Cognition* **4:** 207–220.

Call J. and Tomasello M. 1996. The effect of humans on the cognitive development of apes. In *Reaching into thought* (ed. A.E. Russon et al.), pp. 371–403. Cambridge University Press, Cambridge, United Kingdom.

Call J., Agnetta B., and Tomasello M. 2000. Cues that chimpanzees do and do not use to find hidden objects. *Anim. Cognition* **3:** 23–34.

Call J., Brauer J., Kaminski J., and Tomasello M. 2003. Domestic dogs (*Canis familiaris*) are sensitive to the attentional state of humans. *J. Comp. Psychol.* **117:** 257–263.

Clutton-Brock A. 1999. *A natural history of domesticated mammals.* Cambridge University Press, Cambridge, United Kingdom.

Cooper J.J., Ashton C., Bishop S., West R., Mills D. S., and Young R. J. 2003. Clever hounds: Social cognition in the domestic dog (*Canis familiaris*). *Appl. Anim. Behav. Sci.* **81:** 229–244.

Coppinger R. and Coppinger L. 2001. *Dogs: A startling new understanding of canine origin, behavior and evolution.* Scriber Press, New York.

Darwin C. 1859. *On the origin of species.* John Murray, London.

Frank H. 1980. Evolution of canine information processing under conditions of natural and artificial selection. *Tierpsychologica* **59:** 389–399.

———. 1982. On the effects of domestication on canine social development and behaviour. *Appl. Anim. Ethol.* **8:** 507–525.

Frank H., Frank M.G., Hasselbach L.M., and Littleton D.M. 1989. Motivation and insight in wolf (*Canis lupus*) and Alaskan Malemute (*Canis familiaris*): Visual discrimination and learning. *Bull. Psychonomic Soc.* **27:** 455–458.

Gasci M., Miklósi A., Varga O., Topál J., and Csányi V. 2004. Are readers of our face readers of our minds? Dogs (*Canis familiaris*) show situation-dependent recognition of human's attention. *Anim. Cognition* **7:** 144–153.

Hare B. and Tomasello M. 1999. Domestic dogs (*Canis familiaris*) use human and conspecific social cues to locate hidden food. *J. Comp. Psychol.* **113:** 173–177.

———. 2004. Chimpanzees are more skillful at competitive than cooperative cognitive tasks. *Anim. Behav.* **68:** 571–581.

Hare B., Call J., and Tomasello M. 1998. Communication of food location between human and dog (*Canis familiaris*). *Evol. Commun.* **2:** 137–159.

Hare B., Brown M., Williamson C., and Tomasello M. 2002. The domestication of social cognition in dogs. *Science* **298:** 1634–1636.

Hare B., Plyusnina I., Ignacio N., Schepina O., Stepika A., Wrangham R., and Trut L. 2005. Social cognitive evolution in captive foxes is a correlated by-product of experimental domestication. *Curr. Biol.* **15:** 226–230.

Itakura S. and Tanaka M. 1998. Use of experimenter-given cues during object-choice tasks by chimpanzees (*Pan troglodytes*), an orangutan (*Pongo pygmaeus*), and human infants (*Homo sapiens*). *J. Comp. Psychol.* **112:** 119–126.

Kaminski J., Call J., and Fischer J. 2004a. Word learning in a domestic dog: Evidence for "fast mapping". *Science* **304:** 1682–1683.

Kaminski J., Call J., and Tomasello M. 2004b. Body orientation and face orientation: Two factors controlling apes' begging behavior from humans. *Anim. Cognition* **7:** 216–223.

Kruska D.C. 2005. On the evolutionary significance of encephalization in some eutherian mammals: Effects of adaptive radiation, domestication and feralization. *Brain Behav. Evol.* **65:** 73–108.

Kubinyi E., Miklósi A., Topál J., and Csányi V. 2003a. Social mimetic behaviour and social anticipation in dogs: Preliminary results. *Anim. Cognition* **6:** 57–63.

Kubinyi E., Topál J., Miklósi A., and Csányi V., 2003b. Dogs (*Canis familiaris*) learn from their owners via observation in a manipulation task. *J. Comp. Psychol.* **117:** 156–165.

Leach H. 2003. Human domestication reconsidered. *Curr. Anthropol.* **44:** 349–368.

Markman E. and Abelev M. 2004. Word learning in dogs? *Trends Cognit. Sci.* **8:** 479–481.

McKinley J. and Sambrook T. 2000. Use of human-given cues by domestic dogs (*Canis familiaris*) and horses (*Equus caballus*). *Anim. Cognition* **3:** 13–22.

Miklosi A., Topál J., and Csányi V. 2004. Comparative social cognition: What can dogs teach us? *Anim. Behav.* **67**: 995–1004.

Miklosi A., Polgárdi R., Topál J., and Csányi V. 1998. Use of experimenter-given cues in dogs. *Anim. Cognition* **1**: 113–121.

———. 2000. Intentional behavior in dog-human communication: An experimental analysis of "showing" behavior in the dog. *Anim. Cognition* **3**: 159–166.

Miklosi A., Kubinyi E., Topál J., Gácsi M., Virányi Z., and Csányi V. 2003. A simple reason for a big difference: Wolves do not look back at humans, dogs do. *Curr. Biol.* **13**: 763–766.

Oskina I. 1996. Analysis of the function state of the pituitary-adrenal axis during postnatal development of domesticated silver foxes (*Vulpes vulpes*). *Scientifur* **20**: 159–161.

Osthaus B., Lea S.E.G., and Slater A.M. 2005. Dogs (*Canis familiaris*) fail to show understanding of means end connections in a string pulling task. *Anim. Cognition* **8**: 37–47.

Parker H.G., Kim L.V., Sutter N.B., Carlson S., Lorentzen T.D., Malek T.B., Johnson G.S., DeFrance H.B., Ostrander E.A., and Kruglyak L. 2004. Genetic structure of the purebred domestic dog. *Science* **304**: 1160–1164.

Plyusnina I.Z., Oskina I.N., and Trut L.N. 1991. An analysis of fear and aggression during early development of behavior in silver foxes (*Vulpes vulpes*). *Appl. Anim. Behav. Sci.* **32**: 253–268.

Pongracz, P., Miklósi A., Kubinyi E., Topál J., and Csányi V. 2003. Interaction between individual experience and social learning in dogs. *Anim. Behav.* **65**: 595–603.

Pongracz, P., Miklósi A., Kubinyi E., Gurobi K., Topál J., and Csányi V. 2001. Social learning in dogs: The effect of a human demonstrator on the performance of dogs in a detour task. *Anim. Behav.* **62**: 1109–1117.

Popova N., Voitenko N.N., Kulikov A.V., and Avgustinovich D.F. 1991. Evidence for the involvement of central serotonin in the mechanism of domestication of silver foxes. *Pharmacol. Biochem. Behav.* **40**: 751–756.

Povinelli D.J. and Eddy T.J. 1996. What young chimpanzees know about seeing. *Monogr. Soc. Res. Child Dev.* **61**: 1–152.

Povinelli D., Bierschwale D.T., and Cech C.G. 1999. Comprehension of seeing as a referential act in young children but not juvenile chimpanzees. *Br. J. Dev. Psychol.* **17**: 37–60.

Povinelli D. J., Reaux J.E., Bierschwale D.T., Allain A.D., and Simon B.B. 1997. Exploitation of pointing as a referential gesture in young children, but not adolescent chimpanzees. *Cogn. Dev.* **12**: 423–461.

Reidel J., Buttelmann D., Call J., and Tomasello M. 2005. Domestic dogs (*Canis familiaris*) use a physical marker to find hidden food. *Anim. Cognition* (in press).

Ritvo H. 1998. Foreword. In *The variation of animals and plants under domestication* (ed. C. Darwin), pp. v-xiii. John Hopkins University Press, Baltimore, Maryland.

Savolainen P., Zhang Y-P., Luo J., Lundeberg J., and Leitner T. 2002. Genetic evidence for an East Asian origin of domestic dogs. *Science* **298**: 1610–1613.

Slabbart J. and Rasa O. 1997. Observational learning of an acquired maternal behaviour pattern by working dog pups: An alternative training technique. *Appl. Anim. Behav. Sci.* **53**: 309–316.

Soproni K., Miklosi A., Topál J., and Csányi V. 2001. Comprehension of human communicative signs in pet dogs (*Canis familiaris*). *J. Comp. Psychol.* **115**: 122–126.

Tomasello M. and Call J. 1997. *Primate cognition.* Oxford University Press, New York.

Tomasello M., Call J., and Gluckman A. 1997. Comprehension of novel communicative signs by apes and human children. *Child Dev.* **68:** 1067–1080.

Tomasello M., Call J., and Hare B. 2003. Chimpanzees understand the psychological states of others–the question is which ones and to what extent. *Trends Cogn. Sci.* **7:** 153–157.

Tomasello M., Carpenter M., Call J., and Moll H. 2005. Understanding and sharing intentions: The origins of cultural cognition. *Behav. Brain Sci.* (in press).

Trut L. 1991. Intracranial allometry and morphological changes in silver foxes (*Vulpes vulpes*) under domestication. *Genetika* **27:** 1605–1611.

———. 1999. Early canid domestication: The farm-fox experiment. *Am. Sci.* **87:** 160–169.

———. 2001. Experimental studies of early canid domestication. In *The genetics of the dog* (ed. A Ruvinsky and J. Sampson), pp. 15–43. CABI Publishing, New York.

Trut, L., Plyusnina I.Z., Kolesnikova L.A., and Kozlova O.N., 2000. Interhemispheral neurochemical differences in brains of silver foxes selected for behavior and the problem of directional asymmetry. *Genetika* **36:** 942–946.

Vila C., Savolainen P., Maldonado J.E., Amorim I.R., Rice J.E., Honeycutt R.L., Crandall K.A., Lundeberg J., and. Wayne R.K. 1997. Multiple and ancient origins of the domestic dog. *Science* **276:** 1687–1689.

Viranyi Z., Topál J., Gacsi M., Miklósi A., and Csányi V. 2004. Dogs respond appropriately to cues of humans' attentional focus. *Behav. Processes* **66:** 161–172.

Watson J.S., Gergely G., Csányi V., Topál J., Gacsi M., and Sarkozi Z. 2001. Distinguishing logic from association in the solution of an invisible displacement task by children (*Homo sapiens*) and dogs (*Canis familiaris*): Using negation of disjunction. *J. Comp. Psychol.* **115:** 219–226.

Wrangham R. 2005. *The cooking ape.* Harvard University Press, Cambridge, Massachusetts. (In press.)

The Genetics of Domesticated Behavior in Canids: What Can Dogs and Silver Foxes Tell Us about Each Other?

Anna V. Kukekova and Gregory M. Acland
James A. Baker Institute for Animal Health
College of Veterinary Medicine, Cornell University
Ithaca, New York 14853

Irina N. Oskina, Anastasia V. Kharlamova, and Lyudmila N. Trut
Institute of Cytology and Genetics
Siberian Branch, Russian Academy of Sciences
Novosibirsk 630090, Russia

Kevin Chase and Karl G. Lark
Department of Biology, University of Utah
Salt Lake City, Utah 84112-0840

Hollis N. Erb
Department of Animal Science, College of Veterinary Medicine
Cornell University, Ithaca, New York 14850

Gustavo D. Aguirre
Section of Medical Genetics, Department of Clinical Studies
School of Veterinary Medicine
University of Pennsylvania, Philadelphia, Pennsylvania 19104-6010

DOMESTICATED ANIMALS DISPLAY DISTINCT DIFFERENCES in behavior, morphology, and physiology from their wild counterparts. They usually exhibit reduced aggressiveness, increased social tolerance among conspecifics, and reduced sensitivity to environmental changes (for review, see Sachser 1998; Price 2002). The complex suite of modified social behaviors and cognitive abilities that results from domestication specifically includes tameness, the ability to interact with humans in a positive way. An understanding of the genetics of tameness in canids should provide new insight into the mechanisms underlying social behavior in canids and other species, and into the broader phenomenon of domestication as a whole. Herein, we discuss strategies for identification of the genetic roots of tame behavior in canids. For the purposes of this chapter, we refer to this friendly, dog-like behavior as tameness.

The mutual ability of dogs and humans to communicate, and to interact socially with each other, is so commonplace that it is often not recognized for how strange a phenomenon it represents. The sentiment "She is almost human—just cannot speak" would be immediately understood by anyone familiar with the relationship between dogs and their human companions. Without a word from them, we understand our dogs very well. We can read their body language, facial expressions, habits, and emotional states; and they can read ours. They can understand our mood, intent, tone of voice, and words from our language. Sometimes it is hard to credit that the modern dog (*Canis familiaris*) is such a recent and close relative of the gray wolf (*Canis lupus*), a species that innately and strongly prefers to avoid human contact from a very early age. The domestication of wolves took place at least 12,000 years ago and produced an animal with unique social abilities. Dogs are not just domesticated animals bred under human control, but true companion animals that have a unique relationship with humans.

The roots of dog social behavior must arise from the behavior of wild canids. Members of the family Canidae are territorial animals with well-developed social organizations; their social behavior follows a common basic pattern, but its complexity varies among canid societies (Scott and Fuller 1965). In nature, gray wolves, red foxes (*Vulpes vulpes*), and other canids usually live in family groups that include a breeding pair and often offspring from previous years, although wolf packs including several families are observed (Macdonald 1979; Baker et al. 1998; Mech 1999, 2003; Bekoff 2001; Creel and Creel 2002). Large packs containing 20 and more pack members are well documented for African wild dogs (*Lycaon pictus*) (Creel 2001). In most cases, it is a dominant couple that reproduces while other group members participate in pup rearing—protecting pups from intruders and helping with feeding and teaching. Living in groups helps pack members to protect territory and food, and to kill larger prey through cooperative hunting than less social dogs could (Creel and Creel 2002). Although it is still controversial whether cooperative hunting is one of the biggest advantages of pack formation, pup rearing definitely benefits from having helpers. In long-term studies of African wild dogs, no pups were successfully reared to adulthood if raised by parents alone (Creel and Creel 2002). Cooperative rearing of pups is also observed in other canids (Macdonald 1979; Macdonald and Carr 1995; Bekoff 2001). The social organization of wolves, coyotes, foxes, and other canids can vary from year to year and with living conditions, but the advantages of pack living are greater in difficult years (Bekoff 2001). In contrast to the very common cooperative living in canids, 85% percent of species in the order Carnivora are solitary (Baker et al. 1998; Creel and Creel 2002). It seems reasonable to propose that the preexisting high level of sociality and plasticity of canid social organization and similarity of canid pack structure to human families probably played an important role in the process of canine domestication.

The adaptability of canids to different ecological conditions also deserves consideration in this context. As a group, canids have achieved global distribution. They inhabit all continents except Antarctica, in ecological zones ranging from tundra to desert. The red fox (*V. vulpes*) is probably one of the most widely distributed mammals (Sheldon 1992). Furthermore, foxes live not only in wilderness, but also in settled urban zones (Harris and Smith 1987; Deplazes et al. 2004) and suburban areas (Doncaster and Macdonald 1991; Baker et al. 2000). Microsatellite analyses revealed significant genetic differentiation between populations of red foxes from urban Zurich and adjacent rural areas; the structure of urban fox populations suggested a founder effect and multiplex admixture events between urban and rural foxes, with a much lower immigration rate of urban foxes to rural areas than vice versa (Wandeler et al. 2003). Lack of competitors, reduced exposure to predation, and the availability of deep food resources around human settlements form a beneficial niche for species that can live in close proximity to humans. Not surprisingly, this niche is, even today, often filled by free-living dogs (Fox 1978; Boitani et al. 1995; Macdonald and Carr 1995; Coppinger and Coppinger 2001).

The dog was certainly the first animal domesticated and, unlike the later domestication of livestock, its "complete" domestication probably represents a secondary human-directed selection, superimposed on a longer process of adaptation of dog ancestors to the newly developed ecological niche provided by human presence. Humans and dogs probably first shared a common environment when fully modern *Homo sapiens* migrated from Africa into Eurasia, around 100,000–150,000 ybp (years before present). This is also the earliest date supported by molecular dating for the divergence of dogs and wolves (Vila et al. 1997). Neither *Homo erectus* nor *Homo neanderthalensis*, both of whom were established previously in at least parts of this range, appear to have had canine companions. Several scenarios have been proposed for how, where, and when dogs became adapted to life around human settlements (see, e.g., Coppinger and Coppinger 2001).

Molecular analyses of mtDNA support a common origin for both pre-Columbian American dogs and Eurasian dogs from old world gray wolves (Leonard et al. 2002). Archaeological evidence for the coexistence of dogs with humans in the old world has been identified from as early as 14,000–17,000 ybp (Nobis 1979; Sablin and Khlopachev 2002). A diverse radiation of canine morphotypes took place in early human civilizations (Scott and Fuller 1965; Clutton-Brock 1999); several general types of modern dogs are recognizable in Babylonian, Assyrian, and Egyptian records (Zeuner 1963; Acland and Ostrander 2003). Still unanswered is whether canine domestication took place once, or repeatedly at different times and locations, and perhaps from different ancestral populations. If once, it suggests that tameness of all modern dogs is determined by the same loci (Scott and Fuller 1965). Presumably, domestication was tightly connected with the development of human society, and the spread of dogs around the globe by human migration and trading habits. Although domestication might favor behavioral features that would be disadvantageous for wild canids, domestication has been an extremely successful survival strategy for dogs. The current dog population worldwide vastly exceeds any other canid species in number and distribution. There are at least a thousand-fold more dogs in existence today than gray wolves (Coppinger and Coppinger 2001).

Inherently social behavior is probably necessary, if not sufficient in itself, to permit successful domestication. However, it is also apparent that domestication then leads to dramatic changes in social behavior in animals, and these changes are undeniably genetic. A large part of the difference between dog and wolf behaviors is the ability of dogs to respond positively to cues from humans (Hare et al. 2002; Gosling et al. 2003; Kaminski et al. 2004; Viranyi et al. 2004). Hare and colleagues proposed that social cognition of dogs was changed during domestication (Hare et al. 2002, 2005). In wild canids, social play is mostly exhibited in juveniles, but in dogs, playful activity often continues into adulthood. Thus, the social behavior of adult dogs resembles that of juvenile wolves (Fox 1978). Artificial selection focused on variations in behavioral and morphological traits has led to the development of hundreds of dog breeds, selected to be shepherds, hunters, sled dogs, guardians, and simply companions.

The genetics of dog social behavior was pioneered by Scott and Fuller and their colleagues at Jackson Laboratory in Bar Harbor (Scott and Fuller 1965). Their fundamental research in canine genetics provided a foundation of knowledge that has gained renewed significance with the recent development of canine molecular genetics. Scott and Fuller observed the development of behavior in dogs from birth to one year of age, and identified multiple behavioral characteristics that vary among individuals and dog breeds. They produced experimental pedigrees and confirmed resegregation of numerous behavioral phenotypes in informative generations. They proposed a scheme to account for the remarkable variation in dog morphology and behavior and demonstrated that the inheritance of complex behavioral phenotypes in dogs often can be explained by a small number of genes. Factor analysis to score dog behavior was first introduced in their experiments (Scott and Fuller 1965).

Canine pedigrees similar to those developed by Scott and Fuller would be an invaluable resource for identifying the genetic basis of canine behavioral phenotypes. Until recently, mapping behavioral quantitative trait loci (QTLs) in mammals was explored mostly in rodents (for review, see Flint 2003). Extensive analyses of behavioral variation in dogs and the development of robust methods for scoring dog behavior (Coppinger and Schneider 1995; Serpell and Hsu 2001; Hsu and Serpell 2003) provide a foundation for future mapping experiments in canids. The recent availability of molecular genetic tools, including the canine genome sequence, now enables analyses of complex phenotypes in dog populations (Chase et al. 2002, 2004; Chapter 5) and provides an opportunity to shed light on the genetic architecture of canid social behavior.

This raises the issue, however, of how to identify the genetic changes that took place during domestication of the dog. Newly developed tools of canine molecular genetics could shed light on this question, but the key events took place at least 12,000 ybp, and informative pedigrees resegregating behavioral differences between dogs and wolves are not currently available.

This chapter provides an overview of model experiments aimed at identifying such genetic mechanisms. Our model is a strain of domesticated silver foxes developed at the Institute of Cytology and Genetics of the Russian Academy of Sciences (for reviews, see Trut 1999, 2001). Domesticated foxes interact with humans in a friendly manner, and this behavior is extremely similar to that of domestic dogs.

GENETICS OF TAME BEHAVIOR IN FOXES

The silver fox, a color variation of the red fox (*V. vulpes*), is another member of the Canidae family that recently has been domesticated in a controlled experiment. A tame strain of silver foxes demonstrating behavior like that of the domestic dog was developed selectively at the Institute of Cytology and Genetics (ICG) in Novosibirsk, Russia, in the second part of the 20th century (Trut 1999). This robust experimental model offers a unique opportunity to study a complex behavioral phenotype. The fox-domestication experiment has been well described previously (for reviews, see Trut 1999, 2001; Trut et al. 2004), and we provide herein only a brief outline.

History of the Fox-domestication Project

The silver fox, a species taxonomically close to the dog (Wayne 2001a), has been bred in captivity since the beginning of the 20th century (Nes et al. 1988). Under standard farm conditions, however, foxes normally exhibit distinct patterns of aggressive and fear-aggressive behavior to humans. Belyaev and colleagues hypothesized that selection of farm foxes for less-fearful and less-aggressive behavior would yield a strain of domesticated fox. Starting in 1959, and selecting solely on behavioral criteria for more than 40 generations, a population of foxes demonstrating friendly responses to humans was produced (Belyaev 1969, 1979; Trut 1999, 2001). The selectively bred tame foxes show behavioral patterns extremely similar to those of domestic dogs: By one month postnatal they become eager to establish human contact, whimpering to attract attention and sniffing and licking at humans, just like puppy dogs. Tame foxes exhibit highly social behavior both with other members of their own species and with humans, in a playful, friendly manner (Trut 1999, 2001; http://cbsu.tc.cornell.edu/ccgr/behaviour/index.html). To ensure that tameness results from genetic selection, foxes are not trained and are only allowed brief "time dosage" contacts with humans.

In parallel with this selection for tameness, a second strain of foxes was bred selectively to preserve wild-type ("aggressive") behavior (Trut 1999, 2001). This aggressive strain was developed because foxes exhibiting wild-type behavior can be difficult to handle and, under farm conditions, even unselected foxes are subject to some degree of less-deliberate selection for relatively tame behavior.

Although selection in foxes was strictly limited to defined behavioral criteria, further developmental and behavioral differences emerged in tame foxes. Opening of the eyelids and external auditory canal accelerated; the sensitive period for socialization persisted past 60 days old (compared to less than 45 days in unselected foxes); and play activity (normally seen in infantile wild-type foxes) extended into adulthood (Belyaev et al. 1984/85; Plyusnina et al. 1991; Trut 2001). Ability to read human cues, like the abilities of domestic dogs, was observed in tame foxes (Hare et al. 2005). Significant differences in corticosteroids and neurotransmitter levels were also found between tame and control foxes (Oskina and Tinnikov 1992). Lower plasma cortisol and ACTH levels in tame foxes than in the control population were observed (Gulevich et al. 2004). Significantly lower density of serotonin 5-HT_{1A} receptors was observed in the hypothalamus, and significantly higher levels of serotonin and tryptophan hydroxylase were detected in the midbrain and hypothalamus of the domesticated strain of foxes (Popova et al. 1997; Trut 2001). These results have clear parallels in the recent demonstration of differential hypothalamic expression of two neuropeptides, CALCB and NPY, between dogs and wolves (Saetre et al. 2004).

As an unexpected consequence of selection for tameness, several morphological changes emerged in these foxes, including floppy ears, rolled tails, shorter tails, and changes in skull shape, all of which resemble differences between domesticated dogs and wolves (Wayne 2001b; for review, see Chapter 6). Neoteny, retarded development of adult characteristics, has been proposed as one of the possible mechanisms underlying domestication (Trut 1999, 2001; Wayne 2001b; Coppinger and Coppinger 2001; Trut et al. 2004). The persistence of juvenile morphological characteristics, in foxes selected solely on the basis of behavioral criteria, suggests that the genes under selection may have a broader regulatory role in development (Trut 1999, 2001; Trut et al. 2004).

Established Scoring Systems for Tame and Aggressive Behavior

A rigorous testing system for assessment of behavioral phenotypes, established at ICG to enable selection of foxes for behavior (Tables 1 and 2), required development of two systems to score fox behavior; one for the "tame" and one for the "aggressive" population. This was because,

Table 1. Established system for scoring tame behavior

Animal reaction	Scores
Passive-protection response; fox avoids experimenter or bites if stroked or handled, comes if offered food.	0.5−1.0
Fox lets itself be petted and handled, but shows no emotionally friendly response to experimenter.	1.5−2.0
Fox shows emotionally positive, friendly, response to experimenter, wagging tail and whining.	2.5−3.0
Fox is eager to establish human contact, whimpering to attract attention and sniffing and licking experimenters like dogs. Foxes start displaying this kind of behavior before 1 month of age.	3.5−4.0

This table summarizes the scoring method used for ranking foxes in the tame population, and for selecting the most tame foxes for the breeding program. Tame behavior is scored from zero (representing "neutral" behavior; an absence of both actively aggressive and tame responses directed toward the observer) to 4 (representing the most-tame behavior). (From Trut 1980b.)

Table 2. Established system for scoring aggressive behavior

Animal reaction	Scores
Fox shows teeth, snarls, growls *at first sight of human*. When experimenter is near *closed cage* fox attacks experimenter and other objects in field of view. Bared teeth and fixed dilated pupils.	−4.0
When experimenter is near *closed cage*, fox shows teeth, snarls, growls, tries to attack both the experimenter and other objects in field of view. Bared teeth and fixed dilated pupils.	−3.5
When experimenter is near *open cage*, fox shows teeth, snarls, growls, attacks experimenter and other objects in the field of view. Bared teeth and fixed dilated pupils.	−3.0
When experimenter is near the *open cage*, fox growls but does not attack.	−2.5
When experimenter, near the *open cage*, moves protected arm toward fox, it growls and tries to bite.	−2.0
As experimenter opens cage, fox is calm, but attempts to touch the fox provoke it to show teeth and snarl.	−1.5

The score is based on the critical distance between the experimenter and the caged fox when the animal *first* demonstrates an aggressive reaction to the experimenter's presence. Originally the scoring system was based on a range from zero (least aggressive) to −4 (most aggressive). However, after multiple generations of selection, no current individuals exhibit behavior scoring between −1.5 and 0. (From Trut 1980b.)

although the original farm fox population showed a continuous variation in behavior, very quickly the phenotypes in the selected tame and aggressive populations no longer overlapped. To measure variations in "friendliness" in the "tame" population, and "aggressiveness" in the "aggressive" population, an objective, quantitative approach was instituted (Trut 1980a, 1999; Vasilieva and Trut 1990).

Experiments undertaken in the 1970s included experimental cross-breeding of tame and aggressive animals, cross-fostering of newborn pups, and transplantation of embryos, and demonstrated the genetic basis of the tame and aggressive phenotypes (Trut 1980a,b, 2001). New experimental pedigrees have been developed more recently (2001–2004) by crossbreeding tame to aggressive foxes, and then back-crossing F_1 animals to tame parents. Clear resegregation of the tame behavioral phenotype in the informative generation was observed (Fig. 1).

Development of Molecular Tools for Genetic Mapping of Fox Behavioral Phenotypes

To undertake genetic mapping of behavioral QTLs in foxes, it is first essential to develop a meiotic-linkage map of the fox genome. Dogs and

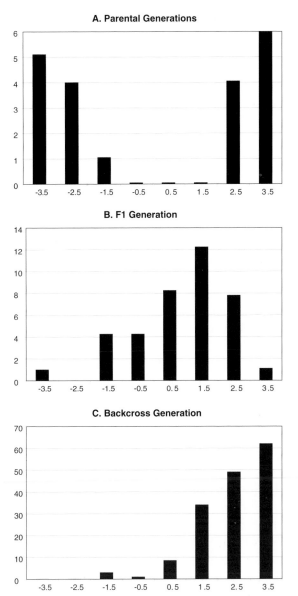

Figure 1. Analysis of behavioral phenotypes in fox populations. Behavioral scores are based on the historically established system developed at ICG. Scores are continuous from most aggressive (-3.5) to the most tame (3.5) phenotype. (**A–C**) Distribution of behavioral phenotypes in parental (tame, aggressive), F_1, and backcross pedigrees, respectively. Scores of parental tame, and aggressive, populations are distinctly different from each other (**A**); the F_1 generation (**B**) is intermediate between that of the tame and aggressive populations, and the backcross population (**C**) demonstrates resegregation of scores typical of the tame population.

foxes are taxonomically close species (Wayne et al. 1997), but karyotypes of domestic dog and red fox are markedly different in chromosome number and structure. Although the dog has 78 mostly acrocentric chromosomes, the red fox has 34 metacentric chromosomes and 0–8 small B chromosomes. Concordance of the dog and fox karyotypes, as defined by reciprocal chromosomal painting, suggests that the fox karyotype evolved from a more-ancient dog-like karyotype by 26 fusion and 4 fission events (Yang et al. 1999, 2000; Graphodatsky et al. 2000, 2001, 2002). This understanding of the homology among chromosomal segments of the two species lets us predict, to a first approximation, how dog linkage groups should be arranged in the fox genome. Because canine microsatellite markers have been amplified successfully from DNAs of several canids (Kohn et al. 1999; Wandeler et al. 2003), it became reasonable to adapt microsatellite markers from the canine genome map to build fox linkage groups.

Initially, 700 of the most informative dog microsatellites (Breen et al. 2004) were tested to identify those that robustly amplify fox DNA, and a primary set of 400 microsatellite markers to map the fox genome was developed (Kukekova et al. 2004). A subset of PCR products (30 markers) amplified from fox DNA were sequenced, and in each case, the microsatellite repeats expected from canine data were identified. The number of fox alleles varied among the microsatellites, with a mean of 5 for 30 markers genotyped for 32 individuals from the fox colony. From the canine MSS-2 marker set of 340 fluorescently labeled markers (Guyon et al. 2003; Clark et al. 2004), a further 100 polymorphic markers were identified that robustly amplified from fox DNA. Selected markers were combined in new multiplexes and used for fluorescent genotyping of fox samples.

Thirty five three-generation fox pedigrees comprising 265 individuals were selected for construction of a meiotic-linkage map of the fox genome. Blood samples were collected at the fox farm in Novosibirsk, Russia, and DNA was extracted for each fox. These pedigrees were genotyped at the Mammalian Genotyping Service of Marshfield Laboratories (Madison, WI) with 249 canine microsatellites and at Cornell University with a further 100 markers.

Proof of Principle

Preliminary analyses of fox genotyping data were undertaken to begin construction of fox linkage groups. Using MultiMap (Matise et al. 1994; Mellersh et al. 1997), marker genotypes were checked for Mendelian segregation (*prepare option*) and two-point recombination fractions and associated LOD scores calculated for all marker pairs. Markers were assigned to linkage groups (*find-all-linkage-groups function*) if linked to at least one other marker in the group with a recombination fraction ≤0.4 and a LOD score of ≥3.0 (equivalent to odds of 1000:1 in favor of linkage). A sex-averaged, framework map then was constructed for each linkage group. Consistency between linkage data and cytogenetic predictions identified by reciprocal chromosomal painting (Yang et al. 1999) was observed. For example, microsatellite markers from canine chromosomes 1, 12, and 33 map to fox chromosome 1 (Vvu1), and the orientation of canine linkage groups within Vvu1 is also consistent with cytogenetic data (Fig. 2).

The high level of nucleotide identity between the dog and fox genomes enables the density and resolution of the fox map to be increased by adapting microsatellite markers from the canine genome map and sequence. The newly available 7.5× sequence of the canine

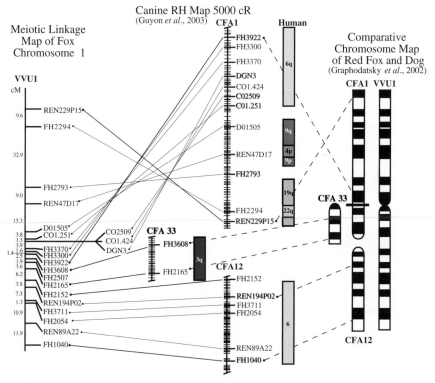

Figure 2. Comparative map of fox chromosome 1 with dog and human genomes. Fox chromosome 1 (VVU1) corresponds to a fusion of canine chromosomes 1, 33, and 12 (CFA1, CFA33, CFA12). The VVU1 meiotic map demonstrates inversion of the segment corresponding to CFA1, representing centromeric fusion (or fission) of the segments representing CFA33 and CFA1 compared to VVU1.

genome (http://www.ncbi.nih.gov/genome/guide/dog/) will greatly facilitate the identification of additional microsatellites and incorporation of gene-specific markers into the fox meiotic map. Because we anticipate that genes important in neurotransmitter metabolism may contribute to fox behavioral traits, we are specifically including multiple such genes in the fox genome map. Similarly, we are deliberately including many genes previously implicated in human and/or rodent behavior to optimize the fox map for identifying and/or excluding candidate genes for behavioral phenotypes. When complete data for both genotyping marker sets and gene-specific markers are merged, we will be able to build a meiotic linkage map of the fox genome with a predicted maximal distance between adjacent markers of 15 cM (which should be sufficient for QTL mapping).

FOX EXPERIMENTAL PEDIGREES FOR MAPPING BEHAVIORAL QTLs

Strains of tame and aggressive foxes currently are maintained at the experimental farm in Novosibirsk, Russia. Construction of fox experimental pedigrees was initiated by crossbreeding tame and aggressive foxes, then backcrossing the F_1 progeny to tame parents, to produce litters informative for differences in behavior between the founder populations. Ten crosses between extreme representatives of the tame and aggressive populations were produced in 2002. Behavior of 45 F_1 pups has been assessed using the tests described above and videotaped. Eight of these F_1 foxes exhibited aggressive behavior; 31 were tame but not friendly (scores 0.5–2.0), and 6 were tame and friendly (scores 2.5–4.0).

POWER STUDIES FOR DETECTING QTLs

The heritability of fox behavior has been estimated from behavioral data measured in F_1 crossbred foxes and their tame and aggressive parents. Regression of observed offspring phenotypes (offspring behavior score) on midparent phenotypes (Fig. 3) indicates an additive heritability (h^2) close to 1. In a simulation study, we estimated the power of a backcross population to detect QTLs associated with behavior score. Assumptions included (1) additive heritability for QTLs (h^2); (2) a backcross population ($F_1 \times$ Tame) containing either 50 or 100 animals; (3) recombination fraction of 0.0, 0.1, or 0.2 between marker and QTL; (4) a target significance level of 1/10,000 (LOD 4) corresponding to a p-value of 0.05 for a multiple-test genome-wide screen (e.g., 500 markers). For all

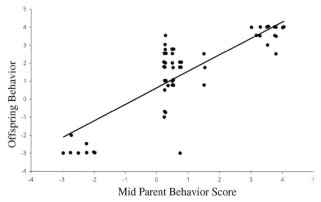

Figure 3. Heritability of fox behavioral phenotypes. Regression of observed offspring phenotypes (Offspring Behavior Score) on midparent phenotypes indicates an additive heritability (h^2) close to 1.

modestly optimistic scenarios, the simulation data (Fig. 4) predict power levels >0.8. Only for the most pessimistic scenarios (very small QTL effect size, and/or small population size, and/or large marker QTL distance) is the predicted power discouraging.

Three hundred backcross foxes were generated during 2003 (160 animals) and 2004 (140 animals). All breeding and assessment of phenotypes was undertaken at the ICG in Novosibirsk, Russia. Because the fox colony continues to be maintained on the same farm, environmental influences are minimized, and behaviors are evaluated under

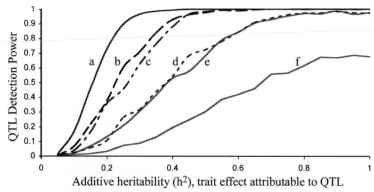

Figure 4. Simulated power to detect a QTL. Significance threshold of 1/10,000 (LOD 4.0) for QTLs of varying strength (h^2). For all modestly optimistic scenarios, the simulation data predict power levels >0.8. Only for the most pessimistic scenarios (very small QTL effect size, and/or small population size, and/or large marker QTL distance) is the predicted power discouraging.

consistent, standardized conditions. Behavior of all backcross animals was or will be tested in the standard test at 3–3.5 months, at 5–6 months, and at 16–18 months of age. Foxes were tested in their home cages at a standard time period during the day. The standard test includes five steps: (1) observer approaching fox cage; (2) observer stays near closed cage; (3) observer is near open cage but does not initiate tactile contact; (4) tactile contact; (5) observer stays near closed cage. Each step (except step 1) is 1 minute long. Fox behavior during the entire test is videotaped to preserve a permanent record and to allow subsequent detailed analyses of fox behavior during the test. Fox behavior is tested at least twice at each age point by the same observer. Distribution of fox behavioral phenotypes estimated using the traditional system developed at ICG is presented in Figure 1.

ASSIGNMENT OF FOX BEHAVIORAL PHENOTYPES FOR QTL MAPPING

Measurement of behavior presents several challenges, particularly in attempting to map the underlying genetic loci. Because behavior is clearly phenotypically complex, it is likely to represent a complex interaction of multiple genetic loci and environmental factors. To identify genes underlying behavioral variation, it is first highly desirable to identify, among the richness of behavioral expression, those specific, independent, and presumably simpler aspects that can be measured objectively and quantitatively, and can be demonstrated to be inherited.

To these ends, we developed a high-resolution, objective, quantitative method for behavioral evaluation of fox populations selected for, and segregating, differences in behavior. A comprehensive primary set of objective measures was developed for scoring the physical manifestations of fox behaviors. Statistical analysis, including principal-components analysis (PCA; Jackson 1991; Chase et al. 2002; Afifi et al. 2004) was used to dissect out the independent, resegregating traits underlying the phenotypic variation expressed in these multiple correlated observations. To develop and test the method, we used a preliminary subset (n = 95) of the foxes in the complete data set.

Behavior of a subset of foxes, including 20 randomly selected animals from each population (tame, aggressive, unselected, and backcross) and 15 F_1 animals (all F_1 foxes of the same age that were available in 2004), was analyzed from videotapes. Behaviors were observed in five test steps, representing different interactive states between the fox and the observer. A total of 319 observations of fox behavior were identified that could be scored simply by their presence or absence. Traits requiring subjective interpretation of animal actions (for example, "Fox is afraid" or "Fox demonstrates submissive behavior") were avoided. Traits selected describe specific fox actions, location in the cage and time spent there, body postures, positions of particular parts of the body, and sounds. Initially, traits were tested for informativeness and the number of factors was reduced to 50 by eliminating traits that either failed to distinguish among the reference populations, or were redundant.

PCA using the 50 retained traits identified two PCs which, between them, explained more than 50% of cumulative variance in the fox behavioral data set, and clearly distinguish the founder fox populations (Fig. 5). PC1 explains 47.3% of variance and distinguishes tame and backcross populations from aggressive, unselected, and F_1; and the F_1 and backcross from tame, unselected, and aggressive (Kruskal-Wallis statistic 53.59; $p<0.0001$). PC2 explains 6.4% of total variance (12.1% of remaining variance) and distinguishes the tame population from the backcross (16.41; $p=0.0025$) (Acland et al. 2004; A. Kukekova et al., in prep.). We also have demonstrated reproducibility of the behavior scores, and of the contributions of traits to the principal components.

We now have very high confidence that our analytic methods dissect the inherited components of fox behavior into independent underlying traits that clearly resegregate in the backcross population. This dramatically improves our power for mapping underlying genes/QTLs.

Heritability of PC1 was estimated for a subset of foxes including 20 randomly selected animals from each population (tame and aggressive)

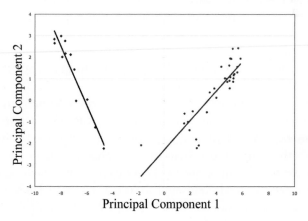

Figure 5. Comparison of behaviors scored as principal components. Principal component 1 (PC1) scores for tame and aggressive foxes (*diamonds* and *circles*, respectively) exhibit no overlap. The relationships (*straight lines*) between PC1 and PC2 for the two populations are also clearly different.

and 15 F_1 animals. Analysis of variance for PC1 was undertaken, and heritability was estimated as the ratio of variance (mean square) between groups to total variance (mean square between groups plus mean square within groups). When all three populations were included in the ANOVA, a heritability estimate of greater than 0.9 was observed. When only tame and F_1 populations were compared, the heritability estimate was lower ($\sim$0.5), as expected. These values are very encouraging for studying resegregation of quantitative behavioral traits in the backcross to tame informative generation.

DISCUSSION

Genetically determined behaviors in the silver fox offer a unique opportunity and resource. The genetic basis of these fox behavioral phenotypes has been clearly and carefully demonstrated (Trut 1980a), with extensive and accurate documentation, including not only scrupulously maintained pedigree data and records for these behavioral phenotypes, but also comprehensive data from biochemical, morphological, and other studies as reported in a large body of literature (for review, see Trut 2001). Because new foxes are born at approximately the same time every year (March–April), all foxes live under consistent conditions and have similar interactions with humans, and fox behavior is tested at precise time points using standard tests under consistent conditions, environmental factors that might influence behavior are held to a strict minimum. For at least 40 generations, well-characterized tame and aggressive fox populations have been bred separately and selectively to exhibit distinctly different and precisely measurable behavioral traits. These selectively bred populations have thus provided ideal founder populations for the newly developed crossbred and backcross populations, and an unprecedented opportunity to map and identify the genes responsible for resegregation of behaviors.

In the history of this project, selection of foxes for amenability to domestication has been intense. Selection started with 130 farm-bred foxes that demonstrated the least avoidance behavior toward humans. Subsequently, only the tamest individuals in each generation have been allowed to breed, while at the same time a deliberate effort was made to avoid inbreeding (Trut 1999, 2001; Trut et al. 2004). The tame population developed relatively fast in response to selection, 18% of foxes from the tenth generation demonstrated extremely tame behavior, and little behavioral variation was observed by 1985 and thereafter (Trut 1999, 2001; Trut et al. 2004). Unlike the modern dog, the population of domestic foxes was created in a relatively short period of time by selection focused only on specific behavioral traits. The speed with which the tame fox population was derived suggests that the imposed selection may be acting on a few genetic loci influencing behavior (Trut et al. 2004).

The resegregation of behavioral phenotypes in these fox pedigrees provides an opportunity to identify loci involved in tameness through meiotic linkage mapping, and power and heritability studies provide strong support for success. Once behavioral QTLs are detected, it will be necessary to better localize and validate the identified regions of interest. Comparative alignment of fox and canine genomes will allow selection of further canine microsatellite markers to refine the identified intervals in the fox genome. The canine integrated meiotic, and radiation hybrid maps (Guyon et al. 2003; Breen et al. 2004), and the newly available canine genome sequence (Kirkness et al. 2003; http://www.ncbi.nlm.nih.gov/genome/guide/dog/) provide essential tools for this process. Additional markers adapted from the canine genome will be genotyped on an expanded and expanding set of informative fox pedigrees to define and confirm QTL intervals. Once the marker frame for a region of interest is defined by linkage analyses, single-nucleotide polymorphisms (SNPs) will be developed for this region and used for linkage-disequilibrium (LD) mapping of distantly related foxes from the tame population. We anticipate that haplotypes from the least-related tame foxes will allow some further reduction of the candidate regions, allowing concentration to be focused on a reduced number of positional candidates within the absolute LD region. The high nucleotide identity observed between the coding sequences of fox and dog genes makes the canine genome sequence highly advantageous for cloning fox genes.

When QTL intervals in foxes are well defined, it will become possible to test whether homologous chromosomal regions influence behavior in dogs and other canids. An integrated comparative map of the dog and fox genomes, currently under development, will allow identification of regions in the dog genome which are homologous to the QTL intervals in foxes. Comparison of SNP haplotypes among dogs and wild canids can then test whether any of the chromosomal regions identified in foxes are distinctive among wolves, dingoes, and dogs with different temperaments. Identification of regions in the fox genome implicated in the development of friendly fox behavior also will provide the opportunity to define new candidate genes for autistic and other human neurological disorders that are accompanied by impaired development of social reciprocity.

Until recently, it would have been an overwhelming task to construct a fox genetic map and to study the molecular genetics of this unique fox population. However, the task has been substantially simplified by the development of tools for the dog genome over the last decade. The fox model of domestication thus offers an accessible opportunity to identify genomic regions involved in social behavior of mammals.

SUMMARY

Dogs and other canids exhibit a fascinating range of behaviors that are clearly heritable. Probably the most commonly recognized examples of this are the very different, breed-specific behavioral repertoires seen in different breeds of dogs. However, one of the most dramatic, although often taken for granted, behaviors that characterize domestic dogs is the way they interact with humans in a sociable, friendly, or tame manner. In essence, dogs and humans accept each other into a mutual social structure. This stands in stark contrast to the behavioral responses that the gray wolf, the dog's immediate ancestor, demonstrates to humans.

The process of domestication, whereby gray wolves gave rise to the domestic dog at least 12,000 years ago, has been recapitulated experimentally in a program to domesticate the silver fox. Over approximately 40 generations, wild-type farm foxes that naturally are aversive to human interaction (much like gray wolves), have been selectively bred to interact socially with humans in a tame, friendly manner, much as dogs and their puppies do.

The recent rapid development of molecular tools for investigating the canine genome can now be exploited to study genetic traits in the silver fox. A set of mapping markers has been adapted from the canine genome to the fox, and a first-generation meiotic linkage map of the fox genome is now in development. Three-generation fox pedigrees informative for behavioral traits have been developed, and resegregate these behavioral traits. Quantitative methods for analysis of objectively scoreable fox behaviors have been shown to reliably distinguish among tame, wild-type, F_1, and backcross foxes, and they yield very encouraging estimates of heritability and mapping power. Once loci responsible for these traits are identified in foxes, the role of these loci in dog domestication and dog social behavior can be explored.

ACKNOWLEDGMENTS

We are grateful to Irina V. Pivovarova for coordination of experiments on the fox farm and assistance with behavioral data analyses; to Jennifer Johnson, Svetlana Temnykh, and Sue Pearce-Kelling for their efforts in developing a fox meiotic map; to Dasha Terehova, Grisha Temnykh,

Sarah Pinkney, and Anna Stepika for fox behavioral data collection; to Tatyana Semenova, Rimma Gulevich, Svetlana Shikhevich, Vasiliy Ivaykin, Vera Vladimirova, Tatyana Konovalova, Tamara N. Kuzhutova, Lyudmila V. Parafienko, Oleg Trapesov, Vera Haustova, and all the animal keepers at the ICG experimental farm for technical assistance; to Sergey N. Borchsenius for his support in the development of this research, to Simon Kizhner for excellent editorial assistance in construction of the fox video library; to Kathleen E. Whitlock, H. Kern Reeve, Alexander S. Graphodatsky, Simon Klebanov, Irina Z. Plyusnina, Brian Hare, and David Carrier for critical discussions; and to Keith Watamura for graphics support. We express our profound gratitude to Marshfield Laboratories Mammalian Genotyping Service for genotyping fox pedigrees. This work was supported by National Institutes of Health grants MH069688-01, EY06855, and EY13729 and GM63056; NATO grant LST CLG.979216; grant #2303.2003.4 from the Russian Fund of Basic Research (Scientific School) and grant of the Program of the Basic Research of the Presidium of Russian Academy of Science: "Dynamics of gene pools of plants, animals and human."

REFERENCES

Acland G.M., Erb H.N., Chase K., Lark K.G., Aguirre G.D., Trut L.N., Oskina I.N., Kharlamova A.V., and Kukekova A.V. 2004. Resegregating behaviors in the Silver Fox. A model system for mapping sociability. American Society of Human Genetics, 54th Annual Meeting, Abstract 2300, p.417. American Society Human Genetics, Bethesda, Maryland (http://www.ashg.org/genetics/abstracts/abso4/f2300.htm).

Acland G.M. and Ostrander E.A. 2003. The dog that came in from the cold. *Heredity* **90:** 201–202.

Afifi A., Clark V.A., and May S. 2004. *Computer-aided multivariate analysis*, 4th edition. Chapman & Hall/CRC, Boca Raton, Florida.

Baker P.J., Robertson C.P.J., Funk S.M., and Harris S. 1998. Potential fitness benefits of group living in the red fox, *Vulpes vulpes. Anim. Behav.* **56:** 1411–1424.

Baker P.J., Funk S.M., Harris S., and White H.C.L. 2000. Flexible spatial organization of urban foxes, *Vulpes vulpes*, before and during an outbreak of sarcoptic mange. *Anim. Behav.* **59:** 127–146.

Bekoff M. 2001. Cunning coyotes: Tireless tricksters, protean predators. In *Model systems in behavioral ecology* (ed. L.A. Dugatkin), pp. 381–407. Princeton University Press. Princeton, New Jersey.

Belyaev D.K. 1969. Domestication of animals. *Science* (Russ.) **5:** 47–52.

———. 1979. The Wilhelmine E. Key 1978 invitational lecture. Destabilizing selection as a factor in domestication. *J. Hered.* **70:** 301–308.

Belyaev D.K., Plyusnina I.Z., and Trut L.N. 1984/85. Domestication in the silver fox (*Vulpes fulvus Desm*): Changes in physiological boundaries of the sensitive period of primary socialization. *Appl. Anim. Behav. Sci.* **13:** 359–370.

Boitani L., Francisci F., Ciucci P., and Andreoli G. 1995. Population biology and ecology of feral dogs in central Italy. In *The domestic dog: Its evolution, behaviour and interactions with people* (ed. J. Serpell), pp. 217–244. Cambridge University Press, Cambridge, United Kingdom.

Breen M., Hitte C., Lorentzen T.D., Thomas R., Cadieu E., Sabacan L., Scott A., Evanno G., Parker H.G., Kirkness E.F., et al. 2004. An integrated 4249 marker FISH/RH map of the canine genome. *BMC Genomics* **5:** 65.

Chase K., Lawler D.F., Adler F.R., Ostrander E.A., and Lark K.G. 2004. Bilaterally asymmetric effects of quantitative trait loci (QTLs): QTLs that affect laxity in the right versus left coxofemoral (hip) joints of the dog (*Canis familiaris*). *Am. J. Med. Genet. A.* **124:** 239–247.

Chase K., Carrier D.R., Adler F.R., Jarvik T., Ostrander E.A., Lorentzen T.D., and Lark K.G. 2002. Genetic basis for systems of skeletal quantitative traits: Principal component analysis of the canid skeleton. *Proc. Natl. Acad. Sci.* **99:** 9930–9935.

Clark L.A., Tsai K.L., Steiner J.M., Williams D.A., Guerra T., Ostrander E.A., Galibert F., and Murphy K.E. 2004. Chromosome-specific microsatellite multiplex sets for linkage studies in the domestic dog. *Genomics* **84:** 550–554.

Clutton-Brock J. 1999. *A natural history of domesticated mammals*, 2nd edition. Cambridge University Press, Cambridge, United Kingdom.

Coppinger R. and Schneider R. 1995. Evolution of working dogs. In *The domestic dog: Its evolution, behaviour and interactions with people* (ed. J. Serpell), pp. 21–30. Cambridge University Press. Cambridge, United Kingdom.

Coppinger R. and Coppinger L. 2001. *Dogs*. The University of Chicago Press, Chicago, Illinois.

Creel S. 2001. Cooperative hunting and sociality in African wild dogs, *Lycaon pictus*. In *Model systems in behavioral ecology: Integrating conceptual, theoretical, and empirical approaches* (ed. L.A. Dugatkin), pp. 466–490. Princeton University Press. Princeton, New Jersey.

Creel S. and Creel N.M. 2002. *The African wild dog: Behavior, ecology and conservation*. Princeton University Press, Princeton, New Jersey.

Deplazes P., Hegglin D., Gloor S., and Romig T. 2004. Wilderness in the city: The urbanization of *Echinococcus multilocularis*. *Trends Parasitol.* **20:** 77–84.

Doncaster C.P. and Macdonald D.W. 1991. Drifting territoriality in the red fox *Vulpes vulpes*. *J. Anim. Ecol.* **60:** 423–439.

Flint J. 2003. Analysis of quantitative trait loci that influence animal behavior. *J. Neurobiol.* **54:** 46–77.

Fox M.W. 1978. *The dog: Its domestication and behavior*. Garland STPM Press. New York.

Gosling S.D., Kwan V.S., and John O.P. 2003. A dog's got personality: A cross-species comparative approach to personality judgments in dogs and humans. *J. Pers. Soc. Psychol.* **85:** 1161–1169.

Graphodatsky A.S., Yang F., O'Brien P.C., Serdukova N., Milne B.S., Trifonov V., and Ferguson-Smith M.A. 2000. A comparative chromosome map of the Arctic fox, red fox and dog defined by chromosome painting and high resolution G-banding. *Chromosome Res.* **8:** 253–263.

Graphodatsky A.S., Yang F., O'Brien P.C., Perelman P., Milne B.S., Serdukova N., Kawada S.I., and Ferguson-Smith M.A. 2001. Phylogenetic implications of the 38 putative ancestral chromosome segments for four canid species. *Cytogenet. Cell Genet.* **92:** 243–247.

Graphodatsky A.S., Yang F., Perelman P.L., O'Brien P.C.M., Serdukova N.A., Milne B.S., Biltueva L.S., Fu B., Vorobieva N.V., Kawada S.-I., et al. 2002. Comparative molecular cytogenetic studies in the order Carnivora: Mapping chromosomal rearrangements onto the phylogenetic tree. *Cytogenet. Genome Res.* **96:** 137–145.

Gulevich R.G., Oskina I.N., Shikhevich S.G., Fedorova E.V., and Trut L.N. 2004. Effect of selection for behavior on pituitary-adrenal axis and proopiomelanocortin gene expression in silver foxes (*Vulpes vulpes*). *Physiol. Behav.* **82:** 513–518.

Guyon R., Lorentzen T.D., Hitte C., Kim L., Cadieu E., Parker H.G., Quignon P., Lowe J.K., Renier C., Gelfenbeyn B., et al. 2003. A 1-Mb resolution radiation hybrid map of the canine genome. *Proc. Natl. Acad. Sci.* **100:** 5296–5301.

Hare B., Brown M., Williamson C., and Tomasello M. 2002. The domestication of social cognition in dogs. *Science* **298:** 1634–1636.

Hare B., Plyusnina I., Ignacio N., Schepina O., Stepika A., Wrangham R., and Trut L. 2005. Social cognitive evolution in captive foxes is a correlated by-product of experimental domestication. *Curr. Biol.* **15:** 226–230.

Harris S. and Smith G.C. 1987. Demography of two urban fox (*Vulpes vulpes*) populations. *J. Appl. Ecol.* **24:** 75–86.

Hsu Y. and Serpell J.A. 2003. Development and validation of a questionnaire for measuring behavior and temperament traits in pet dogs. *J. Am. Vet. Med. Assoc.* **223:** 1293–1300.

Jackson J.E. 1991. *A user's guide to principal components.* Wiley, New York.

Kaminski J., Call J., and Fischer J. 2004. Word learning in a domestic dog: Evidence for "fast mapping." *Science* **304:** 1682–1683.

Kirkness E.F., Bafna V., Halpern A.L., Levy S., Remington K., Rusch D.B., Delcher A.L., Pop M., Wang W., Fraser C.M., and Venter J.C. 2003. The dog genome: Survey sequencing and comparative analysis. *Science* **301:** 1898–1903.

Kohn M.H., York E.C., Kamradt D.A., Haught G., Sauvajot R.M., and Wayne R.K. 1999. Estimating population size by genotyping faeces. *Proc. R. Soc. Lond. B Biol. Sci.* **266:** 657–663.

Kukekova A.V., Trut L.N., Oskina I.N., Kharlamova A.V., Shikhevich S.G., Kirkness E.F., Aguirre G.D., and Acland G.M. 2004. A marker set for construction of a genetic map of the silver fox (*Vulpes vulpes*). *J. Hered.* **95:** 185–194.

Leonard J.A., Wayne R.K., Wheeler J., Valadez R., Guillen S., and Vila C. 2002. Ancient DNA evidence for Old World origin of New World dogs. *Science* **298:** 1613–1616.

Macdonald D.W. 1979. "Helpers" in fox society. *Nature* **282:** 69–71.

Macdonald D.W. and Carr G.M. 1995. Variation in dog society: Between resource dispersion and social flux. In *The domestic dog: Its evolution, behaviour and interactions with people* (ed. J. Serpell), pp. 199–216. Cambridge University Press. Cambridge, United Kingdom.

Matise T.C., Perlin M., and Chakravarti A. 1994. Automated construction of genetic linkage maps using an expert system (MultiMap): A human genome linkage map. *Nat. Genet.* **6:** 384–390.

Mellersh C.S., Langston A.A., Acland G.M., Fleming M.A., Ray K., Wiegand N.A., Francisco L.V., Gibbs M., Aguirre G.D., and Ostrander E.A. 1997. A linkage map of the canine genome. *Genomics* **46:** 326–336.

Mech L.D. 1999. Alpha status, dominance, and division of labor in wolf packs. *Can. J. Zool.* **77:** 1196–1203.

———. 2003. *The wolf: The ecology and behavior of an endangered species.* University of Minnesota Press, Minneapolis.

Nes N.N., Einarsson E.J., Lohi O., and Jorgensen G. 1988. *Beautiful fur animals—And their colour genetics*. SCIENTIFUR, Denmark.

Nobis G. 1979. Der älteste Haushund lebte vor 14,000 Jahren. *Umschau* **19**: 610.

Oskina I.N. and Tinnikov A.A. 1992. Interaction between cortisol and cortisol-binding protein in silver foxes (*Vulpes fulvus*). *Comp. Biochem. Physiol. Comp. Physiol.* **101**: 665–668.

Plyusnina I.Z., Oskina I.N., and Trut L.N. 1991. An analysis of fear and aggression during early development of behavior in silver foxes (*Vulpes vulpes*). *Appl. Anim. Behav. Sci.* **32**: 253–268.

Popova N.K., Kulikov A.V., Avgustinovich D.F., Voitenko N.N., and Trut L.N. 1997. Effect of domestication of the silver fox on the main enzymes of serotonin metabolism and serotonin receptors. *Genetika* **33**: 370–374.

Price E.O. 2002. *Animal domestication and behavior*. CABI Publishing, Wallingford, United Kingdom.

Sablin M.V. and Khlopachev G. 2002. The earliest ice age dogs: Evidence from Eliseevichi 1. *Curr. Anthropol.* **43**: 795.

Sachser N. 1998. Of domestic and wild guinea pigs: Studies in sociophysiology, domestication, and social evolution. *Naturwissenschaften* **85**: 307–317.

Saetre P., Lindberg J., Leonard J.A., Olsson K., Pettersson U., Ellegren H., Bergstrom T.F., Vila C., and Jazin E. 2004. From wild wolf to domestic dog: Gene expression changes in the brain. *Brain. Res. Mol. Brain Res.* **126**: 198–206.

Scott J.P. and Fuller J.L. 1965. *Genetics and the social behavior of the dog*. The University of Chicago Press, Chicago, Illinois.

Serpell J.A. and Hsu Y. 2001. Development and validation of a novel method for evaluating behavior and temperament in guide dogs. *Appl. Anim. Behav. Sci.* **72**: 347–364.

Sheldon J.W. 1992. *Wild dogs: The natural history of the nondomestic* Canidae. Academic Press, San Diego, California.

Trut L.N. 1980a. The genetics and phenogenetics of domestic behavior. *Proc. Int. Congr. Genet.* **2**: 123–136. Mir. Moscow.

———. 1980b. "The role of behavior in domestication-associated changes in animals as revealed with the example of silver fox." Doctoral (Biol.) dissertation, Institute of Cytology and Genetics, Novosibirsk, Russia.

———. 1999. Early canid domestication: The farm fox experiment. *Am. Sci.* **87**: 160–169.

———. 2001. Experimental studies of early canid domestication. In *The genetics of the dog* (ed. A. Ruvinsky and J. Sampson), pp. 15–43. CABI Publishing, Wallingford, United Kingdom.

Trut L.N., Pliusnina I.Z., and Oskina I.N. 2004. An experiment on fox domestication and debatable issues of evolution of the dog. *Genetika* **40**: 794–807.

Vasilieva L.L. and Trut L.N. 1990. The use of the method of principal components for phenogenetic analysis of the integral domestication trait. *Genetika* **26**: 516–524.

Vila C., Savolainen P., Maldonado J.E., Amorim I.R., Rice J.E., Honeycutt R.L., Crandall K.A., Lundeberg J., and Wayne R.K. 1997. Multiple and ancient origins of the domestic dog. *Science* **276**: 1687–1689.

Viranyi Z., Topal J., Gacsi M., Miklosi A., and Csanyi V. 2004. Dogs respond appropriately to cues of humans' attentional focus. *Behav. Process.* **66**: 161–172.

Wandeler P., Funk S.M., Largiader C.R., Gloor S., and Breitenmoser U. 2003. The city-fox phenomenon: Genetic consequences of a recent colonization of urban habitat. *Mol. Ecol.* **12**: 647–656.

Wayne R. 2001a. Phylogeny and origin of the domestic dog. In *The genetics of the dog* (ed. A. Ruvinsky and J. Sampson), pp. 1–15. CABI Publishing, Wallingford, United Kingdom.

———. 2001b. Consequences of domestication: Morphological diversity of the dog. In *The genetics of the dog* (ed. A. Ruvinsky and J. Sampson), pp. 43–61. CABI Publishing Wallingford, United Kingdom.

Wayne R.K., Geffen E., Girman D.J., Koepfli K.P., Lau L.M., and Marshall C.R. 1997. Molecular systematics of the *Canidae. Syst. Biol.* **46:** 622–653.

Yang F., O'Brien P.C., Milne B.S., Graphodatsky A.S., Solanky N., Trifonov V., Rens W., Sargan D., and Ferguson-Smith M.A. 1999. A complete comparative chromosome map for the dog, red fox, and human and its integration with canine genetic maps. *Genomics* **62:** 189–202.

Yang F., Milne B.S., Schelling C., Dolf G., Schlapfer J., Switonski M., Ladon D., Pienkowska A., Bosma A.A., Sargan D.R., and Ferguson-Smith M.A. 2000. Chromosome identification and assignment of DNA clones in the dog using a red fox and dog comparative map. *Chromosome Res.* **8:** 93–100.

Zeuner F.E. 1963. *A history of domesticated animals.* Harper and Row, New York.

APPENDIX

AKC Recognized Dog Breeds

HERDING GROUP

Australian Cattle Dog

Australian Shepherd

Bearded Collie

Belgian Malinois

Belgian Sheepdog

Belgian Tervuren

Border Collie

Bouvier des Flandres

Briard

Canaan Dog

Cardigan Welsh Corgi

German Shepherd Dog

Old English Sheepdog

Pembroke Welsh Corgi

Polish Lowland Sheepdog

Puli

Rough Collie

Shetland Sheepdog

Smooth Collie

HOUND GROUP

Afghan Hound

American Foxhound

Basenji

Basset Hound

Beagle

Black and Tan Coonhound

Bloodhound

Borzoi

English Foxhound

Greyhound

Harrier

Ibizan Hound

Irish Wolfhound

Longhaired Dachshund

Norwegian Elkhound

Otterhound

Petit Basset Griffon Vendéen

Pharaoh Hound

Plotthound

Rhodesian Ridgeback

Saluki

Scottish Deerhound

Smooth Dachshund

Whippet

Wirehaired Dachshund

NON-SPORTING GROUP

American Eskimo Dog

Bichon Frise

Boston Terrier

Bulldog

Chinese Shar-Pei

Chow Chow

Dalmatian

Finnish Spitz

French Bulldog

Keeshond

Lhasa Apso

Löwchen

Miniature Poodle

Schipperke

Shiba Inu

Standard Poodle

Tibetan Spaniel

Tibetan Terrier

SPORTING GROUP

American Water Spaniel

Brittany

Chesapeake Bay Retriever

Clumber Spaniel

Cocker Spaniel

Curly-Coated Retriever

English Cocker Spaniel

English Setter

English Springer Spaniel

Field Spaniel

Flat-Coated Retriever

German Shorthaired Pointer

German Wirehaired Pointer

Golden Retriever

Gordon Setter

Irish Setter

Irish Water Spaniel

Labrador Retriever

Nova Scotia Duck Tolling Retriever

Pointer

Spinone Italiano

Sussex Spaniel

Vizsla

Weimaraner

Welsh Springer Spaniel

Wirehaired Pointing Griffon

TERRIER GROUP

Airedale Terrier

American Staffordshire Terrier

Australian Terrier

Bedlington Terrier

Border Terrier

Bull Terrier

Cairn Terrier

Dandie Dinmont Terrier

Glen of Imaal Terrier

Irish Terrier

Kerry Blue Terrier

Lakeland Terrier

Manchester Terrier (Standard)

Miniature Bull Terrier

Miniature Schnauzer

Norfolk Terrier

Norwich Terrier

Parson Russell Terrier

Scottish Terrier

Sealyham Terrier

Skye Terrier

Smooth Fox Terrier

Soft Coated Wheaten Terrier

Staffordshire Bull Terrier

Welsh Terrier

West Highland White Terrier

Wire Fox Terrier

TOY GROUP

Affenpinscher

Brussels Griffon

Cavalier King Charles Spaniel

Chihuahua

Chinese Crested

English Toy Spaniel

Havanese

Italian Greyhound

Japanese Chin

Long-Coat Chihuahua

Maltese

Manchester Terrier (Toy)

Miniature Pinscher

Papillon

Pekingese

Pomeranian

Poodle (Toy)

Pug

Shin Tzu

Silky Terrier

Toy Fox Terrier

Yorkshire Terrier

WORKING GROUP

Akita

Alaskan Malamute

Anatolian Shepherd Dog

Bernese Mountain Dog

Black Russian Terrier

Boxer

Bullmastiff

Doberman Pinscher

German Pinscher

Giant Schnauzer

Great Dane

Great Pyrenees

Greater Swiss Mountain Dog

Komondor

Kuvasz

Mastiff

Neapolitan Mastiff

Newfoundland

Portuguese Water Dog

Rottweiler

Saint Bernard

Samoyed

Siberian Husky

Standard Schnauzer

OTHER BREEDS

Beauceron

Redbone Coonhound

Index